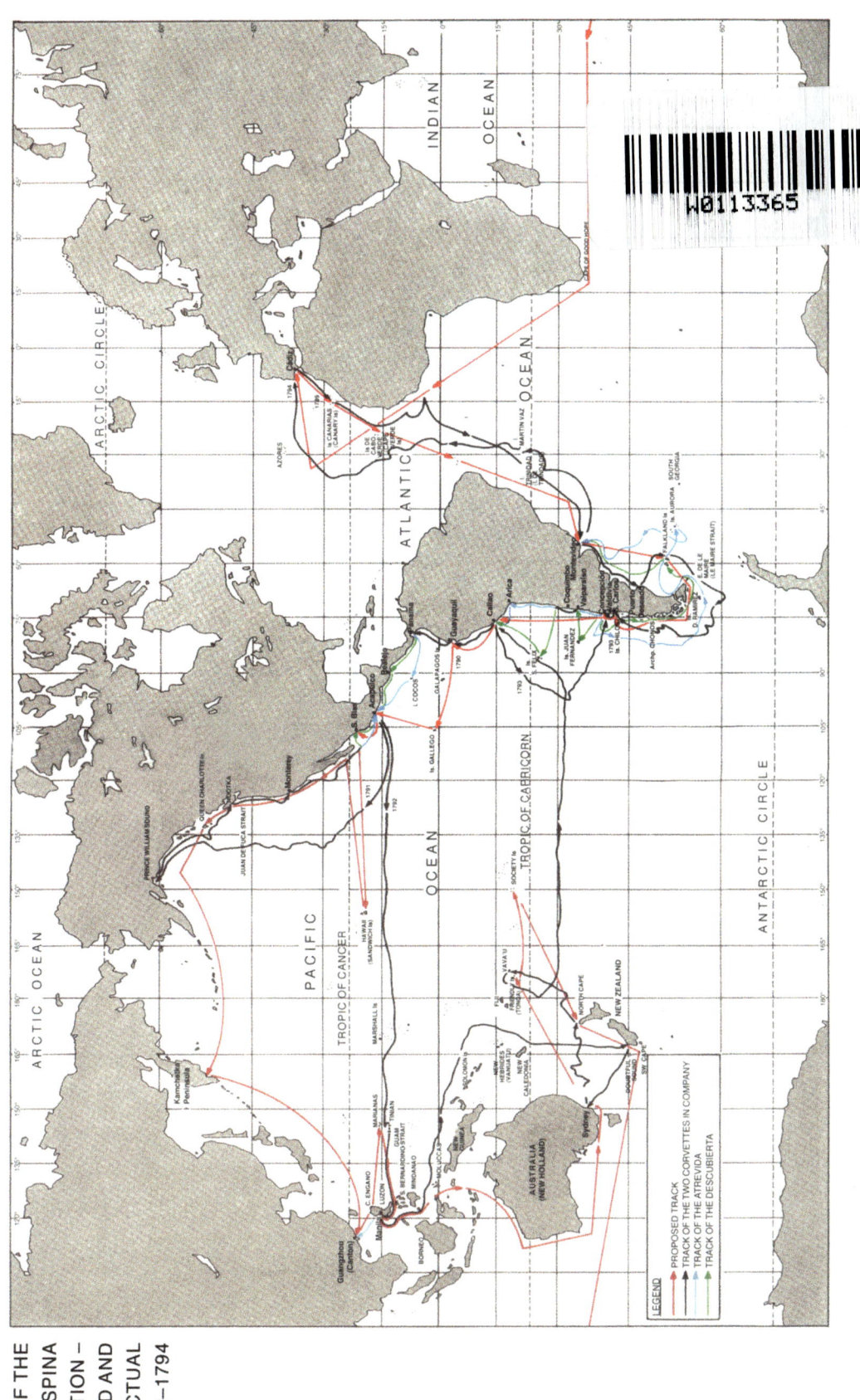

TRACKS OF THE
MALASPINA
EXPEDITION –
PROPOSED AND
ACTUAL
1789–1794

W0113365

WORKS ISSUED BY
THE HAKLUYT SOCIETY

———————

THE MALASPINA EXPEDITION
1789-1794

VOLUME I
CADIZ TO PANAMA

THIRD SERIES
NO. 8

THE HAKLUYT SOCIETY
Council and Officers 2001–2002

PRESIDENT
Mrs Sarah Tyacke CB

VICE PRESIDENTS

Professor R. C. Bridges

Lt Cdr A. C. F. David

†Professor P. E. H. Hair

Professor John B. Hattendorf

Professor D. B. Quinn HON.FBA

Sir Harold Smedley KCMG MBE

Professor Glyndwr Williams

COUNCIL (with date of election)

W. L. Banks CBE (1999)

Dr Michael Brennan (2000)

Tony Campbell (1999)

Dr Margaret Deacon (1999)

Stephen Easton (co-opted)

R. K. Headland (1998)

Bruce Hunter (co-opted)

Jeffrey Kerr (1998)

Professor Bruce P. Lenman (2001)

Professor P. J. Marshall (2000)

Rear Admiral J. A. L. Myres CB (2000)

Carlos Novi (1999)

Anthony Payne (1999)

Michael J. Pollock (2001)

Paul Quarrie (1999)

Royal Geographical Society
(Dr J. H. Hemming CMG)

A. N. Ryan (1998)

Dr Suzanne Schwarz (2001)

Mrs Ann Shirley (1999)

Dr John Smedley (co-opted)

TRUSTEES

Sir Geoffrey Ellerton CMG MBE

Dr J. H. Hemming CMG

G. H. Webb CMG OBE

Professor Glyndwr Williams

HONORARY TREASURER
David Darbyshire FCA

HONORARY SECRETARY
Dr Andrew S. Cook
The British Library, Oriental and India Office Collections,
96 Euston Road, London NW1 2DB

HONORARY SERIES EDITORS
Professor W. F. Ryan FBA
Warburg Institute, University of London, Woburn Square, London WC1H 0AB
Professor Robin Law FBA
Department of History, University of Stirling, Stirling FK9 4LA

ADMINISTRATIVE ASSISTANT
Mrs Fiona Easton
(to whom enquiries and application for membership may be made)
Telephone: 01986 788359   Fax: 01986 788181   E-mail: office@hakluyt.com
Website: www.hakluyt.com

Postal Address only
The Hakluyt Society, c/o Map Library, The British Library, 96 Euston Road,
London NW1 2DB

Registered Charity No. 313168

VAT No. GB 233 4481 77

INTERNATIONAL REPRESENTATIVES OF THE HAKLUYT SOCIETY

Australia: Ms Maura O'Connor, Curator of Maps, National Library of Australia, Canberra, ACT 2601

Canada: Dr Joyce Lorimer, Department of History, Wilfred Laurier University, Waterloo, Ontario, N2L 3C5

Germany: Thomas Tack, Ziegelbergstr. 21, D-63739 Aschaffenburg

Japan: Dr Derek Massarella, Faculty of Economics, Chuo University, Higashinakano 742–1, Hachioji-shi, Tokyo 192–03

New Zealand: J. E. Traue, Department of Librarianship, Victoria University of Wellington, PO Box 600, Wellington

Portugal: Dr Manuel Ramos, Av. Elias Garcia 187, 3Dt, 1050 Lisbon

Russia: Professor Alexei V. Postnikov, Institute of the History of Science and Technology, Russian Academy of Sciences, 1/5 Staropanskii per., Moscow 103012

USA: Dr Norman Fiering, The John Carter Brown Library, P.O. Box 1894, Providence, Rhode Island 02912 *and* Professor Norman Thrower, Department of Geography, UCLA, 405 Hilgard Avenue, Los Angeles, California 90024–1698

Plate 1. Alejandro Malaspina in the uniform of a Brigadier de la Real Armada. Anon.
Museo Naval, Madrid

THE

MALASPINA EXPEDITION

1789-1794

Journal of the Voyage by

Alejandro Malaspina

VOLUME I

CADIZ TO PANAMA

Edited by

ANDREW DAVID, FELIPE FERNANDEZ-ARMESTO,
CARLOS NOVI, GLYNDWR WILLIAMS

Introduction by

DONALD C. CUTTER

PUBLISHED BY ROUTLEDGE
FOR
THE HAKLUYT SOCIETY, LONDON
in association with
THE MUSEO NAVAL, MADRID
2001

First published 2001 by The Hakluyt Society
c/o Map Library
British Library, 96 Euston Road,
London NW1 2DB

Published 2018 by Routledge for The Hakluyt Society
2 Park Square, Milton Park, Abingdon, Oxon OX14 4RN
605 Third Avenue, New York, NY 10017

First issued in paperback 2022

Routledge is an imprint of the Taylor & Francis Group, an informa business

© The Hakluyt Society 2001

All rights reserved. No part of this book may be reprinted or reproduced or
utilised in any form or by any electronic, mechanical, or other means, now
known or hereafter invented, including photocopying and recording, or in any
information storage or retrieval system, without permission in writing from the
publishers.

Notice:
Product or corporate names may be trademarks or registered trademarks, and
are used only for identification and explanation without intent to infringe.

Publisher's Note
The publisher has gone to great lengths to ensure the quality of this reprint but
points out that some imperfections in the original copies may
be apparent.

ISBN 13: 978-1-03-229401-8 (pbk)
ISBN 13: 978-0-904180-72-5 (hbk)
ISSN 0072 9396

Typeset by Waveney Typesetters, Wymondham, Norfolk

SERIES EDITORS
W. F. RYAN
ROBIN LAW

British Library Cataloguing-in-Publication Data
A catalogue record for this book is
available from the British Library

MINISTERIO DE ASUNTOS EXTERIORES
DIRECCION GENERAL DE RELACIONES CULTURALES Y CIENTIFICAS

CONTRIBUTING EDITORS

W. R. P. Bourne

J. S. Cummins

Laurio Destéfani

Iris Engstrand

†Eduardo Estrella

Phyllis Herda

María Dolores Higueras Rodríguez

Robin Inglis

John Kendrick

Robert J. King

Manuel Lucena Giraldo

Jorge Ortiz Sotelo

Juan Pimental Igea

Blanca Sáiz

CONTENTS

BOOK FOUR

FROM COQUIMBO TO CALLAO

BOOK FIVE

FROM CALLAO TO ACAPULCO

ILLUSTRATIONS AND MAPS

Appendixes

SKETCH MAPS

JOINT FOREWORD

On 1 September 1794 Cádiz saw the safe return to their home port of the corvettes *Descubierta* and *Atrevida*, commanded respectively by Alejandro Malaspina and José Bustamante. These officers of the Spanish Royal Navy had been entrusted five years earlier with a scientific and political expedition to the Pacific Ocean, and their return was celebrated as the culmination of the most ambitious voyage of exploration and surveying undertaken by Spain in the eighteenth century.

Most unfortunately, the sad circumstances which surrounded Malaspina's life soon after his return, as well as the political upheavals and economic impoverishment of Spain resulting from the prolonged wars of the period, prevented the immediate publication of Malaspina's *Diario del viaje* and that of the additional narratives, sailing directions, astronomical observations, physical descriptions and political reports drafted during those five years of almost ceaseless navigational activity. It was necessary to wait until 1824-7 for the first edition of the *Diario*, published in Russian at St Petersburg by Admiral Ivan (Adam) Fedorovich Kruzenshtern of the Imperial Russian Navy, and until 1885 for a Spanish edition by the distinguished mariner and historian, Pedro de Novo y Colson. Comprehensive publication of the massive documentation generated by the expedition was finally undertaken as a Museo Naval initiative in 1987, and it naturally included Malaspina's *Diario general del viaje*, printed in two volumes in 1990.

The importance of a work of such scope could not go unnoticed by the Council of the Hakluyt Society, known to Spanish scholars as the most prestigious of British institutions devoted to the publication for the benefit of the international community of voyages which have contributed historically to the advancement of our knowledge of the physical, political and human aspects of the world in which we live. Indeed, the idea for the publication in English of Malaspina's *Diario* in association with the Hakluyt Society was supported with such enthusiasm by successive Presidents of the Society, and by a team of generous contributors, that we are now able to submit this scholarly edition in the knowledge that English-language readers will appreciate at its true value the account of a Spanish maritime undertaking whose contribution to a better knowledge of the globe may deservedly be compared to the achievements of the contemporary expeditions commanded by Bougainville, Cook, La Pérouse and Vancouver of which we Europeans are legitimately proud.

José Ignacio González-Aller Hierro
Almirante Director del Museo Naval, Madrid (1991-2000)

The Hakluyt Society has been publishing accounts of voyages and travels since 1847, and among its volumes are translations of important narratives in languages other than English. Its second and fourth volumes were translations of early Spanish accounts of the Americas, and of the volumes published to date thirty-seven have been based on Spanish sources; but if the Society's edition of Malaspina's journal takes its place in a well-established tradition, in terms of size and importance it is a landmark in the history of Anglo-Spanish scholarly collaboration both in Europe and in the Americas. This edition would not have been possible without the enthusiastic help of several institutions and individuals. The Museo Naval gave permission to the Society to use the Museo Naval/Lunwerg publication of the *Diario general del viaje por Alejandro Malaspina* (Madrid, 1990), which formed Volume II of the multi-volume *La Expedición Malaspina 1789-1794* as its basic text; supplied reproductions of drawings, charts and other materials from its holdings free of all charges; and gave a grant towards initial editorial costs. The Museo de América and the Real Jardín Botánico also supplied reproductions of paintings and drawings free of all charges. Vital to the progress of this edition has been the support of the Ministerio de Asuntos Exteriores, which gave a most generous grant towards the heavy translation and other costs of this edition. The Society is also deeply indebted to grants from Dr Alex Malaspina and the Lampadia Foundation. As President of the Society during the years when Volume I of its edition was being completed, I am aware of the unstinting help given by scholars, libraries and institutions in many countries. These will be acknowledged individually in the Preface and in the list of Contributing Editors; here I would like to thank them collectively on behalf of the Council of the Hakluyt Society. They have helped to make this edition a truly international enterprise.

Sarah Tyacke
President of the Hakluyt Society (1997-)

PREFACE

Among the voyages of exploration and surveying in the eighteenth century that of Alejandro Malaspina was remarkable in terms of its objectives and ambitions. Perhaps more than any other expedition of the period it represented the high ideals and scientific interests of the Enlightenment. In the last forty years the Hakluyt Society has published texts of some of the most significant Pacific voyages of the eighteenth century: John Byron, Philip Carteret, James Cook, Jean de Surville, Johann Reinhold Forster, George Vancouver, Jean François Galaup de la Pérouse, and (forthcoming) Louis Antoine de Bougainville. The Malaspina expedition remained the most serious gap in the Society's publications on the Pacific in this period, and for some years the Society was interested in publishing an English-language edition of his journal. The main obstacle in the way was the lack of a definitive Spanish text which could be used as the basis for such an edition. The publication of the Museo Naval/Lunwerg multi-volume edition of the *La Expedición Malaspina 1789–1794* (1987-99) has solved the problem, for Volume II (Parts 1, 2), *Diario general del viaje por Alejandro Malaspina*, published in 1990 under the editorship of Ricardo Cerezo Martínez, has as its main text Malaspina's autograph *Diario* or journal. This manuscript is held in two volumes in the Museo Naval (MSS 610, 423) and, supplemented by other documentation in MSS 92, 429 and 751, runs to approximately 450,000 words. It is distinct from two slightly later manuscript versions in the Museo Naval: one in three volumes (MSS 749, 750, 751), and a single-volume version (MS 753), which was edited by Pedro Novo y Colson and published at Madrid in 1885 under the title of *Viaje político científico alrededor del mundo por las corbetas Descubierta y Atrevida*. This last version was also published by Ediciones el Museo Universal, Madrid, in 1984 under the editorship of Mercedes Palau, Aránzazu Zabala and Blanca Saíz with the title of *Viaje científico y político, a las costas del Mar Pacífico y a las Islas Marianas y Filipinas verificado en los años 1789, 90, 91, 92, 93 y 94 a bordo de las corbetas Descubierta y Atrevida de la Marina Real, mandadas por los capitanes de navío D. Alejandro Malaspina y D. José F. Bustamante*. We have found the footnotes and biographical entries in this edition of much assistance.

The entries from February 1790 onwards in the version of the journal in MSS 610, 423 were divided into books and chapters by Malaspina. The editor of the 1990 Museo Naval/Lunwerg publication, inserted similar divisions in the earlier section of the journal from July 1789 to February 1790, and this has resulted in fourteen books containing seventy-four chapters in all, together with two appendices. That arrangement has been followed here, but in other ways this edition of Malaspina's journal differs in several respects from the Spanish edition of 1990. Dates which are represented there (as in the original) simply as numbers in the margin, have here been extended and shown as headings in the text; while isolated sentences and parts of sentences have been brought together into paragraphs for easier reading. Our edition will be published over a period of time in three volumes, not two, and will have a considerable amount of extra material. In Volume I this includes an Introduction by Donald C. Cutter, a translation of Malaspina's own *Introducción*, and an appendix containing a selection of Malaspina's correspondence with the Ministro de Marina, Antonio Valdés y Bazán, during the period covered by the volume. Additionally, there are guides appropriate to an English-language edition. These include an essay by Carlos Novi on 'Translating Malaspina', a Spanish-English glossary of geographical terms, and notes on Spanish naval ranks, compass directions, and weights and measures. The text itself is fully annotated, with footnotes explaining technical terms, identifying persons and places, and adding information from other journals and documents on the voyage. An appendix by Andrew David explains Malaspina's surveying methods. A list of works cited in this volume will be found at the end of the text. A full bibliography together with an index to all three volumes will be included in Volume III. Much of the material used in this edition comes from the various volumes in the Museo Naval/Lunwerg edition, of which the most recent, the *Diario* of Malaspina's fellow commander, José Bustamante y Guerra, was published in 1999 under the editorship of Dolores Higueras Rodríguez. These volumes are cited in the footnotes to this volume, as well as many other works on Malaspina that reflect the growth – almost an explosion – of interest in the navigator and his voyage in recent years. Blanca Sáiz's *Bibliografía sobre Alejandro Malaspina* (Madrid, 1992) contains 1134 published items, and many more have appeared since. There is a Centro di Studi Malaspiniani 'Alessandro Malaspina' at Mulazzo in Italy under the directorship of Dario Manfredi; a whole series of Malaspina conferences have been held in recent years in Spain, Italy and Canada; and the worldwide academic interest in the voyage is shown in the list of Contributing Editors printed after the title-page of this volume. These are scholars who have freely given of their specialized knowledge and time to enable us to provide the annotations to this edition, and it is difficult to overestimate our debt to them. In Volume I the annotations on hydrography and astronomy have been supplied by Andrew David, with help from John Kendrick and Richard Campbell; while those on ornithology have been provided by Bill Bourne. The annotations on the many other subjects of interest in this volume have been mainly the responsibility of Felipe Fernández-Armesto, working with material supplied by Laurio Destéfani, the late Eduardo Estrella, Manuel Lucena Giraldo, Jorge Ortiz Sotelo, Juan Pimentel Igea, and Blanca Sáiz. The translation of Malaspina's text in this volume, which readers of Carlos Novi's piece on 'Translating Malaspina' will surmise was not a totally straightforward business, has been shared between several different translators. Here we would like to thank all of them: Edward Ewing, Sylvia Jamieson, Zoë Petersen, Paul Rankin and David Sutcliffe, together with Philip Grundy who was responsible for the overall revision of the translation.

In the joint Foreword to this volume the President of the Hakluyt Society has expressed the Society's gratitude to those institutions whose support has made this edition possible. As editors we would like here to thank those individuals in those and other institutions of learning whose help and advice have been so important to the making of this edition. María Dolores Higueras Rodríguez at the Museo Naval and Mercedes Palau Baquero at the Ministerio de Asuntos Exteriores are in a sense its godparents: at every stage they have encouraged and helped it on its way. Their support was shown in a more personal way during the '150 Years of the Hakluyt Society' celebrated at the Museo Naval in December 1996, when many of the scholars involved in this edition gathered together. We should also like to thank Estado Español, Dirección General de Relaciones Culturales y Científicas, Ministerio de Asuntos Exteriores, Madrid; Doña Paz Cabello, Director of the Museo de América, Madrid, and María Pilar de San Pío Aladrén, Archivist of the Real Jardín Botánico, Madrid, for their kindness in supplying illustrations. We are grateful for the help we have received at different times from: María Victoria Ibáñez Montoya; Luisa Martín-Merás, Curator of Cartography at the Museo Naval; Nieves Amunátegui, Librarian of the Museo Naval; Hugo O'Donnell y Duque de Estrada, historian, for his generosity with his time, and his readiness to give advice; the present Director of the Instituto Hidrográfico de la Marina, Cádiz, Juan M. Nodar Criado, and his predecessor, José María Fernández de la Puente y Ferrera de Castro; the Director of the Servicio Hidrográfico y Oceanográfico de la Armada de Chile, Rafael Mackay Bäckler; Carlos Parra Merino and Sergio Carrasco Delgado of the Universidad de Concepción; Carlos Tromben of the Biblioteca de la Academia de la Historia in Santiago de Chile; Enrique Trucco Delepine, Executive Director of the Liga Marítima de Chile in Valparaíso; Rosa Zeta de Pozo, University of Piura, Peru, who provided issues of the *Mercurio Peruano* with notices about the Malaspina expedition's visit to Peru; Alexei Postnikov (Institute of the History of Natural Sciences and Technology, the Russian Academy of Sciences, Moscow); Peter Barber (The British Library), and Ann Shirley. The Department of History at Queen Mary, University of London, has been generous in its provision of accommodation and other forms of support for much of the editorial work on this volume. Finally, the Society owes a special debt to Dámaso de Lario, who during his years as Cultural Counsellor at the Spanish Embassy, London, did much to secure the funding without which this edition would not have been possible.

Andrew David
Felipe Fernández-Armesto
Carlos Novi
Glyndwr Williams

EQUIVALENTS AND ABBREVIATIONS

SPANISH NAVAL RANKS

Ranks in the *Real Armada* correspond with those in the eighteenth-century British Royal Navy as follows:

Capitán General	Admiral
Teniente General	Vice Admiral
Jefe de Escuadra	Rear Admiral
Brigadier de Real Armada	Commodore
Capitán de navío	Post Captain
Capitán de fragata	Commander
Teniente de navío	Senior Lieutenant
Teniente de fragata	Junior Lieutenant
Alférez de navío	No exact equivalent
Alférez de fragata	No exact equivalent
Contador de fragata	Junior Paymaster and Purser
Capellán	Chaplain
Cirujano	Surgeon
Guardiamarina	Midshipman
Práctico	Pilot
Cartógrafo	Cartographer
Piloto	Master
Segundo piloto	Second master
Pilotín	Master's mate
Director de historia natural	Director of natural history
Botánico	Botanist
Pintor	Artist
Dibujante	Draughtsman
Guardián	Quartermaster

ABBREVIATIONS

AGI	Archivo General de Indias, Sevilla
AMN	Archivo del Museo Naval, Madrid
AMNCN	Archivo del Museo Nacional de Ciencias Naturales, Madrid
ARJB	Archivo del Real Jardín Botánico, Madrid
UKHO	United Kingdom Hydrographic Office, Taunton

WEIGHTS AND MEASURES

Spanish terms which have a direct English equivalent have been translated into English; terms for which there does not appear to be a direct equivalent have not been translated.

Spanish terms which have been translated into English

braza	5·48 feet = one fathom
cable	one-tenth of a sea mile = one cable
cuartillo	0·9 pints = one pint
legua	three sea miles = one league
milla	the internationally accepted unit of distance at sea, one-sixtieth of a degree of latitude = one [sea or nautical] mile
nudo	the internationally accepted unit of speed at sea of one sea mile an hour = one knot
pie	0·91 feet = one foot
pulgada	0·91 of an inch = one inch
línea	one-twelfth of a *pulgada* = one line (an obsolete English term for one-twelfth of an inch)

Spanish terms which have not been translated

arroba	approximately 25 lbs
codo	18 to 22 inches
toesa	approximately 6·4 feet – the French *toise*
quintal	approximately 102 lbs
vara	approximately 3 feet

18TH-CENTURY SPANISH COMPASS DIRECTIONS AND ENGLISH EQUIVALENTS

Spanish	English	Spanish	English
N¼NE	NbyE	S¼SW	SbyW
NE¼N	NEbyN	SW¼S	SWbyS
NE¼E	NEbyE	SW¼W	SWbyW
E¼NE	EbyN	W¼SW	WbyS
E¼SE	EbyS	W¼NW	WbyN
SE¼E	SEbyE	NW¼W	NWbyW
SE¼S	SEbyS	NW¼N	NWbyN
S¼SE	SbyE	N¼NW	NbyW

SPANISH-ENGLISH GLOSSARY OF GEOGRAPHICAL TERMS USED IN THIS WORK

Spanish word	English meaning	Spanish word	English meaning
alto(s)	height/heights	golfo	gulf
archipiélago	archipelago	isla	island
arena	sand	islote	small island, islet
arrecife	reef	islita	small island, islet
arroyo	stream	lengua	tongue
bahía	bay	monte	mountain
bajo	shoal	morro	headland, bluff
banco	bank	morrito	small headland
batería	battery	nuevo/a	new
boca	mouth	pan de azúcar	sugar loaf
bodega	warehouse	piedra	rock, stone
cabo	cape	playa	beach
caleta	cove	península	peninsula
camino	road	promontorio	promontory, headland
canal	channel	pueblo	town, village
casa	house	puerto	port
castillo	castle	punta	point
cerro	hill	quebrada	gorge, ravine
cerrillo, cerrito	small hill, hillock	río	river
cordillera	mountain range, especially the Andes	roca	rock
		sierra	mountain range
ciudad	city	silla	saddle
ciudadela	citadel	teta	nipple, pap, breast
ensenada	bay	tierra	land
estrecho	strait	torre	tower
estero	creek, inlet	vigía	lookout
farallón	small needle-shaped rock	volcán	volcano

TRANSLATING MALASPINA

Carlos Novi

The original text

The original of the *Diario* or journal here translated into English is Alejandro Malaspina's holograph manuscript (AMN MS 610 and MS 423), to which an appendix of five folios on the veracity of Ferrer Maldonado's 1588 voyage is inserted between Books V and VI. Malaspina's own *Introducción* comes from a separate manuscript (AMN MS 753).[1] Since working directly from these manuscripts was impracticable, the master original adopted was an electronic copy of the printed version of the *Diario* published in 1990 by the Museo Naval.[2] This made it imperative for the editors and the translating team to take into account the possibility of transcription lapses or errors in the Spanish printed text used in lieu of the manuscript original. In the event the deficiencies fell into four main categories: errors detected in the original text; errors of transcription; printing errors in the published version; and interpretative errors (misreadings) made by the Spanish transcribers.

An added difficulty arose from the graphic presentation of MS 610 adopted by the Museo Naval in its publication. This point requires clarification. In a effort to provide readers of the Spanish text with a fair 'view' of the original manuscript, the Museo Naval editors included in the printed version all those parts of the manuscript that had been deleted by Malaspina. The unusual circumstances in which the author of the narrative had to work are worth recalling. During the fourteen months that elapsed between his return to Cádiz (21 September 1794) and his sudden arrest in the night of 23 November 1795, Malaspina does not seem to have had adequate time to put into order the mass of documents related to the expedition and proceed calmly from a first draft to a final fair copy. The holograph manuscript is not the neat affair that might have been produced by a well-chosen amanuensis. By design, this material deficiency has been faithfully reflected in the Museo Naval publication. All the author's deletions are shown in the text within quotation marks and accompanied by a footnote confirming 'deleted in the original'. In establishing the Hakluyt Society's English text a decision had to be taken regarding this convention. The option of leaving untranslated all passages deleted by Malaspina was discarded for two reasons. Firstly, many corrections were matters of form which in any case had to be taken care

---

[1] See below, pp. lxxviii–xcviii.

[2] Ricardo Cerezo Martínez, ed., *La Expedición Malaspina 1789–1794*, Tomo II: *Diario general del viaje por Alejandro Malaspina*, 2 parts, Madrid, 1990.

of in translation. Secondly, there was often in the 'deleted' portions substantive information which could usefully be incorporated in the English text without in any way distorting the intention of the passage in question. Therefore, the complete Spanish original has been treated as 'a whole' to be translated, while editors saw that duplications or redundant passages were suitably eliminated. This convention was subordinated to the general assumption that English-speaking scholars who might wish to study the English narrative with its Spanish original at hand must find that both texts are parallel.

Conventions and methodology

The eighteenth-century Spanish original has been rendered into modern English avoiding the use of strident anachronisms. The aim has been to preserve as much of the character, style and general flavour of the original as possible while ensuring accuracy and consistency throughout the English text. This has been achieved, hopefully, through a systematic but subtle use of the 'one-to-one' rule in the choice of words, i.e. the avoidance of synonyms if the author does not resort to them, while respecting the idiomatic conventions of both languages. To this end an electronic Spanish-English glossary of technical terms and idiomatic expressions was compiled as part of the translating exercise. All members of the translating team have had access to the same historical marine dictionaries[1] and to updated copies of the electronic glossary. Its entries (over 1,000) are led by a headword and include numerous examples taken from the Spanish original with their agreed English translation. Special collocations and textual references are easily accessed. Since virtually all entries result from difficulties encountered by individual translators in the course of their work, the glossary complements general and technical dictionaries, and serves as a record of translations and idioms already used in other parts of a lengthy text. For specific subjects agreed bilingual lists of nautical instruments, ranks, boats, birds, plants, sea states, wind scales etc. have been compiled.

Linguistic difficulties

An immediate hurdle encountered by English translators in handling a faithful transcript of the eighteenth-century manuscript is the fact that in those days Spanish spelling had not yet been standardized. Erratic spelling as well as archaic forms of capitalization and punctuation made reading difficult, to say the least. For instance a common Spanish nautical term like *varada*, for 'grounding', is hardly recognizable if spelled *barrada*. Obsolete spelling, unusual scrawl and the use of abbreviation in the manuscript could and did cause confusion. For example, since the word *Estrecho* in a particular passage of the printed version did not make sense, the authentic text had to be checked. Verification showed that this particular word is in fact abbreviated as *Estr°* in the manuscript. Yet it had been transcribed in full as Estrecho (= Strait) certainly a possibility in isolation, but making no geographic or hydrographic sense in the context. By interpreting *Estr°* as the abbreviation of *Estremo* (i.e., spelling 'Extremo' with

---

[1] Timoteo O'Scanlan, *Diccionario marítimo español*, Madrid, 1831 (Museo Naval facsimile, 1974) and Juan José Martínez Espinosa, *Diccionario marino español-inglés/inglés español*, Madrid, 1849 (Editorial Naval facsimile, 1989).

an *s* instead of *x* in eighteenth-century phonetic spelling) it became possible to convert two geographic features nowhere to be found on the relevant chart into the two 'extremities' of an island.

Another problem that the translators have faced is the lack of consistency in the spelling of proper names. A place name, for example, might have several different spellings in the manuscript. One reason for these inconsistences is the number of different reports that Malaspina was using as he composed his journal, and the lack of time that he had to prepare a fair copy. More often than not, these matters would be settled at the hydrographic office when the detailed charts were engraved, but Malaspina's journal did not reach that stage. In this edition, when faced with variant spellings of a proper name, we have accepted the one we feel is the most authoritative, and silently amended the others to agree. As far as place names are concerned, we have used as our authority contemporary Spanish charts, while we have taken our modern spellings from the latest *Admiralty Sailing Directions*.

Readers of sea voyage accounts will be familiar with the peculiar features of this type of literature, particularly when they include extracts from logbooks and records of scientific observations. Malaspina's journal abundantly reflects the great wealth of tradition preserved in the Spanish seafaring language for which there is, naturally, an equally rich English tradition. This has required familiarization with a considerable diversity of scientific jargons, relating both to the substantive description of matters observed and to the operative aspects of the act of observation, such as the handling of instruments like chronometers and the adoption of particular procedures, for example to measure a base for surveying purposes. One may find the term pendulum in any dictionary, but the idiomatic manner of expressing its uses has to be found elsewhere. The accuracy of meaning in the mariner's vocabulary is inextricably linked to the proper choice of terms. So every line of the journal's English translation has been checked meticulously against the Spanish original by a Spanish-speaking editor and then checked by one or more English-speaking counterparts. Only after full agreement on both sides of the linguistic divide was the text allowed to go as approved for further stylistic revision.

A curious and rewarding example of lexical investigation arose in connection with the proper designation of one of the types of boat included in the complement of boats carried by the corvettes. In the Spanish original the word *bombo* is used. Yet, in accordance with the best Spanish marine dictionaries this term would not be acceptable as descriptive of any of the regulation boats carried by naval vessels. It is generally defined as 'a flat-bottomed boat of shallow draught used for working cargo and for crossing shoal water'. But contextual uses of *bombo* in the Spanish narrative suggested that this was the second boat in each of the corvettes. The matter was studied in Madrid by looking through the contemporary documents relating to the building of the two corvettes and their equipment. The search was rewarding. Two documents were found which showed that the *bombo*, with that designation, was included in the list of the corvettes complement of boats.[1] It would appear that this is the first time

---

[1] These are: (1) Doc 18 of AMN, MS 426, f. 54, concerned with the fitting out of the corvettes and their boats (Cádiz 1789), particularly revealing because the *embarcaciones menores* are listed with their respective names (*lancha, bote, bombo, botecillo* and *chinchorro*). (2) AMN, MS Doc 92*bis*, ff. 79-80 which relates to the period of refitting at San Blas mentioned in the *Diario* (12 April 1791). In it the term

the word *bombo* has been used in this context, possibly an innovative idea of Malaspina and Bustamante.

The linguistic idiosyncracies of Malaspina

Anyone reading Malaspina in Spanish, let alone attempting to translate his writings, will soon discover that his prose, particularly in his moments of esoteric rumination, is difficult to digest. The combination of ingredients that made up his practical use and knowledge of the Spanish language merits, therefore, special consideration. Born and bred in Italy, Malaspina was twenty when he entered the Spanish Navy. He was forty when the expedition returned to base in 1794. By the time he received his momentous commission his command of the Spanish language must have been more than adequate. On his aptitude for the job the Secretary of State for the Navy was convincing: 'his knowledge, birth, noble personality and elegance of manners, proud bearing, affability, resolute character and social gifts made him the first in our navy and a unique choice for that commission'.[1] Although co-signed by Bustamante, the plan for a voyage submitted to Antonio Valdés bears the unmistakable characteristics of Malaspina's style of drafting. There is no denying that the nautical parts of the journal are written with ease and professional competence, though he may be less *castizo*[2] in his philosophical extemporizations. The question is whether Italian was impinging on his command of Spanish or whether his manner of writing was attributable to his character and extraordinary personality.[3] Paradoxically, the deceptive similarity of the two languages involved[4] may be part of the problem. This similarity is apt quickly to erode the command of the mother tongue for an Italian fully immersed in an all-Spanish environment, as midshipman Ali Ponzoni[5] experienced just fourteen months after his arrival at Cádiz. Even Malaspina, with false modesty perhaps, hinted at such an erosion in writing to one of his correspondents at the height of his Spanish naval career.[6]

Undoubtedly, the native tongue is bound to brand an individual linguistically for life, but that need not be a negative factor. It has been suggested recently that 'a poor command of the Castilian language' by the author was to be blamed for a number of obscure passages found in an Italian version of the memorandum written by

---

bombo, used three times, obviously refers to the second boat but is treated as distinct from the first *bote* by the Arsenal's master shipwright since it deserves its own separate mention in the listings of the volume of timber (5 quintals) used for its repair.

[1] Quoted by Cerezo in *Diario por Malaspina*, Pt. 1, p. 16, from Marcos Jiménez de la Espada, 'Una causa de estado', *Revista contemporánea*, t. XXXI, vol. IV, February 1881, p. 409.

[2] Linguistically pure-blooded.

[3] See John Kendrick, *Alejandro Malaspina – Portrait of a Visionary*, Montreal and Kingston, 1991.

[4] There are a number of recurrent Italianisms in the *Diario*, such as the use of *ocurrir* as meaning 'to need'; *sistemar* in lieu of *arreglar*; *destruir* for *cancelar*. They are real booby-traps for the unwary translator, but they are irritants rather than problems.

[5] 'You won't believe it but I have so forgotten the Italian language that I can hardly write a letter in Italian', Ali Ponzoni to Ramón Ximénez, Letter 4, from Isla de León (27 November 1788). Italian original in Dario Manfredi, *Alessandro Malaspina e Fabio Ala Ponzone – Lettere dal Vecchio e Nuovo Mondo (1788–1803)*. Bologna, 1999, p. 146 (henceforth Manfredi, *Lettere*).

[6] Writing to Gherardo Rangoni from Acapulco (19 March 1791) '… and the rare chances to practice our sweet language may by now have badly affected the clarity and full expression of my thinking'. Italian original in ibid., p. 247.

Malaspina on the need of peace between Spain and France.[1] This rather dim view may be a cautious explanation by the Italian translator. Malaspina himself has left us in no doubt about his sense of harassment while in the process of drafting the *Diario* in less than ideal conditions. The dramatic circumstances in which that memorandum was written could well be blamed for lapses in drafting or some inappropriate use of terms.[2] This is not to say that translating Malaspina is a suitable task for the uninitiated. As already noted, his style was at best unusual. In his narrative Malaspina jumps without warning from the forbidding tints of a cloud formation and the prescription of down-to-earth sailing directions to the convoluted philosophical abstractions of an enlightened thinker. Having just recorded with telegraphic formality a daily entry in the logbook he may be moved to sentimental strictures as he faces the inexplicable actions of a lowly sailor. These lurches have demanded on the part of translators sudden shifts of register and sensitive changes of mood. But their greatest difficulty has been the need to render into intelligible vernacular English Malaspina's bizarre brand of classicism. When untethered by naval convention and scientific protocol his discourse becomes rich in lapidary sentencing and long-winded periods often preceded by a dreaded ablative absolute, a sure sign of an imminent syntactical labyrinth and a welter of subordinate clauses. The only safe course left to the translator is to proceed with the strictest application of grammatical and logical analysis and search for verb, subject and predicates as with a Latin text. A Latinist, Father Manuel Gil,[3] whose choice as a writer to help draft his journal Malaspina came to regret,[4] is probably objective enough. 'With the exoticism of some of his ideas and, worse still, the words he used to express them – for the truth is that he had not mastered the Spanish language with the perfection that he was persuaded he had – he had come to create a language of his own of which the least one can say is that it was quite extraordinary. We even used to imitate him in his presence, just for fun, resorting to his own high-flown mannerisms ... which reflected extravagant political thinking'.[5]

Publication
Publication and dissemination were implied in Malaspina's plan for a scientific and political voyage to the outer limits of the Spanish empire and beyond.[6] Publication of the results of the expedition was one of the aims stated in his introduction to the journal of his voyage, this time extolling the benefits of public debate for the Spanish nation and for Spain's reputation in the world.[7] It is reasonable to presume that if his

---

[1] *Sull'arresto di Alessandro Malaspina: vecchie certezze e nuove ipotesi* (Appendix: *Riflessioni relative alla pace della Spagna con la Francia*) p. 211, in R. Giura Longo and P. Rossi, *Con Malaspina nei mare del sud*, Bari, 1999.

[2] While Malaspina was trying to order his papers for the narrative he was negotiating who was to cover the publication expenses and was also involved in producing a peace plan with France. See letter to Paolo Greppi from Aranjuez (17 February 1795) in Manfredi, *Lettere*, pp. 333-6.

[3] At the proposal of Malaspina, by virtue of a Royal Order issued at San Ildefonso on 26 July 1795, the Reverend Father Manuel Gil was given the task of drafting and editing the outcome of the voyage.

[4] He used to call him 'le citoyen noir': Pedro Novo y Colson ed., *Viaje político-científico alrededor del mundo por corbetas Descubierta y Atrevida ... desde 1789 a 1794*, Madrid, 1885, introducción histórica, p. XV.

[5] Ibid.

[6] See 'Plan for a Scientific and Political Voyage Around the World', pp. 312-15 below.

[7] See p. lxxxi below.

narrative was to be made available to the world, Malaspina must have also entertained the thought of having his journal translated into other languages. Tragically, this never happened in his lifetime. We know, with the benefit of hindsight, that the proper time to discuss such a project never came. But it causes no surprise to learn that the Italian-born Spanish navigator, Latin scholar, and keen user of the French language, actually spent much of his sixth and, in the event, last year at the prison castle of San Anton in La Coruña translating from French into Spanish a philosophical essay that had been awarded a French Academy prize in 1755.[1] Significantly, Malaspina prefaced his own translation with a discussion on the substantive aspects of the translator's job. From his experience translations of French texts into Spanish rarely reflected accurately the spirit of the original. Each language, he argued, mirrors the soul of a given people. He was referring to what modern linguists have called the 'transcultural transfer' factor in translating. If a translator is not to betray the author's intent he must seek to produce in the reader's target language an equivalent response to that intended by the author in the source language. This is in substance what a dedicated team of translators and the Society's editors have endeavoured to achieve. It is with a sense of restitution owed to the unfortunate Malaspina that his Journal is now offered to an English-speaking readership.

---

[1] Manfredi, *Lettere*, p. 101, n. 209; the full title of this translation is given as *Discurso del Padre Guenard jesuita sobre la questión "en que consiste el carácter de la filosofía" según los consejos de San Pablo en la Epístola a los Romanos. Cap. XII V.III. "Non plus sapere quam oportet sapere". Premiado por la Academia Francesa en 1755, traducido a el idioma castellano por D. Alexandro Malaspina.*

INTRODUCTION

Donald C. Cutter

The white city of Cádiz was sparkling in late July's sunshine and its ancient harbour was particularly active because Spain's greatest scientific expedition was about to depart. It was the culmination of eight months of intensive preparation for a planned round-the-world voyage of two new naval corvettes manned by the finest available crews. Though different in character, this major undertaking was intended to equal or surpass the three well-publicized voyages of Captain James Cook. Now it was Spain's turn, and at 9 o'clock in the morning of 30 July 1789 preparations ceased and the voyage began.

It is tempting even after the lapse of two centuries to compare this Spanish voyage commanded by Alejandro Malaspina[1] with those of Cook or the contemporary French scientific explorer Jean François de Galaup, Comte de la Pérouse, especially since Malaspina and his consort vessel commander, José Bustamante y Guerra, had initially indicated that the scientific aspects of the expedition would be modelled on those of the British and French navigators. But from the beginning there were differences, which had been included in the original 'Plan' of the voyage,[2] although these were probably overlooked in the excitement of the moment. Some of these differences are noted later, but the most significant is evident from the outset. The Malaspina voyage was more than an expedition of scientific exploration, for it had another important, perhaps overriding, purpose – that of imperial inspection. This included making a series of detailed reports on various parts of the far-flung Spanish overseas empire, with emphasis on southern South America and on the Pacific Ocean ports of the Americas. Malaspina's instructions to gather information about the geography and economies of these regions implied the making of recommendations for the improvement of unsatisfactory conditions. As a result, more than half the time of the Malaspina expedition was spent in the role of imperial inspection, and to this its commanding officer devoted himself wholeheartedly.

While acting as investigators in settled areas, the expedition members had the great advantage of access to the experience and knowledge of local informants, as well as

---

[1] Malaspina's name is frequently misspelled in the documentation, appearing as Malespina, Mala Espina, and even as Malaspiña. After arrival in Spain he used the Spanish spelling of his given name, Alejandro, rather than the Italian spelling Alessandro, but never followed the Spanish practice of adding a maternal surname, nor was one ever applied to him as a second *apellido*.

[2] See pp. 312–15 below.

unlimited access to regional archives and map collections. This did not mean that original scientific investigation was neglected, for the naval visitors had resources and equipment unavailable to local residents. In addition, Malaspina made exceptionally detailed reports, an ability he had developed in his earlier career, but which he applied with enthusiasm during the long voyage. Malaspina and Bustamante were not only captains of their vessels, but were commanders of the entire expedition force. There was no division between the military and scientific contingents except as regards the nature of the work expected and accomplished. The only civilians aboard were the artists, the naturalists, and six stowaways who escaped detection at the time of departure (four on the *Descubierta*, two on the *Atrevida*). Save only for royal or ministerial changes to the original orders, the expedition was under the overall command of Malaspina.

Although in time the scientific exploring expedition that sailed from Cádiz in late July 1789 became known as Malaspina's, in concept, preparation and overall plan it was a joint project proposed by fellow officers of equal rank, Malaspina and Bustamante. Yet naval protocol required one captain to serve as commanding officer, and that officer was bound to be Malaspina, senior by two years. Because of this seniority, his capacity for putting his thoughts and plans in writing, and the preponderance of surviving manuscripts in his hand, the older officer took the lead. He also had the advantage of his close personal connection with the Ministro de Marina, Antonio Valdés y Bazán. In general terms Malaspina had greater responsibility for the scientific aspects of the voyage, whereas Bustamante was was more concerned with the handling of the vessels. But there was never any sharp division between them, and any such separation of duties seems relative. On the other hand there is no evidence that even before Malaspina's fall that he expected to share authorship and publication duties with the officer who had earlier been his co-petitioner. Once the voyage was over, it was Malaspina who was busy turning the mass of documentation into a form suitable for publication, whereas Bustamante was destined for return to regular naval duties (and later to important overseas military and civil government responsibilities). With Malaspina's fall and disgrace, it was decided that his name should not be mentioned in publications relating either to the main voyage or to the subsidiary exploring efforts. Instead, with only a few inadvertent exceptions, the singular term 'expedition commander', or the plural form 'commanders of the expedition', were substituted for the personal names of Malaspina and even of Bustamante in drafts being prepared for publication. The elimination of both names is noticeable in 1802 when the journal of the subsidiary expedition involving the sloops *Sutil* and *Mexicana* was published.[1]

In the enthusiasm of recent Malaspina scholarship it is sometimes overlooked that the expedition was a major royal effort supported and carried out by the Spanish navy, not the singular achievement of the foreign-born Malaspina. He was born on 5 November 1754 in Mulazzo, Lunigiana, in northwest Italy, of a Parmesan mother of noble lineage and a Lunigianian father, Carlo Morello Malaspina, Marchese of Mulazzo. The latter was reportedly more progressive than his many cousins and relatives who were among the declining lesser nobility of a feudalistic rural area. The Italian biographer of Malaspina, who takes a rather romantic view of the explorer,

---

[1] *Relación del viage hecho por las goletas Sutil y Mexicana en el año de 1792*, Madrid, 1802.

credits him with having been destined to compensate for the mediocrity into which the family had fallen some years earlier.[1] As the third son of a minor noble, Alejandro was not likely to inherit the family lands or title, so, typically, in his youth he was prepared for an alternative career. To avail his family of important connections, in 1762 Malaspina's father took them to Palermo where his wife's uncle was Viceroy of Sicily. The young Malaspina was only seven years old when he left his native Lunigiana, and at his new destination lived and studied in the protective shadow of the viceregal court for three years. Through family connections, Alejandro was next sent for further study to the Pontificio Collegio Pio Clementino (or Collegio Clementino) where he enrolled on 15 May 1765. There he continued his education for the next seven years, and to this period Malaspina owed much of his future interest in linguistics, geography and physical science. At the Collegio Malaspina first wrote research papers, one of which survives as his earliest known work, *Theses ex Phisica Generali* (1771). It was also at this prestigious school that he met some fellow students who would become important figures in late eighteenth-century Europe.[2]

By 1773 Alejandro Malaspina's formal education was over, and it was time to choose his future career. Contrary to family desires, he had earlier decided against a religious calling. Instead he chose to enter the military, with a preference for the Spanish naval service, a career to which his family background facilitated access; but before any active duty in the Spanish navy Malaspina followed the precedent of his relatives and went to nearby Malta to become a member of the Order of St John whose headquarters were on that island. The military brotherhood of Malta had its own small navy, and in it Malaspina had his first sea experience and also made his first brief visit to Spain. On 18 November 1774 he entered the Spanish navy as a midshipman. His prior period of service as a member of the Order of St John perhaps explains why in less than two months Malaspina was advanced to the rank of *alférez de fragata*,[3] and so missed most of a midshipman's normal training. Certainly at this time and later Malaspina's foreign background did nothing to affect his career adversely. Regardless of his place of birth, Malaspina became a highly successful officer in the Spanish navy, ready to lead in 1789 an expedition that was strictly Spanish and that was to reflect national skills and ambitions.

For his part, Bustamante has until recently only had a meagre share in the interest surrounding Malaspina and the great voyage.[4] No writer has made any comparison between Cook and Bustamante, or indeed between Malaspina and Bustamante, whose life lacks the melodramatic elements of his colleague's. Born on 1 April 1759 at

---

[1] Dario Manfredi, director of the Centro di Studi Malaspiniani, Mulazzo, and author of a sympathetic biography, 'Alejandro Malaspina. Una biografia', in Blanca Sáiz, ed., *Alejandro Malaspina: La América imposible*, Madrid, 1994, pp. 19-133.

[2] Among these was Federico Gravina, later to become Capitán General in the Spanish Navy, and to command the Spanish fleet at the Battle of Trafalgar. Among other fellow students were two future cardinals, a general, an admiral, and an ambassador.

[3] For a list of ranks in the Spanish Navy see p. xx above.

[4] Bustamante's journal of the voyage from 30 July to 21 May 1792 has now been published: María Dolores Higueras Rodríguez, ed., *La expedición Malaspina 1789-1794*, Tomo IX: *Diario general del viaje corbeta Atrevida por José Bustamante y Guerra*, Madrid, 1999. It includes a biographical sketch of Bustamante by the editor at pp. 14-19.

Ontaneda in Spain's northern Santander province, Bustamante enlisted at the age of eleven as a midshipman in the Spanish naval department of El Ferrol in Galicia. His first appointment at sea came on 22 March 1773, when he was a month short of his fourteenth birthday. While still a youth saw action in several battles, and became an *alférez de fragata* when only fifteen. His early career included a voyage to the Americas, and then to the Philippines, Later he was wounded in action, taken prisoner by the British, and after his release saw action both at the siege of Gibraltar, and in a battle against Admiral Richard Howe's fleet at the entrance to the Strait of Gibraltar in 1782. A year later he was placed in charge of a shipment of mercury to Veracruz in New Spain, and on the return trip carried safely to Cádiz an immense cargo of bullion. This transAtlantic round trip gained him promotion to *capitán de fragata* in 1784, the same year in which he was inducted into the prestigious knighthood Order of Santiago. By the time he joined with Malaspina in 1788 in drawing up the 'Plan' for a major scientific expedition, Bustamante was a seasoned naval officer, although he had not yet reached his thirtieth birthday.

As far as the outside world was concerned, the scientific motivation for the voyage was emphasized. The Enlightenment goal of advancement of knowledge provided an attractive rationale for the heavy royal Spanish expenditure involved, and for the deployment overseas of a select group of officers on board a pair of specially-constructed vessels with chosen crews. It also provided a fitting motive for asking assistance and gaining information from other European countries that might well have been unwilling to assist a military tour of inspection.[1] That such a major expedition was entrusted to the Spanish navy, long in the forefront of voyages of exploration, is not surprising since 'the Navy as an institution and the great mariner scientists of the period were the prime movers behind the great scientific movement of the Enlightenment in Spain and of its European character.'[2] As Alexander von Humboldt later pointed out, no European government invested more heavily in scientific research in the late eighteenth century than that of Spain. The expedition was a feat of Spanish science, and essentially one of the Navy. Among the fields to be investigated, the most important was that of hydrography with emphasis on cartography, with the many astronomical observations forming the basis of these activities.

It is doubtful whether any of the inner circle at the Ministerio de Marina in Madrid felt that exploration, rather than inspection of the empire, was the primary aim of the expedition. Amid their solicitude for Spain's old and exposed empire, and the increased threat of foreign challenges, royal advisers would have placed defence of Spanish possessions ahead of contributions to mankind. It was evident that a largely unfortified colonial coastline thousands of miles long was badly in need of better charts and maps, and that a naval exploring expedition carrying on board expert chart-makers, men of recent training and experience, would be of immense benefit

---

[1] For institutions and individuals consulted in France, Italy and Britain see María Dolores Higueras Rodríguez, 'The Malaspina Expedition (1789-1794): A Venture of the Spanish Enlightenment', in *Spanish Pacific from Magellan to Malaspina*, Madrid, 1988, pp. 147-63. Full listings are in María Dolores Higueras Rodríguez, ed., *Catálogo crítico de los documentos de la Expedición Malaspina en el Museo Naval*, III, Madrid, 1987, Appendix, 'Fuentes documentales utilizadas por la Expedición Malaspina'.

[2] Higueras, 'The Malaspina Expedition', p. 147.

to such regions. Besides the stated objectives of the expedition, there were secret tasks of a more sensitive nature. These included the preparation of comprehensive reports on the new Russian settlements in northern latitudes, and on the rumoured English settlements in that quarter and in the southwest Pacific. Thus Alaska, Nootka Sound and Botany Bay were regions of special interest. As it turned out, although the expedition's work was conscientiously accomplished, neither scientific nor imperial goals were well served by the final results of the Malaspina voyage, but it was not for lack of effort in the coming months as time was divided between scientific study and the more utilitarian duties outlined in the final instructions.

Some clarification of motives can be gained from the letter written by Malaspina and Bustamante on 10 September 1788 to the Ministro de Marina, Antonio Valdés y Bazán.[1] With exceptional brevity (considering Malaspina's inclination to verbosity), the letter introduced the projected voyage, and while outlining the contributions which the expedition would provide for the scientific world, quickly moved to additional motives for royal consideration. The first of these was to make much-needed charts of the farthest reaches of America and thereby permit the drawing up of sailing directions for the guidance of Spanish mariners. Beyond this task was the need to investigate the political and economic status of the Spanish colonies in the Americas, relative to the mother country and to foreign nations. This would include the trade situation, defensive and offensive capacity, the adequacy of ports and shipbuilding facilities, resources needed for further development, a survey of natural resources, and an analysis of the colonial system of government. It was this last, to which Malaspina devoted much of his time and attention, that eventually brought a premature end to his career. Finally, it is difficult to see how these grandiose objectives could have been crammed into the three years initially proposed for the expedition.

The outline of the voyage called for departure from Cádiz on 1 July 1789, with Montevideo the first port of call. After investigation and resupplying there, the ships were to visit first the Falkland Islands (Malvinas), and then Bahía del Bueno Suceso inside the Estrecho de Le Maire as a preliminary to rounding Cabo de Hornos (Cape Horn). Once on the Pacific side of South America the ships were to follow the coast from Cabo Victoria northward via the Archipíelago de los Chonos to the large Chilean offshore island of Chiloé. Arrival there was scheduled for the end of 1789. The whole of 1790 was to be devoted to a reconnaissance of the west coast from Chiloé in the south to San Blas in Mexico. Lima and Acapulco were to be important stops on the way north, while a search for the Islas del Gallego (thought to lie far to the west of the Galapagos Islands) plus an extended visit by members of the expedition to Mexico City, were also proposed. From Acapulco 1791 would begin with a three-month reconnaissance of the Hawaian Islands, followed by a visit to the coast of 'California', which was to be traced almost endlessly for more than 3500 miles north-westward as far as the ice of Bering Strait. At that point the ships would cross the Pacific Ocean at its shortest distance between America and Asia, and call at Kamchatka. The Asiatic phase of the voyage would continue with a visit to Canton (now Guangzhou), where it was felt that the furs traded in the Pacific North-west could be

---

[1] See pp. 311-12 below for the text of this letter.

sold advantageously. From Canton the vessels would sail by way of northern Luzon in the Philippines to the Marianas, and back to Manila via the San Bernardino Strait, following the route used for more than two centuries by the Manila galleon. A long stay at Manila, capital of the Philippines, was to be followed by a short stay at Mindanao before the ships passed between the Celebes (now Sulawesi) and the Moluccas until they reached the northern coast of Australia. This they would trace westward to the Indian Ocean, and then follow the coastline of the southern continent more than halfway round the continent to Botany Bay. From there the ships would call at Tonga and the Îles de la Société, both visited by Spanish navigators in the 1770s and 1780s. In October or November 1792, the height of the southern hemisphere's summer, the coasts of New Zealand would be followed from north to south. From that last point of exploration, the ships were to skirt Australia, cross the Indian Ocean by established sailing routes, round the Cape of Good Hope, and sail back to Cádiz with an estimated date of arrival of April or May 1793.

This elaborate itinerary underwent many minor alterations and some major changes, the most radical of which was indicated in a letter from Malaspina to Valdés on 15 September 1790 that resulted in the expedition not being involved in a round-the-world voyage after all, even though it was long referred to as such.[1] If a journal entry by Malaspina on 30 November 1789 represents his feelings at that time, he had given up the idea of a circumnavigation long before his letter to Valdés.[2] Rather, the expedition followed a more southerly track on both its outward and homeward Pacific crossings than originally laid down, and left out visits to Hawaii and elsewhere. It then repeated (in the opposite direction) the first ten months of the voyage between Cádiz and Callao, revisiting several of the places touched on during the outward voyage. For reasons that were not foreseen at first, and others that emerged from changing world affairs, the total time spent from departure until the return of the ships to Cádiz was sixty-two months, almost a year and a half longer than originally anticipated.

It is uncertain whether the original 'Plan' or any of the subsequent documentation ever received any serious study from the reigning monarch, Carlos III. Submission of the project came close to the time of his death and therefore near the date of succession to the throne of his son, Carlos IV. Although the change of kings brought no immediate turnover in royal ministers and advisers it eventually proved to be an unfortunate factor in the months that followed the return of the Malaspina expedition. Carlos III had been a farsighted, energetic king, one in touch with the ideas of the Enlightenment and an exemplar of benevolent despotism, although he had not reigned without some opposition. His successor was in no way similar and soon fell under the influence of his principal minister, Manuel Godoy, who viewed the French Revolution with great alarm and resisted any governmental change felt to jeopardize the King's position, or anything that smacked of progress. If Godoy and Malaspina were on the same track of promoting Spain's welfare, they were going in different directions, one regressive and the other progressive. For Malaspina the long-term

[1] See p. 322 below.
[2] See p. 80 below.

consequences were disastrous. After a secret trial he was stripped of his rank and distinctions, and spent almost half of the rest of his life in prison. Of the expedition's records and collections some were scattered and many were impounded and left largely unpublished, unstudied, and almost totally forgotten for a century.

The brevity of the 'Plan' presented by Malaspina and Bustamante suggests that its approval was a foregone conclusion, and that there must be elsewhere substantial supporting documentation spelling out the project in more detail. This assumption would seem to be supported by the fact that within five weeks of initial composition of a document originating in Cádiz, the project received royal approval nearly 350 miles away in Aranjuez, and was almost immediately ready for implementation. This allowed little time for consideration of the implications of the 'Plan', and suggests that the King's approval was purely formal. Although there is no positive evidence, it seems likely that the proposal never advanced further than the desk of the Ministro de Marina, Antonio Valdés, who was not only in great favour with Carlos III, but who also held Malaspina in high regard (possibly because both were members of the Order of St John). It was from Valdés that on 14 October 1788 Malaspina received news that the plan had received royal approval.[1] Exactly two months later Carlos III died, but the project did not suffer any delay from the succession of his son, Carlos IV.

Although approval was granted for the voyage, many details remained to be settled, the most urgent of which was to decide on the vessels to be used. The plan called for two ships which would normally sail in consort, but which, when circumstances required, might sail separately on special missions, and this turned out on occasion to be the case on the long voyage. Ideally, two naval vessels sailing together on the same mission ought to have similar operating characteristics so that their speeds and responses could be predicted throughout the voyage.[2] Even so, to avoid problems when sailing long distances, it was necessary to establish rendezvous points in case of unplanned separations. For example, upon leaving Acapulco in December 1791 as the ships headed for the Orient, crossing an ocean with few intermediate landfalls, any one of which would have been hard to find, the rendezvous set by Malaspina was Agaña, the capital of Guam. If separated, the first ship to arrive would wait fifteen days and then continue to the next rendezvous at Cape Bojeador at the northwestern extremity of the main Philippine island of Luzon. As it turned out, the vessels did not become separated, and despite the ease with which Agaña was identified, the less established but more convenient port of Humatac was selected as the expedition's anchorage. An obvious bonus of having identical vessels in the hands of skilled navigators was that during the entire sixty-two month voyage in good weather and bad, in uncharted seas, and faced with many obstacles, not a single day was lost in one vessel searching for the other.

[1] See p. 315 below.
[2] Some scientific discovery expeditions made use of vessels with markedly different characteristics, one being essentially the exploring vessel and the other being used for support purposes. Others had a large and a small vessel, the latter to be used in areas where manoeuverability was of paramount importance, such as in detailed charting of coastal areas. For example, the plan for the near-contemporaneous voyage of Captain George Vancouver (1791-94) assumed that the close coastal work would be done by the small 131-ton tender, *Chatham*, rather than by the 330-ton *Discovery*. For this assumption, and the problems that followed, see W. Kaye Lamb, ed., *George Vancouver, A Voyage of Discovery to the North Pacific Ocean and Round the World*, Hakluyt Society, 2nd ser., 163-6, London, 1984, I, pp. 36-7.

When the time came to decide on the size, strength and capacity of the ships to be used for the voyage, there was a distinct difference of opinion between Valdés and the captains-designate. Considerations of economy suggested the adaptation of a *bombarda* (a small, mortar-throwing vessel), the *Santa Rosa de Lima*, and matching it with another to be built with the same characteristics. But Malaspina, supported by Bustamante, countered with a proposal for the construction of two state-of-the-art corvettes. Their opinion seems to have been endorsed by the King, who did not want anything to stand in the way of complete success for the expedition. Having vessels built especially for the expedition was felt to be one of Malaspina's great advantages over other contemporary explorers.

Capitán de Navío Tomás Muñoz, commandant of the naval shipyard of La Carraca in Bahía de Cádiz, and well-versed in the art of ship construction, was entrusted with the building of the vessels. The result was identical twin corvettes, christened *Santa Justa* and *Santa Rufina*. Almost immediately these names were dropped, and the vessels were soon sailing under designators more appropriate to their forthcoming mission. The command vessel became *Descubierta* (Discovery) and the consort *Atrevida* (Daring), commanded by Malaspina and Bustamante respectively. Each of these strong little vessels displaced 306 *toneladas*, with an overall length of 120 *pies* and a beam of 31 *pies*.[1] Depth of hold was 15 *pies* with a draught of 14 *pies* when laden. If finished as corvettes they would have carried twenty guns, but for the time being armament gave way to considerations of the special mission for which they had been constructed, bearing in mind that later conversion to normal fighting ships could be easily accomplished. But for the great expedition the corvettes had to serve as transportation, as living quarters for an extended period, as a laboratory and storehouse for specimens waiting to be shipped back to Spain, as a depository for stores, equipment and provisions, and as a workshop for the multiple functions of the expedition. Their slightly reduced firepower included fourteen 6-pounders and two 4-pounders, though they carried a further eight 6-pounders in the hold.

When building was completed only six months after laying the first keel, the corvettes were subjected to the customary but in this case extremely brief shakedown cruise. The vessels left harbour together on 5 July 1789 to test their handling characteristics. Muñoz and the famous Spanish naval officer and scientist, Teniente General de la Real Armada Antonio de Ulloa,[2] went on board the *Descubierta* with Malaspina, while other dignitaries accompanied Bustamante on board the *Atrevida*. Off to an early morning start, the two new naval vessels carried out joint exercises, combined drills, and individual manoeuvres. The result of the one-day trial cruise was that the corvettes proved both seaworthy and manoeuvrable, to Malaspina's great satisfaction. With the sea trials accomplished, the rest of July was spent in preparing for sea, receiving on board stores and additional personnel to bring each complement up to its authorized total of 102, and making final preparations including formal leave-taking.

---

[1] The Spanish *pie* = 0.91 feet. The Spanish *tonelada* was a measure of both displacement and carrying capacity, for which an English equivalent has been notoriously difficult to establish.
[2] See p. xciv below for a biographical sketch.

No expedition such as this stands alone in time or history. There were both distant and proximate precursors, but this is not the place for a treatise on man and exploration. More to the point were those voyages that shaped and gave focus to the Malaspina-Bustamante expedition. First of these was Malaspina's earlier circumnavigation as commanding officer of the Spanish naval vessel *Astrea* from 1786 to 1788, and although that voyage lacked the investigatory element of the 1789-1794 voyage it was Malaspina's training ground. It not only established the general goal of global navigation for the future, more prestigious expedition, but taught Malaspina useful logistical and nautical lessons that he was to apply on the later project. The voyage of the *Astrea* owed much of its *raison d'être* to Spain's newly-revived commercial interests in the Pacific, and was a joint venture between the navy and the Real Compañía de Filipinas. It was not scientific in nature, but one of its specific purposes was to study the logistics and duration of the outward and return routes from Spain to the Philippines going east or west. It also gave Malaspina an opportunity to visit much of the general area he later touched on during the 1789-1794 voyage, and in particular the coasts of Chile and Peru. An immediate foreign precursor to the Malaspina expedition was the voyage of La Pérouse that left France in August 1785, and had been last reported at Botany Bay in January 1788. The course of the French expedition was known in general outline to Malaspina, who attempted to discover more about its final outcome.[1] At several places where he stopped, particularly at those also visited by the French, Malaspina inquired about the activities of La Pérouse, whose reticence about his exploratory efforts was thought to be a cloak for French political and territorial designs. Malaspina made little headway in obtaining worthwhile information from those who had encountered the friendly but secretive Frenchman. La Pérouse was a good listener, but was reluctant to discuss his own activities. While in Chile in 1790 Malaspina took on board several deserters from La Pérouse's ships, but in view of their humble status they were able to shed little light on the motives of the French exploring expedition. Of greater importance as a precursor both through mutual interest in the Pacific and because of the publication in 1784 of the official account, was the third voyage of Captain James Cook.[2] Malaspina was familiar with all three of Cook's Pacific voyages, both from the published accounts, and possibly from stories passed by word of mouth, private letters of inquiry, and other means. A measure of Cook's importance to the Spanish expedition is obvious in the number of times that the British captain's name is found in the version of his *diario* or journal that Malaspina was preparing for publication. If we add to this, the frequent comparisons and incidental references in portions of the archival record that were never intended for publication, the references to Cook reach the hundreds. By contrast, La Pérouse is mentioned only occasionally, and seldom with the same precision that Malaspina could bring to bear on Cook's activities.

It had initially been expected that the most suitable crew members would be recruited from the northern Spanish regions of Galicia, Asturias and La Montaña,

---

[1] In April 1789 Valdés sent Malaspina a summary of La Pérouse's voyage from Manila to Kamchatka that the minister had obtained from Paris. See AMN, MS 278, f. 44.

[2] James Cook and James King, *A Voyage to the Pacific Ocean for making Discoveries in the Northern Hemisphere*, London, 1784.

where it was felt there would be enough seamen inured to the hard life of prolonged voyaging. It had been decided that the crew members should be young men with nautical experience, with probably none over thirty-five years of age. The officers and scientific staff were not subject to a strict age limitation, but most of them were also young (though there was at least one exception, the botanist Luis Neé). Of the 168 crew members authorized for the two corvettes only seventy-two (including fifty-six able seamen and twelve landsmen[1]) were persuaded to join the first recruiting officer. Almost one hundred additional crew members had to be recruited later from the Cádiz area and from nearby parts of Andalusia. Even there, and despite appeals as far as the distant Naval Department of Cartagena on the Mediterranean coast, there was a noticeable scarcity of volunteers for so long a voyage.

Each corvette's complement of 102 included fourteen officers and two staff members: the captain, three senior and three junior officers, the Chief of Charts and Maps and the Director of Natural History (both with Malaspina on board the *Descubierta*), a midshipman, chaplain, paymaster, surgeon, two *pilotines* (equivalent to master's mates in the British navy), and an artist. The *Atrevida* had in lieu of the Chief of Charts and Maps and the Director of Natural History a *piloto* (equivalent to master in the British navy), a botanist and an artist. Each corvette had fourteen petty officers (*oficiales de mar*), fifteen marines (*tropa de marina*), four marine gunners (*tropa de brigada*), fifteen able seamen (*artilleros de mar*), ten landsmen or apprentice seamen (*grumetes*), and eight servants (*criados*) who were assigned to various duties such as cabin boy, orderly, or page.[2]

Selection of officers was not a difficult process, since Malaspina and Bustamante were given a free hand in their choice and already had in mind the men they wanted. They were a select group, with specialities in the theory and practice of astronomy, cartography, hydrography, and natural science. When the expedition leaders were permitted considerable latitude in their choice, they picked men with whom they had served on board ship on hydrographic surveys or on regular naval duties, or in some cases those who had impressed them at the midshipmen's school in Cádiz (Real Compañía de Guardias Marinas) or at the nearby and recently established Escuela de Altos Estudios del Observatorio Astronómico (School of Advanced Studies of the Astronomical Observatory). Several of the officers chosen for the expedition, as well as Malaspina himself, had been involved in a large-scale charting project of Spain headed by the respected naval hydrographer, Don Vicente Tofiño de San Miguel.[3] As

---

[1] Landsman: a person aboard a ship who had not previously been to sea.

[2] See pp. xcvii–viii below for Malaspina's crew lists, which give the individual names of the officers and staff.

[3] This major hydrographic project embraced the entire Iberian peninsula and resulted in the *Atlas marítimo de España*, the first official charting. It began under the direction of Vicente Tofiño in 1783. His survey methods were used by the Malaspina expedition for good reasons, for Malaspina and five of his officers had worked with him. As early as 1778, Alcalá-Galiano had been with Tofiño in the Portuguese Azores, served with him again in 1784, and finally in 1787 was on a Tofiño-led party in Asturias and Vizcaya. Salamanca was also with Tofiño in the Azores. Bauzá served with the great cartographer on the coasts of Portugal and Galicia in 1786; and Espinosa was with him in 1783-84 in the *Santa María Magdalena*. Vernacci spent a tour of duty with the cartographical project. Most important of all, Malaspina was with Tofiño in 1785 and 1787 (on board the *Vivo* along with Alcalá-Galiano). The latter had also served under Antonio de Córdoba y Lazo in the *Santa María de la Cabeza* during a surveying voyage in 1785-86 to the Estrecho de Magallanes.

Plate 2. Real Colegio de Guardamarinas de Cádiz, 1850. Museo Naval, Madrid

Plate 3. Model of the *Descubierta*, by Pedro Sansó Juan, 1967. Museo Naval, Madrid

a result they had gained just the type of practical surveying and charting experience that would be needed on the forthcoming voyage. A key appointment was that of Felipe Bauzá y Cañas, born in Mallorca, who became Chief of Charts and Maps, in a sense the expedition's geographer. He was not previously known as a scientist, but had worked his way up from humble birth as son of a bricklayer, via the Pilotage Corps, to commissioned officer in the Spanish navy. At the time of his recruitment Bauzá was a teacher of fortification and drawing at the naval school in Cádiz, from which he was given leave of absence. During the voyage, besides his custody of charts and maps, he continued to exercise his chart-making ability, and occasionally turned his hand to artistic representations. While on board the *Descubierta* he also kept an account of the voyage.[1] After the voyage Bauzá became conservator and custodian of much of the surviving Malaspina expedition material. In later life he was promoted to *capitán de navio*, but became involved in political affairs and spent the last ten years of his life in exile in England, where he died in 1834. Among the accounts kept by the other officers that written by Bustamante's first officer on the *Atrevida*, Antonio de Tova Arredondo, has material not found elsewhere but remained undiscovered for

[1] 'Diario del viaje alrededor del mundo desde la salida de Cádiz hasta Puerto Jackson', in AMN, MS 479, ff.1-112.

Plate 4. Felipe Bauzá, shown in naval uniform. Anon. Museo Naval, Madrid

almost 150 years.[1] With an interest in indigenous culture in Patagonia, Chiloé, Port Mulgrave, Nootka and Tonga, his manuscript is of exceptional value. Interspersed among observations which a later generation would class as anthropological, Tova presented comments on the history and geography of the various areas visited. A careful study of his account ranks him as one of the expedition's best natural scientists, even though his initial appointment was based on his reputation in ship handling.

In line with Carlos III's strong preference for Spanish citizens, several foreign candidates were rejected for the role of botanist, though as a last resort the Bohemian-born Tadeo Haenke was engaged. The King had antipathy even towards *criollos*, that is Spanish subjects born in the New World, for only two important members of the party were *criollos*, rather than native-born Spaniards. They were the head of the natural history effort, Teniente Coronel Antonio Pineda, born in Guatemala; and Alférez de Navío Francisco Viana, a native of the corvettes' first port of call, Montevideo. Both were children of Peninsular parents who at the time of their birth were serving on tours of duty in the Americas. Of the two, Antonio Pineda played a more important role as head of the three branches of natural history, and his sudden death of apoplexy at Badoc in the Philippine Islands while on an inland excursion was a severe loss to the expedition and was grieved by all. His death was commemorated by a burial mausoleum in Manila with a fitting inscription, but his work, which fell in some measure to his brother, Alférez de Navío Arcadio Pineda, was never brought to full fruition. One of Viana's greatest contributions was to keep an account of the expedition. This was published rather obscurely in Uruguay in 1849, considerably in advance of the official account, and is the first published account in Spanish of the complete Malaspina expedition.[2] Among the junior officers was a midshipman, Jacobo Murphy, born in Cádiz of Irish parents. He was an asset when proficiency in English was needed, although several officers were known to have rudimentary knowledge for reading and elementary conversation. Murphy's value became most obvious during the expedition's call at Port Jackson, New South Wales, in March and April 1793, by which time he had been promoted to *alférez de fragata*.

Tadeo Haenke joined the expedition after a whole series of adventures and misadventures. The young scientist in his late twenties agreed to join the expedition from his home in Vienna, virtually at the last minute because several previous nominees for the position had disappointed the Spanish government. Haenke had been trained in the universities of Prague and Vienna, and was ideally suited for participation in the great expedition, being highly recommended by persons of status in his field of specialization. Having signed his contract, Haenke packed his considerable personal belongings and scientific gear and started as soon as possible for the departure point of Cádiz, where he arrived on the proper date, but just two hours too late for the

---

[1] His account was published by Lorenzo Sanfeliú Ortiz, ed., *62 meses a bordo: La expedición Malaspina según el diario de Antonio de Tova y Arredondo*, Madrid, 1943, 1988.

[2] Or almost the complete expedition; the last entry is at Montevideo on 14 February 1794. The book has two title-pages, giving different end-dates for the journal. The first title, with correct dates, reads *Diario del viage explorador de las corbetas españolas Descubierta y Atrevida, en los años de 1789 á 1794, llevado por el Teniente de Navio D. Francisco Javier de Viana*, Cerrito de la Victoria, 1849. Viana's jounal was republished in Montevideo in 1958.

Plate 6. Luis Neé, probably by García Condoy, c. 1949.
Museo Naval, Madrid

Plate 5. Antonio Pineda, by García Condoy, c. 1949.
Museo Naval, Madrid

Plate 7. Tadeo Haenke (likeness in V. R. Grüner,
Jindy y Nyní, Prague, 1829)

morning sailing. Not wishing to be left out of the excitement of such an important mission, he took the first available ship so as to intercept the corvettes at their first port of call at Montevideo. Almost within sight of his destination, his vessel was caught in a storm and slowly went down, taking with it most of Haenke's belongings, both personal and scientific. However, clutching what he could of his precious possessions including, by tradition, his copy of Linnaeus, the scientist-turned-swimmer was rescued in the nick of time. Again, his arrival on the coast of South America was just eight days too late to achieve his goal of joining the corvettes. Undaunted, and with help from the Viceroy, Haenke made the long trip across the Pampas, over the cordillera of the Andes, and down to the coast of Chile where he at last joined the *Descubierta* at Valparaíso.

Assigned to the *Atrevida* and placed under Antonio Pineda's orders, Haenke was a great asset, performing his duties with professional skill and personal pleasure. He was a jack-of-all-trades, and although listed as a naturalist he was also a good musician playing the harpsichord, a skilled botanist, an astronomer and mineralogist, a medical assistant when called upon, an accurate observer and careful recorder of many of the resources of the areas he travelled through. He knew and wrote in several languages, sometimes changing language several times in the same sentence, and since penmanship was not one of his assets he has left researchers with challenging puzzles. During the final part of the voyage, while the vessels were on their second visit to Callao, Haenke persuaded Malaspina to give him permission to make a cross-continental journey in order to carry out additional scientific studies, and from this he never rejoined the expedition. When the corvettes reached Río de la Plata on their way home, Haenke was still in Upper Peru (today's Bolivia), accompanied by Gerónimo Arcángel, a Filipino gunner who had been assigned to him as a dissector, and merely wrote to say that he was labouring mightily.[1] Some of Haenke's work died with him, but he also sent specimens, artifacts and collections to Europe, both to Spain and elsewhere. He was never careful to distinguish between his private collections and those of the scientific group on the expedition, to the augmentation of the former and the detriment of the latter. One result of this laxity was that he sent many gifts and small collections to friends rather than to the Gabinete Real de Ciencias Naturales in Madrid to which they ought to have been shipped.[2]

Also with the expedition as a botanist was Luis Neé, who had none of the versatility of Haenke and was referred to in disparaging terms by him as a mere gardener. By now well into his fifties, Neé was probably by far the oldest member of the expedition, but this in no way diminished his enthusiastic dedication. Born in France of French parents, he moved to Spain and became a Spanish citizen. He worked under Casimiro Gómez Ortega of the Real Jardín Botánico in Madrid, and spent much time in avid collecting. Before joining the Malaspina expedition, Neé had botanized at length in Navarre and the Basque region. He had also founded the botanical garden in Pamplona, and just before the departure of the corvettes he had submitted a report on the vegetation of Andalucia. As evidence of his tireless enthusiasm, after returning to Spain he submitted a list of 10,622 plants that he had collected on the Malaspina expedition, many of which are still to be found in the Real Jardín

---

[1] Although Haenke lost contact with the expedition, he never lost touch with the royal paymaster in Cochabamba where he settled on a long-term basis, and even disregarded a royal order in 1810 to return to Spain, as if duties with the expedition required his continued investigations in Bolivia. Instead, he pointed out the impossibility of travel overseas at that time of revolution. He lived at his estate at Yuracaré near that Bolivian city until he died from accidental poisoning in 1817, having being paid regularly during the last 22 years of his life despite the official oblivion of the expedition which he first joined in 1790. For some of the newer details of Haenke's life see María Victoria Ibáñez Montoya and Robert King, 'A Letter from Taddeus Haenke to Sir Joseph Banks', *Archives of Natural History* 23 (1996), pp. 255–60.

[2] From this liberality has resulted the recent discovery of artifacts and specimens in the Czech Republic's Bohemian Museum (the Náprstek in Prague). For this, and much more on Haenke's career, see María Victoria Ibáñez Montoya, ed., *La expedición Malaspina 1789–1794, Tomo IV: Trabajos científicos y correspondencia de Tadeo Haenke*, Madrid, 1992.

Botánico.[1] Some crew members also became interested in the scientific aspects of the voyage, including fishing, hunting and botanizing under supervision. The surgeon of the *Atrevida*, Pedro María González, was so interested in zoology that he was relieved of some of his medical duties to allow him to spend more time on it. His counterpart on the *Descubierta*, Francisco Flores Moreno, took over the care of those patients who could not have regular attention from the *Atrevida's* surgeon. Some of the crew turned to as self-trained taxidermists, but with less than total success. Some of their specimens took on strange shapes as the result of the unanticipated settling of the materials with which these enthusiastic amateurs had filled them.

Making their contribution both to the recording of natural history and to much else were the expedition's artists, six in all. The two original artists, José del Pozo of Sevilla and José Guio of Madrid, were replaced, the first at Lima for not having lived up to expectations, and the second at Acapulco because of ill health and his inability to do anything except technical drawings. Pozo was replaced by José Cardero, originally a *criado* or servant on the expedition, and Guio by Tomás de Suria, an artist-engraver from Mexico. Both contributed significantly to the pictorial record of the voyage, as did the Italian-born artists Juan Ravenet and Fernando Brambila, who joined the expedition at Acapulco. Additional artists were recruited locally, especially in Peru and Mexico, to serve with detachments from the expedition involved in field trips ashore.

During the five years and more of the voyage, two developments were noticeable among the officers and men. First was the increasing competence of most of those who joined the newly-launched corvettes in July 1789. Experience at sea had a positive impact on the seamen, whether inexperienced novices or salty mariners. Secondly, the officers not only advanced in terms of their practical skills, but given the long periods together on board ship the junior officers had more than usual opportunity of learning from their seniors, under whose tutelage some of them had already studied at the midshipmen's school or the observatory in Cádiz. This increasing competence among all personnel was reflected in advances in rating among the crews and in promotions for the officers. In the former case the fact that competent individuals were already on board reduced the need for recruitment of specialized personnel to fill vacancies. In the latter case the same applied, and it is evident that some officers were promoted to higher rank than was normally called for in the billets to which they were assigned. As an example, Cayetano Valdés and Dionisio Alcalá-Galiano were promoted to *capitán de fragata*, even though they continued filling posts normally assigned to a lieutenant or *teniente de navío*. Cayetano Valdés, it should be noted, a young officer on the expedition, was nephew of the Ministro de Marina.

A rather different effect of the long absence of the corvettes from their home port of Cádiz was the gradual attrition of the all-volunteer original crews, sometimes by illness, often by desertion when temptation presented itself in port, and occasionally by death. Because of their unsuitability a few others were lost by transfer ashore. It has been estimated that only about forty per cent of the original crew completed the

---

[1] He listed grasses, algae and fungi separately, bringing the overall total to 15,990 dried specimens. See AMN, MS 1407, ff. 79-81, and Iris H. W. Engstrand, *Spanish Scientists in the New World*, Seattle and London, 1981, p. 106.

voyage, or approximately eighty individuals, a figure that includes twenty officers, scientists and artists. Among the crews, the petty officers remained constant, whereas the seamen and orderlies showed a more frequent turnover. The fact that the ships were distant from any normal pool of recruitment resulted in less select replacements. In the original crews one problem was soon all too clear as even before departure from Cádiz venereal disease was evident among the crew, the result of their 'passions and vices'.[1] By the time of the expedition's first port of call at Montevideo the malady was so widespread that a separate ward was set up ashore for those infected. The first efforts to find extra personnel came from the need to replace some of the sick, and also to obtain substitutes for those who had deserted. Probably some, but perhaps not all, of the newcomers realized that they were going on board for the duration of the voyage, however long that might turn out to be. Later, other crew members were obtained from available Spanish naval vessels, Malaspina and Bustamante using their superior rank to acquire such 'volunteers'. As early as Montevideo the turnover of personnel was almost twenty per cent. On the *Atrevida*, twenty men were lost: ten from desertion, four discharged for bad conduct, and six discharged sick. It should be remembered that although some desertions were permanent, at other times the errant personnel rejoined the corvettes, either voluntarily or as the result of bounties paid for their return.[2]

To reduce the inroads of desertion at successive ports of call, shore patrols combed the streets just before the ships' departure in search of those who might have yielded to the temptations of the land. As the expedition progressed, and as it left the more Hispanicized and better organized areas of Spain's empire, replacement of crew members became a matter of marketplace availability. No documentation has survived concerning the terms of enrolment for substitutes, but it is clear that many were foreigners. Several were deserters from the ill-fated expedition of La Pérouse; others were identified by their lack of Spanish, and when on board must have been given orders in the international language of the sea. Yet others included Filipino seamen who joined in Acapulco in an effort to return home sooner than would have been the case had they awaited the annual departure of the Manila galleon upon which they had arrived in New Spain at about the time of the corvettes' first arrival there. An international atmosphere resulted from the presence of Italians, Portuguese, Filipinos, Frenchmen, Irishmen, and persons whose nationalities are not always identified but whose surnames were Wilson, Diser, Maguire, Kliesner, Cline, Redding, Green, Peters and Loftus. Of the many replacements some served only for brief periods, and it can be presumed that none was much interested in the scientific aspects of the voyage. Nor is there any evidence that any such late recruits were rewarded upon completion of their terms of enlistment with promotion or with bonuses for work well done. This is not to say that these persons were

---

[1] See pp. 3, 8 below.

[2] Statistics on desertion are difficult to compile and of doubtful validity. Absence without leave (or even absence over leave) was frequent, and three days absent without leave was considered desertion according to the existing norms. See AMN, MS 427, 'Apuntes, noticias y correspondencia pertenecientes a la espedicion de Malaspina'. Even allowing for these considerations, desertion remained a serious problem. By April 1791, crew-lists drawn up at Acapulco showed that 93 crew-members had been lost to desertion out of a total complement for the two corvettes of 204. See Higueras, *Diario por Bustamante*, p. 220.

Plate 8. Cayetano Valdés y Flores, by José Roldan. Museo Naval, Madrid

Plate 9. Dionisio Alcalá-Galiano. Anon. Museo Naval, Madrid

incompetent, but as replacements of necessity and not of first choice, their best interests had been served by whatever special inducements to enlist they could obtain in distant ports at a time when other replacements were unavailable.

In preparation for departure special care had been taken to ensure plentiful supplies, with store rooms constructed to carry sufficient provisions for up to two years. Space allocated for fire wood could hold enough fuel for six months. In addition to a considerable supply of potable water in casks, each corvette carried an apparatus for providing fresh water, but more natural supplies were not ignored. They watered ship at every opportunity, and on one occasion in the south-west Pacific the crew of the *Descubierta* caught rainwater as it fell on deck and channelled it into containers. In addition to water, an adequate supply of wine was embarked at Cádiz. The common wine, one hundred barrels of it, came from nearby San Lucar, while the wine for special occasions was obtained from distant Málaga. To help keep the crews healthy, the corvettes carried enough sauerkraut, vinegar and olive oil for two to three years. The olive oil, 600 *arrobas*, was obtained from Morón, not far from Sevilla.

Malaspina also had on board orange and lemon juice, although he seems not to have been aware of the growing reliance on lemon juice as an antiscorbutic on British naval vessels in this period. Daily routine on board ship was punctuated by special precautions aimed at preventing contagious diseases. The captains often ordered their crews to clean ship, and to take clothing, bedding and other impedimenta from between decks and air them on an upper deck. Among various checks carried out the surgeons tested the malignity of shipboard air from the hold upwards by means of a eudiometer, the invention of the Italian physiologist Felix Fontana.

Live animals with their fodder were on board from the beginning, and replacement animals were obtained en route whenever possible. Livestock was kept on deck except in extremely cold weather, and the bovine population could be as high as ten head or more. During long stays in harbour animals were sent ashore to graze, and as a result some of them had a long shipboard life. Another important source of protein for the crews was fish. Sometimes fish were bartered from local tribes, at other times they were purchased from suppliers in port, and at still other times they were caught by sending out small boats. In the latter case, fishing did double service, that of adding to the general mess, and providing specimens for the naturalists, who frequently had the artists draw colour illustrations of any special fish caught. Despite the care taken with the provisions, biscuits became infested with grubs in the first weeks of the voyage, and measures had to be taken at the corvettes' first port of call at Montevideo to prevent a recurrence. One result of the amount of provisions taken and the need for their careful stowage was that, as on most expeditions of this nature, accommodation for even the officers was limited.

The resupply of food items, especially live animals and semi-perishable staple commodities, was carried out as far as possible at all the ports visited. In isolated places such as Nootka, where the local Spanish garrison had a thriving vegetable garden and supplied the visitors with a daily supply of freshly-baked bread from their ovens, or in California, where there was a surplus of livestock, reciprocal courtesies were extended. In New South Wales and Vava'u such items as those places had in plenty were acquired as space and need allowed. In turn the paymasters of the *Descubierta* and *Atrevida* were at times ordered to off-load items of which they had plenty or where they could anticipate resupply in the reasonably near future. At any stop, whether at a busy port or an uninhabited harbour, the corvettes took on water and firewood. Away from areas under European control, care was taken that any supplies were obtained with the consent of the indigenous inhabitants, not always a straightforward matter since it was often uncertain who had the authority to permit wood to be cut, grass to be harvested, or water to be carried off. The fact that local guidance was frequently sought in finding the best places for supplies was one way of ensuring tacit approval of the visitors' activities. Although it is not certain whether or not Malaspina recognized that some of the items taken were in public ownership and that granting permission for their use was the prerogative of tribal leaders, he seems not to have had supply problems even at harbours outside Spanish control; but he did encounter other difficulties with aboriginal people, for example with the Tlingit at Yakutat Bay.

At the time of departure from Cádiz, special care was taken to have on board the right items for the long voyage. Replacement clothing was taken such as kerchiefs, flannel and stockings for the crews, together with soap, cigarettes, shot and fishing tackle. For trading with native peoples there were knives, scissors, jack-knives, brightly-coloured cotton and silk handkerchiefs, packages of jet and glass beads, imitation coral, and small mirrors.[1] Time soon proved that these items were not the most attractive merchandise, and new lines were introduced such as abalone shells, blue cloth from Puebla (about 65 miles east-south-east of Mexico City), common cloth from Querétaro (about 130 miles north-west of Mexico City), copper, and even glass window panes. In general, the Spaniards were often a step behind in their attempts to meet the requirements for successful trade with aboriginal peoples.

Part of the customary expedition routine both in areas being inspected as well as those being explored for the first time was to set up the observatory tent at a proper spot ashore. In settled areas it was centrally located in a level place that was usually set aside by the expedition's local hosts. At the first stop in Montevideo the observatory was promptly sent ashore and installed in a house next to the Fort of San José. Soon a second observatory was erected across the estuary of the Río de la Plata in Buenos Aires. In remote areas the observatory was set up in a defensible position near the shore for convenient setting up and taking down. When forges were sent on shore for the blacksmiths they were situated near the observatory, as was the flag-pole for ship-to-shore signals. The observatory thus became the hub of shore-based activities.

Routine observations and measurements were frequently extracted from the notebooks in which they were originally entered and placed in the official journal at appropriate places. Pendulum observations for gravity were not, however, part of Malaspina's initial instructions, but orders were issued on 22 December 1790 by Valdés, and brought to Acapulco by *tenientes de fragata* José Espinosa y Tello and Ciriaco Cevallos, together with a specially designed pendulum. This instrument, which was probably based on one used in 1773 in Spitzbergen by the British naval officer

---

[1] For a list of these goods see AMN, MS 2513, Document 47.

Captain Constantine John Phipps,[1] was sent from England by Teniente de Navío José Mendoza y Ríos, a noted mathematician and astronomer as well as a naval officer, who was in London at the time.[2]

Land expeditions were carried out by small detachments which were given a specific length of time to carry out their tasks. These special trips were often for purposes of natural science, and therefore were accompanied by one or more of the scientists who at times led the group. The detachments were usually accompanied by a military escort from the corvettes, and often took along with them knowledgeable local residents. Areas investigated included the Pampas, the interior of Peru as far as the headwaters of the tributaries of the Amazon, parts of New Spain (where long and detailed investigations were carried out), and several undeveloped regions in the Philippines. As we have seen, Haenke was given permission at Lima to cross the continent on a botanizing trip to Buenos Aires, while Espinosa and Bauzá made their way across the Andes to Montevideo, taking observations as they went for a map of the area. In yet another long overland excursion Luis Neé crossed from Talcahuano in Chile to Buenos Aires, and added to his impressive collection of plants more than five hundred obtained on the Pampas. The reports made by these small groups added much to the mass of detailed data already collected. Such interior reconnaissance was in marked contrast to other oceanic discovery voyages of the period.

Communication between the vessels at sea was normally carried out by brightly coloured signal flags, but when fog, early morning or late afternoon haze, or darkness, obscured these then acoustic signals such as pistol or cannon fire, or (at night) lanterns were used. Rockets were fired for long distance communication. When groups were ashore, a different set of both acoustic and visual signals was employed. Once the corvettes were at sea, if conditions were relatively calm the ships' boats could be used to keep contact. Tomás de Suria's account kept on board the *Descubierta* has an example of such an operation on the high seas.

> On the 21st [May 1791] the ship was hove to and the commander ordered the boat to be put in the water. This was done and Don Cayetano Valdés, Don Secundino Salamanca, the purser, Don Rafael de Arias, and Don Tadeo Kaienk [Haenke], the botanist, went alongside the *Atrevida*. From the *Atrevida* the same proceeding was followed and Don Fabio Aliponzoni, the father chaplain and the surgeon came to our ship. Our commander did this for the purpose of giving out various instructions on the procedure of our voyage to the commander of the *Atrevida*. For this purpose they exchanged some geographical maps of the regions to the north to which we were going and also instructions were issued about how to combine the daily observations for latitude and longitude of the two corvettes.[3]

On other occasions the boats were used to exchange key personnel on the high seas,

---

[1] See Constantine John Phipps, *A Voyage towards the North Pole undertaken by His Majesty's Command 1773*, London, 1774. Malaspina had a copy on the *Descubierta* of this account, which contains an illustration of the pendulum used by Phipps's astronomer, Israel Lyons.

[2] See AMN, MS 1826, f. 112. A full explanation of pendulum observations for gravity is contained in Appendix II below.

[3] Donald C. Cutter, ed., *Journal of Tomás de Suria of his Voyage with Malaspina to the Northwest Coast of America in 1791*, Fairfield, WN, 1980, p. 29.

such as when the astronomical skills of Alcalá-Galiano were needed on board the *Descubierta*, and a mid-ocean transfer was made from the *Atrevida*.

For various reasons, the most prominent of which was the necessity of preventing unauthorized disclosure of information, accounts of the voyage were written only by the officers among the ships' companies. Although these are in large measure parallel narratives, each has something to add to the story of the expedition. Not surprisingly, although disappointing to historians seeking a balanced account of activity, no account exists written by ordinary crew members. Few of them would have been literate enough to keep a journal, and their life on board ship can only be glimpsed through the filter of the officers' journals. In these there is little information on daily life aboard the corvettes, for this was so routine that it was rarely thought worthy of record. Sunday was a day of rest except for essential operations. On holidays Mass was always said, following which the naval penal code was read to the assembled but perhaps inattentive crew. Only occasionally are there comments which tell us something about the social history of the expedition. One early event was recorded when on 28 August 1789, just four weeks out of their home port, the corvettes crossed the equator in 16° west of the Spanish prime meridian of Cádiz. This led to the traditional inititation of neophytes who had never crossed the Line, and were subjected to a masquerade involving practical jokes administered by veteran seamen.[1] On the three subsequent occasions of crossing the Line during the long voyage, there is no mention of any such celebration, probably because of a shortage of the uninitiated. Another recorded social event came when in honour of the visiting crews the local California hosts held a *corrida de toros* using a young bull, and allowed any one to join in. In another of the infrequent references to recreation, on 28 April 1790 Pedro Clain (Cline?) from the *Atrevida* won a contest in marksmanship as the only participant to hit a target at 120 paces.[2]

A major difference that sets the expedition apart from others of the period is that at many of the places visited Malaspina had special status and responsibility. Either as the King's delegated agent or as senior officer present, he had jurisdiction over and responsibility for all Spanish vessels, both merchant and naval. When he was in port, all vessels reported to him their arrival and departure, and were guided while in the area by his instructions. When necessary they gave support to and accepted assistance from his corvettes. At only a few places visited were Malaspina and Bustamante likely to find more senior naval officers present. Even though the Viceroy in Buenos Aires, Lima, or Mexico City, had more status, such viceregal authority was conditioned by royal orders to aid the corvettes in every way. Supplies, payments for officers and crew, and collections of specimens were sent to Spain without hesitation through local governmental channels, while supplies and payments for officers and crews were received from Spain through the same channels. In the matter of forwarding to proper authority artifacts, charts and maps, statistics and reports, the original itinerary gave the Malaspina expedition a great advantage over other contemporary explorers. None of the items being sent had to pass through foreign hands, as had been the case

[1] See Sanfeliú, *62 Meses a bordo*, p. 47.
[2] See AMN, MS 755, Libro de Guardias, *Atrevida*.

with Cook and Clerke, or La Pérouse, when they forwarded reports and other items from Kamchatka. Except for some duplicates of charts and scientific observations carried on board the British merchant vessel *Kitty* when she sailed from New South Wales in June 1793, all Malaspina shipments during the long voyage were sent home in Spanish vessels. Whereas letters and reports followed normal government channels, the shipping of artifacts and specimens was a more difficult business. Identification, labelling, packaging, and precautions against deterioration became a time-consuming business. Although a small initial consignment of boxes was sent home on 8 October 1789 by mail packet from Montevideo, together with news of the expedition, major shipments of artifacts and specimens were dispatched only from Lima, Acapulco and Manila. Each large crate normally contained items representing the work of a single natural scientist.

An additional duty of Malaspina's command was in large measure a result of his mission as explorer-inspector of empire. On several occasions, he had to provide assistance to other ships facing potential disaster. His intervention was required at Acapulco, for example, where at a moment's notice the expedition was called upon in December 1791 to attempt the rescue of a dismasted and rudderless merchant vessel, the *Sacramento*. Only rarely were there on board the corvettes any persons not officially part of the ships' company. The six stowaways at the time of departure from Cádiz were an exception, as was a naval officer, Francisco Quesada, who embarked as a passenger bound for Montevideo. Some five months later he came aboard again while the *Descubierta* and *Atrevida* were at Talcahuano in Chile. On this occasion he came bearing letters and news, the most welcome of which was that Malaspina had been promoted to the rank of *capitán de navío*, and that some of the other officers had also been promoted – not as a reward for their services but in celebration of the coronation of Carlos IV.[1] A second time when passengers were on board came when the outgoing President of the Audiencia of Quito, Juan de Villa Lengua, his wife, family, and all their personal possessions needed transportation from Guayaquil to take up a new post as Regent of the Audiencia of Guatemala. Faced with the vigorous intervention of Governor José Aguirre, accompanied by assurances that the distinguished visitors would be no bother, Malaspina agreed. Their arrival, however, brought some inconvenience insofar as Malaspina and Bustamante had to vacate their own cabins for two months to make room for the guests. Their arrival in Realejo in today's Nicaragua was a personal relief for the commander, and probably others in the officer chain of command who must have been transferred domino-like to inferior quarters. A more useful class of temporary visitors were the occasional experienced pilots who were assigned to assist in navigating unfamiliar waters. One was Piloto José de la Peña, who using his own vessel and at times going on board one of the corvettes, lent his expertise along the coast of Patagonia. Another was Domingo Velázquez who came on board in Valparaíso. He acted as pilot as far as Callao because of his experience in command of merchant ships during which he had become expert in coping with the fogs and calms of the long coastal stretch.

Time in port was hardly one of leisure. Small boat duty included loading and

---

[1] See pp. 154–5 below.

unloading, sounding the harbour, and making charts and plans of local areas and anchorages. There was also the work of procuring supplies, seeking helpful records, gathering information from local residents, making drawings, and cultivating the friendship of the populace. Periods in port provided the opportunity to scrape, paint, repair, replenish, refurbish, and improve the vessels for the tasks that lay ahead. The first time the corvettes availed themselves of local facilities was immediately on arrival at Montevideo. Repairs to both vessels were urgently needed, and were more extensive than might have been expected for newly-built ships. For several days as many as thirty carpenters and twenty-five caulkers (this number included those from the ships' companies) were hard at work caulking seams, making alterations, and effecting other repairs. It was the first of several times that the expedition had the advantage of local resources to improve the corvettes' seaworthiness. Cook and La Pérouse, on the other hand, had only very limited repair resources, with most jobs usually carried out by their ships' own crews.

Stops at established ports brought the greatest disciplinary problems, absence over leave and desertion being the most frequent offences. The extremely long stay of over four months at Callao in 1790 was brought about by the need to await the end of the rainy season along the planned route to the north. Throughout the long wait only skeleton crews of one officer and four or five men were kept aboard each of the corvettes. Most of the remainder of the ships' companies were assigned to duty ashore where they were billeted in various accommodations, and some were even given leave of absence. In more remote locations or during the infrequent occasions when the corvettes visited non-Spanish ports disciplinary offences were almost non-existent. Although the overall voyage lasted sixty-two months, much of that time was spent in port or at anchor off some known coast. In view of the special requirement to enquire closely into the condition of Spanish overseas possessions, it is not surprising that approximately fifty per cent of the expedition's time was spent in harbour. About another ten per cent was spent in anchorages with limited or non-existent harbour facilities, leaving the corvettes actually at sea for slightly more than forty per cent of the voyage. The voyages of Cook and La Pérouse had shown a very different proportion of time at sea.[1]

In order to create a body of documentation helpful to the expedition's mission, several officers were given the duty both before and during the voyage of searching all available archives for appropriate reports, charts, maps and other papers. The Archive of the Indies in Sevilla was the first repository visited. Subsequently, additional researchers were sent to obtain useful documentation and charts. For example, while in Chile Cayetano Valdés was sent to search the regional archives at Santiago. Some documents were copied *in toto*, abstracts were made of others, while yet others seem to have been bodily removed from the archives. By combining archival sources, local inquiries and personal observation, expedition members provided brief histories of many of the regions they visited, beginning with southern Argentina, the Falkland Islands (Malvinas), Chiloé and Juan Fernández.

---

[1] For example, on Cook's second voyage of more than three years the *Resolution* was in harbour for only about thirty per cent of the time.

Local informants such as the Mexican savant Father José Antonio Alzate (1739-1799) were generous with both time and information, and their ideas became incorporated into expedition reports.[1] In a long stay on shore in Mexico by a party from the corvettes, Antonio Pineda in his position as head of the three branches of natural history on the expedition ordered his brother, Aracadio Pineda, to make extensive copies of materials concerning New Spain's frontier provinces. These included regions not visited by any detachment from the expedition such as areas of present-day Texas and New Mexico, Baja California, and Sinaloa and Sonora. This extended archive of material made available to Malaspina's officers a mass of information that they did not personally gather, and it covered areas well outside the itinerary of the expedition. It explains why today among the holdings of the Museo Naval, Madrid, there are materials which would not normally be expected in a maritime museum, many of them being part of the legacy of the scientific exploring groups from the expedition. It also explains why on occasion the writings of Malaspina and his companions were at considerable variance from reality since at times both archival and printed materials were accepted as true without checking, or were taken to apply to a wider geographical area than intended by the original authors. An example of this was the idea that all the Indians of both Alta and Baja California were a single cultural group. This was based on the earlier descriptions of the Jesuit Fathers Venegas and Burriel of the Indians of Baja California (whom Malaspina had not encountered), and were applied by him to the natives of Alta California, a few of whom had been observed during a two-week stay by the expedition in 1791.[2]

Of special concern to Malaspina while in New Spain was the fact that the entire viceroyalty suffered from a poorly defined position on the globe, its capital of Mexico City being mislocated by what seemed to be as much as three to five degrees of longitude. It was there that Malaspina sent the person he considered to be his best astronomer, Dionisio Alcalá-Galiano, to make precise observations. If these could be coordinated with others made at Veracruz, Puebla, Tehuantepec and Coatzacoalcos, then with all positions correctly determined a satisfactory coordination between the Atlantic and Pacific Oceans could be made.

Unfortunately, there is scant documentation recording the impact of the Malaspina expedition on those areas which it visited. At first glance it is quite surprising, almost incredible, that such an important group would not have left a more obvious paper trail amid the archival records of viceroyalties, captaincies general and provinces visited. On reflection, it is not surprising for in one capacity the expedition members were investigative reporters, making inquiries, having access to confidential files, and

---

[1] There was a close relationship between Antonio Pineda and Father Alzate, who became the inseparable companion of the chief of natural history during his excursions in New Spain. Beyond this, many of the documents copied in Mexico reflect Alzate's participation. He was the source of much of the material copied into the expedition record concerning Mexico City which is in AMN, MS 568, 'Virreinato de Mexico', Vol. II.

[2] Although never a resident of Baja California, Father Miguel Venegas, who lived in Mexico City, collected much primary material from that Jesuit mission field which he incorporated into a lengthy manuscript. This original was sent to Spain where it was shortened and given focus by Father Andrés Burriel. It was published at Madrid in 1757 under the title *Noticia de la California*. Neither of the Jesuit Fathers had ever been to Baja California.

obtaining information to be taken back to Spain. In the case of brief visits only the facts of arrival and departure, and lists of the items given to or by the corvettes seem to have been recorded. During extended stays, more detailed documentation was kept, often in order to balance local ledgers. Later searches of local archives for documents that would have been of most interest to the expedition scientists reveal that such items are sometimes missing. These were presumably the records removed in their original form by Malaspina's officers instead of making copies or abstracts of them. Such was the free hand that had been given to the visitors by royal authority that they seem not to have hesitated to make use of their privileged position. As a reciprocal courtesy, Malaspina gave local officials duplicate copies of charts made on board, thus providing a tangible benefit of the expedition's visit.

Subsidiary explorations were sometimes more detours of one or the other of the corvettes rather than detachments. During the voyage the corvettes operated separately on six different occasions, spread over more than nine months and representing fifteen per cent of the total time away.[1] The longest separation was 102 days in 1791 when the *Descubierta* upon departure from Panamá made an independent visit to Central America, stopping at Realejo from which she sailed to Acapulco to wait for her consort. Meanwhile, the *Atrevida* headed first to the Cocos Islands and then to Acapulco. From there she went northwest to San Blas for two weeks before returning to Acapulco where she rejoined the recently-arrived *Descubierta*. A second lengthy separation was for seventy-two days from early December 1793 to mid-February 1794 during the final stages of the voyage. The *Descubierta* carried out running surveys of the coasts of Tierra del Fuego and Patagonia, made a return visit to Port Egmont in West Falkland to make gravity observations in the farthest point south visited by the expedition, and then headed for a rendezvous at Montevideo. The *Atrevida* spent that time fixing the position of Islas Diego Ramírez, which was in some doubt, visiting Puerto de la Soledad in East Falkland to report on the Spanish settlement there, and seeking the non-existent Islas de Aurora, before making her scheduled rendezvous with the *Descubierta*. Besides these independent excursions, there were occasions when side trips were carried out in vessels acquired on a short-term basis.

In those places where the expedition's scientists encountered native peoples some attention was paid to indigenous culture, but it was not until they came face to face with groups that had little or no contact with Europeans that they could be considered as ethnographers (a term not yet invented by the scholarly world). After departure from the Río de la Plata, first contacts were made with the Patagonians, and this gave officers a chance to observe one Amerind culture at first hand.[2] Of special importance, the artists were given an opportunity to record for posterity the visual details of a group that was soon to undergo major changes in its cultural orientation. Another ethnographic investigation in the first year of the voyage was at Chiloé, where new information was obtained about the little-known Vilches Indians.[3] In these inquiries Malaspina's scientists were able to develop techniques that would stand

---

[1] See the route map on end papers.
[2] See pp. 84-90 below.
[3] See pp. 137-8 below.

them in good stead when later in the voyage they encountered native peoples outside the normal sphere of Spanish influence.

The first visit to an area that for many years to come would remain the subject of international dispute came early in the voyage when from Puerto Deseado the expedition paid a mid-summer visit to Port Egmont on West Falkland Island and spent a week there in December 1789. This visit was prompted by the need to take on water, to rate the chronometers, and to survey the anchorage, one already partially known to the maritime world. The corvettes anchored off the remains of the ill-fated British settlement on Saunders Island. As legacies of that brief occupation, European scurvy grass and wild celery grew in abundance.[1] Pineda and Neé went ashore to collect plants and became discoverers of several new species. Game hunters were also sent ashore, as well as a recreation party. The latter group started a fire which got out of control and spread rapidly despite all efforts to extinguish the blaze. The result was severe ecological damage to the island's fragile botanical cover. A related short-term problem was the resulting smoke cover which greatly complicated the celestial observations of the astronomers trying to determine the exact position of their anchorage. The stay of the corvettes in the Falklands and their departure date were being timed to coincide with the proper season for negotiating the difficult passage around Cabo de Hornos, which was easiest in the southern hemisphere summer months of December and January. By February 1790 the corvettes had arrived safely on the Chilean coast, and Malaspina was ready to resume the expedition's work.

From time to time the expedition collected information of current political or economic importance, for example on the sequence of events that led to the Nootka Sound crisis. Information was also gathered on the potential role of forward bases in the maritime commerce of the Pacific, and the relationship of these to the larger picture of world trade. The original 'Plan' put forward by Malaspina and Bustamante had included a three-month visit to the Hawaiian Islands between January and March 1791. The motive for such an extended visit was nowhere stated, but both political and scientific considerations may well have been involved. In retrospect, the change of orders that cancelled Malaspina's visit to the islands denied the Spanish government a final chance to lay claim to them or at least to investigate the possibility. It seems hardly likely that if the Spanish corvettes had been in the Hawaiian Islands for ninety days, that Malaspina would have resisted the impulse to take symbolic possession of at least one of the islands. Such an act would probably have been considered as a reaffirmation of whatever claim had been made in 1555 when Juan Gaitán explored and charted some islands in a similar latitude. Although the longitude of Gaitán's reputed discoveries was far distant from Hawaii, it is interesting that Malaspina thought that his calculations

> strongly supported our suspicions that Islas Sandwich of Captain Cook were those named as Monge, Ulua, etc. on Spanish charts. They were discovered by Juan de Gaitán in 1555 and located some 10° east of the new position determined by the English.[2]

[1] European Scurvy Grass (*Cochlearia officinalis*), an introduced species, and Wild Celery (*Apium Australe*).

[2] Malaspina's journal entry, 10 January 1792 (to be printed in Vol. II of this edition); see also Donald C. Cutter, 'The Spanish at Hawaii: Gaytan to Marin', in *Hawaiian Journal of History*, XIV, 1980, pp. 16-25.

If a Spanish takeover was still a possibility, however remote, it would have been based on the long-held assertion that Spain possessed exclusive sovereignty over the Pacific Ocean. This pretension was about to be rejected, but it was a claim that Malaspina assumed was well-founded and defensible. Had a visit to the Hawaiian Islands been an important matter, Malaspina when leaving Acapulco in December 1791 after his second stay there could have followed a course to reach those islands on his westbound track across the Pacific. It would have added to the distance to be sailed, but it would have been possible. Instead the corvettes followed the centuries-old sailing route of the Manila galleons from Acapulco to the Philippines by way of Guam. Spanish interest in Hawaii was satisfied by Manuel Quimper's one-month visit in 1792 to the island of Oahu in a vessel from San Blas. He gathered some political and natural history information, and made some charts; but his observations were not in the same category as those by Malaspina's well-trained scientists and officers would have been.

That science could co-exist with national political interests was shown on those occasions when acts of possession were carried out. Clear evidence that eighteenth-century exploring expeditions had dreams of empire can be found in the acts of possession carried out on the Alaskan coast both by Malaspina and by two of his non-Spanish predecessors, Cook and La Pérouse.[1] These provide good examples of the ancient rites, with European exponents of the Enlightenment apparently having no qualms about their cavalier extinguishment of Native sovereignty by such minimal acts as they performed. Malaspina's act of possession at Port Mulgrave (Yakutat Bay) was one of only two he carried out on the voyage. Malaspina cut short what had been a lengthy and tedious ritual for over three centuries in the New World. Stones were gathered to form a pyramid, at whose base was deposited a bottle containing a new silver peso and a document noting that Malaspina and Bustamante had 'discovered this port on 20 June 1791, and named it Desengaño, taking possession in the name of His Catholic Majesty'. Appended were the geographical co-ordinates which, combined with the newly-minted coin, served as proof of place and date. Almost two years later Malaspina performed a rather different act of possession in the Tongan Islands in the distant South Pacific. Whereas the ceremony at Yakutat Bay was mentioned only briefly in Malaspina's journal, that of May 1793 in Vava'u was recorded in some detail. There was no reference this time to any authenticating coin, but an emphasis instead on the perceived enthusiasm that the inhabitants had for their new political status. There was a large Native attendance, and the inscription indicated indigenous approval, so fulfilling the oft-repeated requirement of the consent of the governed. The document recorded that Malaspina and Bustamante,

> having explored the entire archipelago around Vavao, they took possession of it in the name of His Catholic Majesty, unfurling the flag at the site of the observatory, and with this solemn act accompanied by not only both crews saying seven times *Viva el Rey*, but also by the assembled natives who followed the chief Vuna repeating an equal number of times, *Vavao foxa España*, that is, 'Vavao is a child of Spain'.[2]

---

[1] Cook took possession of Cook Inlet in June 1778, and in July 1786 La Pérouse 'bought' an island in Lituya Bay from a native chief, and carried out an act of possession there.

[2] Malaspina's journal entry, 30 May 1793 (to be printed in Vol. III of this edition).

By the end of March 1791 Malaspina in the *Descubierta* had reached Acapulco, while Bustamante in the *Atrevida* was farther north at San Blas. After visiting the Viceroy in Mexico City, Malaspina had intended to rejoin Bustamante, and sail to the Hawaiian Islands. Time did not allow, Malaspina considered, for the exploration north along the coast from California to Bering Strait that had been envisaged in the original 'Plan'. Then everything changed as instructions reached Malaspina from Madrid to sail immediately to the northwest coast to latitude 60°N and search for the passage between the Pacific and the Atlantic claimed to have been discovered by Lorenzo Ferrer Maldonado in 1588.[1] It was following its departure from Acapulco headed northward that the Malaspina expedition was able to give most attention to those scientific goals that have often been put forward as the motive for the entire expedition. For almost two years up to this point the expedition's efforts had been directed towards those aspects of the 'Plan' that can most properly be classified as inspection of empire. And although from this time on there were periods during which imperial goals remained important, much of the remainder of the long voyage was focussed less on inspection and more on those forms of scientific exploration associated with the Enlightenment. Prominent here was investigation of the life style of the native peoples – on the Alaskan coast, at Nootka, in the Philippines, at Port Jackson in New South Wales, and at Tonga. Tova and Cevallos took special interest in linguistic and other anthropological aspects of these investigations.

Supplementing this field work were the drawings by the ships' artists: José Guio, José del Pozo, José Cardero, Tomas Suria, Fernando Brambila, Juan Ravenet, and even the expedition's geographer, Felipe Bauzá and the naturalist Antonio Pineda. One of the most enduring legacies of the expedition was the copious artistic archive amassed during the voyage, as the artists drew both places and people.[2] They took the opportunity to sketch both colonial residents and little-known native peoples, their ceremonials and everyday activities, their dress and their artifacts. Had it not been for the emergence of José Cardero from the humble status of orderly to that of unofficial artist, we would be without dozens of detailed drawings of the Pacific coast of the Americas. After Cardero's detachment from the main expedition to serve on board the schooners *Sutil* and *Mexicana*, Juan del Río Miranda, a carpenter on board the *Descubierta*, turned his hand to drawing. The many drawings give a visual dimension to the anthropological details, and help to identify the scenes described in the journals. Since further details were often added in the finished versions, the drawings generally are most reliable in their original state than when efforts were made to meet artistic expectations. We have examples of drawings being 'completed' by another, more skilled artist, and of more alterations made in the process of preparing a drawing for copper plate engraving. So, Brambila finished a Cardero drawing of Salish Indians in their canoes suspiciously following the schooners *Sutil* and *Mexicana* to which he added an intrusive kayak, a native craft several hundred miles distant from its home area. Occasionally, different versions were prepared by the same artist, as

[1] The new instructions are contained in a letter from Valdés to Malaspina, 22 December 1790 (AMN, MS 583, f.112). The text of this letter will be printed in Vol. II of this edition.
[2] A careful and near-exhaustive study of the graphic aspects of the Malaspina expedition is Carmen Sotos Serrano, *Los pintores de la expedición de Alejandro Malaspina*, 2 vols, Madrid, 1982.

when Brambila during the ships' stay at the new British settlement in New South Wales drew two versions of a scene. One, which was sent to Spain, showed some of the convicts as a chain gang; the other, presented to the governor, Philip Grose, omitted the chains.[1] These and other variations clearly indicate that the artist did not always function as a camera to capture what was being involved, but that the artistic representation might be completed in stages. For example, Suria was given six months following his detachment from the *Atrevida* to complete the drawings that he had made during the expedition's five-month stay on the Pacific Northwest Coast and in California. This time frame for completion was then extended to eight months, after which his work was shipped to Spain where items were being selected for making copper plate engravings for the planned publication of the expedition's findings.

In addition to his artistic work, Suria kept a journal of part of the voyage, and this differs from Malaspina's as well as from the journals kept by other officers such as Tova and Viana. An example of the difference can be seen in the disparity between Suria's personal journal and the official one when a week-long storm was encountered in early August 1791 off the Queen Charlotte Islands. Suria described it as so strong that the ship had to be handled with every care, with tremendous rolls and terrifying darkness making for conditions that Malaspina considered to be the worst he had experienced since the ships left Spain. Suria wrote that the commander tried to hide his anxiety, but he himself thought that had it not been for

> the wonderful construction of our vessels...we would have without doubt perished...
> There was not a man who could keep his footing, simply from the violence of the wind, so that besides the mountains of water and foam which swept over us, there arose from the surface of the water small drops of spray forming a strange and copious rainfall never before seen.[2]

By contrast, Malaspina's journal covers in two sentences the continuous storm, during which, he noted, the corvettes 'had opportunity to prove their outstanding characteristics'.

During his time on board Suria attempted to obtain geographical information from a *pilotín*, Joaquín Hurtado, only to be told that the Ministro de Marina had given special orders that no information about the voyage should be disclosed until the minister himself should authorize it. It was reasoning easily explained since premature leaks of information would have prejudiced the overall impact of the voyage. Similarily, the British Admiralty at the time of Cook's voyages had attempted to prevent the publication of unauthorized accounts by ordering that all journals and logs kept on board the discovery vessels should be handed in and sealed at the end of the voyage. Unfortunately, failure to disseminate widely the results of its discoveries robbed the Malaspina expedition of much of the credit that otherwise would have accrued to it. This is particularly so in botany, zoology and ichthyology, though less so in geography, cartography and toponomy where some names bestowed by or in

[1] See Peter Barber, 'Malaspina and George III Brambila and Watling: Three Rediscovered Drawings of Sydney and Parramatta', in *Malaspina '92 – Jornadas Internacionales*, Cádiz, 1994, pp. 357-69.

[2] Cutter, *Journal of Suria*, p. 71.

honour of the expedition have survived. Along the Pacific Northwest Coast especially, place-names from Malaspina's visit still exist. It is noticeable that many of them are personal names – a departure from the long-standing tradition of using the liturgical calendar as a source of toponyms. In addition to honouring individual expedition members, Malaspina or his officers also commemorated leading naval figures of the day, often those associated with the Naval Department of Cádiz. A point of land with its now-associated town, stretching far north into Clayoquot Sound and shielding the deep fiord known as Tofiño Inlet, honours the great Spanish cartographer. José de Mazarredo, Miguel José Gastón, Francisco Javier Winthuysen, and Juan de Lángara had their names attached – at least temporarily – to geographical locations, as did Antonio Porlier and the Conde de Florida Blanca. Other names were placed on the map later in remembrance of the activity of the Spanish explorers. Malaspina's name is remembered in the Pacific Northwest, especially in the awesome Malaspina Glacier, given that name in 1874. There is an institution of higher learning in Nanaimo, British Columbia, called Malaspina University College, in an area never visited by the expedition proper. Some names can be identified as commemorating incidents of exploration during the voyage, although most such names disappeared as later explorers substituted their own place-names. The location of one short-lived toponym was northwest of Mount St Elias which was named Valle de Ruesga after the home town in La Montaña of Antonio Tova, second-in-command of the *Atrevida*.

The expedition's visit to Port Mulgrave (Yakutat Bay) provided an occasion to see an Indian group at close quarters. It provided the opportunity to visit natives who had experienced only minimal contact with Europeans. They had briefly met Russian traders, the Englishman James Colnett, and most importantly George Dixon. Dixon, who had already been on the coast as an armourer on Cook's third voyage, entered the bay in 1787 in command of a fur-trading vessel. It was he who was responsible for the name Port Mulgrave and other non-Hispanic names that Malaspina used in the area.[1] The Malaspina expedition had arrived at the opening of Yakutat Bay in latitude 59°15'N. because it had orders to search for Maldonado's strait in or near latitude 60°N. In anticipation of a potentially hostile reception, and the possibility of longer than normal trips by the ships' boats, some significant changes had been made. The limited capacity of the most-used boats was increased by rebuilding them, one at Guayaquil and the other at San Blas, so that they could carry two months of supplies if necessary. Whereas the original boats were like longboats, the new ones were intended for heavier duty, and included a small shelter at the stern to permit escape from the weather and even provide room for sleeping. During the fifty-day voyage north, out of sight of land, preparations were taken against possible native hostility on the northwest coast. At weekly inspections all hands were reviewed on the quarterdeck with their arms and clothing. Cayetano Valdés and a select group of men, mostly Galicians, were chosen to man the special longboat which had been modified at Acapulco to withstand attack. Each member of the detachment was equipped with a gun, two pistols, a cutlass and a knife, and this safeguard was not without its subsequent value. Monday was usually set aside for small arms drill, while

---

[1] See [William Beresford], *A Voyage round the World 1785–1788 by Captain George Dixon*, London, 1789.

as the time of first contact drew near, the master gunner exercised the cannon on almost a daily basis.

The Malaspina expedition had no commercial motives for its visit to Port Mulgrave, and there was careful observation of the culture of the local Tlingit inhabitants during the nine-day stay. Short word lists were made, while the artists Suria and Cardero recorded events, drew various natives, and illustrated customs that were at the same time being recorded in the various expedition journals. Even so, relations were not uniformly harmonious. As on other voyages, cultural misunderstanding was a potential source of trouble, and some actions by the inquisitive strangers were not always appreciated by the Indians, who in turn demonstrated a not easily satisfied curiosity about the visitors. One such case concerned the interest of the natives in what they assumed was an Indian slave – in fact a Filipino crew member – whose liberty they wanted to purchase. Since he looked like one of them, they were reluctant to believe that he was really a member of the visitors' crew. It was at Port Mulgrave that Malaspina and his men came closest to armed conflict with any native group. Malaspina's journal gives little space to a nearly fatal confrontation, but Suria goes into some detail. The episode was illustrative of what could happen when cultural differences, misunderstood signals and mutual distrust coincided. A signal from Juan Vernacci that he was preparing to bring the portable observatory on board was taken to be a sign for help against the natives. An armed party led by Malaspina and Valdés went ashore and confronted a large group of natives at gunpoint, while Bustamante fired a cannon from the *Atrevida*. In Suria's version of events, violence was narrowly averted when natives paddled past making signs of peace while hoisting on high a pair of trousers that had been stolen from the expedition, and which they assumed was the cause of the strangers' show of force.[1] Much was accomplished in the short stay at Port Mulgrave, but Malaspina, who took personal charge of the exploration by boat of the inlet, soon discovered that it offered no way into the interior.

After the stay at Port Mulgrave, the expedition had only brief contact with the natives of the coast as it trended north-west. Even so, this fleeting proximity led to a realization that the native peoples of the Alaskan coast possessed distinct patterns of life and were much more culturally diverse than had originally been thought. It was also apparent that unidentified ships had been in the area, since the natives tried to entice the Spaniards to come ashore to trade. The general aspect of the Alaskan coast discouraged Malaspina from making any prolonged attempt to find Maldonado's strait. The ships headed for Prince William Sound, but strong winds prevented them entering it, and they bore away back down the coast on their way to their next intended port of call at Nootka. The great peak of Bering's Mount St Elias, frequently mentioned in the journals, and depicted in several of the artists' drawings of the region, was a strong indication that if any strait existed it would have to be farther south. Although the ships kept a respectful distance off the coast until they neared Nootka, a series of interesting coastal views survive from this stage of the voyage,[2]

---

[1] See Cutter, *Journal of Suria*, pp. 41-2. Suria's description will be given in full at the appropriate place in the text in Vol. II of this edition.

[2] These seem to have been drawn by Bauzá. See Donald C. Cutter, *Malaspina and Galiano: Spanish Voyages to the Northwest Coast, 1791 and 1792*, Vancouver, 1991, p. 65.

more artistically attractive than the simple coastal profiles that were often made as guides to navigators.

Nootka Sound was the next stopping-place where native culture was seen at close range. All the expedition's activities during its sixteen-day stay there were greatly helped by the existence of the small Spanish establishment of Santa Cruz de Nutka founded in 1789 at Friendly Cove. This first non-native settlement in the region was on a small island just off the central west coast of Vancouver Island. It stood as evidence of an effort to strengthen Spanish claims in the Pacific Northwest in the face of Russian and British intrusions, as well as the interests, little-heeded as yet, of the young United States. At the time of Malaspina's visit its future was precarious, for British and Spanish emissaries were on their way to Nootka to put into effect the terms of the Nootka Sound Convention of October 1790 which laid down that the Spaniards should restore to Britain land and buildings allegedly seized the previous year. Among the garrison were those who were of considerable assistance to the expedition. One, who had already become proficient in the Nootkan language, acted as interpreter, while Pedro Alberni, the army officer in command of the First Company of Catalonian Volunteers at the establishment, had gone far to re-establish relations with the chief at Nootka, Maquinna. These had been strained since an incident in July 1789 in which a local chief had been shot and killed by an over-eager Spanish sentry. Malaspina acknowledged how useful Alberni's experience was, as well as the fact that many of the garrison possessed rudiments of the local language. All this permitted a fuller enquiry into both the native culture and the natural history at Nootka than the expedition had time or knowledge for elsewhere. In gratitude for the help he had received, and for the daily supply of vegetables and freshly-baked bread, Malaspina and his officers were generous in their gifts: cloth, flannel, tar, medical supplies, and broth tablets. To these were subsequently added more medicines, four casks of San Lucar wine, and a month's supply of foodstuffs, these including biscuit, dried vegetables, bacon, and vinegar. Then came a whole miscellany of useful items – tobacco, tackle, plates, cups, clay pots, iron hoops, hammers, nails, paper, metal plates, flags, lead lines, and gun parts. Alberni was given cigarette papers, files, flints, caps and cloth. Finally, in order to assist Alberni's daily meteorological observations the scientists left behind a Reaumur scale thermometer. During the stay, crew members repaired firearms and farm tools, while the surgeon of the *Descubierta*, Francisco Flores Moreno, showed the garrison how to make a beer of pine needles and other ingredients, a brew widely thought to be antiscorbutic.

Malaspina also gave a series of presents to Maquinna, including two canoe sails, four glass window panes, a sheet of copper, some yards of greatly-esteemed blue cloth, and a few articles of hardware. As a return, Malaspina suggested that the chief should re-establish his headquarters near the Spanish post, and told him that he would prohibit Spanish sailors going to the dwellings of Maquinna's people. He also promised that when the Spaniards retired from the area they would turn over their establishment to the inhabitants, and that the commandant's large quarters (then under construction) should become Maquinna's house. In response to these promises the chief assured Malaspina of his friendship and indicated that the Spaniards would always be owners of the area which they occupied at that moment. For his part,

Malaspina was reassured that he was successful in his attempts 'to ratify our friendship with him [Maquinna] and to demonstrate, by means of gifts of great value, our desire for a solid and lasting peace'.[1]

Although mentioned only briefly in his journal, one development at Nootka that made Malaspina uneasy was the fact that Father Nicolás Loera, the chaplain from the naval vessel *Concepción* stationed at Nootka, had over the months purchased from Maquinna and other chiefs some twenty-two captured slave children. The price per child was one or two sheets of copper, or a musket, or a few yards of cloth. The chaplain had bought these children in order to supervise their manners and social behaviour, and in particular their religious training. From Nootka they would be sent for their clothing, food and instruction to military persons in San Blas who would care for them. Although Malaspina understood that the adopted children might be bettering their lot, his personal view was that of many in the Enlightenment when he wrote that

> the natural repugnance toward slavery of one's fellow humans and the fear that trustees of these children might under the cloak of religion try to justify a type of permanent dominion over these unfortunate beings, induces me to desire either that a limit be placed on these acquisitions or that the law concern itself with their future well-being with attention to the inclinations that motivate them, to the purity of our religion, and to the inalienable rights of man

Despite the enslaving of children, in many ways Malaspina considered the inhabitants of Nootka to be 'quite advanced in civilization' compared with other peoples of the Pacific coast north of the Tropic of Cancer.[2]

Although it is a subject not mentioned in the expedition's journals, Malaspina also spent considerable time pondering whether or not the Nootkans were cannibals. Maquinna had been stigmatized as a cannibal by John Meares[3], the half-pay British naval officer who had arrived at Nootka in 1788 in command of a trading vessel, and whose account of his voyage, published in 1790, Malaspina now had on board. The latter's enquiry into the matter was not made simply out of curiosity, scientific or otherwise. If the British assumptions about Maquinna's taste for human flesh, regardless of whether this was in the context of a ceremonial occasion, were correct, then this would have implications for the Spanish presence at Nootka, and for the whole question of relations with Maquinna and his people. In an involved discourse, Malaspina proved to his own satisfaction that the chief was not, and never had been, a cannibal.[4] This is a view strongly supported by anthropologists, and by the

---

[1] See Malaspina's journal entry for 27 August 1791 (to be printed in Vol. II of this edition); and, for similar statements by two of Malaspina's officers, Francisco Javier de Viana, *Diario del viaje explorador de las corbetas Descubierta y Atrevida en los años 1789 a 1794*, Montevideo, 1958, p. 222, and Ciriaco Cevallos, 27 August 1791 in AMN, MS 775, Libro de Guardias, *Atrevida*, f.70v.

[2] See AMN, MS 330, 'Examen Politico de las Costas N.O. de América'.

[3] Meares wrote: '... from his manifest confusion in conversing on this subject, and various other concurring circumstances, which will be related hereafter, we were very much disposed to believe that Maquilla [Maquinna] himself was a cannibal.' John Meares, *Voyages made in the Years 1788 and 1789, from China to the North West Coast of America*, London, 1790, p. 125. For more on this contentious issue see Christon I. Archer, 'Cannibalism in the Early History of the Northwest Coast: Enduring Myths and Neglected Realities', *Canadian Historical Review*, LXI, 1980, pp. 453-79.

[4] Also in AMN, MS 330.

Mowachaht-Muchalaht people who are today's descendants of inhabitants of Nootka in the late eighteenth century.

The visit to Nootka was satisfactory in several different ways. In terms of anthropology and ethnography it was perhaps the most productive stay on the long voyage. Despite the shortness of the visit, expedition members had been able to pursue their researches into the life style of the natives, and one result of these was Secundino Salamanca's dissertation on the customs, usages and laws of the area's inhabitants.[1] Additionally, there had been hydrographic surveys of the sound and some of its interior channels, much collecting of botanical specimens, and some superb drawings by the artists. On the practical side, the ships had been resupplied with wood, water and fresh provisions. Finally, the garrison had been supported, and relations with Maquinna put on a firm footing.

The expedition's next call, at Monterey, was primarily for rest and recuperation after the difficult navigation of the North Pacific. The ships' two-week stay in September 1791 provided a glimpse of a new province in the process of development, since Upper California had been founded only twenty-two years earlier, and some of the original settlers were available as sources of information. The first visit in 1791 was supplemented by a second and longer visit by members of the original expedition who a year later returned to Monterey in the schooners *Mexicana* and *Sutil*, commanded by Cayetano Valdés and Dionisio Alcalá-Galiano. The two visits allowed for an independent assessment of the progress of two of the most important institutions of Spanish colonial control, particularly ones used in areas of recent occupation. These were the military *presidio*, as the frontier army posts were called, and the mission, paramount among Spanish efforts to convert and civilize indigenous peoples on the frontiers. Although in great measure these institutions were symbiotic, the limited resources of the frontier areas forced them to compete for government support. The naval visitors observed both the *presidio* and the mission not only closely but sympathetically, and produced a balanced evaluation of them. The journal-keeper of the 1792 visit described the varied activities of the garrison of sixty-three men, most of them married. 'The lack of colonists of any other kind has obliged these soldiers to employ themselves in all of the occupations necessary for a civilized population. As a result, one can be seen acting as a sentinel of the guard; another herding livestock, roping an animal, or driving a cart; still another building a wall, making a door, or sewing shoes; yet another arming himself to go into the interior along the roads to carry information to other presidios or missions. 'A handful of soldiers', the journal continued, was enough to cope with any hostile Indians.[2] The writer was no less enthusiastic about the Franciscan missionaries from the College of San Fernando

---

[1] Ibid., under the title of 'Apuntes incordinados a cerca de las Costumbres, usos y leies de los Salvajes havitantes del Estrecho de [Juan] de Fuca'.

[2] Donald C. Cutter, *California in 1792: A Spanish Naval Visit*, Norman and London, 1990, pp. 121, 124. The authorship of the 1792 journal is not clear. Part of it was included in the two-volume work published at Madrid in 1802, *Relación del viage hecho por las goletas Sutil y Mexicana ... .* The much longer manuscript version, and specifically the portion concerning California, is in the handwriting of the artist José Cardero (see AMN, Vargas Ponce, MS 1060), but logically the account should have been the responsibility of the expedition's commander, Dionisio Alcalá-Galiano.

de México. He praised them for 'their modesty, piety, austerity of customs, diligence in providing for the spiritual good of the natives, and their graciousness of conduct'. One of the missionaries was 'in charge of instructing the converts and the children in the Spanish language and in the dogma of our religion, and the other of the direction and instruction in cultivation of the soil, the mechanical skills, and domestic service'.[1] All this activity was carried out under the direction of the Father President of the missions, Fermín Francisco de Lasuén (referred to in Malaspina's journal and in other expedition documents as Father Matías Lasuén). In 1792 he supplied the visiting officers with an unusually interesting document. This was a catechism, plus a 'dictionary' in the form of word lists of the local Rumsien and Esselen languages, which since these native groups were soon to become culturally extinct forms the most reliable written access to their language and culture.[2]

Malaspina's visit to Monterey gave the opportunity for field work and collecting by the expedition's botanists. Although such activity is scarcely mentioned in Malaspina's journal, here and at other ports of call it resulted in large collections of specimens being sent to the Real Jardín Botánico and to the Gabinete Real de Ciencias Naturales in Madrid. They were accompanied by detailed, precise descriptions of flora, often complete with coloured illustrations done by the artists. During the 1791 visit, Haenke is credited with the botanical discovery of the *Sequoia sempervirens*, the majestic California coast redwood, as well as for its transportation to Spain, where, planted in the area above the Alhambra of Granada and in the gardens of the Casita del Príncipe at El Escorial, the species found a suitable climate. Sometimes the botanists received specimens – roots, branches, leaves of plants and trees – without ever having seen or been near the living organisms. From specimens obtained for him by expedition members in California, Luis Neé, who had been assigned to the Mexican detachment, was the first to scientifically describe two of the local oak species which were subsequently given the names *Quercus agrifolia* Neé and *Quercus Lobata* Neé for the Coast live oak and the White oak respectively. It was unfortunate that because of Malaspina's disgrace on his return, and the collapse of plans to publish a comprehensive account of the expedition and its findings, most of its descriptions did not find a place in the nomenclature of the scientific world.

California also provided an example of Malaspina's intention to provide evaluations of the timber available at his various ports of call. Since Upper California was largely unknown, an extended document was prepared about the timber resources of Monterey and its environs. Usefulness was determined by a timber's commercial value. The document, entitled 'Report on lumber produced at Monterey and useful for shipbuilding and for houses',[3] described the salient characteristics of some fifteen local trees, and also indicates the use that the expedition made of local timber resources. A topgallant yard, a studding sail boom, and (for the *Atrevida's* launch) two masts, were made from Bishop pine or Douglas fir. The whole report reflects the interest in Monterey as a potential ship repair base, a factor that from first contact had

---

[1] Cutter, *California in 1792*, pp. 129, 131.
[2] An English translation of this document is in ibid., pp. 147–55.
[3] AMN, MS 126, 'Relación de Maderas', in Pacífico América, tomo 1.

been a motive for any proposed Spanish occupation of the coast of Upper California. Despite the timber report, Malaspina's stay at Monterey fell somewhere between inspection of empire and exploration. The area was being developed, but was not yet a viable part of the Spanish colonial empire, and almost nothing was known of the interior. In consequence, Malaspina treated this distant province of New Spain more as a region to be explored than to be evaluated.

The *Sutil* and *Mexicana* of the 1792 excursion were tiny, 46-ton schooners that had recently been built at San Blas, and had been placed under Malaspina's orders. Their visit to Monterey had come at the end of a detailed survey of almost three months of the inner recesses of the Strait of Juan de Fuca. Carrying orders both from Malaspina and the Viceroy at Mexico City, the Conde de Revilla Gigedo, the schooners had left Nootka in June to complete the surveys of the strait started by Quimper in 1790 and by Eliza in 1791, and in a wider sense to complete that search for the entrance of the Northwest Passage that had sent Malaspina north into Alaskan waters the previous year to look for Maldonado's strait. Their survey, Malaspina told the commanders, would 'decide once and for all the excessively confused and complicated question of the communication or proximity of the Pacific Ocean and the Atlantic'.[1] It would have been a blunder if Malaspina had not ordered this final effort, only to have a foreign expedition uncover the fabled seaway. This point was given added emphasis when the two schooners, while engaged in their explorations off the mainland coast of the Strait of Georgia behind Vancouver Island, unexpectedly met British naval vessels under the command of Captain George Vancouver, a skilled hydrographer who had been on the coast with Cook in 1778. Both commanding officers followed their instructions to co-operate with any foreign vessels that they encountered, and in addition to working briefly together, an exchange of charts made it unnecessary for the Spanish vessels to enter the complex southern arm of waterways which was to be named Puget Sound after one of Vancouver's officers.

On 31 August the *Sutil* and *Mexicana* arrived back at Nootka to complete the first continuous circumnavigation of Vancouver Island. With this detachment from the main Malaspina expedition was José Cardero, whose work, drawn with great attention to detail, provides many of the early visual images of the northwest coast. One particularly significant contribution was the series of native types drawn by him in his characteristic painstaking style, which preserved for posterity aspects of native culture before it was altered by intensive contact with the European world. Most of these 1792 drawings were not seen by the main group, and only became part of the collections of the voyage after they had been taken to Spain by Cardero himself, or sent there by the Viceroy in Mexico City. Another major contribution of the 1792 expedition was its completion of the charts of the extended Strait of Georgia that had been started by the 1790 and 1791 partial surveys. They provided definitive evidence that the idea of a navigable Northwest Passage was a myth. Although the entry that they had followed was called the Strait of Juan de Fuca, this was merely the name

---

[1] Malaspina's instructions are in Archivo General de la Nación, Mexico City, Marina 82, ff.193 et seq.; the Viceroy's are also in Marina 82, ff.204 et seq., with a copy in AMN, MS 619, ff.1-7v.

given in hopeful anticipation that it would prove to be the broad, penetrating water-way described in the account of Fuca's apocryphal voyage of 1592. Unlike other materials from Malaspina's voyage the journal of the voyage of the *Sutil* and *Mexicana* was published within a relatively short time of the expedition's return. It was printed at Madrid in 1802, together with a handsome atlas; but with no mention of Malaspina's part in organizing the voyage.

Not the least significant aspect of the voyage of the *Sutil* and *Mexicana* was its effect on the complements of the *Descubierta* and *Atrevida*. Four officers from the corvettes were reassigned to the schooners in what was was expected to be a permanent change of duty. Dionisio Alcalá-Galiano and Cayetano Valdés were given command of the schooners, with as their seconds-in-command two other officers from the main expedition, Secundino Salamanca and Juan Vernacci. José Cardero was on the *Mexicana*, as artist and much else, because the arrival to join Malaspina of two skilled artists, Fernando Brambila and Juan Ravenet, recruited in northern Italy, made young Cardero surplus to requirements and for the first time permitted his appointment as official artist. All except Alcalá-Galiano had been with the corvettes on the northwest coast in 1791; he had spent that time on expedition business in the Viceroyalty of New Spain. In addition to the four officers and Cardero, at least nine other crew members from the corvettes were detached to help man the schooners. The detachment of these fourteen men from the *Descubierta* and *Atrevida*, combined with the recurrent problems of illness and desertion, resulted in a thinning of the original ranks, and an urgent need for replacements. Partly compensating for the loss of four experienced officers had been the arrival of Tenientes de Fragata Ciriaco Cevallos and José Espinosa y Tello, who had joined the ships at Acapulco in March 1791. They had arrived from Spain bringing with them two Arnold chronometers as well as the specially designed pendulum obtained from London and mentioned earlier.[1]

While on the northwest coast of America, the expedition members had learned more about the potential of the maritime fur trade and the outlet for pelts in China. Spain had obvious advantages in the future development of the trade. Not only were its nationals near the fur-bearing areas, but it had ready access to great amounts of the beautiful abalone shells on which the trade depended, together with Mexican copper, and cloth from New Spain for warm outer clothing. What was not well understood was that other trade goods of Spanish and Mexican manufacture did not have the same appeal to the natives of the northwest coast as those of competing nations, and there was also a reluctance by the Spaniards to use firearms and alcohol in their trading operations. For these and other reasons, Spain never entered successfully into the fur trade, and a visit by the *Atrevida* to Macao in 1792 also showed the uncertainties of the Asian market.[2]

Malaspina left Acapulco on 20 December 1791 after his second visit there, and headed west across the Pacific to the Philippines. The length of the expedition's stay of almost ten months in the islands was partly explained by weather conditions, but it

---

[1] See p. li above.

[2] See James R. Gibson, *Otter Skins, Boston Ships, and China Goods: The Maritime Fur Trade of the North-west Coast 1785–1841*, Seattle, 1992, especially Chapter 1.

gave the opportunity for a separate voyage by Bustamante in the *Atrevida* to Macao. This was the first and only call by the expedition to that half of the world which shortly after Europe's discovery of America had been assigned by papal decree to Portugal for evangelization and occupation. The reasons for the visit included the desire to check the longitude of Macao, obtain some new instruments as well as supplies for the artists, inspect the operations of the Philippines Company, and to dispose of several hundred furs that had been obtained the previous year by the crews of both corvettes while on the Pacific Northwest coast. Since no regular Spanish naval vessel had been seen at Macao, it was essential for Bustamante to observe the fullest protocol, and he was careful to make friends in all possible quarters. Rainy weather during the ten-day visit hindered some of the crew's activities, but they obtained much local help, took on water, and charted the harbour, while Brambila drew views of the city. Unfortunately, the efforts to obtain a good price for the furs were not successful, and since these efforts were being made in the interests of the crews as a whole, this failure was felt throughout the expedition.

In the Philippines there was time for various detachments to carry out surveys of the maze of islands that had been under Spanish sovereignty but not complete Spanish control for more than two centuries. This led to increased security measures being taken during the expedition's stay, but the long stay allowed for extended anthropological investigations and for yet another of Malaspina's 'reflexiones políticas'.[1] The visit began at Palapa on the island of Samar, and ended at Zamboanga, the principal settlement on the southern island of Mindanao, with much of the time being spent on the main island of Luzón. The time spent in the Philippines provided reminders of the various forms of help that Malaspina received during his expedition: from viceroys and governors at the top levels of colonial bureaucracy, to local scientists and ordinary wellwishers, and perhaps above all from priests, both secular and regular. This latter aid is not surprising. The priests were part of the Spanish system of control over its colonial empire, and this was particularly so of the missionary fathers. Because they were usually the best-educated persons in isolated areas, with a strong feeling of duty to church and state, the priests were in an ideal position to help. Their isolation from civilized society made them the more ready to share their knowledge with members of the expedition. A particular recipient of such aid was Tadeo Haenke, who several times was accompanied on his field excursions by local priests. With them he enjoyed collecting specimens and visiting sites of natural phenomena. On one occasion, for example, a local priest, Rafael Benavente, OFM, went with Haenke and Pineda to the summit of an active volcano, Mount Mayón (also known as Mount Albay), because they all had an interest in volcanology. It was on one such excursion that Pineda died, in June 1792. The special relationship which Haenke developed with the local priests probably owed something to his Jesuit-oriented scientific training. During one lengthy trip on Luzón, Haenke made friends with the Bishop of Bigán, who gave him letters of assistance to subordinates, who were ordered to render assistance to the botanist. Similar messages were sent to secular clergy, and to Augustinian and Dominican missionaries who were active in the Ilocos

---

[1] See AMN, MS 621, ff. 170-231v.

province in northern Luzón. From such helpful contacts Haenke and other expedition members developed an almost universally favourable opinion of the priests, if not always of the mission system itself, which was out of tune with the Enlightenment concepts to which most of them were dedicated. Besides the logistic support, the knowledge of local history, and the field companionship that they provided, the priests were frequently able to supply reports, maps, and rudimentary studies of the native languages that they had made during their attempts at religious instruction.

From Mindanao the *Descubierta* and *Atrevida* sailed for the south Pacific, but their anticipated visit to the southwestern extremity of New Zealand's South Island met with complete disappointment. Following in the wake of Cook's first two voyages, Malaspina had intended to call at Dusky Sound, there to make gravitational observations with the simple pendulum. Contrary weather made the vessels anchor off nearby Doubtful Sound instead. Even there observations were truncated after only a few hours ashore in what appeared to be an area uninhabited by humans but alive with very hungry mosquitoes (or, more likely, sand flies). The permanent record shows only a good chart of part of Doubtful Sound, made quite rapidly by Bauzá. In recognition of the Spanish contribution of 1793, most of the names given by the Malaspina expedition at that time have been officially reassigned to the map of New Zealand.[1] Continued bad weather caused the visitors to abandon their efforts after a long afternoon and head away, with the result that the New Zealand stage of Malaspina's agenda was a nearly fruitless two thousand mile detour on the long run between Zamboanga and Port Jackson in New South Wales.

Although when Malaspina's ships reached Port Jackson the visiting Spaniards might easily have been viewed as agents of an alien power, on the surface they were received most civilly in acknowledgment of the expedition's role in the advancement of science. A despatch from London ordering the governor of the infant colony to afford assistance to Malaspina if he reached New South Wales, had been received almost three years earlier,[2] and it helped that the visit coincided with a brief period when Great Britain and Spain were allies. Malaspina's journal entry on the day of arrival, 12 March 1793, revealed only that the purpose of his call was to take on wood and water, give relief to the ships and their rigging after ninety-seven days at sea, add to their botanical and zoological collections, continue the gravity observations, and allow the crews some rest. All these were valid reasons, but they helped to conceal the additional motive of collecting data about the strength and purpose of the British colony. In the 'Plan' of his voyage Malaspina had alluded to this, and in his confidential reports there is little doubt but that he was engaging in what at a later date might be called espionage.[3] Without saying anything that sounded remotely like a protest against the British colonization of New South Wales, Malaspina told his superiors that the real objective was not to establish a penal colony to rid major cities at home

---

[1] For this, and much else to do with Malaspina's brief visit, see John Hall-Jones, *Doubtfull Harbour*, Invercargill, 1984, and John Hall-Jones, *Fiordland Explored*, Invercargill, 1990.

[2] Robert J. King, *The Secret History of the Convict Colony: Alexandro Malaspina's report on the British settlement of New South Wales*, Sydney, 1990, p. 2.

[3] See his 'Examen político de las colonias ynglesas en el Mar Pacífico', and 'Apuntes sobre la colonia inglesa de Puerto Jackson', translated and printed in ibid., pp. 92-125, 132-50.

of unwanted residents. 'The transportation of the convicts', he wrote, 'constituted the means and not the object of the enterprise. The extension of Dominion, mercantile speculations, and the discovery of Mines were the real object.'[1] It is reasonable to assume that the local British officials were aware of Malaspina's hidden agenda, but they seemed to have given no hint of this. On the contrary, the acting governor, Major Francis Grose, provided an area for the expedition's observatory to be set up, together with the brick hut nearby that had been built for Bennelong, the Aborigine befriended by the colony's first governor, Arthur Phillip. This small building, situated conveniently near the corvettes' anchorage, became the headquarters for the expedition's activities on shore. In return, Malaspina hosted a dinner for Grose and his officers, and it was a special treat for them that the main course was a heifer that had been taken on board in California in 1791. It was only the second time that fresh beef had been served since the colony's establishment five years earlier. Gifts were exchanged; one of the corvettes' chaplains gave the colonists some viticultural tips that he had learned in his youth in Málaga, that wine-producing region of southern Spain, and Malaspina left with the colony's surgeons some Guatemalan quinine that had been taken on board more than two years earlier. As in the *Atrevida*'s visit to Macao, the officers took great care to observe proper decorum, while the crews maintained a subordination and discipline considerably superior to their levels of deportment when in areas under Spanish jurisdiction. It was unfortunate that when the expedition's main tasks had been completed, and the crews had free time on their hands, they succumbed to the wiles of the local convict women, who were described by Malaspina as being more dissolute even than the prostitutes of Tenerife in the Canary Islands encountered by the colony's surgeon, John White, on his outward voyage[2]. As far as the native Aborigines were concerned, the expedition members seem not to have much direct contact with them, and they were scarcely mentioned in Malaspina's journal. In his 'Apuntes' Malaspina devoted more space to them. He described their physical appearance in some detail, but took much of his comment on their customs from Cook's report of his contact with them in 1770, agreeing with him that they were 'the most miserable and least advanced nation which exists on earth...without agriculture and industry, and without any product which would attest their rationality'.[3]

The final, and one of the most productive, areas of anthropological investigation was at Vava'u in the Tonga Islands, where Malaspina's ships arrived in May 1793. While in the Tongan Islands on his third voyage, Cook had heard about Vava'u ('Vaughwaugh' in Cook's phonetic spelling) in 1777, but had not reached the island itself. Four years later Francisco Mourelle[4] had visited the Vava'u group which he named the Archipelago of Mayorga; but in his journal Malaspina used the native name, shown as 'Vavao'. Despite the relative shortness of the visit (thirteen days) and the lack of any Spanish residents, much useful work was done by the members of the

---

[1] Ibid., p. 96.
[2] John White, *Journal of a Voyage to New South Wales* [1790], Sydney, 1962, p. 54.
[3] King, *Secret History*, especially pp. 144-5.
[4] Mourelle's Galician name was often misspelled as Maurelle, a version closer to its pronunciation.

expedition. This was especially so of the ever-curious Ciriaco Cevallos with his linguistic interests, and Malaspina's journal entry on the visit concludes with a substantial word-list. Earlier experience at Nootka, the Philippines and elsewhere stood the ships' scientists in good stead. Relations with the natives, and particularly with the chiefs, were good, and the visit occupies what at first sight seems a disproportionate amount of space in the journals. This may be in part explained by the fact that the stay at Vava'u represented the final visit by the expedition to a place of unique interest, almost unknown to Europeans, and this may have been very much in Malaspina's mind when he prepared his journal for its hoped-for publication. Apart from his detailed journal entries, Malaspina also composed a long manuscript report on Vava'u, its products and its inhabitants. This fulsome treatment of such a brief stay is a reflection of Spanish scientific curiosity, of the general friendliness and co-operation of the native inhabitants, and perhaps an indication of possible official interest. It will be remembered that it was here that Malaspina carried out only the second act of possession performed on the voyage.[1]

Because of the outbreak of war in Europe the last phase of the Malaspina expedition was in large measure anti-climactic. Instead of a triumphal homeward leg of the long voyage, the *Descubierta* and *Atrevida* were delayed in the Río de Plata area so that along with several other warships they could serve as escorts for a convoy of merchant ships bound for Europe. The risk that Spanish vessels with their important cargoes ran of being taken by privateers or enemy warships while on the high seas, although not great, resulted in the corvettes' final duty being a strictly military one. The shadows of European conflict cast gloom over the homecoming of September 1794, particularly since after a deserved period of leave most from the corvettes' complements were reassigned to normal duty on board other ships or were held in a state of readiness at one of the Spanish naval departments. The casualty figures sustained by the crews on the long voyage are surprisingly difficult to obtain. There are records that show that twenty-two named individuals died, mostly on board the corvettes, while ten others were left behind at local hospitals and were said to be gravely ill.[2] In addition, a considerable number of deserters were known to have had health problems, and possibly did not survive.

With the conclusion of the expedition, Malaspina took time to summarize the contribution of each officer. These evaluations were intended for their service records, and included recommendations for promotion. Malaspina and Bustamante had few problems with their officers. The only officer with whom Malaspina had less than cordial relations was José Espinosa, who joined the expedition at Acapulco in 1791, but details about the exact nature and result of such problems as existed are inconclusive. Ciriaco Cevallos was another who was not highly praised by Malaspina at the end of the voyage, being criticized for lacking tenacity and tending to laziness; even so, his contribution to the study of native peoples was second to none among the officers. Evidence of the skill with which Malaspina and Bustamante selected

---

[1] See p. lix above.
[2] The totals come from the Libros de Guardia and the chaplains' list of fulfilment of Easter communion.

their officers for the voyage can be seen in the number who later held high rank in the Spanish navy. Galiano, Viana and Valdés all commanded ships-of-the line at the Battle of Trafalgar, where Galiano was killed. Later, Bustamante became Director General of the Navy, and Valdés was appointed Captain-General of the Navy. Espinosa became head of the Depósito Hidrográfico, while Bauzá was responsible for preserving much of the material from the expedition even though he spent his last years in exile.[1] Apart from Malaspina and Bustamante, who were promoted soon after their return from the voyage, five other officers who had served on the *Descubierta* or *Atrevida* reached flag rank or its equivalent – Galiano, Valdés, Espinosa, Tova, and Viana. The careers of others from the expedition also prospered. Brambila became an official court painter, head of an art school, and an artist of repute many of whose later efforts are on display today in Spanish museums. Guio did fine work with a Spanish botanical expedition to the Antilles in 1797 led by the Conde de Mopox y Jaruco. Cardero became a successful bureaucrat in the Cádiz area. The surgeon of the *Atrevida*, Pedro González, later published an important treatise on health at sea much of which reflected his participation on the Malaspina voyage.[2] As far as the *Descubierta* and *Atrevida* were concerned, on their return from the voyage they were refitted and placed in regular naval service. For the 20-gun *Atrevida* this lasted until 1807 when it caught fire and burned while in harbour at Montevideo, by coincidence its first port of call after its commissioning in 1789. The *Descubierta* had a much longer post-expedition life, being retired from naval service after almost thirty-seven years, and sold for salvage to Felipe Biera for 151,011 *reales de vellón* at Cádiz in 1826. By then Malaspina's corvette had outlived most of the men who had sailed on its maiden voyage.

Although Malaspina denied that his expedition was intended to rival or surpass those of Cook, there is no doubt that Cook's voyages were the yardstick against which Malaspina measured his own. Cook and Malaspina were alike in emphasizing the scientific aspects of their expeditions, although their voyages were also accompanied by strong politico-economic motives. From the beginning, Malaspina realised that there was more to Cook's voyages than a quest for scientific knowledge, as could be seen by the several acts of possession performed by the British explorer. In the same way that other exploratory ventures had multiple motives, so did that of Malaspina and Bustamante. At one level there was the advancement of science in line with the spirit of the Enlightenment, that international movement which had been slow in reaching the Iberian peninsula. This was certainly the most widely-publicized reason for Spain setting on foot so ambitious an expedition, and its seriousness can be seen in the amount of correspondence between expedition members and savants from several European countries. A second reason, and the one which involved the greater amount of time and effort by the expedition, was to check the status and condition of Spain's overseas possessions. Malaspina and his officers spent months both consulting with local officials and independently investigating various aspects of regional history, economy and society, while at the same time the scientists pursued

[1] See Andrew David, 'Felipe Bauzá and the British Hydrographic Office 1823-34', in *Malaspina '92*, Cádiz, 1994, pp. 235-41.
[2] Pedro María González, *Tratado de las enfermedades de la gente del mar*, Madrid, 1805.

their own researches. It was his ideas on Spain's colonial empire and his outspoken advocacy of change that was to make Malaspina the subject of criticism over what were considered his liberal, even seditious, views on colonial reform. In other ways, too, the voyage of Malaspina and Bustamante, differed from Cook's voyages. Those had been followed by swift and widespread publication of their results, whereas there was little dissemination of the records of the Malaspina expedition. This was despite the fact that it would be hard for any other exploration venture of the period to match the materials associated with the Spanish expedition. In the Museo Naval alone, there are 1284 documents on the organization and progress of the voyage, together with more than 300 journals, 450 notebooks of astronomical and hydrographic observations, 1500 hydrographic surveys, 183 charts, 361 coastal views, and more than 800 drawings of places and persons, animals and plants. All this is evidence that when the *Descubierta* and *Atrevida* returned to Cádiz they brought with them 'one of the largest stocks of news and knowledge ever assembled by an expedition'.[1]

The corvettes were at Zamboanga, far removed from world events of importance, halfway round the world in distance and perhaps eight months in travelling time when a political change took place in Spain that eventually resulted in Malaspina's downfall and in the expedition's nearly total eclipse. In a fateful development on 15 November 1792 the Spanish chief minister, the Count of Aranda, was suddenly relieved of his duties. If this had been known to members of the exploring expedition, there might have been a reaction of some surprise, and considerably more at the news that Aranda's place had been taken by Manuel Godoy, a young court favourite not yet thirty years of age. The change marked the end of the stable and progressive era initiated by Carlos III, and ushered in a new regime guided by the inexperienced and frequently venal Godoy. It led to a scramble for preference, and even for survival. The resultant uncertainty led to rumours, intrigues, plots, distrust and secret deals which all helped to produce the atmosphere of unpredictability that greeted the corvettes on Sunday, 21 September 1794. It turned out to be a stormy future into which Malaspina sailed without chart or pilot.

Malaspina, although generally informed of conditions at home, clearly did not pay sufficient heed to the dangerous cross-currents of the Spanish royal court. With almost childlike simplicity he became embroiled in an attempt to replace Manuel Godoy as chief minister.[2] His misjudgments at this time suggest that once ashore he was incapable of navigating with the same skill that he had shown at sea. He had spent many hours on board ship compiling what should have been secret, or at least confidential, recommendations, and that he should have circulated such material to various officials after his return seems in retrospect to have been almost a death wish.[3] It is not clear whether Malaspina fell into a trap carefully laid by Godoy, or whether it was the navigator who, with his openly declared wish to be useful to his adoptive

---

[1] Higueras, 'The Malaspina Expedition', p. 156.

[2] The replacement proposed by Malaspina seems to have been the Duke of Alba. See Eric Beerman, *El diario del proceso y encarcelamiento de Alejandro Malaspina (1794–1803)*, Madrid, 1992, pp. 62, 72, 84, 89, 91.

[3] Some indication of the critical nature of many of those recommendations can be glimpsed in the *Introducción* with which Malaspina intended to preface the account of his voyage. See pp. lxxxi–xc below.

country, left Godoy with no alternative but to strike first.[1] It was while he was busy preparing the results of his voyage for publication that his ultimate misfortune occurred. He was planning to return to his native Italy on leave when he was seized under cover of darkness as he returned from a party, was searched, and imprisoned. He was held incommunicado, his papers were scrutinized to find incriminating material, and the evidence was presented to the Consejo de Estado. Although the hearings were interrupted before a final verdict was reached, Malaspina was stripped of his rank and benefits, and sentenced to life imprisonment (which at this time was ten years and a day). He had depended too much on his friendship with Antonio Valdés, soon to be dismissed as Ministro de Marina, and on a fickle public's admiration for him as a successful explorer. It is surprising that no efforts seem to have been made to aid a figure of such importance, but there is no known evidence of any attempt at intervention on Malaspina's behalf by his fellow officers and friends. Such a move would have involved some personal risk for the intercessors. It did not help Malaspina's cause that between his return and his trial both Vicente Tofiño and Antonio de Ulloa had died, for these were senior naval officers of considerable prestige who had been associated with the great voyage. A curious note on Malaspina's service record states that 'This officer commanded a round-the-world expedition and upon his return he wrote the private life of the Queen María de Borbon for which reason his position was taken from him and he was incarcerated in the fort indicated in the margin [San Antón]'.[2] This explanation of Malaspina's downfall is as interesting as it is unlikely, and probably reflects the secrecy of his *proceso*, of whose charges few would have been aware.

Once imprisoned in the Castillo de San Antón at La Coruña, and with other avenues of assistance closed, Malaspina tried to get powerful outsiders to intercede for him. It was a painfully slow business, although it finally brought results when Napoleon intervened on Malaspina's behalf. At this time the Consejo de Estado was reconsidering Malaspina's case, and although it had not yet acted, it fell in with the strong suggestion made by Napoleon's diplomatic representatives, and on 15 March 1803 Malaspina was freed. The rest of his life was anti-climactic, for freedom brought no recognition of his efforts as an explorer. Among the terms of his release was that he was never again to set foot in Spain, which in the circumstances would not seem to have been a punishment. But Señor Alejandro Malaspina (he was no longer a naval officer) was anxious for a complete pardon, one that would re-establish his prestige and restore him to active duty, or failing such vindication at least give him his retirement benefits. Even though his parents had died without his ever having seen them again, his best option seemed to be to return to the land of his birth, there to begin a new life in a region that had much changed. Although he had friends there with whom he had corresponded for many years, it was hardly the scene of his childhood memories, for he had not lived there since the family had left in 1762 when he was

---

[1] Writing eighty-six years after the event, Marcos Jiménez de la Espada referred to 'a voluminous body of papers' relating to the proceedings in Malaspina's case: see his 'Una causa de estado'. These papers have been lost, and in their absence scholars have to work on the basis of flimsy evidence and vague references.

[2] As noted on his service sheet in 'Antigüedades de los oficiales de guerra de la Armada', Tomo 1, p. 538, in AMN, MS 1161.

only seven years old. In the intervening years much water had flowed down the Magra River, and the face and ownership of the land had changed. For Malaspina at the age of forty-eight, but feeling older than his years, the obscurity of his dreary San Antón prison on Spain's Cantabrian coast was exchanged for a gentlemanly but penurious life in a homeland with which he had never been intimately acquainted. He spent his final eight years there until his death at Pontremoli on 9 April 1810, not far from his birthplace. No steps had been taken to recognize his services either to his adopted country or to the Enlightenment to which he had been dedicated since his school days in Rome. Until recently it would have been easy to agree with the famous scientist and traveller, Alexander von Humboldt, whose epitaph for Malaspina was that he was more famous for his misfortunes than for his discoveries. Even though those mishaps still weigh heavily in any assessment of Malaspina's career, it is now possible to balance them against the many discoveries for which Malaspina and his companions were directly or indirectly responsible, but for which they received little recognition in their lifetimes or later.

MALASPINA'S 'INTRODUCCION'

Malaspina's 'Introducción' was intended by him to form the preface to the published account of his voyage.[1] This was to be an account on a grand scale which would dwarf in size and comprehensiveness the narratives of his predecessors in the Pacific such as Cook and Bougainville. In the first instance there were to be no fewer than seven volumes.

Volume 1 An account of the voyage, together with observations on the lands and peoples, and the existing state of the laws.

Volume 2 A more detailed examination of the territories visited, mostly taken from earlier accounts, and from the observations of the natural scientists on the expedition.

Volume 3 An assessment of the political state of the overseas empire, with detailed recommendations for change and reform.

Volume 4 The journals of the subsidiary voyages of the *Mexicana* and *Sutil* in 1792 to explore the Strait of Juan de Fuca.

Volume 5 Astronomical and meteorological observations.

Volume 6 Navigational material; observations of winds and currents.

Volume 7 Observations on health at sea by the surgeons on the expedition.

Additionally, there was to be a volume of seventy drawings from the voyage, and an atlas of seventy charts, together with harbour plans and coastal views. Later, Malaspina hoped that there might be additional volumes of natural history observations based on the travels and collections of the naturalists Tadeo Haenke and Luis Neé. Of this original design, only the intended Volume 4 was published in Malaspina's lifetime (and then not until 1802 and without reference to the Malaspina expedition as a whole) as *Relación del viaje hecho por las goletas Sutil y Mexicana en el año 1792*.

The 'Introducción' had a double objective: to outline the contents of the seven volumes of the intended work; and to give some indication of Malaspina's own political philosophy and his views on the significance of his voyage. In the text printed here those passages dealing with contents of the proposed technical volumes have been omitted, but the the sections dealing with the proposed third volume and with the more general purpose of the voyage have all been included. Couched in the

---

[1] This translation is based on the text printed in Mercedes Palau, Arañzazu Zabala, Blanca Sáiz, eds, *Viaje científico y político ... Diario de viaje de Alejandro Malaspina*, Madrid, 1984, pp. 29-62, where it is entitled 'Discurso preliminar'. But in AMN, MS 753, Malaspina headed it 'Introducción', and also refers to it as such in the body of the text. See p. xcvi below.

florid and often convoluted language that was characteristic of Malaspina's written style, often switching from the declamation of philosophical principle to some matter of practical shipboard management, it offers an insight into his character and motives. Many of Malaspina's reflections were on politically sensitive, if not politically dangerous, matters, as he argued for a wholesale readjustment of the relationship between Spain and its colonies.[1] His acknowledgments towards the end of the Introducción form a roll-call of some of the most celebrated names in the European Enlightenment of the eighteenth century, and again indicate the breadth of intellectual interests of the commander of this ambitious voyage of survey, inspection and exploration.

TEXT OF THE 'INTRODUCCION'

Any attempt to compare the voyage of the royal naval corvettes *Descubierta* and *Atrevida* with those undertaken by English and French expeditions since 1765 would surely incur a serious risk of error. However various the points of view from which all these voyages are contemplated, differences between them persist in equal number. If only a few of these are listed here, the reader will be able to infer the rest, with no danger of being misled.

In 1789, the first year of the voyage of which we shall now lay the results before the public, the habitable portion of the globe could be considered as known. The limits imposed on navigation by permanent ice at either pole had been determined. The peoples living on the shores of the Pacific, their customs, numbers and origins, had been described and their products examined. The safest and shortest routes between the most distant corners of the earth had been pieced together. Any attempt at a further voyage of discovery would have invited scorn from scholars and, indeed, ridicule from those few readers who seek, in accounts of undertakings of this kind, distraction from protracted leisure or instruction in original developments in systematic thought, whether political or scientific.

Progress in navigation had led me to consider a further point. Rigging, hull design, the quality and quantity of provisions, the supply of drinking water and, finally, the combined effects of unremitting and unpredictable duties at sea with rapid variations in climate and continual ingestion of foul air: none of these could any longer be seen as obstacles sufficient to prevent direct navigation to the remotest parts of the globe. Modern nautical science had overcome them all. Following a change in our outlook, the problem of keeping seamen alive and in good health – whether in the wilderness or on that very ocean which seemed to surround us with perils – had now become as easy and straightforward as formerly it was difficult, even in settled places, and more particularly, in European colonies in Asia and America.

---

[1] These were set out at more length in his 'Axiomas políticos sobre la América', recently published under the editorship of Manuel Lucena Giraldo and Juan Pimental Igea, *Los 'Axiomas políticos sobre la América' de Alejandro Malaspina*, Madrid, 1991. For a discussion of the relationship between the 'Introducción' and the 'Axiomas', and of Malaspina's political thought generally, see John Kendrick, *Alejandro Malaspina – Portrait of a Visionary*, Montreal and Kingston, 1991, pp. 101-22.

It was not, therefore, merely for advances in hydrography and navigation that the present voyage was undertaken, with hopes of a successful outcome. Rather, a glance, however superficial, at the state of European knowledge and dispositions concerning America and Asia raised, in direct and necessary consequence, other matters of equal or greater importance, such as might provide a worthwhile motive for the voyage, while also offering the nation the practical advantages – to say nothing of glory – to which the most recent foreign seafarers have aspired.

The distinctive nature of our proposed objectives could not fail to have a direct influence on the considerably different methods we adopted in order to achieve them. We were to visit most of our colonies in and around the Pacific Ocean and open the way for easy navigation between them; we were, if possible, to exhaust the available resources of geography and astronomy in an attempt to eliminate the hazards or exceed the traditional limitations of commercial voyages. How could we achieve this without spending an extremely long time on a meticulously detailed study of the coasts, an extended stay in the principal colonies, and a quest for favourable seasons on both sides of the equator? In short, how could we manage without constantly exposing, to the combined influence of vice and adverse climatic conditions, crews likely to have fared better at sea or in the lonely confines of the Malvinas, New Zealand or the north-west coast of America?

The English were inspired, in a word, by the desire to find new possessions and new opportunities for trade in countries not yet well known, and thereby to achieve fame, novelty, economic advantage and a happy triumph over a thousand obstacles to navigation and to the easy preservation of seamen's health. Our sights, on the other hand, were fixed on acquiring thorough knowledge of a range of immense possessions, prudently disengaging from those that proved useless or deleterious, and establishing much-needed communications between various points in so extensive a monarchy. The inevitable consequences with which we were faced included the need to compile a great weight of hydrographical and political information; to proceed slowly; to incur high costs; to undertake navigation at modest risk; thoroughly to thresh out ideas; and above all to face the major difficulty – how to keep our crews in good order and health.

However, as occasion arose, it was necessary to scrutinize, in the light of continuous and conscious experiment, all the various ideas suggested by published French and English accounts[1], concerning voyages of great length and duration It would have been reprehensible to follow such ideas unreflectively and with slavish admiration, born of negligence, if not of ignorance. It would have been equally reprehensible to judge their value solely by our own findings, without making allowances in advance and adjustments with hindsight to take account of arguments relating to. The training, character and constitution of our seamen are very different. So various are the internal arrangements of ships and discipline, and there are so many Spanish colonies scattered over the entire surface of the globe, that most of the prescriptions set down by Captain Cook for voyages of this type will inevitably be either detrimental or impracticable for our navy.

---

[1] Almost certainly a reference to the voyage of Bougainville, published in 1771; and to the three voyages of Cook, published in 1773, 1777 and 1784.

Having indicated the general considerations which suggested that the present voyage would be useful and that it would be wise not to limit ourselves in advance to a slavish imitation of the English voyagers, we can now turn readily to the ideas which shaped the way we carried out our task. The compilation of a hydrographic atlas for long-range navigation by Spanish ships, whether for the exchange of goods between the colonies and the mother country, or for more extensive trade with countries outside Europe, was a good enough reason in itself to send to the Pacific Ocean the ships and men needed for the job. Without much additional cost, it was then easy to supplement this meticulous survey of the coasts with some work towards the advancement of natural history in the regions concerned, in respect, chiefly, of human nature, and, further, of the land and the various animals which inhabit it. But if the results of these enquiries were to remain unpublished, their potential benefits, for Spaniards rather than foreigners, would go to waste. To publish them would be to draw aside at last the thick curtain of mystery which, because of America's very size, had previously concealed the true face of the continent from all. The former case, indeed, would be an implicit avowal of our own weakness and would excite the greed of ever-voracious European powers to attack us on every side, with prospects of success; our resolve to maintain defence on all fronts would become as feeble and faint-hearted in practice as the efforts already made to arouse it in prospect ... How sad that our situation should seem to have led us to prefer chaos, ignorance and lack of method, as if they were more advantageous than a thorough, wide-ranging and scientific study of the limits, the nature and the problems comprised within the Spanish monarchy.

But, after all, could our weakness be kept secret? And even if it could , should we regard it as an irremediable defect? Or did some direct conflict between nature and the principles of society subvert the best laid foundations of our laws? Such a question should, in the end, have persuaded a cautious and reflective ministry that, however deep the defects of the present arrangements by which the monarchy is governed, certainly nothing could be worse than a failure to analyse them according to simple and natural principles. The publication of the hydrographic atlas was decided upon,[1] and therewith , as a necessity, a political study of America, which was to reveal, with philosophic detachment, our ills and our remedies, our weaknesses and our resources, our past errors and the most soundest principles of our present administration.

Such a project might of itself have brought together, at the heart of a virtuous society, the Monarch, his ministers and the various classes formed to obey. Would that the task had fallen to hands better able than mine to treat it as it deserves! A few truths, pure and distinct from the muddle of competing systems which create so much confusion, would perhaps suffice to change the face of the Monarchy. Our common task would then have no other object than our common benefit, in which the proper and attainable desires of every individual are comprised; soils and climates so fertile and varied would yield abundant fruit, to owners and tenants alike; continual struggle within society would cease; in political and economic life, slavery would vanish; content with our own happiness in society, we would no longer regard other

---

[1] But never completed.

peoples' conduct with envy or fear, and this civilized detachment would alone suffice, on the one hand, to bring us the respect of other nations and, on the other, to restrain any misuse of the means of war.

Would that I were able, in the exercise of so great a duty, to apply the utmost energy in philanthropy, the greatest assiduity in the study of nature, the highest impartiality in examining the nature and rights of man, and bring them all to bear on social questions! If only I were capable of measured and reasoned reflection on the sad disorders of our day; if I could fulfil, in short, the obligations which arise from my intense and inextinguishable gratitude to the Monarch who has honoured me and to the nation that has taken me as one of its own! If such were the case, never would my voice fall silent in my endeavour. The diversions of my days and the watches of my nights would be dedicated, equally and entirely, to the happiness of all; I would regard it as a greater happiness to be able to guide public opinion towards a tranquil and prosperous society than to lead an expedition of war against an enemy, who, likely as not, is unaware of his reasons, and our own, for fighting ... . But no! It is vain to aspire to a responsibility of such magnitude. Others, of greater abilities, will soon excel the modest limitations which the deficiencies of my political education and the distractions of my seafaring life impose upon me; I shall be happy, however, if the few truths which I have to set down, and which are the fruit of many years of labour, serve simply as a first contribution towards the lofty aim of raising the nation to power and plenty.

In discussing political and social questions, we could simply set aside the assumptions which, since the conquest of America and part of Asia, have held sway in this Europe of ours. We could thus avoid wasting space and effort against a series of principles which have been made obdurate by time, custom and private interest. On the other hand, we might leave the true cause of our ills in its present obscurity; or, if the principles in question were left untouched, we might be led idly to erect a solid and permanent edifice upon what would really be weak and poorly designed foundations. A new project of a familiar kind, perhaps similar to, but less original than, the writings of Abbé Raynal;[1] would entertain the idle and superficial reader for a short time, while encouraging the government to consider its subjects more as enemies than as members of the same body as itself; and such is the tendency of public opinion, that the very insufficiency of the proposed remedies would serve to discourage the application of others, which might confer greater benefits in the future.

In the proposed study of America it is therefore necessary to abandon arguments traditionally proposed; and, after an instructive and impartial look at this vast continent and the true usefulness of its products and of its communications with Europe, we must turn to a particular consideration of the nature of the Spanish possessions, the range of social conditions which characterize them as a whole, the motives which led to their present condition, and, lastly, the prospects they afford for undisruptive reconstruction and the happiness of their people.

The goal of human societies is doubtless none other than to ensure their own security and defence, and facilitate those reciprocal exchanges which lead, directly or

---

[1] And in particular to Raynal's influential *Histoire philosophique et politique des deux Indes*, Paris, 1770.

indirectly, to a tranquil and pleasant life. The Creator has been bountiful towards man, and although troublesome infancy, sedentary old age, weak defences and delicate skin make man perhaps the animal most exposed to the ferocity of other creatures or the inconstancy of the elements, yet he was given a genius and aptitude for thought, with which, indeed, he was readily able to master the whole of nature. At the same time, however, he found himself disposed to exercise these faculties against his own species, moved by envy rather than necessity. Hence arise the different stages of society: at the beginning, strength and maturity, triumph in felling the forests and overcoming the wild animals that inhabit them; next, the dictates of understanding are exercised for shelter from the elements and the easy acquisition of food; finally, there soon follows the third epoch, which is aimed no longer at triumphing over the obstacles of nature but at subjugating man's fellow men and making them work for his own ends. From this, at different times, according to the constitution that each society happens to have, external wars have arisen for the acquisition of slaves and the extension of dominions; and internal, or civil, wars for the overthrow of factions or opinions; the use of navigation for the transport and exchange of goods in bulk; and the refinement of reasoning to mechanize arts and labour. These have been the causes, finally, of the systematic conquest of distant territories overseas, a system which has brought with it the increase of luxury, and the assimilation of all precepts of government in the all-encompassing laws of commerce.[1]

In our society, this is a defect which, however, today occupies a triumphant place in public esteem and which attracts much eager praise from political writers who repeat each other's opinions. For that very reason it must take priority over all other matters as a subject of discriminating enquiry. Just as an industrious farmer, who knows that he cannot prevent the onset of harsh winter weather, chooses, plants and shelters his various trees so as to resist its effects, so we, tracing the course of this evil to its very source and bearing it constantly in mind, cannot now expect to disrupt nature, or make her break laws which she herself has ordained. Rather, we should ensure that the laws we make for society conform to the proper balance which must always be maintained with the changeable nature of mankind.

This approach to the problems under discussion will not appear improper when a reflective mind contemplates, at any given moment, the state of this Europe of ours, that of the colonies in general, and that of simple peoples at the most primitive stages of social development. These are the invariable elements that nature has given us to study: her customs, her laws, the physical environment she provides and her moral precepts all demonstrate, with ample proofs, that human desires must always contend with nature's supremacy; that human imaginations will never cease, amid the uncertainties of the future, to devise a thousand ways of compensating for the evils which overwhelm us in fact; that our desires will themselves subvert the best social institutions, and make it necessary, in consequence, to reform them from time to time. Finally, if reforms, unsupported by the open and agreed opinion of sovereign and

---

[1] In this paragraph Malaspina is repeating the fashionable philosophical views of his lifetime on the origins of the stages of society, the rise of national rivalries, and the increase in luxury.

subject, are ineffectual or else imposed in fear, those which, on the contrary, rest on the basis of universal consent, are effective and beneficent.

If we were simply to ask Spain, that is, all of the parts that make up the Spanish empire: first, what are her real necessities today; secondly, what are her fears; thirdly, what are the lawful covenants by which her internal parts are bound, we would certainly find her hard put to it to reply. Nor, on the other hand, could we find fault with anyone who required beforehand a clear and general response to these doubts, on which to base his reasoning correctly. Spain would hesitate still further if we persisted in asking what were the real necessities which the power of the state can and should supply or control. If the duties of the individual are unalterable in every condition of society, of what use are trade, industry and the colonies? Finally, what does society take to be the proper level of wealth to which we may practically aspire?

Even the simplest glance at political considerations of this type would convince Spain to concentrate on her own priorities. She would see that in general it is equally misleading on the one hand to fear the supposedly excessive strength of the other powers and on the other hand to wish to imitate or outstrip them in wealth; and that the nature of her overseas possessions, the unrestrained desire to add to them, and an imperfect assessment of their worth have not only weakened and warped the monarchy but have also prejudiced our constant enquiries into the causes of a problem of great gravity.

A plan for the useful reform of our colonial arrangements – should our conclusions make it appear necessary – ought then to follow, in clear and incontrovertible order: first, a thorough understanding of what our colonies are like today, and what they will become in consequence of the slow but deleterious effects of present laws, together with an assessment of the true impact on Spain of conquering, possessing and exploiting them; secondly, a straightforward examination of the dues peculiar to each of the parts of which the monarchy is composed, and more especially of their revenues, both internal and external. We should then attempt to devise a comprehensive programme of legislation for them, taking into account the limitations imposed by the deficiencies original to humankind in general and those, in particular, of men born in different climes and environments, as well as the problems which must inevitably ensue from the excessive enlargement of a domain such as ours. ... What good would it do to hold a true mirror to present laws, which have been devised in response to necessities arising from the distorted objectives of our legislation, and fortified by the very institutions which ought to have restrained them? Or how would it avail to utter vain complaints against our weakness and the supposed strength of the other powers, or exchange with them timorous, suspicious and treacherous glances, with which we seem fated to spy on each other in respect of every inhabitant, every inch of territory and every yard of material which belongs to any of us? The upshot would be another confused mass of mean and tired notions – untimely and unavailing for the government and nation, and, in equal measure, humiliating for anyone charged with putting them into effect. They would bear, unmistakably, the detestable stamp of sycophancy or stupidity and anyone who tried to deceive or seduce the people with them would deserve reproof and punishment.

If the monarchy is studied in this way, it will doubtless display a novel and intriguing prospect to men of public responsibilities in Spain. The range of products available in such varied soils and climes presents endless opportunities to supply wants and luxuries, without the aid of any other nation. All at once, the vastness of our dominions distracts attention from the disharmony caused within them by continuous conflict with the hard-pressed natives who formerly possessed them, and equally, from outside, by the interest, of which we have only a poor understanding, shown by political connexions in Europe. Now that Europe is aware of the customs, the nature, the genius and the legal systems of subject Indian peoples, she looks upon them as a precious part of herself. She bids them wake. She offers them the happiness of alternate work and leisure. She makes them multiply, without fear lest they offend. All the wandering tribes who live in the forests and riversides of the interior give themselves up entirely to the soft blandishments of civil life and follow the missionaries' slow – of course – but peaceful steps. No longer does the spirit of conquest inspire our advance into the interior. It is enough for us to see the inhabitants in peace and with a disposition to resettlement and work. Thus the legislator sees recompense for his efforts and the colonist for his costs. Then can be weighed, in the scale of public happiness, the contribution made by rich metals, dyes, simple medicines, industry, agriculture, fisheries and whatever is or can be supplied by way of tribute, from the continent of America, to this Europe of ours and to all other parts of the globe.

Oh, if only by some happy occurrence, amid the terrible convulsions which disturb the human race and combine to impede its increase, it were to be Spain's role to restrain greater spilling of blood and to cauterize some of the present wounds, the consequences of which may perhaps affect adversely the history of the coming century or more in unhappy Europe! If only, by rising above the clash of principles and interests with sublime indifference, Spain could discover at once a morality of peace, a religion of purity, a means of adhering willingly to the energetic and affective legacy of her past! If she could but wrest a source of far-reaching power springing from her own soil, while the limits of her empire remain stable, without further expansion! These would be invaluable means for her improvement and defence, for she would then be able not only to summon popular opinion once more to love stability and the peaceable effects of limits agreed between the authority of the sovereign and the obedience of the subject; but she would be able also to uproot the evil which has prevented society from achieving a just equilibrium: the insistence that private advantage derives from the weakness or ruin of fellow-citizens. In short, Spain would provide peaceful and equitable protection to whoever wished to accept her laws, her government and her rights as a nation.

After these preliminaries, it is now necessary for us to sketch the method adopted in the present work, which will surely seem unsurprising, and may not prove disagreeable. Bearing in mind the need to divide our political materials into categories of study, we have considered the overseas dominions of Spain under three headings: South America from Cabo de Hornos to the Istmo de Panamá; North America, including the Antilles, from the isthmus to the uncertain limits of the continent in the north; and Islas Marianas and Filipinas in the Asian seas, including with the last named, as appeared natural, our interests in that region of the world and in the Pacific

Ocean, whether in relation to European colonies, or to the independent peoples of those regions. To this preliminary categorization, another was added forthwith: that is, the separation of matters of hydrography from others which bear more directly on the instruction and entertainment of the public. Our treatment of hydrography would be neither useful nor accurate if we were not to focus on it with enough detail and corroborative material to safeguard the lives of sailors and the property of merchants in the most remote parts of the earth. Other matters, by contrast, have to be presented in a way which combines utility with pleasure and which attracts readers by means of novelty of purpose, clarity of method and subject matter of interest. A brief lowering of spirits might ensue when the patriotic reader beholds the navigator contending with tempests, shortages, time-changes and the unremitting anxieties which are bound to arise from his oppressive and monotonous way of life, his uncertainty about his fate and the mistakes of his imagination; but boundless pleasure would surely attend a reader who was then suddenly transported among the most prodigious arenas of nature, with their extremely varied climates, the unceasing effects of the immense ocean which surrounds them and the influence of their unwontedly direct and penetrating sunlight. And would not the reader's interest be even more deeply gratified if he were next to see how man fares amid so many opportunities for increase and self-destruction, itinerant existence and social cohesion, fruitful labour and of endless idleness? And finally, would there be any limit to his satisfaction if he were at once to find these reflections and descriptions related, naturally and in a remarkably direct fashion, to the prosperity of his own nation?

This, then, has been the purpose which we have set ourselves in classifying our subject matter. Hydrographic matters concern only the navigator, the legislator and those who wish, for their own interest, to penetrate the detail of one or other of the sciences involved. The narrative of the voyage and the physical and political descriptions of the countries visited, are addressed to every class of educated persons. They should provide a true, albeit incomplete, impression of our establishments overseas, and should contribute to genuine patriotism, by demonstrating to the reader that some advantages of the monarchy are permanent, that some of the ills which obstruct its progress are weak and that remedies afforded by nature can be put into effect. The entire account, moreover, is constructed along lines corresponding to the basis on which the monarchy is divided into its constituent parts. Thus it is easy for the reader not only to gauge and study the portion of the work useful to him, but also to do so with a fully informed approach, so that, just as it would be misleading, in studying this Europe of ours, to confuse the climate, soil, navigation, peoples and customs of Russia and of Spain, so it would be equally inopportune to confuse, in the present context, the snowy mountains of Chile and of the north-west coast of America with the volcanoes and flood plains of the Kingdom[1] of Guatemala and the Philippines, or with the immense prairies of the provinces of La Plata and the coast of Patagonia.

[1] The patrimonial concept of 'kingdom' to mean dominion was embedded in the *Leyes de Indias* at the behest of Felipe II. It is best expressed in his own words: 'Since the kingdoms of Castile and of the Indies are subject to one crown, it follows that their respective laws and form of government should be as concordant and consistent as possible with each other.' See Salvador Minguijón, *Historia del derecho español*, 3rd edn, Madrid, 1943.

... The description of the voyage – that is, in effect our journal – is by its nature rather wearisome reading; but it was indispensable to provide an idea, however approximate, of the method we followed in our work and particularly of the tasks which have absorbed us for the long period of five years. It was indispensable to acknowledge the parts played by all the capable people who served on the expedition and to accord them the praise they deserved. Finally, it was our duty to specify, with the sort of candour proper to seamen, the problems which arose on our voyage, whether from the general deficiencies of the way the monarchy is organized or, as was more usually the case, from some defect in our equipment or conduct. We shall willingly acknowledge that in no way should we wish to equate our voyage with those undertaken by Captain Cook; our sufferings and perils were much less than those of that illustrious navigator; perhaps, had we been anxious to imitate him more closely, without the assistance of the same good fortune, we should have been led, headlong and profitlessly, into the wake of the unfortunate Comte de La Pérouse on the north-west coast of America, or on the Islas de Navegantes, or on the banks near Nueva Caledonia; or we might have followed Captain Riou, who was almost lost with the *Guardian* as a result of encountering an iceberg, or of Captain Hunter, wrecked with the *Supply* on Isla de Norfolk; or of the *Pandora*, similarly lost in the Tierras de Solomon.[1] Yet we shall repeat it once again: ours was not a voyage of discovery. Its object was to survey America, so as to be able to sail safely and profitably along its very extensive shores, and so as to be able to govern there with equity and utility, by uniform and straightforward means.

... But it is now time to provide a somewhat fuller idea of our third volume, which concerns political affairs: the prosperity and defence of America and her direct and natural connexions with the mother country are the essential points, or, rather, the only ones we should have in mind: the history of the conquest and preservation of our possessions will therefore follow immediately on the description of the areas occupied by other Europeans. We shall see why they were seized from our grasp, what advantages, direct or indirect, their posessors derive from them, and what real damage ensues to us from contact with them. And having demonstrated that all these matters are of small consequence, we shall attempt to formulate our territorial rights in such a way as to prevent, at once and for the future, the disputes which hitherto have been excited in such considerable numbers by questions of this sort, innocently misunderstood or maliciously misinterpreted. Once the frontiers of our empire have been fixed, in relation both to those of the other European powers and to our convenience, it is proper to proceed to an examination of the vast extent of those countries which yet remain to us; we shall then identify those which form an effective part of

---

[1] With regard to La Pérouse, Malaspina was referring in his first mention to the loss of two of his boats with their crews in Lituya Bay, Alaska, on 13 July 1786, and in the second instance suggesting that La Pérouse might have been wrecked on Navigators Island (Samoa) or on the reefs of New Caledonia. Captain Edward Riou was taking supplies to the newly-established convict settlement in New South Wales when the *Guardian* struck an iceberg on 24 December 1789 in the South Indian Ocean. By superb seamanship Riou managed to return to Table Bay, but his ship was found to be beyond repair. It was the *Sirius*, Captain Hunter, and not the *Supply*, (which was in company at the time) that was wrecked on Norfolk Island on 19 March 1790. The *Pandora*, Captain Edward Edwards, was wrecked on Australia's Great Barrier Reef on 28 August 1791 when transporting mutineers from the *Bounty* back to England.

our Monarchy, subject to our laws and capable of contributing in some measure to the defence of the state, as distinct from those which we ought not to count as falling within our control. Would it be suitable to subdue them by force of arms, or would it be better to allow the passage of time and the effects of well conducted missions and of commerce, in alliance with humane practice, gradually to bring about the desired result?

However that may be, each constituent part of the monarchy overseas, defined as above, must be organized in such a way as to provide for its own defence and for a certain moderate increase of its own prosperity, before being called on to contribute to the mother country. The colonies must therefore be placed on a military footing in times of peace and war, against savages and against foreign invaders, before taxes are assessed, while their present net contribution to the mother country is calculated, and reduced, maintained or augmented as their resources, properly assessed, permit.

With these considerations in mind, it will become apparent that the presence of naval forces, assigned directly to the colonies, will not only provide effective means of defence and mutual communication along those very extensive coastlines, but will also stimulate industry in the colonies, as the number of ships grows, thanks to the productive natural environment, while greatly reducing the naval costs incurred by the treasury in Spain.

Having argued that the boundaries and defensive arrangements, both internal and external, of each part of the Monarchy overseas should be disposed in this fashion, and each left to see to the means of its own prosperity and to the basic matters of policy and of the administration of justice – which can never be made to conform to a uniform pattern across every region or constrained to be unchanging over time, since needs and circumstances change constantly – the very logic which orders our ideas prompts us next to expound the legitimate rights of the colonies and their duties to each other and to the mother country. What at present are the mutual sacrifices which have to be made, and what are the reciprocal advantages? Examination of this question, if focussed on the distinction between necessities, utilities and superfluities, and if it takes account of the enormous difference in yield generated by internal or domestic trade on the one hand and external or foreign trade on the other, will reveal, simply and logically, what precious metals are really worth, how their free circulation is inevitable, what real harm they cause on our continent under present exigencies, and the greater damage inflicted in America by the mistaken opinion that they symbolize real wealth. From the same error arise the facts that the law continues to support the bullion industry and that the continent remains indifferent to the way violence is used to engorge it. It is the cause, in short, of political mistakes which have kept us weak for three centuries and which have prevented us either from restraining the actions of our conquistadors or from exploiting them usefully.

There is, then, a means whereby our American colonies may be happy, strong, empowered to defend themselves and – taking an active part among the countries which are joined together under the crown – may contribute in due measure to the security and prosperity of the state. What, in that case, are the proper ties which rightly bind America to Europe – ties which should lead us to be absolutely unconstrained by rival powers, to enjoy structures of government and of taxation more moderate and equitable than theirs, to have a population and a system of education better suited to our present state, and to adopt principles of law and economy which, imprinted with equal force on Spaniards at home and overseas alike, and on legislators and subjects, will exempt our commonwealth from the constant collision of aims, vested interests and the force of individual partiality?

Reflections of this kind, without which it would be profitless or even pernicious to attempt an inquiry into the American polity, lead us irresistibly to inquire into the administration of public affairs in Spain. Once the colonies are emancipated – if we may use that term – so as to be able to think of themselves as having a proportionate, and not merely a secondary, role in the Monarchy, and once their direct influence on the strength and wealth of the continent has begun to take effect, it can be expected that they will naturally and necessarily trade with each other; hence will follow the development of a system of commercial taxation which distinguishes what we disburse and consume domestically, as a nation, from what we disburse and import for consumption from abroad. In this connexion we must proceed rapidly to identify the shortcomings in the present laws of Europe, and of Spain in particular; for , suddenly self-transformed into a colonial, industrial and commercial nation, she has effectively obliged numbers of her people to emigrate; this has meant that the cultivation of the very foodstuffs the country needs has been abandoned or prevented. As a result of the confusion of our interest with those of the colonies, and of those of the colonies with partisan struggles in Europe, infinite problems have arisen in the management of affairs of state. Were it not for this, perhaps the differential in prices would not be such that wheat from Beauce and the Orléanais, a hundred and fifty leagues away by sea, can reach Cádiz faster and with transport costs a hundred per cent [sic] cheaper than the same product, conveyed from Palencia, which is only forty leagues from Santander.[1] Would anyone then imagine that a Spanish soldier or miner, who usually, at home, can barely earn a modest wage by hard and vexatious toil, could arrive in America and immediately become a representative of public authority, rich and exempt from the need to labour?

And above all, once Spain has cast aside the need to make political calculations of the kind in which the constitution, defence and administration of her colonies now variously involve her; once she knows the contributions she can expect from them on a regular or extraordinary basis; once she knows the extent of the military commitments to which she is bound by virtue of her relationship with the colonies; and once she has utterly abjured the causes of conflict, whether in respect of territory, trade or rivalry with other nations, why should she not turn her reflections in tranquillity upon herself, without seeking any condition, be it better or worse, other than that authorized by nature, by her united strength and by the help provided by the various parts of her monarchy? Why should she not be able to balance her revenue, replenish her treasury, devote her resources to her own prosperity and make herself respected, without having to rely on anyone else, by the other powers of Europe?

---

[1] Malaspina adds here a footnote here: 'Memoria MS.de don Gaspar Jovellanos sobre la agricultura de España'. His example seems to be an adaptation of that cited by Jovellanos (*Informe sobre la ley agraria*, Madrid, 1917, I, 176) that although there was a Spanish flour factory less than forty leagues from Santander, Spanish colonies relied on flour from France and Philadelphia.

This would perhaps lead to the formation of a new scheme of public law, which at present, as has already been said, is confuse, with commercial law; the interdependence of colonies and mother country would be established by means as direct, well justified and natural as those which currently prevail are warped, unjust and malignant. Lastly, the favour and good opinion of the people could be secured to the government, year in, year out, by accounting publicly for the level of the national income and assets and the way they are handled. But this project belongs to a particular field which does not fall within our present competence – namely, the internal administration of Spain. For our purposes, it will be sufficient to free Spain from the heavy burden to which she is bound by her overseas possessions, and to present a project for reconstituting the Monarchy, and ensuring the happiness of all its parts, so that they neither fear foreign invasion nor covet foreign wealth.

Our desire to provide a reasonably clear idea of our intentions in undertaking our voyage and publishing its results has made us prolix; but this was inevitable if we were to do justice to the method adopted, while also making clear from the outset the final goal which we have kept before our eyes along the way: namely, the solid prosperity of the Monarchy.

... Now that we have given particular attention to the objectives of the voyage and the method currently adopted for publication of the results, we should with equal clarity describe the preparations and measures which preceded it. These will very plainly demonstrate the extent of the generous patronage the King accords to science and navigation and will make apparent the reasons why, in some instances, we have departed from the practices of the navigators who preceded us in expeditions of this sort, while at other times we have humbly followed their example.

The two corvettes with which the voyage was undertaken were alike in every respect. Brigadier Don Tomás Muñoz, chief engineer and commander of the La Carraca naval dockyard, combined in their design all the qualities which seemed most advantageous, as well for the strength as for the capacity and convenience of each vessel: length more than 120 feet, beam 31½ feet, depth of hold 15 feet, 306 tons. With their seams caulked, they had a second skin which sea water could not penetrate, even if a damaging encounter with an underwater reef broke the outer skin. The bottoms were first sheathed in wood secured with metal nails and then lined with copper sheeting, so as to nullify the harmful effects of the nails on the iron pinnings of the hull. The draught was not more than 13½ feet at the stern, permitting access to any shallow anchorage. With a rig designed to make them stiff but reasonably fast, they had excellent steering and made little leeway, especially if the mainsail was properly set. In their hold the ships could carry two years' provisions for the complements assigned to them, and six months' water and firewood. The quantities of stores of all kinds, especially hardware, canvas and tackle, were set to allow for the fact that we would find these commodities absolutely unavailable in American ports. Replacements of clothing for the seamen and goods for exchange were also carried in large amounts. Each vessel had five boats to perform such duties as taking on water and wood, hunting, fishing, making observations, studying natural history and maintaining constant communication between ship and shore. Once the launches were enlarged and fitted with iron sheathing – operations carried out later in Guayaquil and San Blas – our three larger boats could hold the complements of the ships in case of shipwreck. Furthermore, our two iron stills for desalinating sea-water, the second of which was earmarked for providing water for cooking, produced sufficient water for all. And what was of greatest importance to us: there was no one who was not accommodated on the main deck – that is to say, in a well ventilated space, where the galley kept the air circulating continuously. Nor was there wanting, in the accommodation, the organization and respect for rank which is necessary for the maintenance of good discipline over a long period of time. It would be tedious, though not altogether useless, to list one by one the various considerations we had in mind when deciding how to distribute berths. In the case of the commissioned officers, our object was to give each his own quarters, with the peace and quiet needed for scientific work, and to provide comfortable and accessible space in which to gather, without forgetting companionable pleasures, accompanied by music or readings from instructive and entertaining books.

Brigadier Don Fermín de Sesma, deputy inspector at Carraca, was extremely effective in ensuring that rigging, sails and other stores were all of equally good quality. All was of the best quality and size. Any fellow professional will be convinced of the truth of this assertion by our assurance that on the *Descubierta* we used the same maintop halliards throughout the entire course of five years and two months.

Preparations of both these kinds, on which our safety on board largely depended, were accompanied by others no less important, concerning the health of the crews. We were aware (as has already been made apparent) that our frequent calls at various ports in colonies belonging to our nation, would give us the opportunity of re-provisioning as often as necessary; but we also had to keep in mind the fact that among the provisions taken on recent voyages, there were a thousand varieties which might provide means of varying our diet economically, even if they were unable to assist with the apparently difficult task of preserving health.

With this in mind we took sauerkraut and pork, salted according to both the recipes favoured by Captain Cook and the Comte de La Pérouse. We made great use of the wine of Sanlúcar, for which Chilean wine was used as a substitute, and, lastly, grog – that is, watered spirits. Hot meals and gazpacho were served alternately and the use of alcoholic beverages was varied, as were mealtimes, according to climate and season. Meanwhile, bathing was encouraged, as was exercise provided it was done in moderation; nor did we order work when the sun was at its fiercest in the worst regions of the torrid zone, except when it was absolutely necessary. With the same objective – promoting frequent exercise among all ranks on board – we arranged for the seamen and marines always to be organized in two watches and that, in general, other persons should also be included in this useful routine. When at anchor, we never neglected fishing, hunting and the inclusion in our diet of those health-giving herbs which nature provides for mariners, even in the driest and most infertile places. The account of our voyage will shortly demonstrate how many times our crews had to recover at sea from the debilitating effects of time ashore, rather than restoring in port the usual ravages of life at sea.

But the chief means to which we had recourse to keep our typical seaman healthy was undoubtedly peace of mind. We can by no means assume in the Spanish seaman that same insensitivity so often and – it seems – incorrigibly found in northern mariners. Our men are of a thoughtful disposition; they think ahead; and, keeping in mind all the disasters that have happened , often inflated by the effects of imaginative minds, they have some idea of the misadventures that have occurred in hazardous voyages to the South Sea. Hence their importunate meddling in every precaution we take and each difficulty we encounter on an almost daily basis; hence the indescribable manner in which their moods swing from the utmost resolve to the deepest dejection, as their real or imagined circumstances dictate; hence, finally, the equally abrupt change from robust health to widespread sickness: sickness which is made ever more acute by features of life on board to which we have already referred. Fortunately, their genuine sense of discipline keeps their attention fixed on the naval officers who are set over them. It is enough for the latter to know how to judge the right mixture of rigour and kindness, hard work and rest, compulsion and persuasion, a certain measure of familiarity and an ability to withdraw to a sufficient distance at short notice. Our naturally generous national character is then aroused to its proper level and all due effort is made, first, to instil peace of mind and, consequently, to overcome every obstacle. We shall, however, give due attention to ideas relating to the preservation of health in our seventh volume or medical treatise.[1] For the present, we shall limit ourselves to an endorsement of what Cook has already amply demonstrated: namely, that in what concerns nutrition and the economical maintenance of sailors' health, there is no time or clime or any spot anywhere on earth, where success cannot be easily achieved, as long as the normal rules are adjusted according to the customs and qualities of each nation. As far as our provisions are concerned, the sauerkraut kept well for two years, except, however, for some barrels in which the contents quickly rotted and exhibited an extraordinarily foul stench , because they were inadequately salted or because air got in between boards which were not sufficiently tightly coopered. Pork salted by two different methods lasted three years, if the pickle was replenished from time to time. Dried vegetables did not last so long before being infested by grubs. The same thing happened with the bread, though this was not the case with the flour, especially that from Philadephia, which remained in good condition two years out from Cádiz. We also subjected salt beef from Montevideo to an exacting test. We had some which had been processed in 1786 and after we had been at sea for four and a half years, it remained of good quality in March and April 1794.

To these measures undertaken for the health of the crew, we then added others designed, with good reason, for cases of sickness. Soup tablets were prepared by a variety of methods widely used in Europe. We stowed an abundant supply of orange and lemon juice, and took on board a number of barrels of fermented barley meal. The contents of our medicine chests were very different from those usually carried on Spanish naval vessels. Our methods of nursing care were also innovative. We avoided taking on board perishable foods for the sick rations and appointing permanent bays for the sick. The results of each of these measures will appear in due course,

---

[1] Which was never completed, although the surgeon of the *Atrevida*, Pedro María González, later published a treatise on health at sea, much of which was based on his experiences on the Malaspina expedition. See his *Tratado de las enfermedades de la gente de mar.*

accurately and systematically, when, together with the measures mentioned in preceding paragraphs, they are discussed in the medical volume. Here we shall add that
in the determination of our medical arrangements for the service in question, we
were assisted by the medical officer of His Majesty's navy, Don José Salvaresa, who
intervened at His Majesty's command, and whose recommendations on health at sea
can be found in three letters, written in reply to enquiries of our own in which we
expressed our doubts and uncertainties in following practices so different, in some
respects, from traditional Spanish methods and, in others, from the prescriptions
almost invariably made by foreigners.

It is now time to say something further concerning the scientific objectives with
which the voyage was undertaken and the methods employed to achieve them. They
were many, and they kept us busy incessantly. They were, for the most part, conducted by men well known in the scientific community. And the learned minister
who provided the initial stimulus for the voyage, and who continued to sponsor it
with unfailing constancy and generosity, made himself available for consultation from
the outset, with that obliging character which is inseparable from a truly scientific
vocation, and from the desire to make the broadest contribution possible to the true
benefit of our fellow men.[1] In the same way, the archives of the secretariates of state
for the Indies and the Navy were freely made available to us so that we could extract
any hydrographical information they might contain. Our first researches demonstrated anew the need for a voyage of the kind proposed, since materials, sometimes
valuable, sometimes gravely misleading and often mutually contradictory, lay in a
single confused mass. If on the one hand they displayed the frequent and costly efforts
which the government has made to improve navigation, they also demonstrated, on
the other, how easily those efforts could be rendered ineffective or useless by the
imperfections of the methods applied. Similarly, we carried general orders granting us
access to the archives of the dissolved Jesuit houses in the various main cities of
America,[2] since it was highly probable that we should find in those up-to-date
records of the journeys and expeditions of reconnaissance which members of the
Order made into the interior, during this century and the last, either to assist in the
conversion of unconquered peoples or to aid the government in the study and investigation of such an extensive land. These plans, however, also proved disappointing, in
part because we found those archives badly cared-for and in part because they had
been stripped of their most valuable contents. In the end, we had to turn to published writings, albeit with the advantage, in the various chief cities of our kingdoms[3]
and provinces, of being able to consider the principles on which the history of America might be based, in the light of manuscripts at our immediate disposal and of local
knowledge arising therefrom. Without neglecting our responsibility, as is right and
natural, to make the fullest possible acknowledgement in appropriate places in subsequent volumes, we must not overlook this opportunity to state that in the learned

---

[1] A reference to the Ministro de Marina, Antonio Valdés y Bazán. See p. 311 below for a biographical note.

[2] The Jesuits had been expelled from Spain and its overseas territories in 1767.

[3] For 'reinos' see p. lxxxvi, n.1 above.

persons of America – whatever their specializations – we have found, without exception, allies in our work, who have gone on to make enormous contributions to our endeavours and have concurred both in our call for reform and in our identification of the extremely complicated causes which have led America to the state in which she finds herself today.

These factors were essential for our success, as were the very strict instructions from the government to the viceroys and captains-general of our territories to assist our enterprise with whatever measures might be recommended by their zeal and by local knowledge. To these we may add the advice rendered, with immeasurable benefit for us, in a number of submissions by Their Excellencies Don Antonio de Ulloa,[1] Don Juan de Lángara[2] and Don José Mazarredo[3]: these covered hydrography, the physical geography of South America, the acquisition and use of most of our astronomical instruments, as well as certain experiments concerning sea-level in the Atlantic and Pacific Oceans, and various modifications to the hulls, handling and shipboard routines of our vessels. Teniente General Don Gabriel de Aristizábal[4] was also consulted. The Marqués de Ureña[5] contributed various ideas on how different

[1] Antonio de Ulloa y de la Torre-Guiral (1716-95), FRS, one of the most important figures in Spanish science in the eighteenth century, in terms both of his own work and of his patronage of other scientists. He joined the *guardiamarinas* in 1732, and because of his brilliance in mathematics and science was appointed at the age of nineteen to join Jorge Juan on La Condamine's mission to Quito to measure the distance represented by an arc of latitude at the Equator. With time spent on other duties, which included the isolation of platinum, which he identified as an element, he remained in South America until 1745. Capture by the British on his return voyage to Spain brought unexpected benefits; for not only were his scientific papers given back to him, but he was able to form friendships with leading members of Britain's scientific community, and was elected Fellow of the Royal Society. Later, he was commissioned to make numerous excursions around Europe to bring Spanish practice up-to-date with the latest findings in applied sciences, especially in medical education, metallurgy and mining. From 1766 to 1772 he was Governor of Florida; he became *Teniente General* in 1779; and at his death he was Director General de la Armada. In an earlier section of the 'Discurso' (not included here), Malaspina had paid tribute to Ulloa. 'This philosophical observer, who sought the truth by departing from traditional systems, and studied the inhabitants of America with unflagging assiduity for the lengthy period of twenty years, will perhaps inspire new regard among scholars for the clarity and simplicity of his descriptions. Future ages, long after his writings and ours, will provide new support to the measures he undertook towards the classification of nature in that part of the world.'

[2] Juan de Lángara y Huarte (1736-1806), became a *guardiamarina* at Cádiz in 1750, personally selected by Jorge Juan. He was an early Spanish exponent of longitude by lunar distances, commanded several expeditions to the Philippines and the China Seas, as well as the expedition of the *Rosalía* (1774) to correct errors in Spanish charts, and to try out new navigational techniques.

[3] José de Mazarredo (1745-1812), a former *guardiamarina* who served at Cartagena and, under Juan de Lángara, in the Philippines . His distinguished active service record was interspersed with periods as an instructor in navigation (in the late 1770s) and director of the school of *guardiamarinas* (in the late 1780s). He became *Teniente General* in 1789. He was author of *Rudimentos de táctica naval*, Madrid, 1775 and *Lecciones de navegación para el uso de las compañías de guardia marinas*, Cádiz, 1798.

[4] Gabriel de Aristizábal (1743-1805), a former *guardiamarina* who served at Cartagena and later in the Philippines. He commanded the vessel which bore an embassy to Turkey in 1785, and he made a much-admired series of charts. He became *Teniente General* in 1793, and in 1795 brought Columbus's remains back to Spain.

[5] Marqués de Ureña, a polymath celebrated in Cádiz for his patronage of the astronomical observatory and for his model (constructed in 1777-79) of the city. He was author of *Reflexiones sobre arquitectura, ornato y música*, Madrid, 1785.

infirmities are affected by the direction of the wind and on the best uses for our eudiometers, and Don José Armenteros, Secretary to the Real Compañía de Filipinas in Manila, at the government's request, added to our store of information all the thoughts which twenty years of study had supplied on the geographical and political context of the company's installations. So much assistance was enough in itself to inspire men to make a commitment to our enterprise, even if they were tepid in their enthusiasm or doubtful of their own strength. And how could it be otherwise, when contributions to the same end were forthcoming from a number of learned former Jesuits now resident in Italy – namely, abbés Córdoba de Castro,[1] Jiménez[2] and de Cesaris[3] – marchese Gerardo Rangone[4] and abbé Spallanzani,[5] also from Italy, M. Lalande[6] of Paris and the gentlemen from London, Banks[7] and Dalrymple[8] ? To them we owe timely guidance on matters to which we might, with best prospects of success, dedicate our efforts, or subsequent exchanges of correspondence which helped to clarify particular problems which inevitably arise from observations made in isolation at such a great distance from Europe.

In London, Alexander Dalrymple also took a hand in collecting the greater part of our astronomical instruments, of which a more detailed schedule will appear in the journal of our observations.[9] However, we were not so fortunate in respect of an excellent collection of instruments made in Paris for the advancement of physics. It did not reach Cádiz in time to be loaded aboard the corvettes, and the bills of lading were subsequently muddled with others relating to consignments for the Mexican mining industry: so we were never able to take possession, in spite of using our best efforts in all the ports at which we called.

[1] Properly Augustín Castro (1728-90). He left his native Mexico on the expulsion of the Jesuits, and became Rector of the Jesuit College in Ferrara. He wrote works on the botany, history and literature of the New World.

[2] Leonardo Jiménez (1716-86) worked as inspector of hydraulic installations for Duke Leopold of Tuscany. He was author of *Primi elementi della geometria piana*, Milan, 1751, and other works on engineering and surveying.

[3] Angelo Giovanni de Cesaris (1749-1832) worked at the observatory of La Brera, where he remained after the dissolution of the Jesuit Order, helping to compile the *Ephemerides astronomicae*, Milan, from 1775.

[4] Gerardo Rangone-Machiavelli, a Modenese nobleman who served as an imperial chamberlain and ducal counsellor. A notable patron of scientists, he was the father of the celebrated mathematician, Luigi Rangone-Machiavelli.

[5] Lazzaro Spallanzani (1729-99) held the chair of natural sciences at Pavia from 1769. He worked mainly on reproduction and respiration as well as writing admirable travel books.

[6] Joseph-Jerôme Lefrançais de Lalande, FRS (1732-1805/7), a Jesuit-educated lawyer who studied astronomy privately before becoming professor in that subject at the Collège de France in 1758. Among his works were observations on the Transit of Venus, 1761 and 1769, and studies on the distance of the earth from the sun.

[7] Sir Joseph Banks (1748-1820), who as a young naturalist had sailed on James Cook's first Pacific voyage (1768-71). He became President of the Royal Society in 1778, and was at this time of Malaspina's voayge one of the most influential figures in the world of science.

[8] Alexander Dalrymple (1737-1808), best known as a geographer with a special interest in the Eastern Seas and the Pacific. From 1779 until his death he was retained by the East India Company to publish charts and nautical instructions, while in 1795 he became the first Hydrographer to the Admiralty, a position which he also held until shortly before his death.

[9] Again, this was never published.

Nonetheless, the indefatigable spirit of Coronel Don Antonio Pineda never failed to find means of exhibiting his measureless love for the various branches of natural history and practising those endeavours which, in the end, brought his life to a close.[1] He had at his disposal an excellent library, acquired partly in Madrid and partly in Paris. Those able botanists, Don Luis Neé and Don Tadeo Haenke, while pursuing their main purpose with the greatest possible assiduity, did not fail to support him with every useful enquiry which came to hand, especially in lithology. Artists and taxidermists tried – each in the manner proper to his profession – to preserve a record of the rarest species which nature revealed to their gaze in the various countries we visited. Meanwhile, others took charge of hunting and fishing for specimens. We rewarded lavishly natives who made us presents of items useful to collect and study. Thus at various times we could send back to the Real Gabinete in Madrid some seventy cases of items of this kind. Oh, if only fate had allowed us to bring Pineda himself back to his homeland in safety, how much would Spain have benefited from his work in prospect – in studying the collection with that breadth of knowledge he had striven to embrace – and from his character, equipped equally for research and for philanthropy! We shall, at least, in no wise detract from his renown, in publishing his own findings on the voyage, whether preserved in manuscripts or recalled from his conversation. In time, with the publication in all proper detail of his zoological descriptions and the many particular objects covered by his constant observations, the nation will realize the extent of its loss.

We shall conclude this Introduction, which has been long enough, by reminding the reader of the judicious observation made by M. de Bougainville, when he wrote the narrative of his voyage: that there is naturally a great difference between the coarse, dry language of the seafaring man and the more pleasing, elegant and entertaining style required in travel writing.[2] We shall be fortunate if, in our case, that difference can be bridged, at least in part, by qualities we have never for an instant lost from view: truth, simplicity and love of the common good. We shall be fortunate indeed, if at last, in performing our work and publishing its results, we shall have succeeded in perfectly obeying His Majesty's benevolent commands and the sensible measures of his ministry.

Towards the end of July, the chronometers having been rated at the Observatory in Cádiz, and the corvettes having been supplied with all that was necessary and tested for seaworthiness, we were ready to sail. All on board were volunteers. The carpenters, caulkers, smiths and forty-five seamen came from the Ferrol station, the rest from Cádiz. The complement on departure was as follows:[3]

---

[1] A reference to Pineda's death in the Philippines on 23 June 1792.

[2] This seems to be a reference to Bougainville's comment (as it appeared in the English translation of his *Voyage*) that he did not wish his book to be regarded 'as a work of amusement', and wished that he could 'have learnt to counterbalance the dulness of the subject by eloquence of stile'. Louis Antoine de Bougainville, *A Voyage Round the World ...*, London, 1772, p. xxv.

[3] Changes to this list, in terms of the movement of personnel between the two vessels, arrivals and departures, and promotions, will be covered in the biographical appendix to Volume III of this edition.

Corvette *Descubierta*

Commanding Officer	Don Alejandro Malaspina
Senior Officers	Don Cayetano Valdés
	Don Manuel Novales
	Don Fernando Quintano
Junior Officers	Don Francisco Javier Viana
	Don Juan Vernacci
	Don Secundino Salamanca
Midshipman	Don Fabio Aliponzoni
Officer in charge of charts and plans	Don Felipe Bauzá
Chaplain	Don José de Mesa
Purser	Don Rafael Rodríguez de Arias
Surgeon	Don Francisco Flores Moreno
In charge of natural history	Teniente Coronel Don Antonio Pineda
Artist	Don José de Pozo
Pilotines	Don José Sánchez
	Don Joaquín Hurtado
Total of these ranks	16
Petty Officers	14
Marines, with sergeant and two corporals	15
Marine gunners, with one master gunner	4
Able seamen	35
Landsmen	10
Servants	8
Total	102

Corvette *Atrevida*

Commanding Officer	Don José Bustamante
Senior Officers	Don Antonio Tova Arredondo
	Don Dionisio Galiano
	Don Juan Gutiérrez de la Concha
Junior Officers	Don José Robredo
	Don Arcadio Pineda
	Don Martín de Olavide
Midshipman	Don Jacobo Murphy
Chaplain	Don Francisco de Paula Añino
Purser	Don Manuel Ezquerra
Surgeon	Don Pedro María González
Botanist	Don Louis Neé
Piloto	Don Juan Maqueda
Taxidermist and botanical artist	Don José Guio

Pilotines	Don Jerónimo Delgado
	Don Juan Inciarte
Total of these ranks	16
Petty officers	14
Marines, with sergeant and two corporals	15
Marine gunners, with one master gunner	4
Able seamen	35
Landsmen	10
Servants	8
Total	102

As will be seen in the journal, the expedition was later joined by the botanist Don Tadeo Haenke in Santiago de Chile, Tenientes de Navío Don José Espinosa and Don Ciriaco Cevallos, and the artists Don Fernando Brambila and Don Juan Ravenet in Acapulco; Don José del Pozo left us at the first stay in Lima, Don Dionisio Galiano, Don Cayetano Valdés, Don Juan Vernacci, Don Secundino Salamanca and the artist José Guio in Acapulco, Don Martín de Olavide, Don Juan Maqueda, Don José María Sánchez in Manila, Don Tadeo Haenke at our second stay in Lima, and finally Don Juan de la Concha and Don Juan Inciarte at our second stay in Montevideo, all with different commissions relating to the essential mission of the corvettes, except for the two painters and the pilot Don José Sánchez, who were obliged to leave because of their poor state of health. During the six months that we were on the north-west coast of America, we were also accompanied by the Mexican scholar, Don Tomás Suria, as a painting instructor.

BOOK ONE

FROM CADIZ TO MONTEVIDEO AND AT MONTEVIDEO

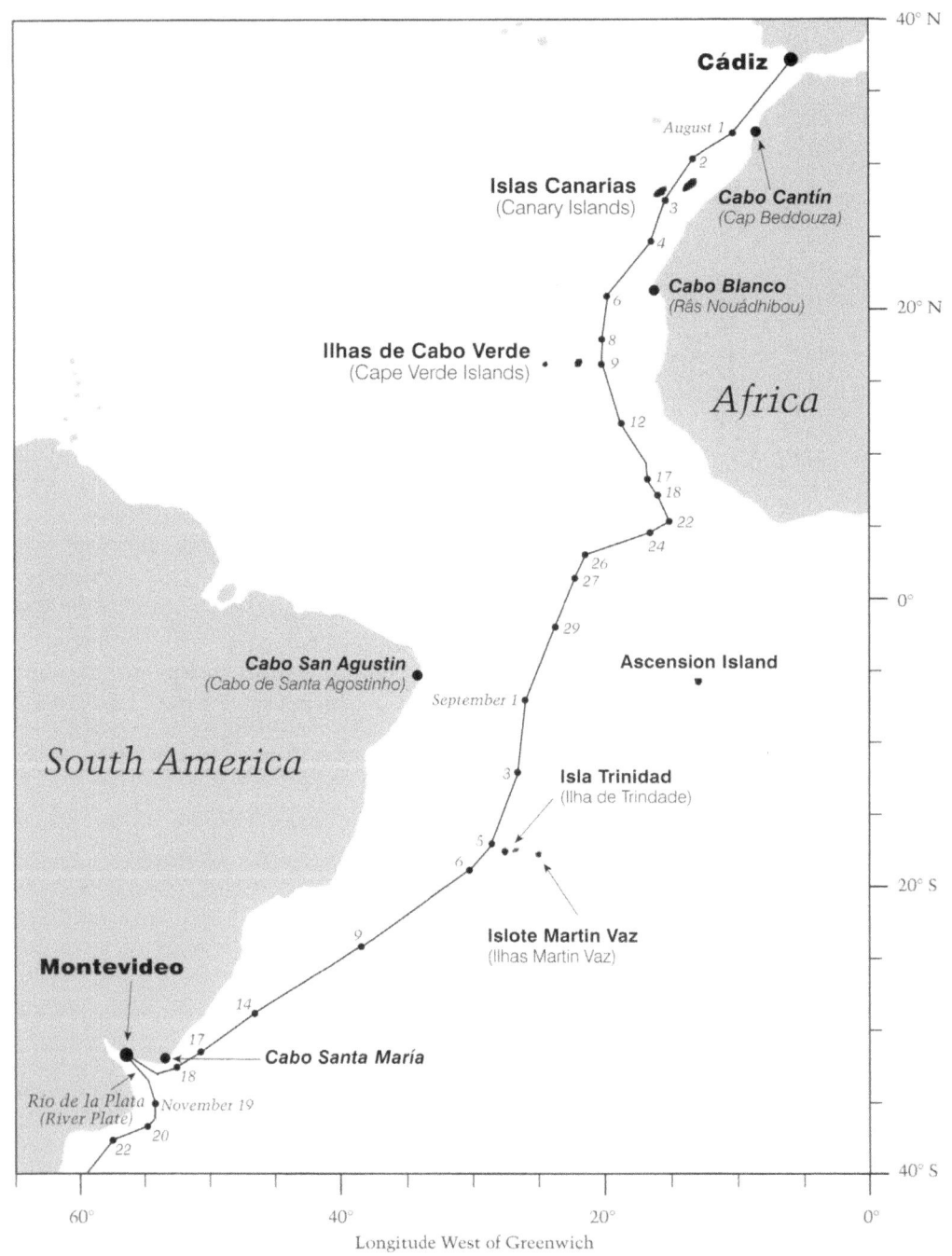

Fig. 1. Cádiz to Montevideo, July to September 1789

CHAPTER 1

From Cádiz to Montevideo

30 July 1789

After the *Atrevida* had taken on board some supplies that she still needed, we[1] set a course under full sail for Punta de Naga on Isla de Tenerife.

I had taken the precaution of handing over to the Harbour Master able seaman Antonio Marcelo who was totally unfit for service, being ravaged by venereal disease, and in accordance with an order previously obtained from the Capitán General de la Armada, two other able seamen were transferred from this corvette[2] to His Majesty's ship *San Dámaso*. Their conduct had been so disorderly after receiving their four months advance that we were obliged to have them draw their wages on board another ship of His Majesty's navy. On the *Atrevida* there was no need for such disciplinary measures. These three individuals, and another who deserted after drawing his pay, were replaced by four sailors who came forward voluntarily. Of these, landsman Agustín Muñiz warrants special mention. He was a member of the Harbour Master's boat who happened to be on board and offered to stay for the voyage, instantly forgetting not only how much this new venture would differ from his present one, but also the fact that committed to such a long voyage he would be abandoning his wife and children, and that without the slightest warning. How many times have fate and the rough nature of the sailor given rise to bold actions which the best tempered heart and the noblest birth might scarcely achieve!

At noon, we were a good three leagues from Torre San Sebastián,[3] bearing N89°E, Rota[4] N58°E, and Altos de Meca[5] S35°E. According to these bearings on Don Vicente Tofiño's new chart[6] our latitude at noon was 36°28′N, and our longitude

---

[1] The *Descubierta* and *Atrevida*.

[2] Malaspina's ship the *Descubierta*.

[3] Castillo de San Sebastián, situated at the western end of the peninsula on which Cádiz stands.

[4] A small town standing at the northern entrance to Bahía de Cádiz.

[5] A level ridge of uniform white appearance, 24 miles SE of Cádiz and close NE of Cabo Trafalgar.

[6] '*Carta esférica desde Punta Candor hasta Cabo de Trafalgar. Año 1787*', plate 9 in Vicente Tofiño y San Miguel, *Atlas marítimo de España*, Madrid, 1789. Tofiño (1732-95) was a protégé of Jorge Juan (see p. 219, n. 2 below), and became Director de las Compañías de Guardias Marinas (in effect, the Spanish navy's intellectual *corps d'élite*) in 1768. While working at the Observatorio de Cádiz from 1773, he promoted the idea of co-ordinating observations made all over Europe. He was employed between 1783 and 1787 surveying the entire Atlantic and Mediterranean coasts of Spain, including Islas Baleares, and became the central figure in the precocious scientific circle to which Malaspina and many of his colleagues belonged. Of those on the Malaspina expedition, Galiano, Espinosa and Vernacci all served

3

0°12'30" west of Real Observatorio de Cádiz, whose meridian we will consistently adopt with the appropriate distinction of longitude west or east, although during our voyage we will of course calculate only in longitude west.[1]

A long swell from south and SE caused us some discomfort throughout the day's run. A fresh breeze continued to blow from ESE to east. We profited from it by putting on all sail, at which to our surprise both corvettes showed the same speed, and raised our hopes that we would be able to keep company without inconveniencing each other. During the afternoon we sighted a brig in the far distance sailing close-hauled towards the NW. At sunset variation was observed to be 21°9'40"[NW].[2]

At this time, four stowaways revealed themselves, having hidden on board and frustrated all our efforts to avoid this problem. The hope of an easy life in the Americas, and the fact that the upbringing of the common people gives them little incentive to work if they have this particular hope, is the real reason for this continuing exodus, which in many ships, particularly merchantmen, can amount to as many as fifty or sixty individuals. Naturally enough, our principal concern this day was with standing orders and discipline. A signal was made to the *Atrevida* for the seamen and marines to be divided into three watches. On this ship, not only were the watches arranged in this way but each was placed under the personal command of a commissioned officer, so that the men would have someone in authority to direct them and be attentive to their hygiene, food, discipline, and their other needs and preferences. Each watch was assigned topmen as well as two specially chosen helmsmen. For the greatest unity and quietude, not to mention vigilance, it was laid down that all members of the watch should station themselves aft apart from two hands forward to watch out for any danger. Lastly, an examination, albeit cursory, by the surgeon revealed without a doubt that all our people were healthy and robust, except for those ailments inseparably associated with seamen.

31 July

The following morning, the weather promised to be more pleasant, not only because of the steady wind which settled at ENE and NE, but because the swell was much less heavy and the outlook fair. At half-past nine we signalled *Atrevida* to observe longitudes by chronometers.[3] Their results (entirely identical to our own) were 2°54' at

---

under Tofiño on the survey of the Spanish coasts. His *Atlas marítimo* was accompanied by *Derrotero de las costas de España*, 2 vols, Madrid, 1789, which was republished in London in 1812 as *España Maritima, or Spanish Coasting Pilot*, with further editions in 1813 and 1814.

  [1] I.e. he intended to continue to calculate his longitudes west of Cádiz on reaching 180° W. Similarly both Cook and Vancouver continued to calculate their longitudes east of Greenwich on reaching 180° E.

  [2] One of two methods of obtaining variation. Variation by amplitude is obtained by observing the magnetic bearing of the Sun when on the horizon at rising (eastern amplitude) or setting (western amplitude) and then, knowing the Sun's declination and the ship's latitude, the Sun's true bearing can be calculated, the difference being the variation. Variation by azimuth is obtained by observing the magnetic bearing of the Sun when it is above the horizon, when its altitude must also be measured in order to calculate its true bearing. Malaspina gave his variation as NE or NW instead of the more usual E or W.

  [3] Malaspina refers to his chronometers in some places in his journal as *reloxes marinos,* i.e. marine watches, and in other places as *cronómetros,* i.e. chronometers; both terms have been translated as chronometers.

noon, at which time we were in latitude 34°36′N by observation. Since this position was considerably ahead of our dead reckoning, we asked the *Atrevida* whether they thought that the latter error could have arisen from the division of our log-line at every fifty and two-thirds English feet to correspond to thirty seconds of the sand-glass.[1] They replied that it was the reliance we placed on longitude by chronometer that caused us to notice the effects of this division which, conforming as it did to the measurement of a degree of the Earth's meridian as provided in the Royal Decree of 1753, was still little followed by Spanish seamen. They told us at the same time that they had found two stowaways hidden away, but had succeeded in finding in good time two more whom the Harbour Master's boat had taken ashore. At half past three in the afternoon, the azimuth compass by Martínez[2] gave the variation as 22°01′NW; the English compass by Gilbert[3] consistently gave a reading one degree less.

On this day, the three officers in charge of the watches carried out a thorough examination of each individual's clothes. Deficiencies were noted, chests and stowage were arranged, and kit put in order, since it was important to take good care of these items, given the inevitable shortage of space on board small ships, with so many things needed on the voyage, and required to be at hand. Administrative arrangements this day included a reallocation of the hours of work and rest for the officers, and the issue for the crew's supper of *gazpacho* (while we still served as part of the ration the rest of the victuals taken on board for consumption in port), and lastly the issue of an extra ration of a pint of white wine and three ounces of sauerkraut for the ships' officers.[4]

Presently we noticed that, as the result of the slight alteration in our trim caused by the necessary changes we had made in our stowage, combined perhaps with some favourable alteration in her trim, the *Atrevida* was sailing considerably faster than we were, and this continued the following day. The night was fine and calm, and with a press of sail we made full use of the favourable NE Trade Wind; the south-easterly sea having moderated considerably in consequence.

1 August

The three chronometers were almost in agreement in giving our longitude as 4°57′ at noon on this day. The latitude by observation was 32°59′. The difference between

---

[1] Malaspina is referring to the interval apart of the knots in his log-line. Placed 50⅔ feet apart in conjunction with a 30-second sand glass represents a nautical mile of 6080 feet, which is close to the international value of 1852 metres. British seamen usually placed their knots 50 feet apart, which represents a nautical mile of 6000 feet, since it was argued that it was safer to have the reckoning before the ship than after it.

[2] A Spanish instrument maker, known only as Martínez; supplied by Arsenal de la Carraca, one of three naval dockyards in Cádiz.

[3] Probably John Gilbert of Tower Hill, London, who was in business from 1751 until his death in 1791.

[4] Sauerkraut was considered as one of the main antiscorbutics of the expedition, based on Cook's favourable opinion of it. On the other hand *gazpacho*, a cold soup made with bread or biscuit, olive oil, vinegar, salt and water, was not considered antiscorbutic, nevertheless it must have been an important source of vitamin C: Julian de Zulueta and Lola Higueras, 'Health and Navigation in the South Seas: the Spanish Experience', *Starving Sailors: the Influence of Nutrition upon Naval and Maritime History*, National Maritime Museum, Greenwich, 1981, pp. 93–6.

these results and the dead reckoning was negligible. Cabo Cantín[1] on the coast of Africa bore E14°30'S, distant thirty-four and a half leagues. From half past nine in the morning we had in sight, some five leagues SSW of us, a vessel sailing on the same course as ourselves. We signalled this to the *Atrevida* and at five in the afternoon, when the vessel bore SE from us about two leagues, we hoisted our ensign and she hoisted the French ensign in reply. The azimuths taken this day with the Gilbert compass gave the variation as 20°4'NW. The consistent value of 18°30' obtained with the two azimuth compasses by amplitude at sunset, when conditions were eminently favourable for this delicate operation, was much more reliable.

The sky was bright but a little hazy. A flat sea allowed the airing and cleaning of the interior of the ship. To this end, all the hammocks were unslung and laid out on deck until sunset. At that time what seemed to be the trade wind sprang up which we took advantage of under full sail. At midnight, the meridian altitude of α Cygni[2] gave a latitude of 31°58'30", which revealed the existence of a current which was in our favour. Throughout the night there was a very heavy dew.

2 August

The *Atrevida* retained some advantage over us in speed, although not as much as on the previous day. The horizon was much obscured by haze and the seas became choppy because of the trade wind, which was strengthening all the time. At noon we were in latitude 31°1' and longitude 7°40'37", this being determined by chronometers 13 and 61, while 72 differed by 3' of longitude. The error in dead reckoning on this occasion was considerable, but not unexpected. By observation we were 26' farther south and 33' farther west than by dead reckoning. How great are the risks to which the navigator is exposed when he lacks the excellent and simple methods on which today's astronomical navigation is based.

With Punta de Naga now bearing sixty-one leagues S38°W from us, we set a direct course to sight it, bearing up one point from the course we had been following, which was WSW by compass. Since the trade wind promised to freshen considerably during the night, we decided to take the fullest advantage of it and signalled the *Atrevida* during the evening to sail close to us that night, keeping a good look-out and observing as many latitudes and longitudes as possible. A large vessel which we had sighted at a considerable distance on our bows at three in the afternoon was much closer at sunset.

3 August

At dawn the next day she was already far astern. Not far off we sighted two small vessels ahead, on the same course as ourselves, which appeared to be Catalan. Indeed when they were close to us they responded to our signal by running up the colours of the Spanish merchant navy.

---

[1] Cap Beddouza, a prominent headland on the coast of Morocco, that rises steeply to 60 m above the sea.
[2] Deneb.

The thick haze, which during the previous night had made any kind of observation impossible, still hid the sight of land from us this morning. We signalled the *Atrevida* to observe longitude with the chronometers. Finally at eleven o'clock she signalled land to SSW, and at noon the small islands off Punta de Naga were bearing S32°W. The latitude by observation was 28°53′30″. This position and the certainty that the chart of Don José Varela[1] was accurate, enabled us to make an exact comparison with the longitudes derived from our chronometers. We found that number 61 showed a discrepancy of only 33″ from the longitude calculated from the bearings. The error of number 13 was 1′52″ to the east, that of number 72 was fully 4′20″ to the west. Daily comparisons with number 61 showed that this error had been progressive and, according to Don Juan Vernacci, the rate of this chronometer was closer to that determined by observation during our last two days in Cádiz, rather than that deduced from observation over eight days. We preferred, however, to determine this matter over a longer period of time; our sole concern during that afternoon was to take full advantage of the favourable weather, coasting at a distance of about one league from land, from Punta de Naga to the town of Santa Cruz, keeping the ensign flying as we sailed past the capital.

Since the previous night the *Atrevida* had shown a tendency to lag behind, and this was further confirmed during the day. We lost sight of the settees,[2] and shortly after another merchant vessel[3] on the same course as us was also left astern. We also saw a schooner crossing from Gran Canaria to Tenerife. By the afternoon the wind began to freshen considerably and a heavy atmosphere with low pressure, shown by the ship's barometer dropping one line,[4] warned us that we might have to adjust our running rigging. Indeed, at nightfall we began to experience strong gusts, with heavy seas from NE. Sailing under foresail, topsails and topgallant sails, we managed to keep twelve (statute) miles offshore. Later we had to furl the topgallants until eleven o'clock that night, when, with the weather turning more favourable, we could again make more sail, and advised the *Atrevida* with the appropriate signal. Bearings of El Pico[5] and of Punta Roja[6] on the island of Tenerife determined our position at six o'clock that evening to be latitude 28°8′ and longitude 10°00′. From this position we resumed dead reckoning. The error [by dead reckoning] on this landfall was 1°15′ to the east, and 51′26″ to the north. This error, accumulated over just four days, ought to have convinced even the most inveterate adherents of traditional methods, had

---

[1] José Varela y Ulloa (1739–94), a colleague of Tofiño, was employed in 1776 in Tenerife, where he established an observatory in Santa Cruz in conjunction with Captain Jean-Charles Borda, a French naval officer in command of the *Boussole*, to test the performance of two chronometers made by the Swiss watchmaker Ferdinand Berthoud. At the same time they established the longitude of Pico de Teide. When Cook called at Santa Cruz in August 1776 Borda and Varela compared their chronometers on board the *Boussole* with their astronomical clock on shore by means of signals, which were also repeated to the *Resolution* to enable Cook to do the same. Borda and Varela also surveyed Islas Canarias and the adjacent coast of Africa.

[2] Sharp-prowed lanteen-rigged sailing vessels with two or three masts, presumably the small vessels sighted earlier in the day.

[3] *Fragata* in the original, with *mercante* omitted – a three-masted merchant vessel.

[4] A twelfth of an inch, a term no longer in general use.

[5] Pico de Teide, 3715 metres high, was at the time of Malaspina's voyage one of several prime meridians from which longitude was measured.

[6] Punta Montaña Roja, the SE extremity of Isla de Tenerife.

they not during the passage taken refuge in the futile conviction that the log-line adopted had not been as officially prescribed.[1]

4 August

When the *Atrevida,* which had fallen some distance astern, caught us up this allowed both ships to make more sail without either getting ahead of the other during the day. The sea was still rough and the weather was very hazy, although the outlook was reasonable.

Almost since departure, signs of venereal disease had appeared, as was natural among seamen, that unhappy class whose life is no more than a series of perils undertaken for little pay, and of misfortunes and infirmities arising from their very endeavours. Men of this class, which Europeans consider valuable, are nonetheless more than any other given up to their passions and vices. Their very upbringing contributes to their miserable lot. In addition to those affected by this complaint, we were concerned this day to find that there were two more individuals on the sick list: the baker had a lung affected by pleuritic pains that had not been properly treated, and a corporal had a liver complaint accompanied by an excess of bile.

The whole morning was employed in arranging our anchors better. The two stern anchors were hung with their flukes inboard of the gangway. The small bower cable was unbent, and the kedge anchors were carried to the foot of the mainmast, leaving only the best bower anchor bent on. Following the usual track, at noon this day we were at longitude 10°43′west of Cádiz. The latitude by observation was 25°49′. The weather, though favourable, gave signs of an imminent change for the worse judging by both the heavy atmosphere and by changes in the wind, which fell almost to nothing during the last hours of the night and shifted to SE and SSE as it did so. Warnings from the barometer ceased the next morning, [5th] when after a few hours of variable light airs a wind got up from NW, and steadily backed to a fresh NE breeze. Our latitude at noon was 23°36′. The longitude according to chronometers 61 and 13 was 12°15′ west of Cádiz. The results from number 72, being the opposite of those obtained when making our landfall at Islas Canarias [Canary Islands], indicated that its rate had not settled down properly, and consequently the instrument inspired little confidence. That same afternoon the Gilbert azimuth compass gave the variation as 15°48′W. We could never obtain from the Martínez compass even average results, although position, observer, circumstances etc were all changed in its favour. The next morning the difference between the two compasses remained the same and no less difficult to explain. The results from the Gilbert compass gave the mean variation as 14°47′NW. This day a considerable mistake in calculating the longitude had persuaded us that there was an easterly current, nothing unusual during these months, and indeed to avoid coming too close to Cabo Blanco[2] we had steered a little to the west. Much later, however, we realized that there was a current in the opposite direction, so much so that at noon, [6th] when our position was latitude 21°10′, longitude

---

[1] See p. 5, n. 1 above.
[2] Râs Nouádhibou, about 45 metres high, situated at the southern end of a long promontory in Mauritania in 20°47′N.

14°1', we found a cumulative error of 40' east of our dead reckoning. The discrepancy in latitude was no less, and this had been shown the night before by the meridian passage of α Lyra.[1] A thick haze had made it totally impossible to observe lunar distances until then.[2] Moderate breezes settled in from the fourth quadrant. The horizon remained hazy and everything warned that the trade winds would soon cease in the same way that they had done when we left Islas Canarias. This should not have surprised us unduly in view of the warning given by Monsieur d'Après in his sailing directions.[3] He clearly states 'that the south and south-west winds also blow in and around the Cape Verde Islands during the months of July, August, September and October' and elsewhere he adds 'that during the months of June, July, August and September the action of the Sun's rays on the land, and indeed on the seas to the north of the Line, alter the state of the atmosphere, making the winds less constant, so that the trade winds drop in June at 10° latitude and in July, August and September between 13° and 14°, and do not return to regular limits until December and January'.[4]

Some azimuths taken during the afternoon gave the variation as 14°48'. Our sick showed an improvement. Clearing the decks and ventilating below deck contributed much to keeping the crew constantly occupied: they had already protected the rigging throughout, with various servings and parcellings.[5]

7 August

This day differed not at all from the one before, except for the gloomier aspect of the horizon and the greater fickleness of the winds. Finally, at a late hour of the night, a steady drizzle set in, with favourable winds from the second quadrant. At noon, latitude was 19°4' and longitude was 14°11' west of Cádiz. As usual, bonito and flying fish began to appear, and we were pleased to see that they were not frightened away by our copper sheathing; and that our sailors even caught a bonito with hook and

---

[1] Vega.

[2] By observing the distance between the Moon and the Sun or between the Moon and a star it was possible for Malaspina to obtain the Mean Time of his observation at Paris from tables in the French Nautical Almanac, *Connaissance des temps,* or at Greenwich from tables in *The Nautical Almanac and Astronomical Ephemeris,* both of which were held on board the *Descubierta.* Local Mean Time, obtained either sometime before or after noon by observation of the Sun, was then carried forward to the time the lunar distance was observed. The difference between the two times was the difference in longitude expressed as time between the place of observation and the meridian of Paris or Greenwich. The known difference in longitude between Paris or Greenwich and Cádiz was then applied to give the longitude west of Cádiz. The first edition of *Almanaque náutico y efemérides astronómicas,* based on the meridian of Cádiz for the year 1792, was not published until 1791. Copies of the French tables were, however, reprinted in *Almanak náutico y estado general de marina,* which Malaspina also had on board.

[3] Jean-Baptiste Nicolas Denis d'Après de Mannevillette (1707-80), astronomer and hydrographer, made a number of voyages to the East and Africa, mostly in the service of the Compagnie des Indes (the French East India Company}. From 1762 to 1780 he was head of the company's chart and map library in Lorient. His *Neptune oriental,* published in 1745, with a revised edition in 1775, was issued in an English version in 1782 as *The East India Pilot.*

[4] *Instructions sur la navigation des Indes Orientales et de la Chine, pour servir au Neptune oriental, par M. d'Aprés de Mannevillette,* Paris and Brest, 1775, pp. 2, 3.

[5] To parcel and serve a rope is to wrap it round with canvas (parcelling) and then bind it tightly in place with spun-yarn (serving).

line. The showery weather caused us, at three o'clock that morning, to furl some of the smaller sails. We signalled this to the *Atrevida* and at dawn the next day [8th] we put on more sail, as the wind had settled in from ESE. We steered close-hauled to the south, not only to keep clear of Islas de Cabo Verde[1] but also to head further east and thus catch the winds which were now almost from south and SSE. At noon, our latitude was $17°56'30''$. The longitude was $14°30'$, deduced from chronometers 61 and 13 since number 72, whose longitude differed increasingly from the other chronometers with every passing day, revealed an alteration in its rate. We preferred not to work this out yet, so as to have more time to check the error against the mean of as many lunar distances as opportunity allowed.

Moderate breezes continued from east to ESE with the horizon and skies constantly overcast. We saw some schools of flying fish and bonito, various common birds, and some small Portuguese men-of-war;[2] while some ominous rain was also in sight. There was an improvement in our sick, among whom was a stowaway who had come on board with an ulcer on his leg which had to be treated with great care. The *Atrevida* hailed us to say they had no sick. Keeping company with her was becoming easier, and thanks to the constant care taken in her handling, she caused fewer delays during our passage, and her speed was virtually the same as ours on most occasions when we came together.

9 August

At dawn we sighted three vessels at a good distance, all in the second quadrant. The weather which had been squally the night before, with some thunder and lightning in the east, improved somewhat during the morning and permitted us to stand to the south, carrying a press of sail. At noon we were able to observe the latitude which was $16°42'$. Chronometers 61 and 13 gave the longitude as $14°16'$; the island of Sal[3] therefore bore forty-four leagues true west. Somewhat later, one of the vessels not far ahead of us challenged us to give chase. We overhauled her at two o'clock in the afternoon, with our ensign already flying. Having hoisted the English ensign, they lowered a boat and their boatswain came on board. He told us she was the merchant vessel *Philips Stevens* of Liverpool, five weeks out of England, bound for Calabar on the Guinea coast to ply the slave trade. We told him our names, and that we were bound for Buenos Aires and also informed him of our longitude, which they were most interested in knowing, since their dead reckoning was a full degree and a half to the west. Finally we offered them some refreshment, and whatever else they needed that was within our power.

Scarcely had the other ship hoisted their launch, and we had set all sail in pursuit of the *Atrevida* (which we had hailed not to shorten sail since we would follow her) than we encountered a fierce squall from the SE, bringing rain and dismal conditions. We were caught by surprise carrying as much sail as possible given that we were close-hauled. Both our seamen and the corvette showed sterling qualities on this occasion,

---

[1] Ilhas do Cabo Verde.

[2] *Galerillas* in the original, which is apparently a diminutive, probably of *galera*, which in this context is a Portuguese man-of-war rather than a small galley: see p. 16 below.

[3] The NE island of Ilhas do Cabo Verde.

the latter withstanding the full force of the wind, the former immediately furling the topgallants, jib and staysails without the slightest mishap. The mainsail was then clewed up, the main topsails were lowered and the mizzen topsail furled. The *Atrevida*, whose crew behaved with the same promptness and exhibited the same excellence in seamanship and ship-handling, kept very close to us. The SWbyW course that we were steering was scarcely favourable, since the wind was blowing almost from the south; it remained fresh and gusting, and the outlook generally dismal. We hauled the main tack on board in the rain, and with the topsails and main topsail staysails hoisted, we set ourselves a double objective: to avoid falling to the west, and to put more and more to the test the corvette's sailing qualities, which we found to be excellent. At sunset variation was 12°44′NW. As night drew on the weather seemed to improve – if a sailor can speak of good weather when a westerly wind has turned to calm. We took advantage of light airs from the east, and at eight o'clock in the morning [10th] these backed to NNE, and finally became a moderate ENE breeze which entirely cleared the skies and horizon. We thought ourselves fortunate indeed (with the reading of the marine barometer further improved) that the trade wind had not entirely abandoned us. On this day too we were fortunate with our fishing: two sharks, two dorados, and sucking-fish[1] clinging as usual to the former, were objects of scientific research for Teniente Coronel Don Antonio Pineda in charge of natural history, and a pleasant dish to serve at our table and in the sailors' mess. The *Atrevida* hailed us to say that they had also been successful with their fishing. For the considerable time that we sailed in company, our conversation was of the particular news of each other, our mutual comradeship, and our hopes for a happy voyage. All the while, the English merchant vessel with which we had spoken on the previous day was just visible from the masthead.

At noon we were in latitude 16°2′, longitude 14°6′. Therefore the centre of the Sun was only 40′ distant from our zenith. The afternoon's run was quiet, but during the night several squalls repeatedly obliged us to trim our sails. The previous evening the variation was 13°7′. On the following day [1th] it proved impossible to observe it, or indeed the latitude, either by the meridian altitude of the Sun, or by double altitudes[2] because the compass needle values were erratic.[3] We therefore abandoned the calculation of longitude, for which purpose hour angles had been observed.[4] As we reached the limit of the trade winds, we recognized that from now on we would be

---

[1] The Remora (*Remora* sp.).

[2] A method of obtaining latitude from two observed altitudes of the Sun and the elapsed time between the two observations, knowing the Sun's declination when the greater altitude was observed and latitude by account. A table to facilitate the calculation was first included in the *Nautical Almanac* for 1771 and in the second edition of *The Tables Requisite* in 1781, with the heading 'For Computing the Latitude of a Ship at Sea from two Altitudes of the Sun, &c.'. This method is now generally known as obtaining latitude by double altitudes. Malaspina's explanation why it was not possible to take these observations ('porque continuaron repetidas las agujas'), which follows, is incomprehensible.

[3] Because the Sun was almost overhead at noon it would be necessary to know exactly the direction of due south to obtain a viable meridian altitude which would not be possible if the compass was behaving erratically.

[4] The angular distance of a heavenly body (in this case the Sun) east or west of the meridian. An integral part of obtaining longitude by chronometer.

subject to the variables, usually blowing to the SW,[1] that would take us to the Line, and we were determined to derive the best advantage from them. So with winds from the second quadrant, which prevailed until midnight, we steered to the south and SW; after which, with the winds now from the third quadrant, we altered course to the second, and the *Atrevida* keeping close to us did likewise.

12 August

In the early morning the weather began to clear, and from eight to ten o'clock we were able to observe distances from the Sun to the Moon which, when carried forward to noon, gave the following results, our latitude by observation being then 13°2′, and the longitude by:

Number 61	12°	40′	52″
Number 13	12	39	5
Mean of 3 sets observed by Don Juan Vernacci	12	4	55
One by Don Fernando Quintano	12	14	52
One by Don Francisco Xavier Viana	12	46	7
One by Don Cayetano Valdés	12	44	22
One by Don Secundino Salamanca	12	39	52
Mean of 3 taken by me	12	38	28

During the afternoon, we observed the variation to be 12°52′ with a number of azimuths as well as by western amplitude.[2] That night when we hailed the *Atrevida*, she confirmed that these observations exactly matched others she had taken. At the same time she told us her longitude at noon had been 12°37′, which agreed almost exactly with our own.

Until now the strong SW winds had only demonstrated the corvettes' sea-keeping and sailing qualities, which we viewed with much satisfaction. We considered that since we were in the region of variable winds, neither the unexpected difference of 38′ to the north obtained at noon,[3] nor finding ourselves being drawn towards the coast of Africa, made us feel that the voyage was in any way tedious. Our four sick seamen showed encouraging signs of a speedy recovery. A number of shirts and shoes had been distributed among the seamen. The whole ship's company evinced the best of humour, intelligence, and rude health, and we knew that they enjoyed the same good fortune on board the *Atrevida*.

13 August

But from early morning, and indeed during the night, we noted from the heavy, confused seas, the occasional flashes of lightning to the east, and more than a little rain, that a violent WSW wind was setting in, whose gusts involved us in constant trimming of the sails. The seas caused us to make considerable leeway, despite all the sail

---

[1] The ships were entering the Doldrums, an area of low pressure between the Trade Winds of the two hemispheres. It is not clear why Malaspina should in this instance give the direction to which the wind was blowing instead of the direction from which it was blowing.

[2] See p. 4, n. 2 above.

[3] The ships had probably lost the influence of the south-going current usually experienced between Ihlas do Cabo Verde and the mainland of Africa.

that we tried to keep on, and the rain was so heavy and persistent this day and the next that the sailors had not a stitch of dry clothing left. Also the high seas made it impossible to ventilate between decks. In the morning we saw a turtle and a number of common birds and in the afternoon of the 14th we sighted a large vessel to lee-ward, sailing close-hauled under the four principal sails,[1] with the main topsail double-reefed. She acknowledged our colours with hers, which appeared to be Por-tuguese. By nightfall she was on our lee quarter, and to give our crew more rest that night we took in a reef in our topsails, lowering them a little.

During the preceding two days we had no opportunity of observing latitudes or longitudes, with the sole exception of a break in the clouds this morning, when chronometer number 61 gave our longitude as 10°53′; although this may have been incorrect as we had not been able to obtain our latitude at the same time. We were forced by a wind from SWbyS to steer either in a SE direction (which would take us too close to the African coast) or towards the fourth quadrant (in which case we would sacrifice some latitude). We chose the former course and sailed close-hauled all night in the second quadrant under the principal sails, although the very rough sea caused us to make considerable leeway.

15 August

We were not therefore surprised, next morning, to see that the sea had taken on the colour of shallow water,[2] nor at ten o'clock to see a number of shearwaters and other birds.[3] However, as a settled and clear day gave us every possibility of determining our position with certainty, we intended to delay changing tack until noon; but by eleven o'clock it became essential to do so as the SSW wind had eventually died down and obliged us to sail close-hauled to the west. At noon our observations placed us in latitude 10°11′N and longitude 9°46′. We were thus 44′ minutes farther to the east and 47′ farther to the north of our dead reckoning. According to the Eng-lish chart from the *Neptune Oriental* (*East-India Pilot*), which we believe deserves our full confidence, Isla Poilon[4] off the coast of Africa bore seventeen or eighteen leagues, and our sounding should have been twenty-five to thirty fathoms. We did not delay to check this, but sailed close-hauled to the west all afternoon under a press of canvas. The wind continued a strong breeze from SSW, and the sea was very rough, but the outlook was much more pleasant.

At about three o'clock in the afternoon we again sighted on our weather side the vessel of the previous day, approaching on an opposite course; soon afterwards she went about to the same tack as us, and the *Atrevida* hailed us to say her ensign appeared to be Portuguese. It was not until nightfall that the sea lost its shallow-water colour. The wind continued strong, the sea falling, and on the following morning

[1] In Spanish usage *cuarto principales* (the four principal sails) refers to the mainsail, the foresail, the main topsail and the fore topsail, while *seis principales* (the six principal sails) refers to the mainsail, the foresail, the mizzen-sail and the three topsails. See O'Scanlan, *Diccionario marítimo*, p. 437.

[2] The normal colour of the sea in the open ocean in middle and low latitudes is an intense blue or ultramarine, but in shallow water its colour is modified to shades of bluish-green and green.

[3] Possibly Cory's Shearwater (*Calonectris diomedea*), the commonest shearwater off Arquipélago dos Bijagós, although other species of shearwater may be encountered as well.

[4] Ilhéu do Poilão off southern Guinea-Bissau.

[16th] the clouds cleared sufficiently for me to observe lunar distances. The results obtained at noon were as follows:

Don Cayetano Valdés	10°	21′	39″
Don Manuel Novales	10	10	00
Don Juan Vernacci	10	30	00
Don Fernando Quintano	10	13	00
Don Francisco Viana	10	32	24
Don Secundino Salamanca	10	18	54
Mean of these six observations	10	21	58
Piloto Don José María Sánchez	11	1	39
Mean of two of my own	11	16	9
Longitude by { number 61	10	58	9
{ number 13	11	3	24

Number 72 had already shown a steady increase in its rate, which required further assessment. Today it gave a longitude which exceeded that of number 61 by approximately fifteen miles. However, it seemed to us to be more sensible to postpone this assessment until the next quarter,[1] which would furnish us with good lunar distances, especially since a longer run of daily comparisons with number 61 that we carried out indicated a much more reliable and well-established rate. Variation today and yesterday was constant at 12°NW, somewhat greater than that shown on the charts for this area, which are invariably based on dead reckoning. With the use of chronometers, it is time that magnetic variation charts cease to depend on this method, if we are to derive any benefit from them, however remote.[2]

All afternoon, evening and night the skies and the horizon retained their customary veil of misty cloud. This veil, if we consider the strength of the Sun in these latitudes,[3] might be taken as a further mark of divine goodness; but the mariner regards it only as a sign of calms and contrary winds in a voyage the duration of which naturally occupies all his concerns and thoughts. The variable and falling light airs from the third quadrant, which had made our passage on the previous night so irksome and unprofitable, gave us on the following morning [17th] the compensation of an exchange of visits by the officers and men of the two corvettes. We were now 500 leagues from where we had last exchanged visits.[4] Furthermore there were close ties of friendship and genuine high regard linking the two bodies of officers, which the seamen naturally copied. It is hardly to be wondered at, therefore, that these two reasons alone, although we had no risks or adventures to recount, and after a passage of time of only eighteen days, should put us all into such uncommonly good spirits.

[1] The Moon was getting too far from the Sun to take lunar distances, which would only be possible again at the Moon's next quarter.

[2] Malaspina appears to be referring to the theory that longitude could be found by an accurate knowledge of variation, which was disproved by the results of Halley's voyage. See Norman J. W. Thrower, ed., *The Three Voyages of Edmond Halley in the Paramore 1698–1701*, 1 vol. and portfolio, Hakluyt Society, 2nd ser., 156-7, London, 1981, p. 366.

[3] I.e. the Sun, whose declination was about 14°N, would be close to the zenith at noon.

[4] The last exchange of visits was therefore before leaving Cádiz.

Until noon the pinnaces continually conveyed first some then others between the two corvettes. The *Atrevida's* seamen presented our crew with a recently caught shark. The agreeable news that everyone was in good health gave new encouragement to us all; and the light airs from WNW that obliged us to separate at noon to stand on our course, gave us hopes of a favourable passage for some days to come.

Having hoisted the pinnaces just after noon we set a course to the south under a press of canvas. Observations of the meridian altitude of the Sun gave our latitude as 9°42′. Number 61 and number 13 were in almost total agreement, giving our longitude as 11°40′. This differed very slightly from the longitudes given by Berthoud's number 10 and Arnold's pocket chronometer, both of which were on board the *Atrevida*. The two chronometers which had recently arrived from London[1] appeared, therefore, to be the only ones whose rates had not settled down and become uniform.

A disagreeable subject but one that must be discussed in detail, necessarily took up a good part of my conversations with Don José Bustamante, who had joined us on board. When the bread-rooms were opened in both corvettes, seven or eight days after departure when we had finished the daily allowance, we found all the biscuit[2] to be infested with a caterpillar which Don Antonio Pineda, after carefully examining it in all its phases and processes, described as follows.

It is a caterpillar which turns into a membranous, transparent, yellowish chrysalis, from which emerges a small, whitish moth of the type known as a clothes moth, which lays its yellowish eggs stuck together by what look like spider's threads.

Seen with the naked eye the caterpillar seems to be eight or nine lines long and rather more than half a line wide, whitish with protuberances and red markings, which gives it an elegant red tint. The head is chestnut coloured. Small white hairs sprout from the protuberances. With a simple microscope or with the aid of a magnifying glass, two large eyes can be seen in the head, which are in all probability composed of many more since they are clearly nodular; just behind this there are two little patches. The mouth is composed of little cheeks and cusps. The body consists of nineteen or twenty rings. In the forepart of the body there are six legs, distinct from the rest, which end in hooks. Then separated from this by three or four rings are four pairs of legs, and then finally another pair next to the truncated, conical anus, with their feet surrounded by a border of red spots. On the back of this caterpillar there are four lines of nodules arranged lengthwise. These are red in colour and from them sprout bristles or very fine hairs. This insect sheds its skin or sheath when it reaches the stage of greatest growth and turns into a membranous yellow chrysalis with chestnut tints. From this there emerges a little white moth whose antennae taper from the base to the ends, being somewhat longer than the thorax or body. The tongue is coiled and the little spikes or whiskers next to the mouth are feathery. The upper wings are arranged horizontally. They are shorter than the body, whitish in colour with little blackish spots. The lower wings are only half as wide and similarly off-white in colour. The body is large and quite bulky,

---

[1] John Arnold, London, chronometers Nos. 61 and 72. The firm supplied marine chronometers to the British Admiralty.
[2] Flour well kneaded into flat cakes, with the least possible quantity of water, and slowly baked. Also known on board ship as bread and stowed in the bread-room. Malaspina uses both *galleta* = ship's biscuit and *pan* = bread, which have been translated accordingly.

more so than other parts. The anus ends in a sharp point. The abdomen consists of seven rings. The thorax is darker; the legs are blackish. This moth lays a cluster of eggs in the biscuit which are tied together by filaments similar to those of a spider's web. These eggs are yellow and somewhat cylindrical and appear dark against the biscuit. The caterpillar which emerges first makes a kind of spider's web, and then bores small holes within the biscuit, from which it protrudes its head, enlarging them as it goes, and eating until it reaches maturity and undergoes the transformations intended by nature. Apparently this moth belongs to Linnaeus's genus *Tecnes* Geofroi *Tennia*, and the species tends towards Geofroi's 19 but without his description or the one provided by Fabricio tallying with what I have observed in this case. Accordingly I take it to be a variety of or different species from *Tenia granitella* and *farmalis*, given that our caterpillar differs from those derived from those moths.

The introduction of this insect on board, which seemed to be limited to the bread,[1] had not spread to the numerous sacks of vegetables deposited in the same store-rooms, giving us ample grounds for suspecting that when the biscuit was brought on board it contained within it at least the eggs of the grub which the heat below decks caused to hatch and multiply rapidly. But as the properties of this grub were not harmful to health, and since we took every opportunity from the outset to overcome natural feelings of revulsion by our own example, on neither corvette did this problem cause alarm. Indeed, the exchange of visits that day, with both corvettes suffering the same problem, helped to assuage it.

Don Antonio Pineda, ever watchful for ways in which he could contribute to the progress of his favourite sciences, derived two further advantages of some importance from that morning's calm. The first was to have one of our yawls catch two Portuguese men-of-war (Linnaeus's *Holothuria physalis*) which he immediately subjected to the most thorough examination. The second was to experiment for the first time with a new container which he had invented for extracting sea-water from a specific depth. Although the latter was extracted at only ten fathoms below the surface, it nonetheless gave a difference of half a degree of temperature, when the Fahrenheit thermometer was immersed first at one and then at another depth.

Except for a few hours of calm during the night, we had a persistent moderate breeze from NW and west, and we took advantage of this under full sail, despite constant rain which fell almost continuously from sunset until the following morning.

18 August

This morning was no different from any of the others in its overcast appearance. The smooth sea and the weather allowed us to ventilate below decks with the greatest thoroughness. All kit was brought out into the open air, and the less ventilated parts of the ship were sprinkled with vinegar. We could now see that all the crew were the very picture of health, the more so now that quinine appeared to have restored the baker, despite his serious illness with its almost unremitting fever. And the few who were affected by stomach troubles or suffering from venereal disease were now making great strides towards full recovery.

[1] I.e. biscuit: see p. 15, n. 2 above.

At noon the latitude of 8°37′ and the longitude according to number 61 of 11°33′ showed that we were making good progress. By now we had lost sight of the Portuguese vessel which we had noticed was sailing somewhat more to the east than we were. The rain became even heavier and more frequent today than yesterday. The wind steadily shifted round to the third quadrant with strong gusts and so dismal an outlook that we were obliged to sail all night with a reef taken in the topsails, and in addition we displayed lanterns on both ships to help us keep company. At sunset we observed variation, somewhat uncertainly, to be 11°30′NW. Even without help from our chronometers we had to assume that there was an easterly current, close as we were to Sierra Leona [Sierra Leone], from which point a strong current flows toward Golfo de Guinea [Gulf of Guinea]. And indeed this current became evident during the day's run, flowing at a rate of one or one-and-a-half knots in the same direction for the next few days.[1]

19 August
At noon we were in latitude 7°30′ and longitude 10°10′. The sky continued overcast, and a heavy swell from the SW caused us some inconvenience. The direction of the wind shifting somewhat from SSW to SW allowed us to sail on both tacks during the day's run. So, although our gain in latitude was slight, we achieved this without much loss to the east, despite the fact that the current, the heavy swell, and the wind itself, all seemed to combine against the good sea-keeping qualities of this type of ship. Indeed, in these months of July, August and September, when the SW winds are so tenacious as far as latitudes 13°, 14° and 15° N,[2] it is important that if the vessel is not to experience long delays in these latitudes she should display good sailing qualities, especially when close-hauled. No action or thought should then be spared concerning sail trimming and tacking; for the proximity of the African coast and the currents that naturally flow towards it, make it quite dangerous to tack in the second quadrant, which otherwise is the combination that might provide some advantage in latitude.

20 August
We devoted the whole of the morning of the 20th, when the weather dispelled any fears of rain, to cleaning and ventilating between decks. To keep the seamen busy, and at the same time to make provision against the cold which we might encounter even before reaching Montevideo, cloth, baize and duck were distributed, which [the crew] could cut up at their leisure to make warm clothing to suit their fancy. The natural vivacity of Spanish seamen begs that they be kept constantly occupied and that some notice is taken of their passing fancies. These two reasons alone, even if one ignores the unfavourable price and quality of everything the sailor buys on land, should persuade all commanding officers to take on board clothing, preferably in pieces ready to be made up, on His Majesty's account, and to distribute them during the tranquil sailing days within the tropics, for use in the stormy conditions of the South Sea.

At noon our latitude by observations was 7°16′, and our longitude 9°54′. Number

[1] The ships were being set to the east by the Equatorial Counter Current.
[2] North of the equator the South-east Trade Wind is deflected and blows from the SW.

13 maintained the same difference of 5½′ to the west compared with number 61, and was very similar to the chronometers on board the *Atrevida*, as we learned that afternoon As experience had shown the need to make various changes to the signals communicated in Bahía de Cádiz, these corrections or annotations were carried out, and a signal was made to the *Atrevida* that afternoon that we wished to speak with her, and that she should send over the pinnace with an officer. We hove-to until these instructions were delivered and the pinnace hoisted once more; then we were both able to stand again to the SEbyS under full sail. The wind was from SSW and SWbyS with a moderate swell. Azimuths taken by Don Juan Vernacci gave the variation as 12°22′. Azimuths which I took myself at almost the same time gave the variation as only 10°51′, although the same compass was used. Around sunset a signal was made to the *Atrevida* to issue a daily ration of wine to the crew and sauerkraut three times a week. We took this decision in order as far as possible to divert the attention of our susceptible seamen from their natural irritation with a crossing of the Line which had proved tedious up to now, and which still seemed likely to be prolonged.[1] The rather rough seas mostly from SW, and the prospect of somewhat squally weather, persuaded us to pay special attention to our sails this night. The wind did not allow any other tack but in the second quadrant. As a consequence of the wind we found ourselves at noon [21st] in longitude 8°22′ and latitude by observation of 6°24′ N. A little after midday, even though the wind was still from SSW and SbyW, we altered course to the west in order to take us away to some degree from the strong easterly currents close to the African coast in these parallels. These currents had a noticeably adverse effect on our longitude, and were carrying us away from western meridians to that of Tenerife,[2] where the winds are customarily more constant and bring less rain. From midnight onwards, the weather did in fact begin to correspond to our predictions by shifting round to the south and SbyE so that at dawn on the 22nd we had the more pleasing prospect of the SE Trade Wind. The clear horizon, the direction of the clouds, a brownish-grey overcast sky with some breaks, a moderate breeze from SSE, and the sea itself a long, moderate swell from the same quarter, all confirmed our conjectures, especially at this season of the year, and showed them to be well-founded.

Some fifteen or twenty frigatebirds[3] were our companions today and yesterday. The tuna and bonitos had abandoned us. Every now and again we saw a flying fish, which although rare appealed more to our appetites than the bonitos that had left these waters. Don Antonio Pineda noted that the frigatebirds came from the east in the morning, and in the evening flew back in the same direction, without taking a rest, however brief, from flying all day long. If their abode is the rocky islets off the coast of Africa,[4] from which at the time we were distant some seventy leagues, what must be the speed and ease of their flight, and how great their need of food, to take them so far from the tranquillity of their nests?

---

[1] The ships were still in the Doldrums.

[2] Malaspina would appear to be in error here since his longitude at noon of 8°22′ west of Cádiz places him approximately 1°50′ east of the meridian of Tenerife.

[3] The Magnificent Frigatebird (*Fregata magnificens*), which breeds on Bõa Vista in Ihlas do Cabo Verde and possibly on Arquipélago dos Bijagós off the coast of Guinea-Bissau.

[4] Frigatebirds disperse at sea to feed during the day, but roost onshore at night.

We took advantage of the exceedingly pleasant morning to air various personal effects, particularly the bedding and kit of all on board. Our latitude by observation at noon was 5°46'. Our longitude according to number 61 was 9°18', differing from number 13 by the same 5½' to the west. In the afternoon the southern hemisphere trade wind steadily became more noticeable.[1] True, it was rather slight at the moment, but it promised nevertheless to strengthen very soon, gradually swinging from south to east as we approached the Line. We steered WSW under all sail during the night, although after sunset we noticed that the main topmast had sprung a hand's-span above the cap[2] because of the juncture of four horizontal knots. It would be difficult to guess at what point of our recent sailing they had given way under the not excessive stress of the wind.

We had carried a press of sail during the night and consequently we were one or two miles ahead of the *Atrevida* the next morning [23rd]. When we were waiting under easy sail for the other corvette to join us, we set about hoisting a new main topmast to replace the one that was sprung. The men worked as one, busily and intelligently, so that within two and a quarter hours we were again under way under full sail. At the same time the *Atrevida* came up within hailing distance, and consequently did not need to clew up. The wind continued very favourably, although always from the south. However, the chronometers were by now indicating a current setting to the west, not unusual in these latitudes once the coast of Africa is left behind.[3] At noon, number 61 gave the longitude as 10°26', latitude by observation was 4°50'. Variation which we had been able to observe in very favourable conditions on repeated occasions remained constant between 12° and 13°NW. At nightfall on this day we were greeted by the welcome sight of a very large albacore caught by hook.[4] It proved a hard task to hoist it in, but at length this was done and the following day it provided a pleasant meal for the officers and crew.

24 August

We had already begun to use the apparatus for distilling water during the mornings only, while the full allowance of firewood was used for the coppers. Sailing on the same tack as the side where the still was set up was not in truth the best arrangement; thus, on this and the following morning we were only able to distil at the rate of 8½ pints per hour. The *Atrevida* had hailed us on a previous occasion to report that they had distilled up to 12 pints per hour.

This water, whose excellent properties had been tried and confirmed on board the *San Sebastián* last year provides not only a valuable addition to our supply, but is also a means of curing certain illnesses.

Observations at noon gave the longitude as 11°55' and latitude 4°12'. We inferred

---

[1] The ships were now under the influence of the South-east Trade Wind.

[2] A strong thick block of wood with two holes, which holds an upper mast in one hole against the top of a lower mast in the other hole.

[3] They were now under the influence of the South Equatorial Current, even though they were still north of the Equator.

[4] Probably the Atlantic Albacore (*Germo alalunga*), also known as the Long-finned Tunny, which rarely exceeds a length of one metre and a weight of 30 kilos.

from the errors in our dead reckoning that allowance had to be made for a westerly current flowing at approximately one knot. Today we began to obtain distances from the Sun to the Moon, which we were keen to have in order to check the rates of the chronometers, adjusting that of number 72. At the same time we were eager to reassure ourselves on both corvettes of the degree of perfection that we had brought to this most interesting aspect of our tasks. Not surprisingly, two sets by Don Juan Vernacci entirely agreed with number 61, while two others taken by myself and one by Don Fernando Quintano differed by 30' to the east, but this, as we learned later, was the result of a slight error in the sextant. Don Francisco Viana's observations were 36' to the west of the same chronometer.

25 August

However, on the following day, when we were able to make observations very much at our leisure and convenience, the following results were calculated, which of course we took as definitive.

Number 61	13°	17'	19"
Number 13	13	26	41
Four sets by Don Francisco Viana	13	21	34
Three ditto by Don Juan Vernacci	13	16	34
Three ditto by Fernando Quintano	13	23	00
Three ditto by me	13	33	40
One by Don Cayetano Valdés	13	23	00
One by Don Manuel Novales	13	42	34
One by Don Secundino Salamanca	13	32	34
Two ditto by Piloto Don José Sánchez	13	28	19

At noon we were not able to observe latitude. The wind was still strong from SbyE, and it was evident from the very choppy sea this produced, as well as from the brownish-grey clouds with breaks which at times darkened the horizon, that the SE Trade Wind would not abandon us from now on.

At sunset a large vessel was sighted to the SW, sailing north at a distance of two or three leagues. The desire to speak with her and indeed take advantage of the occasion by sending news to Europe, persuaded us (having made the appropriate signal to the *Atrevida*) to keep our ensigns flying until night fell. When she was nearer, at around seven o'clock, the vessel fired a gun and displayed a lantern in her bows. We replied by also firing a gun, and having made a signal to the *Atrevida* to heave-to, we did so ourselves and made ready to lower the pinnace. The firing of the gun and the displaying of the lanterns soon dispelled our ideas of approaching the strange vessel. She brought-to while she was to windward of us and so far off that our mutual efforts to exchange recognition with the help of a speaking trumpet came to nothing. Such an awkward manoeuvre and our concern not to waste precious time naturally persuaded us to leave her. We made the crowd-on-sail signal to the *Atrevida*. It was not without regret that we saw that half an hour on the other tack would have brought us up to that vessel which at first had sought to speak with us as inopportunely as she had afterwards sought to avoid us, and whose sailing appeared to be very poor. We finally learned from the *Atrevida* that the vessel had hailed her with considerable wariness to say that she was a Portuguese ship out of Cabo de Buena Esperanza [Cape of Good Hope], bound for Lisbon.

We passed a very peaceful night and the wind, which on the following day [26 th] began to swing round to NE and SE, revived our hopes of crossing the Line before long, although the quite strong westerly currents seemed intent on making us sail close-hauled without respite, and indeed on driving us towards the coast of Brazil. At noon latitude by observation was 2°50′, indicating that in the last two days' working we had lost 23½′ by dead reckoning, while in longitude 14°55′ we were 44′ more to the west according to number 61, the accuracy of which was once again vouched for by eleven sets of lunar distances whose mean differed by only 7′ to the east from that given by the same chronometer. In the afternoon some tropicbirds[1] and many flying fish were to be seen, but the bonitos and the albacores appeared to have abandoned our vicinity. At nightfall the wind shifted around to SE and sailing close-hauled with all sail set, we were at noon [27th] in latitude 1°14′N, longitude 16°14′ according to number 61, which was also confirmed that afternoon by ten sets of lunar distances from the Moon to the Sun, which only differed from the chronometer by 2½′ to the west. The wind which had already settled at ESE was so favourable that by three-thirty the next morning we were able to cross the Line in longitude 17°5′ west of the Real Observatorio de Cádiz. The variation was 9°20′NW by different azimuths measured at nine o'clock that morning [28th], as well as by amplitude measured the previous evening. Seventy sets of lunar distances confirmed the same chronometer's longitude of 17°00′. By noon we were already in latitude [0°] 39′S and longitude 17°24′. We found the effect of the westerly current which had been forty-two miles over the last twenty-four hours excessive,[2] and all the more surprising because there was no appreciable difference in latitude. Having ascertained the reliability of chronometer 61 with so many observations, and with its almost constant agreement with number 13, we decided to adjust the rate of chronometer 72 by reference to the rate of 61. The following comparisons begun on the sixth of the month had already showed its rate to have settled down and become quite uniform:

Day of Comparisons	daily gain of 61 over 72
6	49″
7	45″
8	43″
9	49″
18.	48″
20	47″
24	48″
27	49″
29	

[1] Both the Red-billed Tropicbird (*Phaethon aethereus*) and the White-tailed Tropicbird (*Phaethon lepturus*) occur in this vicinity.

[2] They were still under the influence of the South Equatorial Current setting to the west at one to one and three-quarter knots (24 to 43 miles a day).

The mean of $47''\cdot75$[1] subtracted from the $56''\cdot95$ daily gain of number 61 gave, over number 72's mean, a daily gain of $9''\cdot20$, greater by $2''\cdot47$ than what had been determined in Cádiz.

29 August

The probability of this interpretation being correct was confirmed by two checks that were as convincing as the ways in which they were obtained were different: in the first place the last two intervals of the comparison in Cádiz had indeed indicated a gain of $9\frac{1}{2}''$, which, however, we had to consolidate at the time into the general mass of observations, since it was too small to serve as a basis for determining its rate; secondly, that once this rate, now shown to be $9''\cdot20$, was carried forward from the day on which we sailed from Cádiz to noon on the 29th, it provided a longitude perfectly consistent with those of number 61 and 13, and also with the lunar distances, to the extent that our position this noon could be agreeably determined in the following manner:

By seventy sets of lunar distances	$17°\ 28'\ 12''$
By number 61	$17\quad 33\quad 12$
By number 13	$17\quad 41\quad 10$
By number 72	$17\quad 39\quad 27$

Latitude by observation $2°58'$S, variation by western amplitude $8°13'$NW. However, Don Juan Vernacci very shrewdly observed that since this was a new adjustment to number 72 deduced from 61's rate, and since the latter's rate seemed to be accurate to the last degree, it would seem justifiable to make the two coincide completely. And thus the absolute gain in respect to Cádiz mean time was determined by the longitude of number 61 for this noon, and henceforth a new period would commence for rating number 72. The officers on board the *Atrevida* (according to what they told us by hailing) were not as fortunate in the accuracy of their chronometers. In the comparison with the lunar distances, all entirely consistent with our own, number 10 showed an error of $30'$ or $40'$ east, and Don Dionisio Galiano, with his usual dedication to science had compiled for it a new table of temperatures, which brought it to the true longitude. Chronometer 71 and pocket chronometer 105, both by Arnold, had repeatedly shown an alteration in their rates. It seemed to them that number 10 in conjunction with the new table of temperatures could in future match their expectations.

The precautions we had taken in strengthening our rigging with preventer shrouds and backstays had clearly not been wasted. These precautions were also taken by the *Atrevida* whom we had advised by signal of this procedure on the 27th. The excessively heavy swell from SSE and SE would have made it rather hazardous to take advantage of the very strong trade wind from ESE to east under full sail, without first providing for the security of masts and topmasts. We saw many flying fish, and an

---

[1] The mean of the eight values tabulated is actually $47''\cdot25$, but was probably $47''\cdot75$ if the ninth value omitted in Malaspina's journal is taken into account.

occasional frigatebird.[1] The bonitos and dorados had completely abandoned us. Obviously, sailing south (an intention which we were following closely) as far as parallels 8° or 9° did not take us off course in the slightest. From that latitude we could steer a course directly to soundings off Cabo Santa María,[2] but in order to be able to identify Isla Ascensión,[3] we would have to steer slightly to the east of the direct course. The position of the island[4] had not only been determined by His Majesty's frigate *Santa Rosalía* in the year 1774 by Señores Lángara, Mazarredo and Varela, the first two now *tenientes generales* and the third a *brigadier de la real armada*;[5] it had also been fixed by La Pérouse and Langle[6] in 1784 during their voyage in the *Boussole* and the *Astrolabe*, frigates of his most Christian Majesty, sent in quest of new discoveries. It seemed, therefore, at first sight, that the advantages of an early fitting-out in Montevideo might persuade us to omit this new survey. However, we did not know the circumstances attending the determination of the position by Comte de la Pérouse, while in the case of the *Santa Rosalía*, whose abstract was kindly sent to me beforehand by His Excellency Don José de Mazarredo, this could have involved distances which, since they depended on the less accurate lunar tables and reflecting instruments of the time, could not be corrected by those officers even if their experience and knowledge of astronomy much exceeded our own. Moreover they lacked chronometers and since these have come into use together with lunar distances, the observations made by Tenientes de Navío Galiano and Belmonte in the frigate *Cabeza*, by myself in the *Astrea*, and by Tenientes de Navío Churruca[7] and Cevallos in the snows *Eulalia* and *Casilda,* suggest that the longitude 24°12′ west of Cádiz determined by the above-mentioned gentlemen should be reduced to 23°0′.

Having considered these circumstances, and given that the position determined by Comte de la Pérouse[8] (according to the abstract appearing in the Gazettes by order of the French government) had been obtained under sail using chronometers and lunar

---

[1] Probably Magnificent Frigatebirds (*Fregata magnificens*).

[2] Regarded by Malaspina as the NE entrance point of Río de la Plata.

[3] A non-existent island with the name Ascençaon shown on a chart by d'Après de Mannevillette 300 miles west of Ilha da Trindade and only 15′ farther south, which La Pérouse searched for unsuccessfully in October 1785.

[4] Malaspina is clearly referring here to Ilha da Trindade, having decided perhaps that Ascençaon and Ilha da Trindade were one and the same island. He continues to refer to Ilha da Trindade as Ascençaon until he has fixed its position: see p. 31 and n. 4 below.

[5] Juan de Lángara y Huarte (for whom see p. xciv, n. 2 above) was in command of the *Santa Rosalía*, with José de Varela and José de Mazarredo serving under him, on a voyage to correct errors in Spanish charts.

[6] Paul-Antoine-Marie Fleuriot de Langle (1744-87), presided over the Académie Marine de Brest from 1783-5, when he joined La Pérouse's Pacific expedition (having already served with the same commander in the Hudson Bay expedition of 1782). He was killed by natives on Samoa in December, 1787.

[7] Cosme Damián Churruca (1761-1805) was a *guardiamarina* at Cádiz from 1776. In October 1788 he embarked with Ciriaco de Cevallos in charge of geographical and astronomical work on Antonio Córdoba's expedition to Estrecho de Magallanes. He returned to work in the Observatorio de Cádiz from 1789 to 1792.

[8] According to La Pérouse the longitude of the SE point of Ilha da Trindade was 30°57′ west of Paris: John Dunmore, ed., *The Journal of Jean-François de Galaup de la Pérouse 1785-1788*, Hakluyt Society, 2nd ser., 179-80, London, 1994, p. 24.

distances, I decided that if the winds would allow us to do so without detriment to our main objective of spending the whole of next summer off the coast of Patagonia and Tierra del Fuego, we ought to investigate the position of the island. We would probably need to modify our previous assumptions in accordance with scientific facts should the results of this new and exhaustive investigation confirm – as well they might – the findings of the year 1774.

30 August

Thus, in spite of the very heavy and long swell from SSE which made it very uncomfortable, even risky, for us to set a course to the south, I determined that such a course should be steered until latitude 10° or 12° was reached, from where a course should be set to take us to latitude 20°30', longitude 22°40'. By taking this precaution we could also pass in sight of Islotes de Martín Vaz [Ihlas de Martin Vaz] and we would thus ensure that we always had a good chance to identify any land seen to windward from east to west. Our latitude was 7°41' and longitude by number 61 was 18°4', although the fact that this differed somewhat from numbers 13 and 72 aroused the suspicion that there had been some alteration in the rate of the first-mentioned chronometer.

We continued to see many flying fish, and a few frigatebirds.[1] A number of showers from ESE and SE inconvenienced us which, combined with the gusty conditions, obliged us to trim our sails. Variation had by now decreased to 6°15'NW.

31 August

The wind was rather light on the following day, compelling us to steer somewhat more to the west. However, we managed to maintain our chosen course, and to that end, despite some gusts and very high seas from SSE and SE, we crowded on sail which in itself said something about the good sailing qualities of both corvettes.

Assuming with Captain Cook that the longitude of Cabo San Agustín on the coast of Brazil was 29°30' west of Cádiz,[2] we were 150 to 155 leagues from it at midnight, bearing true west from us. In this position we could consider the month of August as concluded, seeing with some satisfaction how much it had contributed up to now to the success of our voyage. First we had not been afraid to close the coast of Africa with quite stormy winds from SW, and second that we had committed ourselves to the prevailing trade winds, in accordance with the expert opinion of the best modern navigators, crossing the Line 7° west of the meridian of Tenerife, without the slightest concern that we might be set towards the coast of Brazil, a concern that has made

[1] Probably Magnificent Frigatebirds (*Fregata magnificens*).

[2] Cabo de Santo Agostinho on the coast of Brazil. During his third voyage Cook was in the vicinity of the cape on 8 September 1776 when his longitude 'deduced from a very great number of lunar distances, was 34°16' West; and by the watch, 34°47'... Hence I concluded that we could not now be farther from the continent than twenty or thirty leagues at most': Cook and King, *Voyage to the Pacific Ocean*, I, 34-5. Assuming, as Malaspina appears to have done, that Cook meant he was twenty or thirty leagues from the cape, his longitude for it by lunar distances west of Cádiz would have been 34°16'W +1°30' (for 30 leagues) - 6°11' (the difference in longitude between Greenwich and Cádiz observatories) = 29°35' west of Cádiz.

many voyages very long when they could have been extremely short and consequently pleasant. The daily reckoning left us in no doubt that while we were off the
coast of Africa there was an easterly current, which later changed its direction to west
at a rate of one or one-and-a-half knots as soon as we had entered the open sea, or
begun our approach to the American coast. This current, which from the perusal of
a large number of journals I should venture to call constant, to some extent excuses
the old rule which prescribed that one should take the currents to be east or west
depending on whether one was east or west of the meridian of Tenerife, and should
make corrections accordingly.

An exceptional occurrence (which I have not yet mentioned) has to a large extent
marred our efforts to maintain the good health of all on board this ship, although the
skill of the ship's surgeon Don Francisco Flores may seem to promise a happy conclusion to the very delicate case under his care. A member of the royal marine artillery,
whose conduct for many years had been irreproachable, in a moment of weakness
perhaps more induced by the urgings of a dissolute relative of his than by the prompt-
ings of sensuality, which until then he had known how to restrain, had caught venereal
disease, which his very honourable character caused him to conceal from everyone.
Ten days out from Cádiz, he went to the sick bay with a fever which was soon accompanied by delirium. This made the surgeon suspect that there might be some hidden
cause of this affliction. He found in fact, upon looking into it, that this unfortunate
man's modesty in hiding his true illness had proved even more fatal than his weakness
in contracting it. Gangrene had spread so rapidly that only in the last three days has it
been possible to arrest it, at which stage the patient's extreme weakness gives reason to
doubt whether he can survive the long period of suppuration. Although it may be
widely felt that such a wretched individual, who has suffered not inconsiderable tor-
ment before our eyes, and who now suffers the dread of an early death, deserves no
place in this journal, yet we should mention this man, a serving marine who fell to his
own sense of shame in hiding an illness which in his profession is almost a routine
hazard. In contrast, similar cases in five other seamen, although serious, have progressed towards a favourable conclusion. The use of quinine has finally overcome the
very deep-seated fevers afflicting the baker which had seemed to be turning rapidly
into tuberculosis, and lastly, the judicious use of Dr Masdevall's antimonial compound[1]
has forestalled at their very beginning certain catarrhal fevers which, with the naturally
attendant stomach upsets, appeared to derive from the mixture of heat and humidity
which we experienced during this month.

The remainder of the crew continue to be in good health with the rider which I
must add of sailors riddled with the effects of their vices whenever they leave a port
where they have been able to indulge themselves. I believe that the diet adopted has
not only been useful in keeping them healthy but has actually improved their health:
a pint of wine was issued daily, sauerkraut three times a week, and *gazpacho* every

---

[1] Dr José Masdevall y Terrades (c. 1730-1801), a Spanish physician who practised medicine in
Figueres, where his experiments with chemical treatment for fevers yielded a mixture which proved
remarkably efficacious during the typhoid epidemics in Catalonia in 1783 and in Cartagena in 1786.
His treatise on the treatment of fever, *Relación de las epidemias de calenturas pútridas y malignas*. appeared in
1785. His compound was known as Opiata Masdevall.

evening. The dried vegetables and salt pork could not be better. Even the condition of the bread, the store-rooms for which were frequently aired with the help of a ventilation hose, offered firm grounds for hoping to get rid totally of the insect which had already been banished in part, and which had been present from the very outset. The water, particularly that which had been stored in cured casks with a little quicklime, retained the best taste and indeed the greatest purity.

1 September

The springing of the fore topmast due to an unexpected flurry of wind which struck us at eight o'clock in the morning, contributed somewhat to dampening our satisfaction at making so much progress in our voyage. It might imply that we were too uncompromising in crowding on all sail, which, however commendable on other occasions, might have been quite reprehensible on this one, where the main object should have been to conserve as well as we could equipment which was of such value and was always to be preferred, on balance, to a gain of speed. We should, however, have been reassured by the fact that the *Atrevida* had suffered no damage while always keeping on as much or more sail than ourselves, and that the causes of the double damage were clear. In the first incident (as we have already said) there was a row of four transverse knots and in the second we observed that our having cut to size a rough spar of considerably greater diameter had greatly weakened it, and (according to our carpenters) bared its very heart. It was replaced by a new one and at the same time the foremast shrouds were set up to our satisfaction. At noon we were already sailing with the fore topsail hoisted, although the heavy rolling arising from the rough seas running from south and east caused us considerable inconvenience.

Our latitude at this time was 9°35′ and longitude by number 72 was 19°53′. We had now to rely directly on this chronometer for our position as, in agreement with number 13, it had indicated since yesterday a loss of no less than twelve to sixteen seconds per day in the rate of number 61.

The wind continued strong from ESE with very rough seas. By sunset the sky and the horizon had become quite overcast, and our suspicion that the next new Moon could bring some ominous shift in the weather was confirmed by a signal from the *Atrevida* that her barometer indicated bad weather. As our own barometer still read 30 inches and five hundredths of a line we disagreed in our signal back. Nevertheless, by eight o'clock in the evening we were forced by repeated squalls and some downpours to take in a reef in the topsails.

2 September

The next morning it was still blowing from ESE, but the skies were clearer. The rough sea was falling, and it seemed that our efforts to sight Ascensión would not be frustrated, since the natural effect of the westerly currents was to give us an extra 10′ or 15′ to the south, an advantage which by now we regarded as a daily occurrence. At noon, number 72 gave our longitude as 26°46′, Number 61 continued to lose, and 13 continued to agree with 72. Latitude by observation was 11°35′. Such a thorough and repeated appraisal of the rates of the chronometers should not seem superfluous.

Their intrinsic value and our care in examining them should be seen as the basis on which the reliability of our work will be judged, and consequently the degree of confidence that future mariners will place in it. We believed that it was mainly because we had set up the foremast shrouds yesterday and those of the mizzen mast today, whereas the *Atrevida* had not done so, that she had shown a considerable advantage over us in speed. But she hailed us to say she had already set them up as taut as possible.

We had hoped to measure various lunar distances that night, which although very clear at first, soon clouded over, and we even had some passing showers with gusts from ESE, so that we had to strike the topgallant sails and staysails.

3 September
Nevertheless the weather was fairer at dawn, with a fresh breeze from ESE to east, and we continued to steer on a course close to due south. At sunrise the variation was 3°35'NW. There were a number of common birds, especially frigatebirds.[1]

In the morning we set up the mainmast, foremast and fore topmast shrouds which had on previous days become slack. Afterwards we shook out a reef in the topsails, and running under full sail we found ourselves at noon in longitude 21°4', latitude by observation 13°56'. Although the rolling had not completely subsided, at two o'clock that afternoon we attempted to observe azimuths with the two azimuth compasses. The results were in close agreement: the one from La Carraca[2] gave our variation as 6°20', while the English azimuth compass gave it as 1°41'. The western and eastern amplitudes which were observed very much to our satisfaction, indicated in the former instance a variation of 2°18' and in the latter one of 4°9'.[3] From how much uncertainty is the navigator relieved, for he can now find his longitude without depending on those unreliable or inaccurate observations that have hitherto rendered useless both the care taken by many good observers and the theories propounded by many scientific experts.[4]

A peaceful night allowed us to establish, from four sets of lunar distances to Antares, a longitude that was 39'30" further west than shown at the time by number 72. The wind continued a moderately fresh breeze from east to ESE, with a settled outlook and smooth sea.

We continued to steer close-hauled to the south until noon [4th] when, having observed the latitude as 16°25' and the longitude as 20°52', we considered that Ascensión bore S25°30'W, and accordingly bore away to SSW5°W by compass.

The *Atrevida*'s superiority in speed indicated to us that our stowage needed some adjustment. This necessity (to which we were alerted yesterday by the undue ease with which our corvette luffed when we put the helm amidships in order to overhaul

[1] Both the Great Frigatebird (*Fregata minor*) and the Lesser Frigatebird (*Fregata ariel)* breed on Ilha da Trindade and Ilhas Martín Vaz.

[2] See p. 5, n. 2 above.

[3] These values are suspect and were probably wrongly transcribed when a fair copy of Malaspina's journal was being prepared. The values obtained for variation for several days before and after 3 September suggest that on this day variation was about 2° and the values were probably 1°20', 1°41', 2°18' and either 1°9' or 2°9'.

[4] *See* p. 14, n. 2 above.

the tiller-ropes, and with almost all the after sails either lowered or clewed up) persuaded us this morning to shift weight from the stern to the bows. We did this by moving sixty pigs of iron and all our allowance of shot and bar shot that we had in the shot locker which weighed approximately 120 *quintals*[1] and which we carried forward and placed in the forward hold. The advantages of this alteration were immediately apparent during the afternoon and night when we considerably outsailed the other corvette, even though sailing at the time without studding-sails.

This night, twenty-seven sets of lunar distances to Aquila[2] observed by Don Juan Vernacci and myself, whose means gave a longitude only 3′52″ more to the west than number 72, confirmed the excellent state of this chronometer.

5 September

Since the weather had allowed us to maintain a speed of seven knots all night, which was confirmed at sunrise that this was going to last, we could be confident of sighting Ascensión that same afternoon if we made more sail, without the slightest danger of becoming separated from the *Atrevida*. Having therefore signalled that we were going ahead in search of land, we ran under full sail, and at noon we were already four miles ahead of the *Atrevida* by tangents to the horizon.[3] Latitude by observation was 19°14′, longitude by number 72 was 22°43′. Consequently we luffed round to SbyW by compass.

At a quarter to five in the afternoon we did indeed sight the island bearing SSW 5°W. We bore down on it on this course, and having signalled to the *Atrevida* that we had sighted land, we afterwards advised her to observe longitudes with her chronometers. At quarter past five the island was clearly seen from below[4] and, after some clouds had dispersed, we were able to observe for longitude by chronometer using number 72, at the same time the two extremities of the island bore S20°30′W and S26°W by compass. The longitude was 22°49′16″, latitude was carried forward by dead reckoning from noon,[5] variation from several azimuths and by western amplitude was 1°10′.

It was my initial intention to wait for first light, and be as close to the anchorage as possible by dawn, and if the weather permitted have the two pinnaces employed in fishing and hunting, while on board we carried out multiple observations for latitude and longitude. But I then reflected that both latitude and longitude could be equally well determined if we were far offshore at dawn, provided that our bearing was accurate, that longitude was observed at the same time, and that there were very few errors in the latitude derived from the meridian altitude of the Sun observed the following noon. The commanding officer of the *Atrevida*, after a signal had been made

---

[1] Approximately 5½ tons.

[2] α Aquilæ or Altair, one of ten stars whose lunar distances were tabulated in the *Nautical Almanac*.

[3] It is not clear from Malaspina's remarks exactly what observations he took to the horizon. Perhaps he was referring to a time when the *Atrevida's* waterline coincided with the horizon, when knowing his height of eye above the sea the distance of the sea horizon could be obtained from any manual of navigation. With a height of eye of 16 feet the distance to the sea horizon is 4·6 miles.

[4] The island would in all probability have been sighted by a look-out at the masthead before it would be visible from below - i.e. from the deck.

[5] This corresponds to point C on the diagram, which is oriented north at bottom, which gives *Descubierta's* latitude at the time as 19°45′17″S.

to come within speaking distance, agreed with this opinion. Both of us having considered how important it was for us to reach Montevideo in the shortest possible time, we steered SWbyW by compass all night under full sail. We sighted the island on our beam at eleven o'clock that night, at which time twenty-seven sets of lunar distances observed to Aquila[1] by Don Juan Vernacci and myself determined a longitude 20' farther to the west than that of number 72. Finally, at 6.47 next morning it bore N74°45'E from us by compass. The longitude observed at the same moment according to number 72 was 24°7'23".[2] Variation by eastern amplitude was 1°30'NW.

6 September

The difference in latitude by observation between these noon positions,[3] exactly equal to the difference by dead reckoning, confirmed the reliability of the latitudes we had established by dead reckoning this morning and the previous afternoon.[4] Thus we were able to consider that the position which resulted from our different interconnected operations was very close to the true position of the island, which we lost sight of at nine o'clock. By the following noon we were in latitude 21°41', and longitude 24°25' west of Cádiz.

The adjacent figure shows our different operations individually; the longitude of point A, determined from C with the latitude difference and angle of bearing, is 23°7'45" and from point D, similar data determine it to be 23°7'23"; and the mean 23°7'34" is consequently the true longitude of point A. Finally, the great accuracy of both triangles[5] is vindicated by the very similar values for sides CA and AD, resulting again from the triangle CAD:[6]

The fact that the bearings combine in such a way that differences in latitude would have a marked effect on the distances has also convinced us that the latitude of 20°32' could not under any circumstances be accepted,[7] and it was essential to replace it by a latitude of 20°26'45", based on my observations on board the frigate *Astrea*. We had attempted to tie in the bearings to the new latitudes with the meridian altitude of Lyra[8] observed from the *Atrevida* at seven-thirty that night, and that of Aldebaran observed from here at five o'clock the next morning. But we very soon decided that the two meridian altitudes of the Sun were preferable, because the considerable northern declination of both stars would make their movement exceedingly slow,[9] and because the difference deduced from the dead reckoning agreed exactly with that

---

[1] α Aquilæ or Altair: see p. 28, n. 2 above.

[2] This corresponds to point D on the diagram, which gives *Descubierta*'s latitude at the time as 20°43'56"S.

[3] I.e. between 19°14'S at noon on the 5th and the latitude at noon on the 6th, which is not given.

[4] The latitudes of points C and D.

[5] 'Triangulos parciales' in the original MS, referring to the two triangles CAM and DAS.

[6] I.e. the two slightly different values for these sides given against them in the diagram.

[7] Presumably the latitude of Ihla da Trindade based on Malaspina's original latitudes for points C and D, which he has not given and which he had to revise when he accepted the latitude for the island from his earlier observations in the *Astrea*.

[8] α Lyræ or Vega.

[9] Malaspina's meaning is not clear since neither Vega at 38¾°N nor Aldebaran at 16½°N can be described having considerable northern declinations.

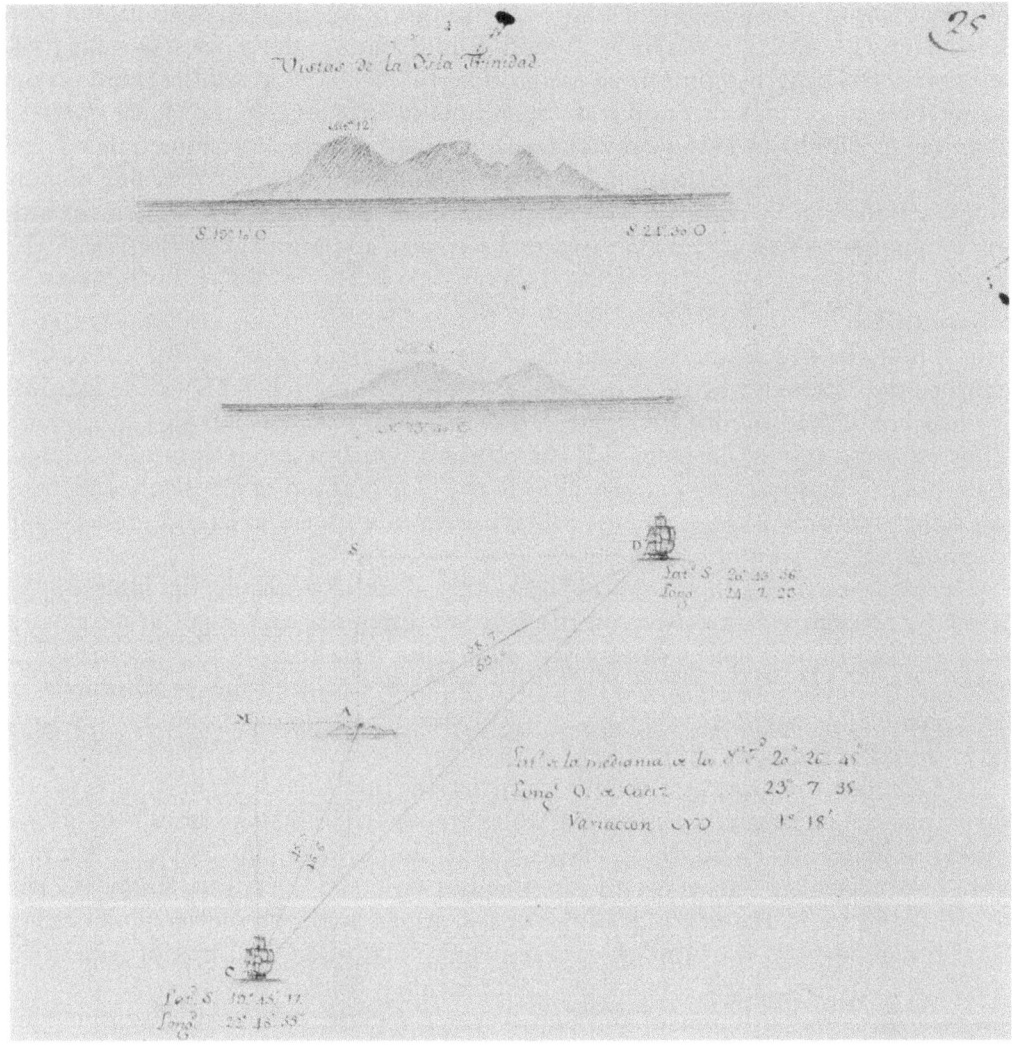

Plate 10. Views of Ilha da Trindade with plan showing Malaspina's method of obtaining the island's position. Museo Naval, Madrid

observed in these twenty-four hours, while the stars, both to south and north, indicated significant differences in the dead reckoning.[1]

For the last eight days we had adopted number 72 as the master chronometer rather than number 61, because number 13 decided in favour of the former, which in any case perfectly matched the longitudes deduced from lunar distances when they had confirmed the accuracy of number 61 and number 13. Consequently, to prefer the said chronometer for determining the longitude of the island would seem all that

[1] Taking altitudes of heavenly bodies at night rarely gives satisfactory results since the horizon cannot be distinguished easily.

was required since the latter's small difference with 13 was long-standing and constant, while number 61 inclined a little to the east by 21′ from the position determined by the frigate *Santa Rosalía*. And lastly, all but a few of the lunar distances observed over several days at this time to Aquila,[1] once again agreed exactly with number 72. In truth, on this occasion with just the minimum of impartiality, I preferred to take without distinction all the lunar distances observed by Don Juan Vernacci and myself, since when taken together they determined the longitude to be somewhat more to the west. However, the distances observed between the Moon and the Sun before and after this all agreed in showing that 13's longitude and therefore 72's, were precisely accurate at the time, and that the result from the lunar distances on this occasion differed somewhat from the correct one because of our cautious policy of adopting any information that would bring our results more in line with those obtained by persons whose knowledge was admired by the entire navy.[2]

The longitudes aggregated with those of number 72 were therefore as follows:

By thirty-three sets observed by me to Antares and Aquila on the
3rd, 4th, 5th and 7th and referred to the chronometer in the same
period . . . . 23° 22′ 19″
By thirty ditto observed by Don Juan Vernacci 23 19 49
By chronometer number 13, according to comparison with 72 at
noon on the 5th, and corrected for the loss indicated in comparison
with the 7th 23 13 12
My observations on board the *Astrea*, with eighteen sets of distances
from the Moon to the Sun agreeing with Arnold pocket
chronometer, number 71 . . . . 23 1 00

Giving a mean of . . . . 23° 10′ 45″

It would be insulting to the meticulousness with which the frigate *Santa Rosalía* compiled the plans and accompanying remarks for this island, if it were thought that these charts might be deficient in any way: thus, solely on the basis of this new astronomical position, we have altered the position of the main islet of Martín Vaz to Latitude 20°24′45″S, and longitude 22°37′35″ west of Cádiz. Its position relative to the island remains the same, and this we shall call Trinidad[3] without hesitation, to differentiate it from the island called Ascension situated in latitude 7°57′S and longitude 16°19′ west of Paris,[4] as well as to dispel the idea of the existence of another island in this immediate vicinity and parallel. This is a point that should be taken as definitely decided by our navy, although Monsieur D'Après appears to maintain the contrary.

---

[1] α Aquilæ or Altair: see p. 28, n.2 above.
[2] I.e. Lángara, Mazarredo and Varela: see p. 23, n. 5 above.
[3] Present day Ihla da Trindade.
[4] The real Ascension Island, over 1,000 miles NE of Ilha da Trindade, in contrast to the non-existent island Ascençaon charted by d'Après de Mannevillette 300 miles W of Ilha da Trindade: see p. 23, n. 3 above.

We have seen very few birds, either boobies or shearwaters,[1] even when in the vicinity of Trinidad and this convinces me (having noticed the same off Avilés[2]) that in these parallels, where easterly winds constantly prevail, almost all birds are careful to make their outward flight in the same direction to ensure the ease and safety of their return. The weather continued fair, with the wind from NE, which allowed us to make six or seven knots, and brought us the following morning [7th] to where we crossed the dividing line between NE and NW variation. We observed to our full satisfaction a variation by eastern amplitude of 0°6′NE.

To steer a course directly towards Cabo Santa María at the northern tip of the mouth of Río de la Plata, we were obliged to adopt a longitude for it. The one indicated in *Conocimiento de Tiempos*,[3] determined in Buenos Aires, differed by only 10′ from that which PilotoTafor[4] had given (perhaps following the observations of Brigadier Don José Varela) on an excellent chart of Río de la Plata. Other charts which he had sent us differed considerably from this longitude. It would have been reprehensible not to place greater confidence in the position given by Tafor, since the difference was negligible, nor would it have been proper to ignore the great experience and zeal with which he had served for a long time in these seas. Consequently, we took Cabo Santa María to be in latitude 34°55′, and longitude 48°19′ west of Cádiz. At noon we were in 26°13′, latitude 22°42′S.

The night provided us with some observations of longitude by distances from the Moon to Aquila,[5] which because of the short interval we could refer to Isla Ascensión.[6] Twelve sets observed by Don Juan Vernacci and myself gave 13′ to the west of number 72. Another twelve observations by Valdés, Quintano and Viana, differed in their mean by only 6′ to the west of number 72, with which number 13 agreed the following noon. Number 61 had differed by 22′ to the east.

8 September

On crossing the Tropic of Capricorn the trade winds began to die away, and the wind tended to swing to the north. The days were exceedingly temperate and pleasant and as we continued on our course this allowed some officers of this corvette to visit those of the *Atrevida*. By noon the pinnace had already been hoisted again. Our latitude was 24°, the longitude 27°37′ and variation by eastern and western amplitudes, was 2°48′NE. In the afternoon the wind got up again, and continued for the next two days, persuading us to keep all sails set, the only exception being a few occasions during the night, when intermittently cloudy horizons to the SW obliged us shorten and trim sail.[9th and 10th] At noon on the 10th we were in latitude 26°28′ and longitude 31°49′. Variation by amplitude the previous evening was 5°0′NE.

---

[1] The Red-footed Booby (*Sula sula*) breeds on Ilha da Trindade, but in this instance Malaspina probably sighted the Herald or Trindade Petrel (*Pterodroma arminjoniana*), which also breeds on this island, and not a species of shearwater.

[2] A town on the north coast of Spain.

[3] The French Nautical Almanac, *Connaissance des temps*, which contains a list of geographical positions.

[4] Piloto Don Bernardo Tafor who had carried out surveys in Río de la Plata and Patagonia.

[5] α Aquilæ or Altair: see p. 28, n. 2 above.

[6] Malaspina has reverted to Ascensión (Ascençaon) instead of IslaTrinidad.

We had taken care to air the bread-rooms as much as possible to rid them of the caterpillar that had got into them, the transformation of which and its arrival once more at the moth stage had shown Don Antonio Pineda that this was Linnaeus's *Tecnia biscotela*. Yet we already knew it was impossible to achieve this, and we were heartily glad that this moth had not spread to other articles. The bread at this time was quite good.

In the last hours of the morning the wind had abated considerably and shifted to NW, by which time the horizons in the third and second quadrant were overcast. A little after noon the wind, after a sudden shift with constant downpours, settled into a fresh breeze from south and SE. With the fore topsail braced aback we went on the port tack, and seeing that the wind was growing stronger we made signals to the *Atrevida* to put the marines and the seamen in two watches, to give them a daily ration of wine, to send down the topgallant yards and topmasts, and to take in two reefs in the topsail, which we performed ourselves shortly after.

With the arrival of the cold weather, which had been very apparent from the moment that the winds from the polar regions settled in, the gunner whose illness we have already mentioned soon received the fatal blow. After a month of unspeakable suffering, all his strength abandoned him, and left him unable to withstand the suppuration of the gangrene, which had now been totally cleansed. He finally succumbed to the results of his untimely silence, frustrating from the start those many measures which we had adopted to avoid the loss in our little society of even one man.

Finally, at nightfall we reduced sail to the foresail and the two topsails so that we could proceed in company, and to further this end we kept a lantern burning in both corvettes.

11 September

The wind, now from ESE, was increasing in strength; there was a high sea and the rain was incessant, so much so that even at noon on the 12th we were unable to observe latitudes and longitudes. Fearing therefore that this wind with the same gloomy weather could take us rapidly to our destination, we had bent on two more anchors and we signalled to the *Atrevida* to take the same precaution, since it would later be more difficult, with the heavy swell. We took in three reefs in both topsails, and the sea, now very high, gave us an opportunity to test the corvettes in a following sea, with the result that both vessels hardly shipped any seas, while neither their hulls, masts, yards, nor their rigging suffered any damage, despite the top-hamper of our boats, and a large load of spare masts and yards.

13 September

The wind did not abate, and the foul weather continued until the morning of the 13th; in fact, during the night the wind had blown so fresh from east and NE that with only the foresail and main topsail, with three reefs taken in, our speed was nine to ten knots by long log-line.[1] The lightning, the foul weather, and the wind, by now

---

[1] A log-line in which the knots are placed 55 *pies* and 5 *pulgadas* or approximately 50⅔ feet apart, which in conjunction with a 30-second sand glass represents a nautical mile of 6080 feet: see p. 5, n. 1 above.

from north, all indicated to us the imminence of a violent *pampero*.[1] We feared that it would arrive (as is usual) with a sudden change in direction. For this reason, together with our desire to try out the corvettes lying-to, we first reduced sails to main topsail and fore topmast staysail, and then finally to the fore topmast staysails, main staysails and mizzen staysails. Extremely heavy seas from the SW, with a wind now almost fallen completely calm, made the next few hours very uncomfortable because of the heavy rolling of the ship.

At noon during some breaks in the cloud (the rain which accompanied us for the last three days having scarcely left off) we obtained a rather doubtful latitude of 31°48' and longitude of 40°2'. For several days we had been accompanied by various species of birds, boobies, Cape Petrels, shearwaters and Wilson's Storm Petrels;[2] and a few hours earlier we had seen a clump of sargasso which in these latitudes (it is commonly said) is seen from 150 to 200 leagues from the mouth of the river.

The following day the weather cleared up. Having had either calms or moderate winds from the third quadrant, we sailed close-hauled in the second and fourth quadrants; finally with all topsail reefs shaken out, and the sea somewhat calmer than it had been, [14th] we spent the rest of the day ensuring we did not approach the other corvette too closely, rather than making good our course. The latitude of 31°32' and longitude of 40°20' which we determined to our satisfaction at noon placed us 150 leagues from Cabo Santa María. New irregularities had appeared in number 72's rate, and today it differed by 22' to the west compared with the other chronometers which mainly agreed with each other. For that reason we ceased noting its results, but we were impatient to see its true rate from observations we were about to take in Montevideo.

During the afternoon the gentle breeze shifted to SSE, and we sailed close-hauled in the third quadrant under full sail except for the topgallant sail, but swaying up the topmasts in the meantime. The night was beautifully fine and peaceful and the wind by now a moderate breeze from east and NE; we crowded on all sail, while waiting only for first light before hoisting home the topgallant yards.

15 September

At sunrise variation was 11°9'NE; and subsequently various sets of lunar distances to the Sun referred to number 61, gave the following results:

Number of sets	Observers		Difference from 61 by comparison
12	Don Juan Vernacci	W	11' 30"
3	Don Franciso Viana	W	9' 45"
5	Don Fernando Quintano . . .	W	4' 30"

[1] A fierce wind from the Pampas which often springs up with great strength and without warning.
[2] Brown Boobies (*Sula leucogaster*), which are the only species of booby found so far south off the coast of Brazil, Cape Petrels (*Daption capense*), various species of petrels and shearwaters, not distinguished from one another in Malaspina's time, and storm petrels, probably Wilson's Storm Petrel (*Oceanites oceanicus*). Malaspina gives the alternative name *martín placas* (whose derivation has not been ascertained) for *pampero* = Wilson's Storm Petrel.

10	Don Secundino Salamanca ..	E	5′ 15″
9	Don Cayetano Valdés	E	16′ 45″
5	Don Manuel Novales	W	14′ 15″
4	Don José Sánchez (*Piloto*) ...	E	13′ 30″
1	D. Fabio Ali Ponzoni (*Guardiamarina*)	E	11′ 30″
7	Self	E	5′ 15″
56			

The latitude by observation was 32°7′S. Number 61's longitude was well-supported by the above observations.[1]

This evening and the following night were very peaceful, even calm. Every advantage was taken of the light airs from north and NE. Several passing squalls from SE and south which appeared on the horizon, blew themselves out. [16th] In the morning we saw a number of common birds. Our latitude was 32°37′ and longitude according to number 61 was 42°9′, variation by observation was 12°40′NE. Very soon afterwards, the same fortune which had accompanied us until now favoured us once again. Moderate breezes set in from the first quadrant and very soon freshened, promising us a prompt end to our voyage. Many birds could be seen. The water began to take on a brighter, shallow-water colour. Variation by observation was 13°. [17th] At noon in latitude 33°27′, longitude 44°9′ we considered ourselves to be N69°15′E from Cabo Santa María distant seventy-four leagues. Nevertheless our navigation was made somewhat complicated that night by an exceedingly thick fog, which required great care with our consort, and made it impossible to make further observations even of latitude. We continued under full sail, steering SWbyW. At nightfall, our hopes of obtaining soundings with eighty-four fathoms of line were frustrated. [18th] Our estimated latitude at noon was 34°27′, and longitude 46°6′, so that Cabo Santa María was now S72°W distant thirty-six leagues. In the afternoon we were already visibly in soundings.[2] We waited, however, until after sunset before testing this, and we made a signal to the *Atrevida* that we intended to continue under way during the night, while one or other of the corvettes would be sounding for two hours, alternately, to reduce inconvenience and save time. At five o'clock in the afternoon, we were surprised to find ourselves in thirty-two fathoms, sand and shell, while the *Atrevida* obtained only twenty-five. We did not know our distance from the coast, which we inferred must be less than it actually was, and considered it wise not to close any further before reaching the parallel of Isla de Lobos.[3] We therefore steered south; at nine o'clock we had thirty-four fathoms, sand, and at one o'clock the next morning, forty-five fathoms, fine sand. At this time, estimating that we were in latitude 35°0′, we crowded on all sail and steered true west at six to seven knots.

[1] According to the version of Malaspina's journal printed in Palau *et al.*, *Diario de viaje de Malaspina*, p. 71, the longitude calculated from these 56 lunars is 41°22′ and by number 61 is 41°24′.

[2] Presumably because of the change in the colour of the water.

[3] An islet about 20 metres high, 4 miles SE of Punta del Este, so named for the *lobos de mar* or sealions which abounded there.

19 September

Despite the extremely thick haze and an unexpected sounding of forty-two fathoms, sand, which we obtained at five o'clock in the morning, and having already run some ten leagues to the west, we were determined not to lose the favourable wind. We made a signal to the *Atrevida* to that effect, instructing her to follow in our wake. The wind, at this time a strong northerly breeze, later threatened to develop at any moment into a *pampero*. It therefore seemed prudent to strike the topgallant yards; and we further sent a signal to the *Atrevida* to run with topsails double-reefed, although soon afterwards we cancelled the order as the wind had dropped, and shifted to the NE once again.

Since eight o'clock, we had sounded twenty-seven fathoms, fine white sand. From nine onwards, we began sounding every ten minutes; at this time we were in sixteen fathoms, black sand and small shells. Our soundings from then until noon were fifteen, thirteen and fourteen fathoms, fine black and white sand, at times with various small shells, some spiral-shaped. As this bottom and that from a twelve-and-a-half fathoms sounding, obtained at around ten o'clock, indicated that we were somewhat to the south, we altered course to NW, first under easy sail, then successively with the topsails and the foresail hoisted.

The haze was still very dense; nonetheless at noon the latitude and longitude, observed by numbers 61 and 13, seemed to me to be fairly reliable. This position, 35°15′S and 48°4′, plotted on Tafor's chart,[1] placed us S59°30′E of Isla de Lobos, distant twenty-nine miles. This being the case, the sounding we had obtained was not surprising, nor was there any doubt about the position of a flat[2] with a depth of ten fathoms, placed by Tafor in almost the same position where, at eleven o'clock in the morning, we had sounded in eleven-and-a-half and twelve fathoms, gravel. We therefore instantly hauled our wind to NWbyN, under a press of canvas, once again squaring the topgallant yards, noting that our soundings gradually increased to eighteen and nineteen fathoms, black sand and ooze, which finally at three o'clock that afternoon to our satisfaction become ooze. Soon afterwards, as the fog lifted somewhat, the port look-out sighted Isla de Lobos. We immediately crowded on sail to close it, observing longitude by chronometers in both corvettes. Since the weather cleared considerably in the afternoon, we passed about a league from it, taking bearings and transits so as to be able to fix its different points to perfection. At sunset, the south point bore N85°E by compass, distant two leagues. We sounded seventeen fathoms, ooze, and we could clearly see all the Sierras de Maldonado[3] as far as Solís Chico, of which we took some views.

The clear, calm skies heralded the continuation of the northerly wind. Accordingly we ran under very light sail to cover the distance needed until next morning, enabling us to sound with ease. Our speed with the topsails lowered was three or four knots. We had already got up a range of cable for the best and small bowers and were steering W5°S by compass.

---

[1] Copies of Tafor's charts are held in AMN, MS 427, ff. 43-4.

[2] A extensive area of much the same depth, usually of mud.

[3] Probably Sierra Ballena, which rises to 146 metres a short distance NW of Maldonado, and Morros de Maldonado, its extension northward.

The errors in our dead reckoning carried forward from Islas Canarias, checked against chronometer readings when we observed for longitude by chronometer in sight of Isla de Lobos, were 2°4′9″ to the north and 3°42′17″ to the east; that is, these were the amounts by which our reckonings of latitude and longitude respectively differed from those obtained by observation and measurement relative to our initial readings. Anyone with the patience to look at the daily differences between the dead reckoning positions and the observed ones, which we present in a separate table,[1] will see plainly that this error can in no way be attributed to the log we adopted, which in any event has proved consistent, time and time again, with the measurement of a degree of latitude.[2]

At nightfall the weather began to thicken towards the west; soon afterwards there were light airs from north and NW with some thunder, considerable lightning, and very forbidding skies. We had no choice but to drop our small bower anchor, and since a tempestuous *pampero* seemed imminent, we double-reefed the topsails and hailed the *Atrevida* to do the same. She had anchored at the same time as ourselves at a reasonable distance from us. Fortunately, the wind shifted rapidly round to south and SbyE, at first with rain, later bringing very gloomy weather, and indeed there were some gusts that caused the sea to rise noticeably. Our anchors held over fifty fathoms of cable. However, at four o'clock in the morning it began to blow very hard, and the sea grew so high that our attempts to weigh the anchor were to no avail. We dragged anchor from five o'clock in the morning onwards, and even though we tried lying-to with topsails, mizzen sail and jibs, we were never able to haul in more than two fathoms of cable, and broke the purchase block in the process. There was a strong breeze, however, and by dawn the skies were clear. As the heavy swell made it unwise to drop a second anchor because of the risk of losing it, we fell away dragging our anchor towards the coast, and the decision to cut the cable became unavoidable, whereupon we sailed close-hauled under the four principal sails, with the topsails double-reefed, to round the immediate headlands. We soon succeeded in doing this, at a distance of about one-and-a-half leagues and the *Atrevida* which had to make the same sacrifice, followed not far behind.

20 September

At dawn the wind began to abate and indeed veered to the east. We shook out all the reefs and, sounding constantly, set a course to come within sight of Isla de Flores. The courses we steered were from west to WbyN by compass, which seemed the most advisable, since we were still obtaining soundings of eight or nine fathoms, ooze, and it was our belief that the freshets flowing from the swelling of the rivers would tend to push us off the land. The mountains, and indeed the coast, were covered with thick haze. As a result we could not be sure either of the trend of the coast or of our position. Thus when we managed to sight Isla de Flores it bore SW5°W by compass,

---

[1] Not present

[2] Malaspina refers here to the results of various scientific expeditions to measure an arc of the meridian (i.e. a degree of latitude on the earth's surface) to determine the size and shape of the globe. Since a degree of latitude is 60 nautical miles its value is directly related to the correct distance apart of the knots on a log-line: see p. 5, n. 1 above.

and the coast was no more than two leagues distant on our beam. We therefore luffed round under full sail, and despite the heavy swell which at times set us considerably towards the land, at twelve-thirty that afternoon we managed to double the island, passing south of it at a distance of a good mile. With its southern end bearing true west, several of our observers were in agreement in observing the latitude to be 34°56′30″, and then with the middle part of it bearing true north its longitude was 48°56′13″ by number 61. Now clear of Isla de Flores, we steered for the port under full sail, with the weather clear and fresh, at a speed of eight to nine knots. We gave a wide berth to Punta Carretas and Punta Brava.[1] Finally at three-thirty in the afternoon we dropped anchor in the port of Montevideo, fifty-two days out of Cádiz. Everybody was unanimous in assuring us that this was faster than any run until now by His Majesty's mail frigates.

His Majesty's ships *Santa Sabina* and *San Gil* were at anchor in this port, the first on service with the small coastguard squadron under the command of Capitán de Navío Don José Orozco,[2] the second about to set sail for the ports of the Patagonian coast under the command of Teniente de Navío Don Pedro de Mesa. Two of His Majesty's brigs commanded by *pilotos* of the Navy are at the immediate orders of the local governor. The mail frigates *Colón* and *Princesa*, the first about to return to Europe, are by custom independent of the navy. The merchant vessel *Dichosa* of the Cádiz trade, and a Catalan settee, are ready to set sail at the first breath of favourable wind. Another six merchant vessels[3] and twenty-one two-masted craft complete the total of vessels in the bay; and all these belong to the Europe trade except one from the Lima trade.

[1] Situated 3 miles SE of Montevideo with Punta Carretas, a name no longer in use, situated a further 5 miles NE.
[2] For Orozoco see p. 39, n. 4.
[3] *Fragata* in the original, with *mercante* omitted: see p. 7, n. 3 above.

CHAPTER 2

At Montevideo

[20 September]

A peaceful night enabled us to complete almost all the operations of mooring according to the custom of this port, running out two long stays of cable from the bows, one to the SW and the other to the SE, securing the stern to the north with a hawser. In this position the summit of El Cerro[1] bore due west of us, its prominent headland and rocky ledges bore WSW[2] and the black stones of the anchorage N3°W.[3] The depth was sixteen feet, soft clay, with southerly winds, reducing to thirteen with the ebb tide from the north. We were about one-and-a-half cables from the *Sabina* and she was two-and-a-half cables from the quay. The *Atrevida* was moored in the same way, a short distance from us. Launches from the *Sabina* and the *Gil* had helped us to manoeuvre into [our berth] with hawsers. The orderly officer, who came on board both vessels in the name of the senior naval officer, Don José Orozco,[4] took charge of the public correspondence and private mail that we had arranged to bring from Cádiz. Don Francisco Viana went on my behalf to present my compliments to the Governor of this place, Don Joaquin del Pino,[5] who was very soon to be succeeded by Brigadier Don Antonio Olaguer, inspector of troops in these provinces, since along with his promotion to brigadier he had been entrusted by His Majesty with the presidency of Charcas. On that same night of the 20th we had the opportunity to observe the occultation of Spica by the Moon from on board, although this was rendered useless by a mistake either in noting down the moment of the observation or in referring the times by the small watch[6] to the chronometers.

---

[1] Cerro de Montevideo, a conspicuous and easily recognized isolated cone-shaped hill, rising to a height of some 140 metres on the western side of Bahía de Montevideo.

[2] Punta del SE.

[3] Rocks extending about 3 cables offshore from Punta de Piedras at the head of Bahía de Montevideo.

[4] *Capitán de navío* in command of the frigate *Santa Sabina* and senior naval officer in Río de la Plata; also senior to Malaspina who was a *capitán de fragata* at the time.

[5] Joaquín del Pino (1729-1804) was a military engineer, assigned to the Río de la Plata region in 1771; in 1773 he became military governor of Montevideo and pursued progressive policies which sometimes brought him into conflict with the municipal council. He sought peaceful relations with the Indians and opportunities to open up the interior for settlement. The promotion and transfer to which Malaspina alludes were gazetted on 2 April 1789.

[6] A hand-held watch with a second hand, sometimes called a hack watch.

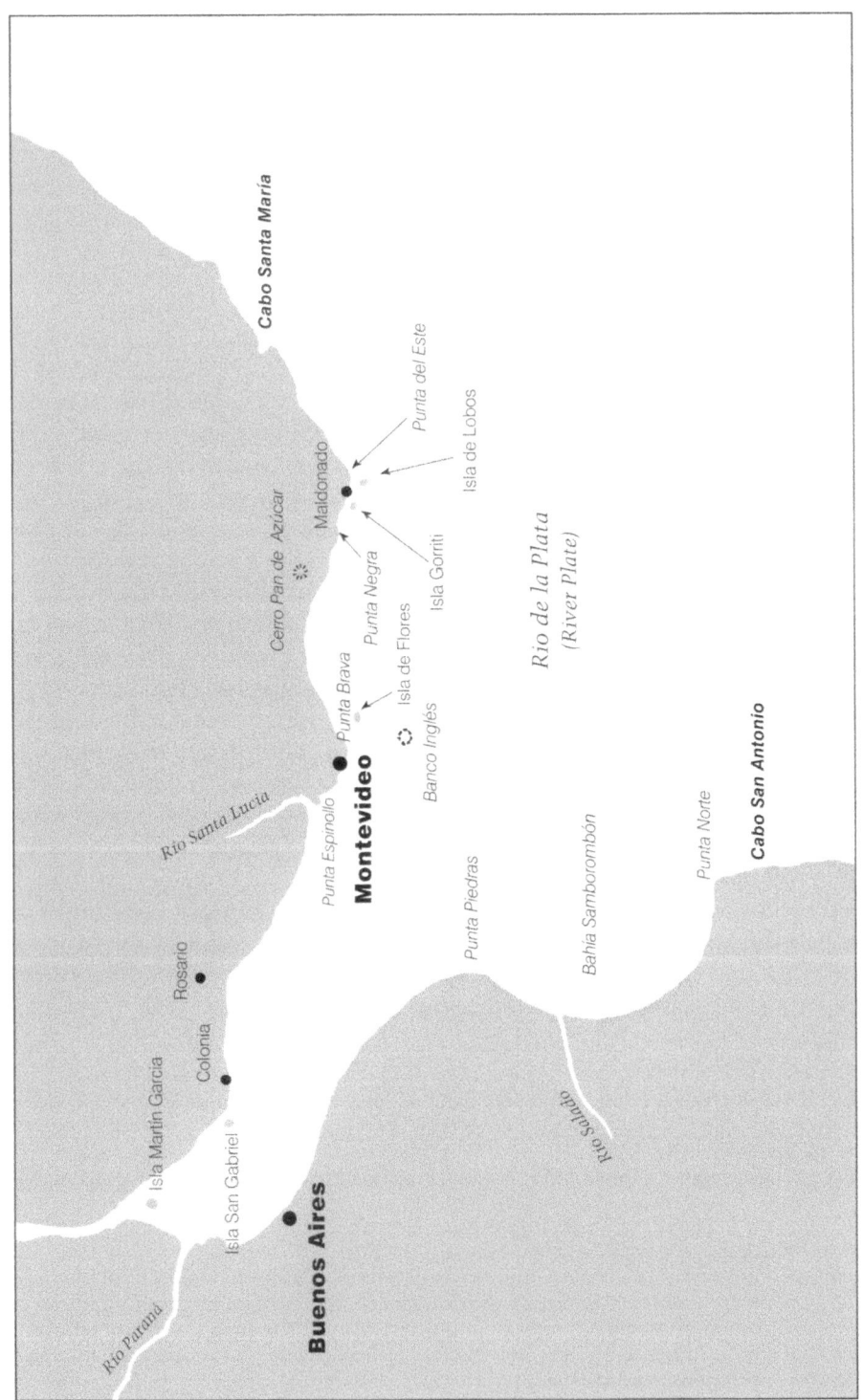

Fig. 2. Río de la Plata, September to November 1789

21 September

The following morning we lowered the topmasts and yards; all sails were got down and sent to the naval warehouses. Don José Bustamante and I presented our compliments to the senior naval officer and to the Governor. Having shown the latter the royal warrant by which he was requested to afford us every assistance, we asked for the use of their revenue *sumaca*[1] to go to Buenos Aires the following day, since on passage its sturdy construction and ease of handling would not only safeguard us from the many hazards commonly experienced in the navigation of this river, but it would afford us the speed without which we could not easily examine such an extensive coastline. It was at our disposal that very afternoon, and at the same time we were assigned a house for our observatory, and some instruments had already been sent to it.

At first sight, producing a chart of this river did not seem feasible for it was a project that would require months rather than days. To undertake it with no hope of finishing it was the sort of idea we should hesitate to carry out; nor on the other hand could we sacrifice to such a task a single day of the coming summer, the whole of which would have to be spent upon the coasts of Patagonia and Tierra del Fuego. However, having considered these obstacles carefully as well as our own strengths, and also the time we must of necessity spend in this port not only because of the earliness of the season but also because of the extensive overhaul which our ships required, the difficulties began to vanish. I saw, with much satisfaction, that an active and intelligent team of officers and the wealth of instruments that we possessed could show our desire to be useful to the nation with a significant task to which we had not previously been committed.

Having set up the observatory in Montevideo where the chronometers were to be compared daily, a series of continuous astronomical tasks was undertaken at the same time, both to determine a reliable longitude, and to make an additional contribution to the advancement of astronomy in an area not much frequented, so far, by scientists. We could also look upon this site as the centre or meeting place for our excursions, its position being almost equidistant from all the important places to be included on our chart and indeed close to the corvettes, which would enable us to work with greater ease without hindering, however briefly, the refit of both our ships.[2]

Thus the following day it was decided that Don José Bustamante, and the junior officers Valdés, Quintano, Concha and Vernacci should cross over to Buenos Aires in the *sumaca*. There, with the assistance that His Excellency the Viceroy would provide, they would undertake a survey of the southern shores of Río de la Plata from that capital as far as Cabo San Antonio.[3] I undertook to deal with the coast as far as Maldonado, but with the wish to begin as soon as I had seen that all was in place to ensure that our refitting would be completed with all possible speed. Only the coast between Montevideo and La Colonia would thus be left, and I believed that it would

[1] A small schooner used in the coasting trade along the Atlantic coast of South America.

[2] Malaspina sent the results of the astronomical observations home on 10 November 1789, with a request, addressed to Ministro de Marina Valdés, that Spanish embassies should circulate them among specialists in Europe: AMN, MS 583, f. 58. See A. Orte Lledó, 'El posicionamiento astronómico de las costas de América en la expedición Malaspina', *La ciencia española en ultramar: Actas de las primeras jornadas sobre España y las expediciones científicas en América y Filipinas*, Madrid, 1991.

[3] The SE entrance point of Río de la Plata.

not be difficult for me to have the opportunity to examine this also, provided that on my return from Maldonado I went to Buenos Aires to discuss with the Viceroy our needs and the other measures which might yield to His Majesty the greatest benefit from this expedition that lay within our power.

The weather did not permit us to sail to Buenos Aires until the 28th, but we took advantage of this time for Don Felipe Bauzá to measure a base at the head of the roadstead and another extending to Punta Carretas so that with the appropriate bearings he was able to begin a plan of this port and [to fix the] the positions of adjacent points. He accompanied me on the morning of the 26th to take bearings with the theodolite from the highest ground in Montevideo to all visible points, among which Pan de Azúcar and the extremities of Isla de Flores were to us of the greatest importance. Don Antonio Pineda and Don Luis Neé also accompanied me on this short excursion. They had already botanized and even hunted in the vicinity of the town, yet nevertheless they encountered much to engage their curiosity and to confirm the first impression that there were many plants unknown to natural history growing in this soil.

Similarly the energy of Comandante Orozco left us nothing to wish for by way of greater or more efficient assistance. I had handed him a separate schedule of the work that needed to be carried out, and finding the thirty-two half-allowances which I had previously requested in Europe awaiting me here, we limited our requests for provisions to a supplementary allowance of bread to replace the badly deteriorated supply we had on board. The statement of work to be carried out, approximately the same for both corvettes, was as follows for the *Descubierta*:

Carpenter
Extend the coach up to the taffrail.
Raise the coamings of the main and forward hatchways.
Enlarge the galley hatchway, and move the galley forward from the stern.
Make a pin-rail to take running rigging on the after side of the foremast.
Place a movable washboard in the prow extending from one foreshroud to the other to give protection from the seas, and make two heads or seats of ease within this.
Shutter the lower half of the windows in the wardroom and quarter galleries and adjust the window lights and sliding port-lids for these windows.
Put guards on the chain-wales forward and amidships.
Remove the division in the after storerooms of the orlop deck and enlarge their door.
Lift a number of planks in the gangways that have come adrift from the beams, and replace two or three that are slightly rotten.
Cut four scuppers in the upper work.
Make two new topmasts to replace the two that are sprung.
Make a hawse plug for the hawse-holes.
Raise two more sills in the petty officers' mess.
Make two snatch blocks for anchor messengers, and two for weighing the kedge anchors.
Strengthen the launch with a number of reinforcing futtocks, fit a removable washboard and two masts and two booms for the spritsails.
Repair the other boats.
Cut a mizzen mast.

Plate II. View of Montevideo, by Fernando Brambila. Museo Naval, Madrid

Blacksmith
Make eight iron knees to strengthen the ship's sides fore and aft to reduce stress when rolling.
Lengthen the pins of the main tack blocks; insert them in line with the waterways, and forge another block of the same type.
Make various hinges, staples, claws, etc. and other odd articles.

Caulker
Overhaul and repair the sides and decks.
Overhaul and repair the boats.
Line the manger[1] with lead.
Lay two tarred canvas coats, one over the mainmast and the other over the foremast.

Boatswain
Overhaul the sails in use.
Supervise watering ship and refitting water casks to keep them in good condition.
Take delivery of six *quintals*[2] of tallow candles and six tallow cakes as spare stores.

Cooper
To reassemble broken casks.

Lantern maker
Repair the fighting, hand and signal lanterns.

The excessively high cost of labour in this country, the unskilled artificers, and the high cost of the materials themselves, particularly the timber, might well have dissuaded me from having all this work done had I not considered on this occasion that part of it was absolutely necessary, both for our own safety and for that of the ship or its equipment, while another part, more trifling, had to be allowed for the sake of comfort in expeditions of this sort.

26 September
The work was begun on the 26th. Our forges had been set up in a building on shore close to the naval warehouses.

First comparisons of the chronometers had shown that their rates were now quite different from what we had determined in Cádiz. Number 61 had lost approximately 3″ daily, number 13 had lost 1′11″ daily, while number 72 had been gaining an average 14″ to 16″ per day. But by reducing their results to Isla de Lobos, whose position had been determined accurately by astronomical observations carried out by Brigadier Don José Varela in Montevideo, it could be conjectured that only 72 suffered this alteration during the period when we had suspected it. The original rates of chronometers numbers 13 and 61 agreed so closely in longitude that the first gave only 4′ less and the other 14′ less than what we had inferred from our astronomical observations. Thus we were confirmed in our assurance that our position as determined at Isla Ascensión,[3] based in particular on number 13, was not very far from the

[1] The space just abaft the hawse holes.
[2] Approximately 600 lbs (a quintal = 101¾ lbs).
[3] Malaspina is clearly referring here to Ilha da Trindade: see p. 23, n. 3 above.

true one. The difference in longitude between Isla de Lobos and Montevideo by number 61 was 1°24'42" with 1°24'8" as the mean of numbers 10 and 105 on board the *Atrevida* which agreed with our trigonometrical observations.

While others were pressing on with our refitting, I could not overlook the extent to which the crews of both corvettes were now suffering from venereal disease, the result of their excesses in Cádiz.[1] It was important to me that they were thoroughly cured, and I was aware of how harmful to the restoration to full health might be the poor nursing that members of crews commonly encounter in hospitals no less than the excesses in which they themselves indulge even during the course of their treatment. Once all the necessary instruments were collected from on board ship or ashore, I arranged for the patients to be treated at our expense in a separate ward of the general hospital,[2] in accordance with the method prescribed by our surgeons, and with their custody and supervision entrusted to a corporal and three trusted marines.

27 September

As weather conditions on the 27th still did not allow Don José Bustamante and the other officers to cross the river to Buenos Aires, they decided to travel overland as far as Colonia del Sacramento,[3] where they would take the *chasquera* or mail boat and immediately cross over to that capital. It was left to Vernacci to oversee the transportation of the *Atrevida's* collection of instruments and our chronometer number 61 by sea; a *pilotin* and a marine were detailed to accompany him. The road to Colonia is said by the natives to be forty-two to forty-four leagues long, as it turns considerably inland from the coast at many points to ford the rivers in safety, although it appears to be considerably shorter on our maps. It passes through El Carrelón, El Campamento,

---

[1] The treatment of venereal disease was a major subject of medical controversy. Until about the last quarter of the eighteenth century, rest, light diet, electuaries to keep the bowels free and even blood-letting were common prescriptions but medication was increasingly favoured – mercury for syphilis and, for gonorrhea, urethral irrigants induced by syringe. Typically, these included dilute solutions of powerful acids, mixed with heavy-metal and salt solutions. The indigenous American herbal nostrum, guaiaca, featured in some patent remedies popular in Europe. Pedro María González, who reported from Montevideo on 3 November, 1789, on the health of crew-members of the *Atrevida* between Cádiz and Montevideo, expected to maintain a healthy crew by lightening duties with the use of three watches instead of the usual two and prescribing an 'orderly and salutary diet ... but we were soon disillusioned.' He diagnosed as gonorrhea most of the cases under his care but, after experiments with milder emulsions, proceeded fairly rapidly to internal and external applications of mercury, attributing cases of non-recovery to 'insufficient introduction of mercury into the bloodstream': AMN. MS. 271, ff. 13-14v. See also A. Orozco Acuaviva, J. L. López de Cózar and J. R. Cabrera Afonso, 'El "Diario medico-chirúrgico" de la corbeta *Atrevida*,' *Malaspina '92*, Cádiz, 1994, pp. 115-25.

[2] There were five working hospitals in Montevideo at the time of Malaspina's visit. The place where Malaspina's men were accommodated is called Hospital de la Plaza in González's report. It was probably Hospital de Marina, on the northern edge of the town, rebuilt in 1781, which had the largest capacity. See F. Guerra, *El hospital en Hispanoamérica y Filipinas,* Madrid, 1994, pp. 520-28.

[3] This settlement, on a promontory overlooking Río de la Plata, 100 miles WNW of Montevideo, was a long standing point of contention between Spain and Portugal. It had begun as a fort erected by the Portuguese in 1679; with the town that grew up around, it changed hands repeatedly from 1680 to 1777, when the Spanish commander, Pedro Antonio de Cevallos, demolished the settlement only to see it arise anew in the following years.

San José, Sufré, El Rosario, and El Sauce,[1] where there are posts of dragoons with horses belonging to the crown. These posts will arrange for a dragoon to accompany the traveller on presentation of a pass or safe-conduct from the Governor of Buenos Aires or Montevideo. In this way the *chasques* or special couriers, the periodical mails, and messages as far as the posts in Río Grande[2] via Maldonado, reach their destination with a speed of which it would be difficult to give a precise idea without fear of being accused of exaggeration. There are a number of villages and many small holdings along this road where the traveller may be well received. The meat and milk here are the fruit of nature rather than the product of industry and may generally be described as of little value in all these small holdings.

I had issued Bustamante with instructions on the essential points he should follow regarding this assignment. I also gave him official letters for the Viceroy suggesting how important it was to us to survey Cabo San Antonio, and to obtain the assistance, and in particular the ready cash, without which this special assignment could not proceed with the speed expected of it.

28 September

Bustamante, Valdés, Quintano and Concha reached Colonia on the night of the 28th and Buenos Aires the next morning, almost at the same time as the *sumaca*, in which Vernacci was bringing the instruments, dropped anchor. During a crossing of a few hours, and as a result of comparisons before and after the crossing, number 61 determined the difference in longitude between our observatory in Montevideo and the Town Hall in Buenos Aires as 2°10'22", which was absolutely identical to that which Brigadier Don José Varela had deduced from his own observations.

The efforts made by Bustamante and the other officers received full support from his Excellency Señor Marqués de Loreto,[3] the present viceroy of these provinces. They set up an observatory, where meridian altitudes taken to north and south by the astronomical quadrant, determined the latitude to be 34°36'39". They measured a base from which they observed a series of triangles as far as Ensenada de Barragán,[4] the terrain not permitting them to carry them any further to the east. The snow *Belén* was put totally at their disposal, together with a shallop. Piloto Juan Bautista Acosta commanded the first vessel and the shallop was entrusted to a *pilotin* who had gained much experience of these waters while sailing with Acosta. Of necessity surveying from seaward was preferred, rather than by land, not only because of the difficulties presented by the distances and the roads, but also because of the great danger in which they might find themselves from the close proximity of the Pampas Indians

---

[1] The route followed that of the modern road through Las Piedras, Canelones, Santa Lucía and San José.

[2] In present day Brazil, 200 miles NE of Cabo Santa María.

[3] Nicolás Francisco Cristóbal de Campos, Marqués de Loreto, followed a military career in Islas Baleares and Andalucía before serving as Viceroy of the recently-created Viceroyalty of Río de la Plata from 1784-99. His policies were to expand the administration, colonize the coast of Patagonia, explore the Falkland Islands, promote fishing and whaling and port construction and delineate the frontier with Brazil. Malaspina, whose policy priorities overlapped but did not coincide, is less warm about him than about most of the other high officials he worked with.

[4] Present day Puerto de la Plata, 27 miles SE of Buenos Aires.

near Cabo San Antonio. Concha and Vernacci took charge of this important operation, embarked in the *Belén*. Bustamante laid down how they were to proceed in the following instructions given to Teniente de Fragata Concha.

> Consequent upon the order I have from Comandante Don Alejandro Malaspina to arrange a survey from this capital of the southern shores of the river to determine astronomically the position of Cabo San Antonio, I have asked his Excellency the Viceroy for all assistance conducive to this objective, so important to our nation's shipping. To this end his Excellency has provided the royal snow *Nuestra Señora de Belén*, and a shallop with the best pilots of this coast as recommended by the captain of this port. The snow is commanded by Juan Bautista Acosta, who assures me he has seen the said cape and obtained a sounding of ten fathoms three leagues from it. This information, communicated to me by a person of good professional standing, confirmed my decision to undertake the operation with every precaution with regard to safety as is due to common sense and to the royal service. To this end His Excellency has also decided that fifteen soldiers should embark in the snow under your command and that of Alferez de Navío Don Juan Vernacci, your second officer in this mission. These will be available to land in case of need or emergency and to assist on shore in the work that you judge opportune, when the weather permits.
>
> The primary aim being to fix the position of Cabo San Antonio by astronomical means, for which chronometer 61 is embarked, you will proceed to examine this point, or the most easterly part of this coast, observing latitudes where possible without becoming involved in other tasks other than those which may be carried out without prejudice to completing this important survey. The shallop may be used for obtaining lines of sounding much closer to land than the snow will be able to go, and for taking you and the instruments ashore when it is necessary.
>
> If the winds should allow you to bear down on the cape and determine its latitude and longitude, you should also endeavour to determine the position of Punta Piedras and examine an anchorage, which I am told lies to the SW or at the entrance to Bahía Samborombón, as well as the anchorage at the mouth of Río Salado which lies halfway along. This task, which will depend on the state of the weather, and on the certainty that there will be no risk in carrying it out, I leave to your discretion, guided by the dictates of prudence and reserving your daring for other more important occasions. This, as well as being the express wish of Comandante Malaspina, obliges me to repeat to you that same instruction with the warning that you should return either here or to Montevideo, depending on the winds, by the 24th of this month, bearing in mind that if you were to return here, you should attempt to take all possible bearings of any salient points from Punta Piedras as well as numerous soundings from the snow and the shallop.
>
> At Buenos Aires, on 7 October: José Bustamante y Guerra.

Don Cayetano Valdés and Don Fernando Quintano, who had worked with the others as far as circumstances would permit,[1] used the remaining time to enquire into the political state of the capital and its dependent provinces.[2] In consequence of a

---

[1] This paragraph and the following one appear to be based on a report from Bustamante to Malaspina.

[2] See Malaspina's 'Descripción política de las provincias del Río de la Plata': AMN MS 590, ff. 14-25v; printed in Juan Pimentel Igea, ed., *La expedición Malaspina, 1789–94*, Tomo VII: *Descripciones y reflexiones políticas*, Madrid, 1995, pp. 37-54. As Pimentel has shown, this document, like most of the rest of the copious political reports Malaspina sent home, was based less on the explorer's own observations than on reading existing publications and, to a lesser extent, on local advice, obtained in person or

royal circular which [Bustamante] delivered to [the Viceroy] to allow us to examine the documents of the expelled Jesuits,[1] Quintano obtained leave from His Excellency to have access to these archives and with commendable diligence he had acquired a substantial knowledge of the country not only from them but also from other reliable sources.

On the 10th [October] the snow and the shallop, aided by a favourable wind, were lost to sight, and on the 12th Bustamante, Valdés, and Quintano returned to Montevideo, crossing in twenty-four hours in the *sumaca.*

From 29 September, having taken, as already mentioned, every measure to ensure a rapid completion of our refitting, I personally undertook, from on land, the survey of the coast from Montevideo to Cabo Santa María. Although I was mounted, I took with me chronometer 105 belonging to the commanding officer of the *Atrevida.* Some sextants, a theodolite, and all the instruments needed for measuring bases and taking soundings followed in a cart. Don Felipe Bauzá and both naturalists went with me. Capitán de Fragata Don Santiago Liniers,[2] second in command of the *Sabina,* and Piloto Don José de la Peña also accompanied me on this excursion. The knowledge which the latter had of these coasts was of as great a value to us as was the skill of the former in procuring, by means of hunting, a thousand objects of interest to natural history.[3]

30 September

On the night of 30 September we were at the foot of the mountain called Pan de Azúcar. The following morning Bauzá and Peña decided to climb to its summit with the intention of taking bearings by theodolite to all points along the coast, accompanied by Pineda, Neé and Liniers with the intention of examining scientifically the mountain's soil, which in country such as this should show a different facet of nature from that which the immense pampas or plains are mostly composed. As for myself, I went on to Maldonado intending to observe the meridian altitude of the Sun with

through questionnaires addressed to officials: J. Pimentel, *La física de la monarquía: ciencia y política en el pensamiento de Alejandro Malaspina (1754–1810),* Madrid, 1998, pp. 184-97.

[1] To the general accusations against the Jesuits from the middle of the century – Pelagianism, moral laxity, abuse of the confessional, temporal ambition and political subversion – were added others peculiar to their missions in America, including contraband, concealment of treasure and the canard that the Jesuits planned a secessionist kingdom of their own. These charges became the pretext for a series of expulsions, beginning in the Portuguese dominions in 1759 and culminating in the general suppression of the order by Pope Clement XIV in 1773. See R. Carbonell, *Estrategias de desarrollo rural en los pueblos guaranies (1609–1767),* Barcelona, 1992, pp. 267-99 and D. Alden, *The Making of an Enterprise: the Society of Jesus in Portugal, its Empire and Beyond, 1540–1750,* Stanford, Ca, 1996, pp. 321-661.

[2] Santiago Liniers y de Bremond (1753-1810) was born at Niort (Poitiers) and joined the Spanish navy in 1775. He was seconded to the *Atlás marítimo* project (p. xxxviii, n. 3 above) at his own request in 1783. In 1788 he was sent to Río de La Plata and played a conspicuous role – crowned with appointment as Viceroy in 1807-8 – as organizer of its defences for the rest of his career. Dalmiro de la Valgoma, *Real Compañía de Guardias Marinas. Catálogo de pruebas de Caballeros Aspirantes,* Madrid, 1943.

[3] The methods by which specimens were acquired were limited by the pressure of time. Hunting in this respect was generally uneconomical. Gathering, barter from Indians and acquisition from local collectors and savants were usually preferred in circumstances, like those of the excursions from Montevideo, in which time was at a premium.

the sextant, but this was impossible because I was unable to travel the great distance from the village to the port before noon.

I spent that same afternoon establishing the position of the port and the adjacent coastline. The instruments arrived at Maldonado at nightfall, and soon after that those who had climbed the mountain arrived in two groups. Lithology and botany had made considerable progress with this excursion.[1] The bearings they had taken had already tied in the main points of the coast, and despite the steepness of the mountain, neither the instruments nor the travellers had come to the slightest harm.

1 October

The 1st of October dawned with the best of prospects. I immediately went down to the port with Bauzá and Peña, the instruments having been sent on before us, and working without pause until five o'clock in the afternoon we measured a base on Punta del Este and, placing to the north another flag, the ends of the base were soon extended by another mile. Using various bearings we constructed a plan of the port and at the same time we fixed the principal points of Isla de Lobos and Cabo Santa María. Having observed on Isla Gorriti its latitude from the meridian altitude of the Sun and its longitude by chronometer number 105 and variation by theodolite, we proceeded to sound Boca Chica[2] from the east, and thoroughly surveyed the shoal off Punta del Este and the rocky ledges off Isla Gorriti.

The naturalists and Liniers, who had spent the whole morning in arranging the many specimens acquired along the way, went in the afternoon to Pueblo Chico, a settlement two leagues from Maldonado, consisting either of Portuguese families exiled from Brazil or of Spaniards brought over in recent years as settlers of the Patagonian coast but placed for the present near Maldonado.[3]

---

[1] Geology, including lithology, and botany were the sciences most prominent in the preparations and instructions for the expedition. In connexion with both, the primary task was the accumulation of specimens for further research, but problems specific to the time dominated the expedition's work in both disciplines. In geology the main objectives were to investigate seismic phenomena and clarify the contemporary debate about whether stratification was the result of inundations in the remote past, including, perhaps, the biblical flood, or of deposits made by volcanic activity: S. Mason, *Historia de las ciencias*, 5 vols, Madrid, 1986, IV, pp. 7–29. In botany, the expedition occurred at a moment of great promise in Spanish history: the Real Jardin Botánico had been transferred to Madrid and re-organized in 1781; in 1788, gardens of acclimatization had been founded in Mexico and at La Orotava on Tenerife in order to provide a framework for the transfer of live specimens and for the controlled adaptation of species. The commission of the botanists of the Malaspina expedition was part of a vast project to establish the taxonomy of the florilegia of the Spanish empire on consistent lines; but they also adopted a practice generally recommended by savants in the New World, of recording native names which illustrated the practical and medicinal properties of plants. From this point of view, the expedition was a pioneering venture in what would now be called ethnobotany. See M. C. Iglesias et al., eds, *Carlos III y la Ilustración*, 2 vols, Madrid, 1989, especially I, pp. 274, 295–301; II, pp. 712–21.

[2] Not identified and not shown on *Carta esférica del Río de la Plata ... levantada de orden del Rey en 1789 y rectificada en 1794 por varios oficiales de su R¹ Armada,* published in 1798; possibly Puerto Viejo close NNE of Cabo Santa María.

[3] According to the chronicler Cosme Bueno there were 1,000 families of Galician origin in the settlement in 1770. The census of 1781 recorded 8,773 Spaniards, 586 Indians, 711 mulattoes, 352 free Blacks and 1,760 slaves: Novo y Colson, *Viaje político-científico*, p. 563.

2 October

On the 2nd, having concluded our operations, and examined the surrounding country, as far as time would allow, we started on our return journey. We took bearings from different points of the coast, such as Punta Ballena, Punta Negra and from the mouth of the Pando,[1] and managed to be back on board on the afternoon of the 4th, noting with much pleasure that the rate of chronometer 105 had not altered, and that its results as well as the latitudes taken during the return journey, differed by only a few seconds from the results of the trigonometrical operations taken from Pan de Azúcar, Maldonado and Montevideo.

On my return I was most gratified to find that Don Francisco Viana, in whose charge I had left the corvette because Don Manuel Novales was ill, had made considerable progress with the refitting. All the tasks in hand were proceeding with equal speed. All our bread had been taken ashore. The bakers were unanimous that its bad quality could be attributed either to fraud or at the very least to negligence, as it was known that all had come from the same bakery and the cause must undoubtedly have been the poor quality of the flour. Both store-rooms had been opened, aired and thoroughly scrubbed. We could now flatter ourselves that we had totally got rid of the caterpillar which would otherwise have continued to spoil large quantities of bread. On board the *Atrevida*, Teniente de Navío Tova had proceeded with just as much energy, while Don Dionisio Galiano, tenaciously pursuing his astronomical observations, had observed early on the morning of the 27th the immersion of Jupiter's second satellite, thereby obtaining fresh rates for the astronomical clock and the chronometers. He had also observed almost daily the inclination and declination of the compass needle, and by different meridian altitudes of stars, well determined in Monsieur Delambre's[2] catalogue, had deduced the observatory's latitude. The daily tracing of the Moon's orbit and the calculation with the aid of diagrams of the timing and positions of the occultations of stars was the kind of work which, although fruitless until now, demonstrated the precision of the work of this officer and astronomer as well as the benefit we would derive in the future from this constant examination of the Moon's movements.

The *guardiamarinas* and the *pilotos* detailed to sound the inner and outer harbour did not neglect this invaluable task for improving the accuracy of our plans, although this was always rendered doubtful in this port by the difference in sea level which usually decreased by four or five feet with winds from NE and NW in contrast to winds from SE, south and SW.

13 October

With Bustamante back in this port, I undertook a trip to Buenos Aires on 13 October. As the weather was unsettled, I chose to travel overland. Pineda and Neé, who were to accompany me, preferred to take the *sumaca*, and had the good fortune to reach Colonia del Sacramento on the afternoon of the next day, several hours before I did.

Having with me a sextant, a compass and chronometer 105, my intention was to examine each day, from the most suitable vantage points, the continuation of the

---

[1] A river, 28 miles in length, which rises on the southern slopes of Cuchilla Grande and enters Río de la Plata 22 miles east of Montevideo.

[2] The French scientist Jean-Baptiste Joseph Delambre, FRS (1749-1822).

Plate 12. View of Buenos Aires, by Fernando Brambila. Museo Naval, Madrid

coast westward so that its alignment along this stretch, although of little importance, could also be recorded correctly. However, since the road turned well inland from the coast, I found it difficult to carry this out without the loss of two or three days. This seemed to me an important consideration in view of the risk that overcast weather conditions might not allow me to take the observations in Colonia, the latitude and longitude of which would eventually allow us to tie in the intervening coastline. With this in mind, I made my way directly to Colonia, only turning aside towards Arroyo de la Cavallería,[1] where I took some bearings and established the trend of the coast to the east, as far as it was visible.

14 October

Pineda and Neé had been botanizing that same afternoon with some success. They were even more successful on the following morning when, having accompanied me to Isla San Gabriel, where I observed the meridian altitude of the Sun and took hour angles with No 105 to determine its longitude, they quickly gathered such a variety of shrubs, plants and flowers that they seemed to be the fruit of the examination of an entire country rather than of one small island. I had taken advantage of the early morning hours to take different bearings from Colonia and from Isla de San Xavier,[2] and so establish the position of the other islands in the immediate vicinity. The desire to obtain a good latitude, and the fact that I was alone and did not have all the necessary instruments, obliged me to omit the measurement of a base. This and the fact that I was using a portable compass may have occasioned some slight error in the position of some of the many islands and reefs in the vicinity.

[15 October]

Shortly after noon I returned with the two naturalists to the *sumaca*, and taking further bearings we set sail for Buenos Aires with gentle breezes from south and SE. For a long time our course was west or WbyS, and having made good four-and-a-half to five leagues we sighted the towers of Buenos Aires to the SW, and were able anchor by sunset close to the city. The current at that time was very slow, setting to seaward.

With every day that passed the Viceroy was more favourably disposed towards our operations, and having learned from His Majesty's warrant, which Bustamante had presented to him, and from my own official letters how important it was for our full funding that the necessary money (twenty-eight thousand *pesos fuertes*[3]) was made available to us, he made such speedy arrangements that on the 19th I was given this amount out of royal funds which I took with me the following day in the *sumaca*. I had the satisfaction of seeing Concha and Vernacci on the morning of the 21st return after completing their mission in its entirety. I was able to make the crossing to Montevideo in a few hours and I immediately took the money to the building of the

[1] Arroya de la Caballada, which enters Río de la Plata close east of Colonia.
[2] An unidentified island off Colonia not shown on *Carta esférica del Río de la Plata*: see p. 49, n. 2 above.
[3] In silver *peso* coins as opposed to *pesos* in notes.

Ministro de Marina. I then examined the rate of chronometer 105 which I found identical to that determined previously.

The eight days of my absence from Montevideo had proved to be most favourable for achieving the twin objective that I had set myself of completing a chart of the river without causing any delay in our refitting or our departure.[1] Bustamante and Valdés had completed almost all the work below decks as well as replenishing provisions and water. In a small sloop chartered for the occasion, Robredo, Bauzá and Peña had taken chronometer 71 with them and obtained the latitude and longitude of Banco Inglés and taking soundings as far as Isla de Flores. On the north bank, Don Antonio Tova, with the *guardiamarina* from this corvette, and Bauzá and Peña, had taken the same sloop and proceeded to Río Santa Lucía to examine that anchorage, look for a shoal close to Punta Espinillo, and extend the triangulation as far to the west as possible. At the same time sounding in the vicinity of the harbour was not abandoned and Galiano continued his astronomical observations in the observatory.

26 October

On the 26th, Pineda and Neé returned from Buenos Aires. The former had made important new acquisitions for natural history on an excursion to Las Conchas.[2] The latter had examined the area around Buenos Aires, and both of them, after landing at Martín Garcia at the mouth of Río Paraná, had in five days carried out a thorough examination of the terrain between that port and Montevideo. Finally, on the 31st, with the return of Concha and Vernacci, we saw the full complement of officers now reunited; henceforth only the demands of service duty would separate them, and that only for very short periods.

Let me summarize briefly the achievements so far on board His Majesty's corvettes *Descubierta* and *Atrevida*. This will be the highest praise I can accord to the officers of the two ships. The coast from Cabo Santa María to Colonia de Sacramento, a distance of sixty leagues, has been surveyed trigonometrically almost in its entirety; plans of Maldonado and Montevideo have been carefully drawn and the whole tied into the established position of Montevideo either by a series of excellent chronometer observations or by latitudes observed most reliably by sextant. Islas de Lobos and Flores, Banco Inglés, the shoals along the northern shore as well as the dangerous reefs off Punta Carretas and Punta Brava, have all been fixed with the greatest accuracy using geodetic and astronomical methods. The meticulous examination of the anchorage off Santa Lucía, the bar of which was found to have less than two feet of water over it, dispelled all hopes making use of it. The shoal off Punta Espinillo was searched for three times in vain. The track followed by Concha and Vernacci and the examination of Cabo San Antonio, combined with measured bases tied to observations of latitude and longitude using good sextants and chronometer 61, fixed the

---

[1] Malaspina had not originally planned so extensive a cartographic programme in this region, but he was led into it by the delays imposed by adverse winds and the need to prepare his ships for the next leg of the voyage. A copy of his survey of Río de la Plata was sent to Spain before the expedition left Montevideo (see p. 60 below). *Carta esférica del Río de la Plata*, see p. 49, n. 2 above, was one of only a few scientific products of the expedition to achieve early publication.

[2] The area south of the mouth of Río Paraná.

true limits of Río de la Plata. These same officers sounded Ensenada de Barragán and carried to there a triangulation scheme from Casa de Cabildo in Buenos Aires, whose latitude and longitude were both duly determined.

Again, at the observatory in Montevideo the rates of the chronometers were checked with the utmost care. The inclination and declination of the compass needle could not have been determined with greater precision. On 27 September and 26 and 28 October the immersions of Jupiter's second, first and third satellites, respectively, were observed. The longitude of the observatory was deduced from 225 sets of lunar distances to the Sun and stars and calculated by three occultations of stars by the Moon, a partial eclipse of the Moon, and the passage of Mercury across the Sun's disc at the beginning of November, all being phenomena of equal importance for the precise determination of this longitude and for the advancement of astronomy.

The natural history collection was even more splendid. Pineda and Neé, covering over one hundred leagues of flat country, several islands, the mountainous terrain of Pan de Azúcar, the vicinity of Maldonado, the pleasant surroundings of Paraná at Las Conchas and Martín García, collected for a first consignment for Madrid a herbarium of almost five hundred plants, at least fifty of which appeared to be unknown to naturalists in Europe. They examined the characteristics of this fertile soil and collected more than fifty species of aquatic or land birds which were either unknown or poorly described until now. There were also some fish, some fossils, and a moderate collection of insects. There were acquisitions like spider silk, partly in the raw as found in a tree and partly woven, that were beautiful to behold, especially when they were accompanied by a scientific explanation which would facilitate their study and understanding in the future, and were realistically depicted by artists with unusual skill and accuracy.[1]

Meanwhile, all work below decks which experience had taught us to be necessary was undertaken in both ships. The sails and rigging were all examined and the exteriors of the corvettes were cleaned, the water was completely replenished, and provisions for a whole year were embarked. Finally, at the muster held that day, those ill or unfit for service and the deserters were replaced, our sick who had recovered full health were re-embarked; the commissioned officers received two payments on account and the petty officers, marines and seamen one, though excluding the unruly and the less reliable. This small reward went to those who had most faithfully borne the daily routine of hard work.

Consequently we were able to fire the signal gun to get under way, and having granted the crew three or four days rest and shore leave, we made unhurried preparations to set sail on or around the tenth of the month once the astronomical operations had been completed. This was conveyed to His Excellency the Viceroy, who after thoroughly familiarizing himself with our assignment, agreed to my proposal to relieve the corvette *San Gil* of her task, she having been appointed by His Excellency to carry out a thorough examination of our harbours on the coast of Patagonia, leaving us to take care of this survey and at the same time draw up the relevant plans, and

---

[1] According to Malaspina, the artists concentrated on botanical and zoological specimens during this leg of the voyage, because the topography lacked interest: Sotos, *Los pintores*, I, 69, nos 3-24, 28-9 in her catalogue.

to assign the brig *Carmen* as our consort. She would be commanded by Segundo Piloto Don José de la Peña, who would then return to this port to report what had been accomplished and with the charts of the coast that had by then been drawn.[1] With this addition to our company and in accordance with the plan of operations that I had set myself, I sent the commanding officers of the *Atrevida* and the brig their respective instructions, so that neither in a chance encounter with foreign ships nor in the event of stormy weather would we be exposed to the slightest misunderstanding. I fixed rendezvous points, the method of placing bottles with information of interest, and the location where I would bury any sealed letters for the Viceroy, should the brig become separated from us and the corvettes unable to wait for her. Lastly, in a brief exhortation, I explained to my officers and to those of the *Atrevida* those areas where the performance of our duty could reach greater perfection, and what should be the basis of peaceful and safe intercourse with the natives of the regions where we might land.

On the point now of departing these shores, where we have travelled so extensively, and where I, like all the other officers, have been accorded not only such a fine and indeed undeserved reception, but also the constant assistance which has so greatly helped us in our endeavours to merit royal approval, I have to recall with gratitude if not every individual person, at least all those whose kindness towards us has had the most beneficial effect for the King's service: His Excellency the Marquis of Loreto not only energetically provided for our needs, but donated various objects for the furtherance of natural history, and showed certain tokens of confidence which were truly appreciated in a person of his distinction, and will always occupy first place in the memories we have of these provinces. The following responded to his orders and at the same time displayed a truly friendly spirit: the Governor of Montevideo, the Governor of Maldonado Don Joaquín del Pino, Don Francisco Glimes, Ministro de Real Hacienda of the same town, Don Rafael Pérez, Teniente Coronel Don [Miguel][2] Echaudi Governor of Colonia, and Capitán Don Joaquín Estrada also appointed there. A particular expression of gratitude must also go to the senior naval officer, Capitán de Navío José Orozco for his efficiency in attending to our almost daily requests. Above all, however, pride of place in our memory will always be given to the Oficial Real Factor [Crown Agent] in Buenos Aires, Don Felix Casamayor. We are obliged to this worthy and generous friend for accommodating us in his own home and looking after us with constant attention, as well as supplying some excellent natural history specimens, and lastly for coming with us on the scientific excursions and taking great pains, either with hunting or with information and inquiries that might most interest us. The knowledge imparted by Brigadier del Ejército José Custodio de Saá y Farria[3] and Don Manuel Lavarden, in charge of the Archivo de

---

[1] The voyages of the *Carmen* and *San Gil* were commissioned by the Viceroy, but Malaspina's voyage, as a commission from the crown, took precedence. Peña continued with a survey of the coast while the *Atrevida* and *Descubierta* visited the Falklands and rounded Cabo de Hornos.

[2] Blank in the MS journal.

[3] José Custodio de Sáa Faria (1733-92), Portuguese architect, captured at the siege of Santa Catalina in 1777. He entered Spanish service and was sent to explore the San José region in 1779, producing two reports illustrated with maps. He lived in Buenos Aires from 1780, where he rebuilt the cathedral and designed the new prison and other important public works. P. de Angelis, *Documentos del Río de la Plata*, Buenos Aires, n.d., pp. 132-7.

Temporalidades,[1] also proved extremely useful. Without such records we should not have known the real extent of these provinces, nor the latest surveys carried out on the Patagonian coast, nor, lastly, their exact political state and the rapid progress of which they are capable.

Losses in Montevideo at the muster of the 31st

Descubierta

Rating	deaths	chronic sick	disorderly	deserters	Total
Commissioned Officers . . .					
Petty Officers...				1	1
Marines.		1	1		2
Marine Gunners.	1				1
Able Seamen.		2		3	5
Landsmen.		2		3	5
Servants.		1		1	2

Atrevida

Rating	deaths	chronic sick	disorderly	deserters	Total
Commissioned Officers. . . .	0			0	0
Petty Officers...	1			1	2
Marines.	1			2	3
Marine Gunners.	1				1
Able Seamen.	4		2	8	14
Landsmen.			1	4	5
Servants.				1	1

Month of November

Having granted the seamen the modest advances I mentioned earlier, it was my intention to reward them for their work, test their conduct and avoidance of disorder again, and lastly prevent their vices (were they inclined to them) from sullying the celebrations being prepared in Montevideo for the swearing of allegiance to His Majesty (long to reign over us) after all the work on board had been finished. The whole company was given three days' shore leave. Those who had joined us to replace the sick and the deserters were detailed to man the boats, as were those who had been sick since our departure from Cádiz and so had not assisted their fellow crewmen with the work, but who were now completely recovered.

On this day our shore hospital was closed and men remaining unfit or not completely recovered were transferred to the station's hospital as members of the local squadron. There were four of these, all from the *Atrevida*, who confirmed the next day how effective had been our efforts to cure them and above all to keep them well behaved. Once removed from the rigorous routine applied to our crew, they were

---

[1] The archives kept by the Jesuits of records, transactions and documents relative to their secular properties and revenues.

immediately to be seen in the streets and their very ailments deterred them from acquiring new ones.

With the object of closing and putting the accounts in order, I stopped all work on the ships and paid off the artificers. Only the smiths continued working until the 6th. We still needed several articles, some for strengthening purposes and some as tools for the ship.

The first days of this month were too favourable astronomically for us not to seek to take advantage of them, all the more because the longitude of Montevideo was still not settled, either because of the poor disposition of the satellites of Jupiter or because of the unusual orbit of the Moon, which had not yet produced a single visible occultation of stars of even the sixth magnitude. Don Dionisio Galiano had prepared all the preliminary calculations with considerable care. The eclipse of the Moon and Mercury's passage across the Sun's disc deserved our full attention. These events possibly might not be observed in Europe because of overcast conditions prevalent in the first days of winter. The egress of the planet would not be visible there, while here it would occur between two and three o'clock in the afternoon.

2 November

On the night of the 2nd, which was completely clear, a partial eclipse of the Moon was observed. It began at 7.41 apparent time,[1] and ended at 9.48. By this time we had already observed the occultation of Mayer's[2] 90th star by the Moon. Its emersion took place at 10.18 in the evening. We then expected to see the occultation of the 93rd star in the same catalogue at 2.14 next morning. Its emersion, which was due at 3.30 approximately, could not be observed as we had planned, because of the excessive brightness of the light. All the officers who were free, attended these observations. In the interval between them, we occupied ourselves taking lunar distances with sextants, their results were added to those which had been observed previously.

5 November

Dawn on the 5th brought us considerable anxiety. A substantial amount of cloud seemed intent on thwarting our efforts and Don Dionisio Galiano was not even able to obtain two consecutive altitudes with the astronomical quadrant for comparison with the corresponding afternoon observations. The heliometers[3] had been got ready. What was very useful to us on this occasion was the graduation determined by Mr Dalrymple[4] for those that had been received recently from England. In the event, a large bank of thick cloud made it absolutely impossible to see the first contact of the planet.[5] But the skies cleared immediately afterwards, and Galiano was able to follow its track with the astronomical quadrant as was Vernacci using the heliometer. The egress was observed to the entire satisfaction of both officers.

That same night it proved possible to observe the immersion and emersion of τ

---

[1] Apparent time is based on each day starting as the Sun crosses the meridian at noon (sundial time).
[2] The Göttingen astronomer Johann Tobias Mayer (1723–62).
[3] A divided object-glass micrometer probably fitted to the equatorial by Dollond.
[4] For Dalrymple see p. xcv, n. 10.
[5] I.e. Mercury.

Tauri when occulted by the Moon. As this was a star of the third magnitude, whose exact position was given in Monsieur Delambre's catalogue, we considered this to be decisive in determining the observatory's longitude. The error of this phenomenon could be easily deduced relative to the lunar tables by the Moon's meridian passage being observed in different observatories in Europe.[1]

Finally, on the night of the 6th we observed the immersion of Jupiter's first satellite at 3h 3′ 11″, which compared with the time given in the nautical almanac gave the longitude of the observatory as 50°5′45″ west of Cádiz. The following day, after we had taken equal altitudes to determine the rate of the astronomical clock, all the instruments were packed away in their boxes and taken on board. It only remained to ensure that the charts were completed and to hasten the embarkation of the natural history collection.

8 and 9 November

On the 8th and 9th we received the gunpowder and were warped farther out to give ourselves more sea-room. The instruments, the forges, and the rest of the implements were brought on board; we swayed up the masts and set up the rigging. On this final day of our stay, aware of the disorderly conduct of our seamen who would in no way want to return on board, embroiled as they were in their vices, we sent out during the evening several shore patrols who seized twelve sailors and harassed many others who immediately came on board.

11 November

It was only on the 11th, when a strong breeze blew from the SW, that were we able to complete our charts and other papers and documents relating to the expedition, which were to be sent to His Majesty by the first mail boat departing for Spain as the first fruits of our mission. This shipment consisted of five chests, numbered one to five, and marked Dª Aª,[2] addressed to his Excellency Señor Bailío Don Antonio Valdés.[3] In number one were the original papers recording our passage to Montevideo; these included one packet from each corvette with the ship's logs, ratings of the chronometers, the meteorological log, the commanding officer's journal and that of the surgeon, and the original calculations of distances, longitude by chronometer and variation. This chest also contained the papers dealing with maritime affairs in the port of Montevideo, comprising the astronomical and meteorological logs, my own and those of the officers who went to Cabo San Antonio, the new rates for the chronometers of both corvettes, the list of latitudes and longitudes needed to construct the Mercator chart of Río de la Plata, the original calculations of lunar distances and a packet detailing the astronomical observations either accomplished or attempted. Another packet contained the journals and useful comments of Don Antonio Pineda and Don Luis Neé on all branches of natural history. Two bundles

[1] If the same phenomenon was observed simultaneously in a European observatory the difference between the local time of the observation at that observatory and the local time in Montevideo would be the difference in longitude between the two expressed as time.
[2] Dª Aª = *Descubierta* and *Atrevida*.
[3] Ministro de Marina, who was also head (*bailío*) of the Order of Malta in Spain.

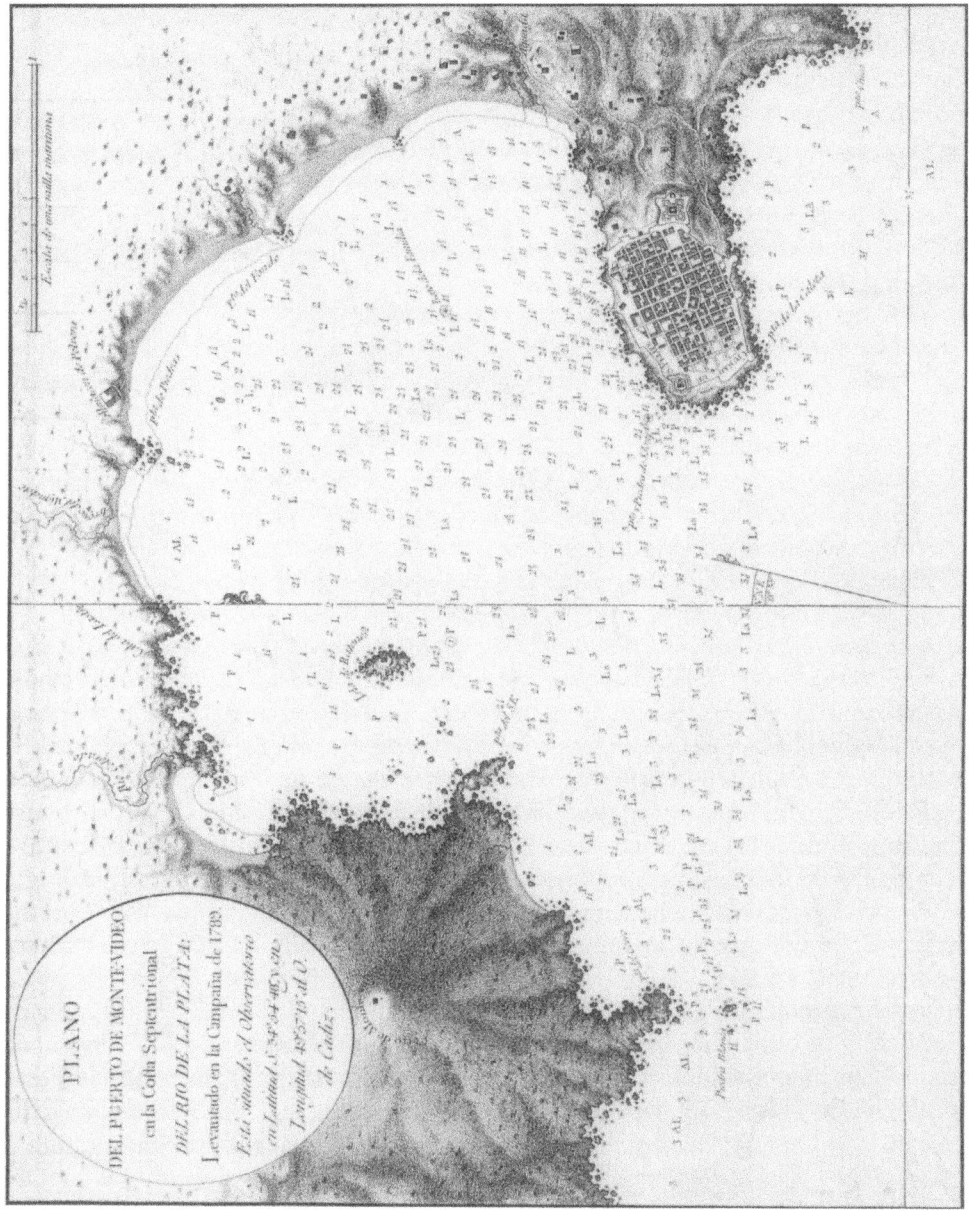

Plate 13. 'Plano del Puerto de Montevideo ... Levantado en el Campaña de 1789', inset plan in *Carta Esférica del Río de la Plata...Año de 1789*; UKHO, t15. Reproduced by permission of the Controller of Her Majesty's Stationery Office and the UK Hydrographic Office.

contained the herbarium consisting of more than 300 plants. A small box contained a sample of Paraná beeswax, and several birds, poorly stuffed but with beautiful plumage, donated by His Excellency the Viceroy. Finally, another small box contained some spider silk, partly in its raw state and partly woven into a web. Doña Tomasa Altolaguirre of Buenos Aires had been promoting this industry in a small holding of hers close to Las Conchas, which Pineda had visited personally.[1]

In chest number two there were up to fifty species of stuffed birds, some fish and insects. The plumage, the description of their characteristics, and even the preparation of this sample of the animal kingdom should be seen as significant steps of progress in natural history.

Of equal importance were the contents of number three, relating particularly to lithology and mineralogy.

Number four contained only a hive of Paraná bees.

In number five there were two large boxes made of tin, contained first the hydrographic charts, that is, the Mercator chart of Río de la Plata and the plans of Montevideo and Maldonado; and secondly all the natural history paintings, of both the animal and the vegetable kingdoms, the work of Pozo and Guio.

Having handed this consignment to Comandante de Marina Don José Orozco, to be sent by the first mail boat to Spain, as I had previously informed His Excellency the Viceroy I would do, I thought it right, and in accordance with His Majesty's intentions for this expedition, to make available without delay to our country's merchant and naval shipping all the information we had recently acquired. Accordingly, at the same time, I sent to the *Comandante* the plans that had to be supplied in duplicate to the Royal Court, charging him first to have copies made for the naval command, and to send one to the Viceroy, to whom I had promised it as a gift to one who had provided most actively for our needs during the whole of our stay; indeed he had just sent a Dollond microscope recently confiscated by His Majesty from private individuals, which, at my request, was added to our collection of instruments.

Our astronomical observations, carried out partly to advance the science of astronomy and partly to increase our hydrographic knowledge, demanded to be brought to the notice of various astronomers in Europe, whose work together with our own would support these operations in a way which could hardly be inferred from the astronomical tables alone.

Galiano, Vernacci and Concha took joint responsibility for this correspondence, sending a copy of our major observations, together with another copy of those that we required to be combined with their [observations], to the officers at the Real Observatorio de Cádiz, to Monsieur Lalande[2] in Paris, and to Señores Oriani, Reggio and de Cesaris[3] at the Brera Observatory in Milan. With these letters I

---

[1] She also gave a collection of dried butterflies to be added to the expedition's specimens: Mª de los A. Calatayud Arinero, *Catálogo de las expediciones y viajes científicos españoles a América y Filipinas (siglos XVIII y XIX): Fondos del Archivo del Museo Nacional de Ciencias Naturales,* Madrid, 1984, No. 563.

[2] For Lalande see p. xcv, n. 6 above.

[3] Angelo Giovanni de Cesaris (p. xcv, n. 3 above) and Francesco Reggio (1743-1804) helped to compile the *Ephemerides astronomicae* from 1775 onwards. Barnaba Conte Oriani (1752-1832) was a poor scholar selected for education by Barnabites. Joining the observatory of La Brera in 1776, he was

enclosed one of my own as I had the satisfaction of having been in correspondence with these gentlemen previously, and I sent these letters open to His Excellency Señor Ministro de Marina, so that he could send them where they would best please His Majesty.

We had been able to devise the best method of measuring bases by means of angles of elevation [from the water-line] to the top of the masthead of the other corvette: Galiano with his accustomed mathematical aptitude devised the following formula which seemed the most ingenious and was accordingly adopted.[1]

On this day, the full account at the Ministerio de Marina for the expenses incurred by the corvettes having been verified and settled, each of the two corvettes received a payment of 800 *pesos fuertes*, the remainder of the 28,000 that the Buenos Aires treasury office had made available to us.

12 November

Having accomplished all the objectives that we could manage in this part of His Majesty's dominions, we hoisted the boats and then unmoored. We had intended to set sail on the morning of the 12th, but the wind was not favourable and several new desertions from both ships were also of concern. I had asked the senior officer of the squadron[2] the evening before to supply us with voluntary replacements. In the event, these had to be taken by force and included some who were unfit for the service, so that the *Atrevida* was scarcely able to make up her complement, while four more men were needed in this corvette. Seeing that it would not now be possible to effect our departure this day, I asked the *Comandante* to proceed, on behalf of the Governor, to round up vagrants (hoping in this way to catch some of our deserters). At six o'clock in the evening we had the five men we needed on board, having rejected one as unfit. The *Atrevida* also completed and indeed exceeded her complement.

In the afternoon the wind freshened considerably from the east and ESE, the water level was falling rapidly, and the *Atrevida* had to warp out some distance, as she was in danger of grounding. The night was quite fair.

13 November

Dawn brought strong and somewhat gusty winds from NNE to NE. I immediately gave the order to get under way, and made the appropriate signal to the *Atrevida* and the brig. We were almost up and down[3] and ready to depart, when the commanding officer of the brig came in person to advise me that the sea was exceptionally low and the very overcast horizon to the SW led him to believe that it would only be a few hours before the weather would deteriorate and become tempestuous. I immediately gave up the idea of departing. I advised the *Atrevida* of this, at the same time warning her by signal to cut off all contact with the shore, to avoid further abuses and

---

appointed to a permanent position as an astronomer in 1778. All three corresponded with Malaspina during the expedition and again during his last years in Mulazzo.

[1] Galiano's formula and explanation, which seem unduly complicated, have been omitted: see pp. 325-6.

[2] Capitán de Navío Orozco.

[3] I.e. the cable was almost vertical immediately prior to weighing anchor.

desertions. We had hardly brought down the topgallant yards and struck the topmasts before a storm got up from NW, NE and east. As the sea level was lower than even the day before, both corvettes grounded, with their bows to NE. We therefore had no reason to fear for our cable, with only twenty five fathoms paid out. With the strengthening wind, which was like a veritable hurricane by the afternoon, came heavy rain and not a few claps of thunder and lightning. It was two o'clock in the morning before the hurricane abated.

14 November

Dawn brought light airs from the fourth quadrant but these very soon gave way to a fresh SE breeze which continued in the afternoon with a promising outlook. The commissioned officers were given leave to go ashore in the afternoon using the *Sabina*'s boat. At the same time her launch replaced the water we had consumed.

That same afternoon the officers of both corvettes, convinced that there was too little space for the crew's comfort because of the number of cattle we had taken on board, asked to be rid of them, leaving only six calves on board each ship. In this way, with common consent, we gave up what was to have been the only relief from our work in the days to come, for the sake of the better health and comfort of the crew. We carried this out immediately in the *Sabina*'s launch, leaving two calves for the crew's use over the next few days.

15 November

The night was calm, and the dawn beautiful, with a moderate breeze from NE and north, which we immediately made use of to get under way. By six o' clock all three ships had done so, and having come together soon after, we stood to the south towards Cabo San Antonio.

BOOK TWO

FROM MONTEVIDEO
TO PUERTO SAN CARLOS DE CHILOE

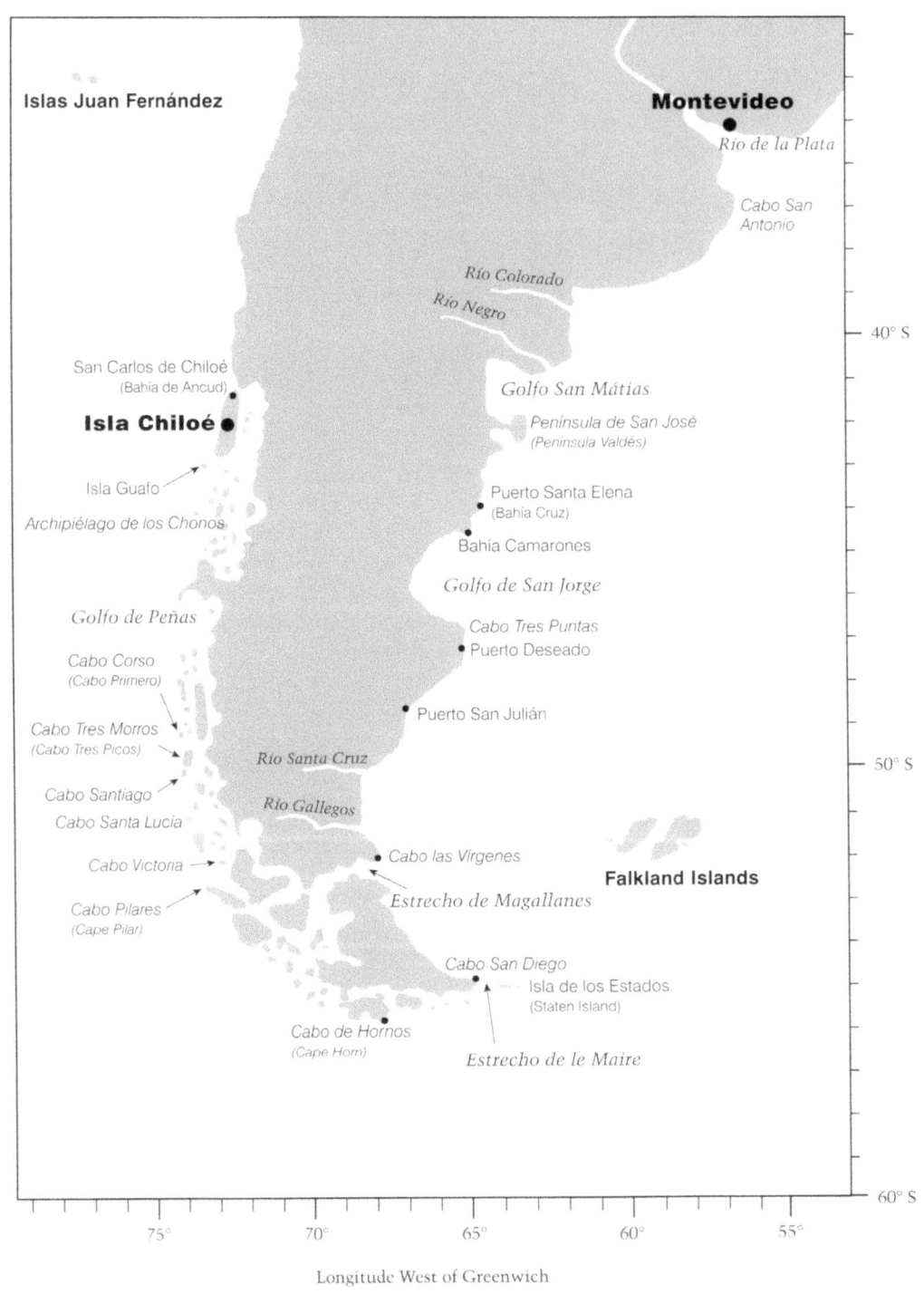

Islas Juan Fernández

Montevideo

Río de la Plata

*Cabo San
Antonio*

Río Colorado

Río Negro

— 40° S

San Carlos de Chiloé
(Bahía de Ancud)

Isla Chiloé

Golfo San Mátias

Península de San José
(Península Valdés)

Isla Guafo

Archipiélago de los Chonos

Puerto Santa Elena
(Bahía Cruz)

Bahía Camarones

Golfo de San Jorge

Golfo de Peñas

Cabo Tres Puntas
Puerto Deseado

Cabo Corso
(Cabo Primero)

Puerto San Julián

Cabo Tres Morros
(Cabo Tres Picos)

Río Santa Cruz

— 50° S

Cabo Santiago

Río Gallegos

Cabo Santa Lucía

Cabo Victoria

Cabo las Vírgenes

Falkland Islands

Cabo Pilares
(Cape Pilar)

Estrecho de Magallanes

Cabo San Diego
Isla de los Estados
(Staten Island)

Cabo de Hornos
(Cape Horn)

Estrecho de le Maire

— 60° S

75°   70°   65°   60°   55°

Longitude West of Greenwich

Fig. 3. Montevideo to Isla Chiloé, September 1789 to February 1790

64

CHAPTER 1

From Montevideo to Puerto Deseado
and at Puerto Deseado

14 November

It was common knowledge that by sailing due south by compass from Montevideo one would come to shallow water in the vicinity of Banco Inglés.[1] We therefore set a course in that direction. In accordance with my instructions the brig was to take the most direct route towards the bank. I hailed the *Atrevida*, instructing her to keep well to starboard of us and begin a separate line of soundings. The bearings of different positions on the coast confirmed in every instance their exact position on the new chart, but at the same time indicated that we were being carried somewhat to the west by the current. We therefore altered course to SbyE and SSE, in order to reach the end of the bank. Indeed, shortly before noon, the depths shoaled from six fathoms, ooze, to five-and-a-half fathoms, sand, gravel and shell. In this depth we were in latitude 35°13′20″ with the Mount [Cerro de Montevideo] bearing N2°E, true, twenty miles distant. Since up to that time the breeze had remained moderate from NNE and NNW, we had not crowded on sail, relying on the breeze continuing, with the intention of waiting for the brig to catch us up, as her known poor sailing was now very evident. When the erratic light airs and excessive heat indicated that the favourable winds would not last much longer, both corvettes crowded on sail, standing to the SSE close-hauled at first with studding sails, and later, as a moderate easterly breeze set in, with all sails set except the mainsail. At two o'clock in the afternoon, with Cerro de Montevideo disappearing from sight bearing true north, we observed longitudes in order to compare the rate of the chronometers with that deduced on the 7th, when the rates were last calculated. It was found that number 61 had kept approximately the same rate, that of number 13 had increased, while that of 72 had slowed considerably. We refrained from adjusting their rates, however, until better data obtained in some harbour could confirm the exact extent of the changes.

At four o'clock in the afternoon, we were dismayed to see that the brig had not only fallen some two-and-a-half leagues astern of us, but was considerably to leeward and even though for the rest of the afternoon we were only under topsails and foresails, she had only gained a mile on us by sunset. Our soundings at this time were

---

[1] A dangerous bank separated from Isla de Flores by a wide channel.

ten fathoms, black sand. As night came on the weather became so threatening that it was no longer possible to concern ourselves with the brig. The wind was gusting from NE and ENE, with very vivid lightning in all four quadrants, considerable thunder to the north and NW and skies so heavily overcast that they threatened a hurricane. In this predicament we judged it best to sail close-hauled to the SE, under foresail and topsails.

To keep clear of Banco Inglés and the vicinity of Cabo San Antonio, we furled the mizzen topsail, to be more in control of the sail in case of a sudden shift in the wind. In the event, the weather held until ten o'clock in the evening with winds from NE, accompanied by a few passing showers. The skies cleared a little, and the lightning was now less frequent. We sounded constantly with depths of twelve fathoms, black sand. The wind now veered to ESE, becoming a fresh breeze. Having made the appropriate signals to the *Atrevida*, which had kept close company with us, we altered course to the first quadrant, and took in two reefs in the topsails. The wind gradually died down and at daybreak it was completely calm. The sky and the horizons had cleared, but the brig *Carmen* was not in sight. [16th] Soon afterwards, a strong wind got up from south and SbyW. We hauled our wind to ESE under the six principal sails[1] and with the reefs in the topsails shaken out in order to get clear of Río de la Plata as soon as possible. Soundings continued to be ten fathoms, black sand. According to our charts and the parallel we thought we were in, we expected the soundings to increase steadily. But in spite of this and steering sufficiently to the south, we suddenly found at half past eight that the depth had decreased from ten to six fathoms and remained at this depth for some time, before slowly increasing to ten. We judged this to be a detached shallow flat rather than part of Banco Inglés, which we thought we were far to the south of by now.[2] As we had already taken the precaution of signalling the presence of the flat to the *Atrevida,* we were later able to advise her that soundings indicated a good depth of water.

By noon, we were in fourteen fathoms, black sand. Our latitude was 35°52′20″, and our longitude according to number 61 was 47′50″ east of Montevideo, both these values agreeing with our dead reckoning, which meant that we had not in any way been affected by currents. The wind was now a moderate gale from the south; the ESE and easterly winds which had prevailed during the night and the sluggish sailing of the brig led us to assume that she was lagging behind to the north and convinced us that in the vicinity of Banco Inglés she had probably decided to run for safety to Montevideo, rather than weather it out to little purpose at sea, without any hope of catching us. This decision must have been a very difficult one for her captain, Don José de la Peña, but it was dictated by prudence and perhaps by necessity. In the same way it was a natural decision for us to make a good offing, setting courses in the second quadrant, in order to get completely clear of the coast and gain as much latitude as possible, so as to reach a position where the SE wind would take us to latitude 38° or 39° where we could start our survey. So, soon after noon we took in two reefs in the topsails, sent down our topmasts and topgallant yards, set up the rigging and secured our anchors, considering that we would now no longer need to drop them.

---

[1] The mainsail, the foresail, the mizzen-sail and the three topsails, see p. 13, n. 1.
[2] They had encountered Banco Rouen.

The wind was still blowing a moderate gale from the south, the sea was very rough and there was no prospect of a break in the bad weather.

At sunset the variation was 14°11′NE and at dawn the next day it was 15°5′. After nightfall we shortened sail somewhat, to allow the *Atrevida* to catch us up. Having achieved this at nine o'clock in the morning, we hauled to the wind under the four principal sails; the soundings, which had increased since noon from fourteen to nineteen fathoms, brown sand, then increased so rapidly that at one o'clock no bottom was obtained with thirty-five fathoms of line and at five o'clock with forty-five fathoms of line. After midnight the wind veered round to SSW and SW and strengthened.

17 November

At dawn the weather outlook was good. We shook the second reef out of the topsails, and with the mizzen topsails and main topmast staysail set we continued close-hauled in the second quadrant as the only option the wind afforded us, even though this took us a considerable way from the coast. We saw many birds of the species known to frequent these waters. The corvettes were evenly matched in speed and in the ability to stand up to heavy seas. At noon our latitude was 36°26′ and longitude by number 61 was 1°58′ east of Montevideo. Dead reckoning gave a latitude greater by 15′ while longitude agreed almost exactly. In the afternoon the weather turned much colder. The wind freshened from the SW, a heavy swell developed, and the horizon was thick with cloud. With the new Moon imminent, we feared that the night would be very stormy. Accordingly we took in two reefs in the topsails, and at five o'clock in the afternoon, having sounded one hundred fathoms, fine sand, we continued with only the foresail and topsail set, to give the crew a rest, since the cold was very noticeable and at all events with the wind in this quadrant our progress was of little use.[1] The night was murky with a very high sea causing us to roll excessively. The wind continued to gust from SSW to SW; consequently we did not make any changes to the sails as set the night before.

18 November

Dawn brought a better outlook, and at that time a vessel was sighted to the east lying-to on the same heading as ourselves, scarcely two leagues from us. There was no doubt that she was one of the many whalers that frequent these waters. We passed this information to the *Atrevida,* who had drawn our attention to this vessel by signal. We hauled home the fore topsail, but left the two main topsails lowered, since the other vessel had gone on the port tack, and crowded on sail, perhaps wanting to speak with us. At almost seven o'clock she hoisted the British ensign. We responded with our own ensigns, but decided not to wait for her, partly because it would be very difficult to communicate by boat in a high sea, which would only make her wary, without our being able to ascertain her true intentions, and partly because we would be wasting another three or four hours unnecessarily, falling off considerably to leeward. A signal was made to the *Atrevida* to crowd on sail. We hauled aboard the main tack and

---

[1] With the wind from SW, a course in the second quadrant was taking the two corvettes away from the direction that Malaspina wanted to steer.

hoisted the topsails. The mizzen topsails and main topmast staysails were hauled home, and we stood close-hauled to the SSE with the wind by then a strong breeze from SW. We soon lost sight of the English vessel, which was still on the other tack. At noon our latitude by observation was 36°49′ and our longitude was 47°8′ west of Cádiz by chronometer number 61.

In the afternoon the wind eased off considerably. We saw near us large flocks of *chorlitos, tableros, carneros, tiñosas, pardelas,* and *pamperos.*[1] Everything pointed to a favourable change in the wind which would allow us to begin our tasks. During the first hours of the night the wind fell to a dead calm. The rolling was considerable. The sea was still high from the SW, and there was a very heavy dew. Only at eleven o'clock, did light airs set in from the north. We lost no time in taking advantage of these, steering SWbyW, with all reefs shaken out, having signalled to the *Atrevida* to put on all sail.

19 November

The favourable breeze did not last long, nor, consequently, did our uncertainty about the best course to steer for the coast. As the Sun rose, the wind freshened to a moderate breeze, backing progressively to WNW, west and WSW. By ten it was from SW, gusting strongly, and the atmosphere was hazy. We were obliged to haul the wind yet again to SSE, and consequently distance ourselves from the coast once more. At noon our latitude was 37°55′ by observation and our longitude by number 61 was 47°35′, which confirmed a favourable current to the SW. After the Sun had crossed the meridian the weather outlook improved greatly, as it had on the previous afternoon. As an unusual advantage in speed had caused us to get well ahead of the *Atrevida,* we used the time spent in waiting for her to sound, obtaining no bottom with 120 fathoms of line. Finally we bore up for half an hour, until we were in company once again.

With the weather looking better, we swayed up the main topgallant mast, but decided not to use the foresail thus lessening the strain on the rigging with a sail that was of little use when close-hauled, and also to meet the impending need to have a man at the masthead at all times. By six o'clock in the evening with light airs from south and SSW, we altered course to the west. In a short while, however, we were completely becalmed and remained so until midnight. The dew was even heavier than the night before. By one o'clock light airs were noticeable from east and NE. These gradually increased in strength, bringing with them fairer weather, so that by dawn with all sails set we were making three or four knots.

With the wind in our favour, I could have steered west and WNW, to start surveying the coast from 37½°, where Concha and Vernacci had concluded their work.[2] However, after carefully considering the essential objectives of our mission and

[1] Possibly small plovers of various species on migration, but more likely prions of several species; Cape Petrels (*Daption capense*); possibly Wandering Albatrosses (*Diomedea exulans*), although any of the larger albatrosses or Giant Petrels (*Macronectes giganteus*) are also possible; possibly terns, shearwaters and storm petrels, possibly Wilson's Storm Petrel (*Oceanites oceanicus*).

[2] From this it is clear that Concha and Vernacci had extended their survey of the south coast of Río de la Plata to a position 70 miles south of Cabo San Antonio.

taking into account the prevailing weather conditions in these latitudes, I realized that altering course now to the north, with the likelihood of a fresh wind from SW ruling out a NW tack, would make it impossible to survey the coast without considerable delay to little or no purpose and also mean wasting fifteen to twenty days. It would be an easy matter for any smaller vessel on her way back from the Malvinas[1] to survey the coast between Río Colorado and Cabo San Antonio. The labyrinth of islands and shoals between the mouths of Río Colorado and Río Negro had already been very accurately surveyed by Pilotos Tafor and Villarino[2] both from the land and the sea. The astronomical position of this coast would therefore be well determined if the southern extremity of the area adjacent to Río Negro could be accurately located. In that case, examining the area around Puerto San José,[3] an important point for whaling, could be carried out speedily, followed by a somewhat longer stay in Puerto Deseado, an ideal place for research into the natural history of the coast of Patagonia, which was of such interest to the advancement of science. This decision would also put us in the pleasant position of being able to progress more quickly with our work on the west coast of Patagonia. This, combined perhaps with some good fortune on our part, and the work of Pilotos Machado and Moraleda[4] (particularly the latter, whose accuracy was unquestioned) could well result in our being able to complete our work on the coasts of Patagonia in just a year. After all, His Majesty had spent large sums of money on different expeditions and truly intrepid, meticulous and intelligent *pilotos* had preceded us.

With all this in mind I altered course to WSW5°W true. To the extent that this would take us toward the coast, it would considerably increase our latitude and leave us free to approach land wherever it was most prudent for us to do so.

20 November

The sea had fallen almost completely calm; at times, however, thick mist hid the *Atrevida* from sight, obliging us to shorten our sail somewhat. Finally, at dawn, the weather set in more persistently from the north and we had the familiar company of many common birds. We had been able to air and clean thoroughly below decks and at noon we were in latitude 38°31' and longitude 48°13'.

At six o'clock in the evening we were still making seven or eight knots, when a dense haze which was beginning to build up to the SW warned us that the favourable wind was about to abandon us. We hurriedly set about taking in all sail, leaving the topsails lowered. Indeed the SW wind was soon upon us and the horizon and skies, by now very squally, threatened a violent storm. Presently there was continuous, fearsome thunder and lightning. In an instant, the wind would shift round from SW to north, and the temperature of the air changed noticeably with each change in the atmosphere. Half an hour of rain seemed to put an end to this struggle, leaving a persistent breeze from north and NNW with much vivid lightning and ugly skies stretching from west round to south.

[1] The Falkland Islands.
[2] Segundo Piloto Basilio Villarino who carried out surveys in Patagonia in 1781-2.
[3] Situated on the south side of Golfo San Matías.
[4] Pilotos Francisco Machado and José de Moraleda; for the former see p. 127, n. 4 below and for the latter see p. 119, n. 2 below.

We hoisted the topsails and altered course to the opposite tack, the *Atrevida* keeping close to us. However, at nine o'clock in the evening the weather was again looking threatening and soon after there was a heavy hailstorm, much thunder and lightning and some strong gusts from SW. These died away, only to be replaced an hour later by even more threatening skies, constantly varying winds and pouring rain which did not abate until five o'clock in the morning, at which time a moderate breeze having sprung up from the north, we were able to resume our course under full sail.

21 November

The *Atrevida* was within hailing distance; accordingly I asked what the longitude had been by her chronometers at noon the previous day. Seeing from what they told me that our chronometers (which did not quite agree amongst themselves) placed us somewhat further to the east, I took advantage of a very quiet morning to make a more exact comparison, independent of dead reckoning. To this end I made a signal at ten-thirty to observe longitude by chronometer, and then signalled the moment to do so for both ships. The results we obtained were as follows:

Descubierta	*Atrevida*
N° 72......0° 21′ 5″ West of Montevideo	N° 71......32′ 10″
61......0° 22′ 58″	10......33′ 00″
13......0° 36′ 21″	105......31′ 15″

Latitude by observation at noon: 39°0′

Early this morning we managed to catch on board three plovers which had doubtless been swept out to sea with many others by the squalls of the night before. In the afternoon, a small falcon was similarly unable to escape the grasp of one of our sailors at the foremast crosstrees. The falcon had taken a small bird and chosen that perch to devour its prey.[1]

From eleven in the morning onwards breaking seas had indicated to us that we were coming into soundings. We checked this at two o'clock in the afternoon, after making the appropriate signal to the *Atrevida,* and found we were in fifty-two fathoms, fine black sand. Twice we had tried unsuccessfully to observe the distance between the Sun and the Moon but we did succeed in obtaining variation by western amplitude which was 15°15′E. Today we distributed new and warmer clothing to the marines and sailors. The marines would need to don another jacket and long breeches. The sailors, naturally enough, lacked everything: the new hands (volunteers and pressed men) because they would not have had any spare clothing for a long time, and our old hands because their recent disorderly behaviour had led them to sell all they formerly possessed.

After nightfall the wind now clearly abating, veered to WNW and west. We hauled to it under full sail, and the soundings which at nine o'clock in the evening and two o'clock in the morning had been forty-five fathoms, black sand, had by six o'clock in

---

[1] Possibly small plovers of various species on migration, but more likely prions of several species. The falcon was probably a Peregrine Falcon (*Falco peregrinus*).

the morning [22nd] become fifty-five fathoms, fine sand, and continued so to nine o'clock. It is impossible to imagine fairer weather than that which greeted us early this morning. Everything promised a pleasant spring. The smooth sea, and the light airs were themselves indicative of alternating land and sea breezes. Variation by eastern amplitude was 15°32'.

At noon, our latitude by observation was 39°33', and the longitude by our chronometers was 1°20' west of Montevideo. Number 13 continued to show a difference of 15' to 16' further to the west, leading us to suspect that the loss detected at the time of the last rating had in fact increased. At the same time the *Atrevida's* chronometers agreed very closely with ours, according to what they told us the following afternoon, and the findings were supported by their lunar distances. As for us, we were mortified to find that all the results from a set of 114 observations taken before and after noon were in error,[1] no doubt because of some optical illusion.

From three o'clock in the afternoon the wind began to strengthen, although the weather remained very fair, and the sea did not change appreciably. We continued making six or seven knots heading SWbyW by compass. At sunset, we sounded fifty-three fathoms, black sand; variation by observation was 17°52'. At the same time we saw in the far distance three or four small whales, and some common birds. At nine o'clock in the evening, fog began to drift in from the west. This caused us to take in the studding sails and booms in order to sound with greater ease. At eleven o'clock that night seeing that we were in no more than thirty-five fathoms, black sand, it seemed prudent to steer SSW for four hours. By three o'clock in the morning we were in fifty-five fathoms, small gravel. From four o'clock we crowded on canvas again, but due to a mistake by the helmsman we steered SbyW until six o'clock [23rd] when we reverted to our original course, steering SWbyW, while sounding fifty fathoms, fine black sand. From dawn onwards small whales, a few common birds and a kind of land plover[2] were in sight not far off; variation was 18°9'. The wind continued from the north and NNW. The weather outlook and the sea were most promising. We were making seven or eight knots, with the *Atrevida* in close company.

At ten o'clock in the morning we observed hour angles for longitude and at the same time sounded again in forty-four fathoms, black sand; by noon this was now forty fathoms, same bottom. At this time our latitude was 40°38', and our longitude 4°0' west of Montevideo. Steering WSW5°W (true) we soon found ourselves in thirty-five fathoms, sand and shell. We had taken forty-eight sets of lunar distances whose mean only differed by 1'30" to the east from the longitude by chronometer 61. The weather seemed set fair, with a strong breeze from the north and NNE.

It was not long before the weather began to deteriorate. By six o'clock in the evening, when we were only eighteen miles from the shoals off Río Colorado,[3] squalls developing from west and SW forced us to trim our sail without a moment's delay. Occurring with sudden shifts of direction, sometimes from SW and sometimes

---

[1] *muy cortos* (very short) in the MS journal.

[2] Could be one of several species of small plovers, possibly the Two-banded Plover (*Charadrius falklandicus*) common in the Falklands Islands.

[3] Possibly Banco Vibora and Banco Intermedio.

from NW and NE, the squalls were accompanied by very vivid flashes of lightning, some rain, and considerable thunder.

Given these conditions, it seemed wiser to steer in a more southerly direction, which would at the same time take us somewhat away from land and give us a little sea room to be able to survey the coast, if, as was very probable, the wind were soon to shift to SW. Contrary to our expectations, however, a strong wind started to blow from NE as night came on. Accordingly, we bore away to WSW once more, taking care to moderate our speed, continuing to sound thirty-five to thirty-three fathoms, gravel and small shells. From eleven o'clock the northerly wind, which had been gusting and variable, died almost completely away. By two o'clock in the morning there were light airs from the third quadrant, but not enough to allow us to maintain steerage way. There was a heavy swell with cross seas from NE and south.

24 November

At dawn the outlook was reasonable. We were sounding thirty-two to thirty fathoms, sand and small shell, and at seven o'clock in the morning when a moderate SE wind set in we crowded on sail, and steered due west (true) with the soundings decreasing to twenty-eight, twenty-nine and twenty-seven fathoms, gravel, sand and small shell. At noon our latitude was 41°24′ and our longitude by number 61 was 5°59′. We were thus only a few miles from the coast, and with the horizon completely free of clouds, we expected to sight it very soon, which we did at two o'clock in the afternoon. The coast seemed to run from north [to south] and then [from east] to west, low lying, but rising gently from the coast. We immediately altered course to NWbyW and at about three-thirty when we were two or three leagues off the coast we observed hour angles and started to measure bases by log-line,[1] to fix the coastline with greater accuracy. There was no doubt that this was the stretch of coast that runs between Río Negro and Punta Belén.[2] But we found it difficult to locate the mouth of Río Negro.[3] Soundings of twenty-five, twenty-four and in some cases nineteen fathoms, gravel and small pebbles, indicated that we were just to the south of its mouth, according to information from our navigators and from Tafor's chart.

The coast ahead of us seemed to be formed of horizontal layers of loose, blackish earth in thin whitish and reddish layers arranged perfectly horizontally. There were perhaps more than twenty of them, probably consisting of sandy marl, clay and so forth. The part of the surface that we could see was dotted with a few shrubs and plants with earth visible between them. Among the pebbles brought up by the lead, Don Antonio Pineda found something to stimulate his never-ending and informed curiosity. A number of sea acorns[4] or marine barnacles[5] of the smallest species stuck to a little shell, when preserved in salt water in a glass looked like a set of feathers or

---

[1] To obtain the distance run between two 'ship stations'.
[2] Promontorio Belén on the north side of Golfo San Matías.
[3] This sentence crossed out in the MS journal.
[4] *Bellotas de mar.*
[5] *Balanos marinos.*

paint brushes, either extended or retracted, exactly like the shape of a parasol, a structure no doubt devised by nature to allow these tiny crustaceans[1] to feed off other smaller ones.

While measuring bases at the same time as us in order to establish the configuration of the coast, the *Atrevida* hailed us to say that their distances agreed very closely with the chronometers. They were somewhat closer to the land than we were and had sounded, fifteen fathoms, small pebbles.

At nightfall we still had not been able to fix the position of Punta Belén. The very dearth of conspicuous features as far as we could see and the total lack of information about this coast, made our position somewhat doubtful; and the many charts we had of these parts seemed to increase rather than decrease our doubts about the part of the coast we had sighted. I therefore decided (at the cost of several hours of fair wind) that both ships should stand off and on under topsails until one o'clock in the morning and keep underway until dawn so as to be north/south with Punta Belén and proceed directly to Península de San José.[2] If we succeeded in getting good latitudes off this peninsula, we would accurately determine the width of the bay[3] and its soundings for the convenience of shipping in the future. The greater or lesser degree of accuracy of my work would depend in large measure on that of the charts made by Tafor and Villarino, although in any event, I would have to sacrifice some precious time particularly with a Moon as favourable as this one promised to be. By late afternoon, the wind had shifted to the east, becoming a fresh breeze, with clear skies and a smooth sea, which continued until midnight. The wind then veered to south before backing once again to the east, with the sky and the horizon becoming overcast.

25 November

However, we sighted the coast in the early morning, taking advantage of the lulls between showers to determine our longitude and to take bearings of Punta Belén so as to fix it to our satisfaction with the bearings taken the previous afternoon. Our soundings were thirty, thirty-five and thirty-two fathoms, gravel and shells. The weather outlook and the elevation of the coast, were the same as on the previous day. By seven-thirty in the morning we were able, as I had originally planned, to set a course to SSW towards the northern tip[4] of Península de San José, whose position I considered particularly important. There was a moderate breeze from NE, with a smooth sea, but the general outlook remained showery, with persistent fine drizzle. At nine o'clock we could no longer obtain bottom with fifty fathoms of line. At ten o'clock we had seventy-two fathoms, mud, but no part of the land was visible. At this time, the wind fell completely, leaving us without steerage way. The persistent drizzle also made it impossible for us to take meridian altitudes of the Sun.

During the afternoon, light airs rose from the third and second quadrants, which we took advantage of with all canvas set in order to reach the peninsula. The weather had cleared, and we set about observing longitudes. It was also our intention to

[1] *Insectos* in the MS journal.
[2] Península Valdés.
[3] Golfo San Matías.
[4] Punta Norte.

deduce the latitude from two altitudes,[1] but the weather thickened with a strong squall from SSW and south, accompanied once again by rain. The skies promised stronger winds that night, which if we penetrated deep into the bay might oblige us to drop anchor in sixty or seventy fathoms. The coast that we had been unable to see had been plotted from land by Piloto Villarino, while Puerto San José[2] had been visited by Tafor and Peña; so this was probably unnecessary work which could cost us the loss of anchors, and perhaps the waste of a great deal of time. With this in mind, I made a signal to the *Atrevida* that I would not now drop anchor in Río de San José.[3] Then with foresail and main topsail somewhat eased, I set a course to SE and east, with the intention of making land by morning on the outer coast of the peninsula at its northern point.

The shoal that extended the length of this coast merited a serious examination, since it could not have been surveyed with great accuracy, for while the latest charts showed it, we saw that Tafor had omitted it.

By sunset the horizon had cleared considerably, when we were lucky enough to sight the high sierras at the inner end of the bay bearing true west. By comparing them with our position we were able to confirm yet again the accuracy of the work of Tafor, Peña and Villarino. The *Atrevida* had sounded seventy fathoms, ooze. After nightfall the wind shifted round to SW and freshened. We were obliged to sail with the foresail and main topsails lowered, and at ten o'clock we hauled the wind to south and SSE, although we later had to lie-to, and eventually bear away to allow the *Atrevida* to come up with us, having clearly failed to see our signal to alter course. At eleven o'clock that night we obtained a sounding of forty-five fathoms, and at four thirty in the morning, fifty fathoms, sand and small shells.

26 November

At dawn, the outlook was extremely pleasing; there was a fresh wind and the sea was rather rough, but it seemed that one or the other would be moderated by the strength of the Sun, as it promised to continue to shine. Península de San José stretched away from west to SSE, no more than two or three leagues distant at its closest points. It was exceedingly important for us to fix the latitude and longitude of its northern extremity to avoid having to tie it in with good bases to the observations which we would make later on. For that reason we put about to WNW, under foresail and main topsails, later going on the other tack to SSE, sounding fifty-nine fathoms, hard mud. Finally, at ten o'clock, in position for measuring a base, following the trend of the coast, both corvettes hove-to and at the timed signal we measured the masthead heights against the horizon to deduce the length of the base,[4] while taking all possible bearings and observing hour angles for longitude. The sounding was then fifty-seven

---

[1] Latitude deduced by taking two altitudes of the same or different heavenly bodies off the meridian, the interval between the observations being taken by chronometer. Also known as latitude by double altitudes.

[2] Golfo San José on the southern side of Golfo San Matías.

[3] Possibly a slip for Golfo San José.

[4] Malaspina appears to mean horizontal and not horizon. The base was obtained by measuring the angle between the waterline and truck of the mainmast simultaneously from each ship. See pp. 325-6.

fathoms, pebbles and small shells. The weather had eased and now looked very fair: the sea was smooth with the wind a moderate north and NNE breeze which we took advantage of with full sail.

The nature of the land, along the coast before us, looked no different from what we had seen on Punta Belén. It too was composed of horizontal layers forming a humpbacked ridge of almost the same elevation. What did apparently differ considerably was its trend when compared with the direction shown on the charts. At noon our latitude by observation was 42°5' and our longitude 7°33' west of Montevideo and so our only thought in the afternoon was to continue taking advantage of the favourable wind and to survey this coast in greater detail. We set to work measuring bases by log-line, and proceeded with caution, looking for a shoal shown in these waters on the national charts. Although its existence was disproved by our previous night's sailing and that of the morning, we found it hard to believe it had been charted without at least some doubt of its existence.

From one o'clock in the afternoon onwards, following the coast at a distance of two or three leagues, we found that depths began to decrease to thirty-seven and thirty-eight fathoms. Accordingly, we altered course somewhat more to the east. Even so, at two o'clock, we found ourselves in twenty-one fathoms. The bottom consisted of small pebbles, sometimes mixed with small shells and at other times with signs of rock. At times we passed through such tide rips that the *Atrevida* found it difficult to keep a steady course even while making five to six knots. We steered to the east and very soon we were in thirty-eight fathoms. Soundings decreased again, and we steered EbyS and ESE in thirty-nine and forty-two fathoms, at a distance of three to four leagues from the coast, the extremities of which stretched from WNW to SSW. A set of thirty distances observed that same afternoon from the Sun to the Moon, all perfectly consistent, gave a mean [longitude] of 3' to the west of [that given by] number 61. There was no longer any doubt that the rates of 13 and 72 had altered since they had been determined in Montevideo. From observations by azimuths and amplitudes, variation was 20°20'NE.

We made very little distance that night: from eleven o'clock onwards we hove-to, and having altered course towards the land at one o'clock we were in sight of the coast at seven, with the two final positions of our previous day's triangulation in view.

27 November

Soundings forty-four and forty-five fathoms, sand, small shells and small pebbles. The land in sight, which was still part of Península de San José, looked the same as that seen on previous days. Having closed the coast to just over a league, we were still sounding thirty-eight, thirty-five and thirty-three fathoms, small pebbles. We measured bases by observing the masthead heights and by log-line; longitudes were also observed. When the wind, which had previously been from the west, backed to SSW we sailed close-hauled under full sail in the second quadrant. Our latitude at noon was 42°31'. The longitude from number 61 was 7°10' west of Montevideo. During the afternoon the wind, now a moderate breeze, backed to the second quadrant. We stood off from the land close-hauled until four o'clock in the afternoon, when we altered course to the south, keeping the coast in sight at a distance of four or five leagues. Its southern extremity,[1] which was a low lying point, bore WSW from us by compass. The favourable wind encouraged us to make the most of the night, taking care to be a short distance from the coast by dawn, altering course towards the point whose bearings we had taken the previous evening. Our soundings continued to be forty-seven and forty-eight fathoms, sand and small shells.

We hove-to for a few hours until dawn, when we were two leagues from the land and at five o'clock we were able to start our geodetic tasks, measuring masthead heights, [28th] which was repeated at seven, ten and twelve o'clock, the intervals being tied in by bases measured by log-line by both corvettes. At the same time we also noted every possible alignment which might be of use in determining the trend of the coast.

From nine o'clock onwards we were in sight of the entrance to Puerto Nuevo,[2] which forms the southern side of Península de San José, and at the back of the harbour of that name.[3] We set about fixing the position of the two extremities of the harbour mouth and devoted a further half hour to measuring the meridian altitude of the Sun, from which we deduced our latitude to be 43°7'30", longitude 7°53'; soundings thirty-seven to forty fathoms, fine sand.

The wind continued to blow favourably from NW to north. The sea was smooth and the horizon fair. For this reason we tried to make full use of the whole afternoon and evening in surveying the southern shore from the entrance to Puerto Nuevo, which all the charts showed as unknown. From Puerto Nuevo the coast turned markedly towards the west, until it formed a considerable inlet, whose inner shores were formed of such low-lying land that surveying it was difficult.[4] Under topsails alone we were making seven to eight knots. The need to sound continuously, and our wish to survey the coast thoroughly had required us to proceed under so little sail. So we hauled the wind to SWbyW and to SW, coasting at a distance of one-and-a-half to two leagues, the soundings gradually decreasing from thirty-five to twenty-seven fathoms, sand, shells and small pebbles.

At three o'clock in the afternoon we hove-to in order to measure bases and to observe hour angles. I found this method preferable to taking masthead heights under way, because the different operations could thus be carried out more slowly and uniformly at the extremities of the base, and also because the actual direction of the bases could in this way be better aligned in accordance with the points along the coast chosen as suitable for taking essential bearings. By about five thirty in the afternoon we could already see the whole stretch of coastline, whose extremities bore from SSW to NNW from us. The soundings had decreased to sixteen and fourteen fathoms, small pebbles. The coast we could now see was identical to that of Península de San José, until it decreased considerably in height and became so low-lying that it seemed to be wetland. Behind it, however, to the west and SW the land rose to a considerable height. Indeed, the country to the west of us featured a wide inlet or large

[1] Punta Delgada.
[2] Golfo Nuevo.
[3] Golfo San José from which Puerto Nuevo is separated by a narrow isthmus.
[4] Malaspina is referring to the southern shores of Golfo Nuevo, which he did not enter.

haven, formed by the low ground which lay before it. The opening to the inlet appeared to trend to ESE, and seemed to extend far inland.[1]

Having surveyed this coast in thorough detail, and having at six o'clock measured new bases and longitudes, we steered SE, in order to be ready at dawn to take as the northern extremities of our work, those which lay to the south of us at that moment. The barometers and the sky itself, announced an approaching change in the weather. For that reason, and also in view of the short distance we had to cover that night, we took in one reef of the topsails, and lowered the topgallant yard. Steering SE we were soon in deeper water, and between ten and twelve o'clock at night soundings were forty-two to forty-five fathoms, small shells, sand and some stones. At one o'clock in the morning, having run the required distance, we hove-to, and at three o'clock we got under way and steered SWbyW, so as to close the coast by dawn.

29 November

By then we had some fine rain, and several flashes of lightning in the second and third quadrants. Soon after the wind shifted to a moderate SW breeze. We stood to the SE close-hauled with more sail and finally with a southerly wind, we went about to WSW, intending to close the coast, which we could see in the third quadrant some five or six leagues off.

At nine o'clock in the morning the wind was still blowing fresh from the south, and there was some cloud on the horizons. We steered ESE and, having measured bases by masthead heights, remained on this course until ten o'clock by which time the sky had finally cleared, and assured now of what seemed to be a steady breeze we once more went about to WSW under full sail. By a quarter to twelve, we were a very short distance from the extremities of the land we had seen to the south the day before. We hove-to in order to measure bases by masthead heights and while we were in this position we waited for the meridian altitude of the Sun, which gave our latitude as 44°2'00", while our longitude by chronometer 61 was 8°52'30".

The land now in sight was no higher, and looked no different from that which we had seen for the last three days. It was undulating and barren looking, but terminated at each end with a more promising appearance. Finally, far away to the south could be discerned the high land around Puerto Santa Elena,[2] an area already frequented by our ships since settlement of the Patagonian coast began.

After midday, with the wind still from SSE, we had to set a course in the second quadrant because the extremities of the coast before us did not give us enough sea room to continue on the other tack. At three thirty we were able to observe some azimuths, from which we found the variation to be 19°NE. Soon afterwards, the wind veered to SE, east and NE, and we could steer NW, a suitable course to survey the next stretch of the coast. Soundings throughout the afternoon were forty-five and forty-six fathoms, blackish sand, with some ooze. At seven forty-five, bases were measured once more, and all the appropriate observations were taken. At nine

[1] Malaspina is mistaken; there is no such inlet.
[2] Bahía Cruz.

o'clock, having finally concluded all our tasks for the day, we steered SSE, under easy sail to take up an advantageous position at first light the following morning.

After running for six leagues on this course, as planned, we lay-to, standing in for the land from two o'clock in the morning with soundings of forty-eight, fifty and fifty-three fathoms. [30th] Eventually we made the land at first light, and at dawn we were two leagues off the coast, in such a favourable position that the entrances to Puerto Nuevo and Santa Elena and the adjacent coast in both directions could all be surveyed and plotted in the greatest detail. The weather was too fair and favourable for us to carry out our idea of anchoring in Puerto Santa Elena. We could make out clearly the entrance to the harbour – Piloto Tafor's work could be considered quite accurate. We had measured some bases at distances of one-and-a-half leagues from the northern extremity, coastal views had been taken of its adjacent coasts, azimuths had been observed and hour angles had been taken. Consequently we continued the survey of the coast, through Bahía Camarones to Puerto San Gregorio.[1] As the weather was so very favourable and clear, we measured bases at different times by masthead heights, and at noon having gone almost round the whole of Bahía Camarones, we sighted on our bows Islas Arce, Leones and Rasa, which lie at the northern end of Golfo de San Jorge. Our latitude was 44°50′, our longitude 9°17′ and our sounding forty-seven fathoms, ooze, having sounded thirty-seven fathoms towards the middle of the bay, a good league from the land on our beam.

The accuracy of Piloto Tafor's work was further confirmed by the sighting of the coast from Santa Elena, through Bahía Camarones, as far as the vicinity of San Gregorio. This *piloto* had surveyed the length of this coast several times in small craft, and if his charts had earned a measure of trust as far as the plotting of the coast was concerned, they had done so all the more with respect to his soundings. The shallow waters with strong tides and their lack of shelter meant that the harbours in this locality did not merit the attention of Intendente Viedma[2] at the time that settlement in the area was being considered. They seem destined by nature to be forever desolate.

Thanks to a strong, favourable breeze we soon reached Isla Rasa. We chose to make our passage between this island and those of Arce and Leones, rather than enter the much narrower but deeper channel between these two islands and the mainland. Our shipping plying between San Gregorio and Santa Elena has always preferred the latter channel. For that reason it seemed more important and safer for us to examine the outer one in detail. We were making five or six knots under foresail and main topsail with the *Atrevida* keeping a constant distance astern of us, after we had signalled her to keep this station as the most suitable one for measuring successive angular bases.[3] As we approached the channel or gut formed by these islands, we began to experience such strong tide rips or overfalls that at first everyone thought they were breakers, and soon our steering, bearings and distances were all thrown out. We

---

[1] South of Bahía Camarones, from which it is separated by Cabo Dos Bahías.

[2] Francisco Viedma, a number of whose papers on the coast of Patagonia were copied in Buenos Aires for Malaspina in October 1789 by Teniente de Fragata Fernando Quintano: AMN, MS 314, ff 159-159v.

[3] Presumably by masthead heights together with simultaneous bearings from the two corvettes to prominent features on shore.

altered course towards Isla Rasa when we were about a mile off and when it was almost abeam we sounded forty-four fathoms, small pebbles. We immediately attempted to measure bases, which the irregular speed of both ships, caught here and there by the tide rips, rendered dubious, if not useless. Suddenly we faced further complications with the discovery of an island which appeared to be flatter and more to seaward than the one marked as Rasa on Tafor and Peña's charts.

Following the channel amidst continous breaking seas, at an equal distance from the large islands on our right, and the known Isla Rasa on our left, the depth increased to fifty-eight fathoms, small pebbles and shell; and having now no doubt that the island which we could see some two or three leagues to SE was not on the charts, we named it Isla Cordero after the boatswain of the *Descubierta,* who was the first to sight it.[1] Clearly, it would be difficult, perhaps even impossible to see it from the deck of a launch or *sumaca,* sailing close in to the main as Tafor and Peña naturally did. They would not have missed it had they passed through the same channel as we had that day.

After surveying the waters adjacent to Puertos San Gregorio and San Sebastian[2] as accurately as we were able, we made for Isla Cordero, from where we could see a reef some way off. We obtained the island's longitude by taking fresh hour angles, and measuring a base at a distance of about two miles from it to determine the length of the reef which stretched away to the SW. Lastly, precise bearings were taken of the entrance to Canal de San Jorge.[3] Although this channel was known to our Río de la Plata *pilotos* from the days when they began to explore the coasts of Patagonia, it was completely unknown to European hydrographers, including our Spanish ones. Our shipping bound for Lima followed Monsieur Bellin's chart,[4] which, like Anson's, did not show the slightest trace of such a channel. Only at Río de la Plata did it come to my notice that Piloto Tafor had sailed from Puerto San Gregorio in a *chalupa*[5] and, having surveyed some thirty leagues of the northern coast of this gulf,[6] he found it strewn with islands, and these, like the coast itself, were very difficult to reach because of the many reefs that surround them. However, in Buenos Aires it was commonly said (on no particular authority) that according to all Patagonians the gulf extended inland to the *cordillera,* and this news had everyone imagining hoards of new riches.

Neither the mission which I was undertaking nor the boats at my disposal, left me in the slightest doubt about whether or not we should undertake the survey of this gulf. I could not see the remotest benefit in it for our national shipping and in any

[1] There is no such island to seaward of Isla Rasa either on modern charts or on *Carta esférica de las costas de la America meridional desde el paralelo de 36°30' de latitud S hasta el Cabo de Hornos,* published in Madrid in 1798, which incorporates Malaspina's surveys. It therefore seems likely that the Isla Rasa shown on Tafor and Peña's charts was the outer of the inshore islands and that subsequently Malaspina rejected the name Isla Cordero and substituted Isla Rasa instead and adjusting the names of the inshore islands accordingly.

[2] Bahía San Sebastian, within Bahía Camarones.

[3] Possibly Canal Leones, ½ mile wide, separating Isla Leones from the mainland.

[4] Jean-Nicolas Bellin (1703-72) was Ingénieur de la Marine et du Dépôt des Cartes, Plans et Journaux to the French government. He made the maps for some of the best-known French collections of voyages and travels, and was responsible for *Le Neptune français.*

[5] A long boat or shallop

[6] Tafor had obviously surveyed the NW side of Golfo San Jorge.

case it could be undertaken from Buenos Aires with more suitable vessels. Alternatively, if it was His Majesty's intention that we personally undertake this task, then the safest way, and the best use of this expedition's time, would be to carry this out during our final year, when we returned to the Atlantic Ocean. In this case, His Majesty's orders would be decisive, with assistance forthcoming; and in the meantime, I would endeavour not to commit the expedition to any objectives other than those set at the time of our departure.

I therefore abandoned any thought of entering the gulf and that same evening, after completing the hydrographic survey of its northern extremity, I decided to sail straight across [the gulf] to its southern extremity, and from there round Cabo Blanco and enter Puerto Deseado, where it was my intention to meet the brig *Carmen* and to check the rates of the chronometers. At six o'clock in the evening, after sounding fifty-eight fathoms, small pebbles and shells, we steered SSE with all sails set, maintaining this course all night, not neglecting to sound every two hours, and to check the bottom, which varied between forty-nine and fifty-two fathoms, fine sand and mud.

1 December

The night was clear, and the wind moderate, becoming gentle the next morning and as the sea was quite smooth, I signalled to the *Atrevida* to approach us and heave-to at the same time as this corvette. I took advantage of this occasion to send Don Felipe Bauzá to the *Atrevida* to collate the many points to which bearings had been taken when measuring bases by masthead heights. This precaution, which we had already taken with respect to Península de San José, was all the more necessary, because the coast we had just surveyed was lacking in conspicuous features easily distinguishable from one another, and the data we were collecting were accumulating with our incessant surveying thanks to such favourable winds.

Bauzá returned at eight o'clock, and after hoisting the pinnace which had conveyed him, we steered SbyE under full sail. The wind was moderate, WSW and west, and the horizon was hazy. We expected to sight land at any time, as we were making some five to six knots, and soundings were already increasing to sixty or sixty-three fathoms. Just before noon, the *Atrevida* signalled land to the south and SSW, six to seven leagues distant, and the depth sixty-five fathoms. At that point our latitude was 46°33' and longitude 9°20'.

A strange mirage similar to that suffered by Byron[1] in these waters deluded us into thinking, soon after noon, that we could see land extending as far as WNW. We signalled this to the *Atrevida* whose officers were equally deceived by appearances in the haze and thought that it stretched as far as north. We then stood to the SW, with a moderate breeze from WNW, and only with a closer approach at three o'clock in the afternoon were we disabused of a delusion shared by almost all on both corvettes. On realizing our mistake, and with the real coast only three or four leagues distant, we

[1] According to Commodore John Byron in the *Dolphin* in 1764 '... what we had taken for land, vanished all at once, and to our great astonishment appeared to have been a fog-bank.': John Hawkesworth ed., *An Account of the Voyages undertaken by the Order of His Present Majesty for making Discoveries in the Southern Hemisphere*, 3 vols, London, 1773, I, p. 10.

bore up and when the mirage finally disappeared, we applied ourselves once more to our usual tasks. The coast was the same as that which the *Atrevida* had pointed out at noon and which Anson had named Cabo Blanco,[1] although it was actually Cabo Tres Puntas.[2] From the real Cabo Blanco, easily distinguished because of a small island lying just off it, which we could see, the coast stretches WNW towards Cabo Tres Puntas and then turns west to form the southern shore of Golfo de San Jorge.[3] A view of this stretch of coast inserted into the *Voyage* of Lord Anson agreed closely with what we could now see, and there was now no doubt that Commodore Byron had named Cabo Blanco that same point which the Admiral[4] had distinguished by this name.

At the said distance of three to four leagues from the coast, we obtained a sounding of fifty fathoms, small pebbles and mud, but quite soon when we steered SbyW to within about two leagues from the coast, it decreased to twenty-eight and twenty-three fathoms, stone. We coasted at this distance, and with the same depth of water for quite a while we managed not only a thorough survey of these shores, but were able to observe on two occasions the longitude at the meridian of Islote del Cabo Blanco. And finally we were able to align the two corvettes in the most convenient direction, to ensure that bearings taken from them would be as reliable as possible.

Having completed all these operations and observed azimuths for variation, as there was still just over an hour of daylight left, it seemed most opportune to make use of it by examining the shoal found by Commodore Byron, whose bearing of Cabo Blanco coincided with our track at that time.[5] We therefore signalled to the *Atrevida* to follow us, trimming our sails with extreme care to proceed along that same track, increasing the distance, which the commodore had indicated, to five leagues. We were sailing under the topsail, and without forgetting to sound. The extremely calm sea made us all the more apprehensive about the possible existence of any danger which would not be revealed by breakers on this occasion. It was not long before we went from twenty fathoms to sixteen, fourteen, and thirteen, small pebbles, from one cast to the next. A NE course was taking us away from the coast, so everything seemed to confirm the existence of the shoal in that area. It was too late, now, to send a boat to carry out a more thorough examination, and it would have been reckless to do so in the corvette, all the more so since, according to Byron, the danger was hidden and soundings did not indicate its proximity.[6] So, persuaded of its

[1] Anson did not in fact name this cape, merely referring to it as 'Cape *Blanco*': George Anson, *A Voyage round the World in the Years MDCCXL, I, II, III, IV.* ed. Glyndwr Williams, Oxford University Press, 1974, pp. 69-70. Its origin has not been determined , but 'C. Blanco' is charted on 'The Mapp of the Streights of Magellan' in *An Account of Several Late Voyages and Discoveries to the North and the South,* London 1694, f.p. 1.

[2] The southern entrance point of Golfo de San Jorge, with Cabo Blanco situated 8½ miles SE.

[3] ENE and E in the MS journal, which are clearly errors for WNW and W as can be seen from any chart or map.

[4] I.e. Anson.

[5] Off Cabo Blanco Byron '... perceived the sea to break right ahead of us; we immediately sounded, and shoaled our water from thirteen to seven fathoms ... so that we went over the end of a shoal, which a little farther to the northward might have been fatal to us. Cape Blanco at this time bore WSW½S distant 4 leagues': Hawkesworth, *Voyages,* I, p. 13.

[6] See n. 5 above; the shoal, now known as Banco Byron, with a depth of about one metre over it, is situated 5 miles east of Cabo Blanco, with breakers 1¼ miles east and 4 miles NE of it. '

existence, not without reason, we resumed our course to east and SE, and were soon in eighteen and twenty fathoms.

Following our routine, our run during the next night was to take us to the south of Cabo Blanco by dawn and at no great distance from it. The weather was still exceedingly pleasant, and there was now some hope that it might continue. We steered SE then SSE, and finally south for some eight leagues; the soundings went almost instantaneously from twenty to thirty, forty and forty-eight fathoms, stone, and soon afterwards to sixty and sixty-six fathoms, sand and ooze. Finally, as we closed the coast on a SW tack, the soundings decreased once more to fifty-five, fifty, and forty-five fathoms, sand and ooze.

2 December

We could not wish to be in a more pleasing situation than that in which we found ourselves at daybreak the next morning. The weather was wonderfully clear, the coast and Cabo Blanco itself were in sight, no more than three leagues distant, the sea pleasantly smooth, with many small whales which cut through the water calmly and majestically. Everything told us that the benign effects of spring were extending even into these desolate climes.

At five o'clock in the morning, with the light airs from the south, we steered still closer in to the land. We were soon measuring bases, and as we still had little wind and were scarcely a league from shore, we sailed close-hauled in the first quadrant, with a gentle SE breeze, which backed progressively to the east and after we bore off from the land for a while, gave us the chance to steer SSE clear of the coast. By six o'clock in the morning the soundings had shoaled from forty-three to thirty fathoms. For the rest of the time until noon the soundings were nineteen, twenty and twenty-one fathoms, small pebbles and small shells. At this time our latitude was 47°29' and longitude 59°0'W.[1]

In the afternoon the wind increased slightly and settled favourably from the north enabling us to keep between one and one-and-a-half leagues from the coast, while the soundings remained unchanged at twenty-two, twenty and eighteen fathoms, pebbles (in contrast to the experience of Commodore Byron who many times while sailing this same passage, and much further out, had shoaled to only seven to eight fathoms). At three o'clock in the afternoon a fresh sea breeze sprang up from east, enabling us to set a course for Puerto Deseado,[2] which would not be far off when Isla Reyes[3] was clearly visible. So we kept a keen lookout for a tower-shaped rock (the *steeple* of the English)[4] which although somewhat difficult to make out serves as a landmark for the entrance to the harbour. We sighted it at four o'clock. and the moment we saw it bearing nearly due west of us we bore down on the coast.

---

[1] Presumably from Cádiz.

[2] First visited by Magellan in in 1520 but named Port Desire by Cavendish in 1586.

[3] Isla Pingüino, 10 miles SSE of the entrance to Puerto Deseado.

[4] El Torreón; as he approached Puerto Deseado Byron saw 'In the evening ... a remarkable rock, rising from the water like a steeple, on the south side of the entrance to Port Desire; this rock is an excellent mark to know the harbour, which it would otherwise be difficult to find.': Hawkesworth, *Voyages*, I, p. 14.

Several sources of information led me to believe that the ebb would turn at around seven o'clock in the evening; but since a survey sheet of Piloto Tafor, which had come into my possession, gave a very different time for the establishment of this harbour,[1] though I later learnt that it had been incorrectly copied, I thought it prudent to drop anchor at the mouth of the harbour and keep an eye on the tide before effecting our entrance the following day. I finally anchored at seven o'clock not far from the southern end of the harbour mouth, in seven fathoms, gravel, with the tide still ebbing at a rate of one knot. The *Atrevida* anchored close to us soon afterwards. It as not long before we saw a launch under oars and sail, coming out of the harbour, and we immediately recognized it as belonging to the brig *Carmen* with its captain, Segundo Piloto Don José de la Peña. He came on board the *Descubierta* at once and, advising us that it was slack water and thus a very favourable moment to enter harbour, he suggested that we should lose no time in doing so, as well as offering to act as our harbour pilot, even though only a few minutes remained before dusk.

The tidal stream in this harbour, especially at its mouth, runs at an almost unimaginable rate, and if one adds to this the many drying reefs and the restricted room available for anchoring, one must certainly consider it to be one of the most difficult harbours to enter.[2] However, in such latitudes fair weather was a gift that could not be relied upon for long. We therefore did not delay for a moment in setting sail, signalling the *Atrevida* which quickly prepared to follow, but was eventually unable to do so, her cable having run out as the anchor was being weighed.

It was nearly nine o'clock in the evening before the *Descubierta* got under way with the fore topsail, mizzen topsail and jibs set. The wind continued moderate from ENE, the sea was smooth and the flood stream was increasing in strength. First we sailed close-hauled to the north to approach the coast and cross the entrance to avoid the reefs projecting from the harbour's southern point. We then bore down on the small rocks in the fairway[3] and on sighting them, left them on our starboard side, soon reaching a suitable place to anchor. At nine-thirty we let go the starboard anchor in six fathoms, paying out the cable to counter the tide pulling us astern, which by then was flowing strongly. Later, little by little, we paid it out in harmony with the tide to drop the western anchor and finally at two o'clock in the morning with the ebb already appreciable we hauled in the bower cable, which gave us a good mooring. So having made use of the tide for the different tasks involved in mooring, by daybreak the corvette was finally secured, riding to two anchors.

Stay in Puerto Deseado

After the incident of the run-away cable, the *Atrevida* had to give up the idea of entering harbour that night since she was unable to follow close behind us. Don José de la Peña, with his usual energy, went out to her that same night to pilot her in, arriving on board at daybreak. [3rd] However, since a fairly strong wind that blew all

---

[1] The time of high water 'full and change' i.e. the time of high water on days of the full or new moon.

[2] In the entrance to Puerto Deseado tidal streams attain a rate of five to six knots at springs and three to four knots at neaps, which would make it difficult for a vessel under sail to enter harbour.

[3] Rocas Dos Hermanas, two large above-water rocks lying near the edge of the coastal reef.

morning from SW, which made her drag her anchor considerably, she was not able to get under way until the afternoon. Even then for a long time she made no headway against the tide until it turned in her favour at eight oclock that evening when finally she was able to enter harbour and anchor close to us.

The brig *Carmen* was riding at anchor about a mile further up the harbour so as to be less affected by the strength of the tide. Her captain informed me that on the night that we became separated in Río de la Plata he had borne down on the coast of Samborombón,[1] and taken shelter there for four days. Then, once the weather had improved, he set a direct course to Cabo Blanco; and, having read the sealed rendezvous instructions which gave this cape as the first point of call, he had preferred, rather than exposing his frail vessel to any further risks, to enter this harbour and wait for us here, which he had done two days earlier. He also informed me that on the very afternoon of our arrival at the mouth of the harbour, he had been visited on board ship by the *cacique*[2] and some other people, most of whom he knew, from a small tribe of Patagonians who were wandering in the area at the time. In fact many of them, men as well as women, had been here at the time of our unsuccessful settlement in this harbour and had lived in the greatest harmony with our people, and they still knew a number of our words and customs.[3] Peña had also taken the precaution of sending two of his crew by land along the shore adjacent to Isla Reyes to see if there were signs of any new foreign settlement in that area since his last visit in the launch from the corvette *Santa Elena*.[4] They both came back the same day and agreed that they were sure that all was in the same state as before, except for a beached canoe on Isla Reyes which appeared to have been left there recently.[5]

It seemed likely that the settled weather would continue and from our position in the anchorage we could see a promising horizon to the east, precisely where the Sun's vertical altitudes could best be sighted to make the most use of its absolute altitudes, measured with the sextant, to check the rates of the chronometers and also to obtain the longitude of Puerto Deseado.[6] This method, which is sufficiently accurate in

---

[1] Between Punta Piedras and Cabo San Antonio on the southern side of Río de la Plata.

[2] Headman of a South American tribe.

[3] There had been a Spanish establishment at this point on the coast from 1778 to 1784. The facts that the inhabitants were accustomed to deal with Spaniards and knew some Spanish words, are understandable against this background. Malaspina's reports home argued strongly for the abandonment of Spanish efforts to control the coast by garrisoning strong points: the better course, he believed, would be to rely on the friendship of native communities and to adopt an attitude of indifference to the activities of representatives of foreign powers. See his 'Reflexiones políticas sobre los dominios de S.M. desde Buenos Aires hasta Chiloé por el cabo de Hornos', in Pimentel, *Descripciones y reflexiones políticas*, pp. 55–74.

[4] Peña was presumably looking for British whalers. In April the *Sappho* and the *Elizabeth and Margaret*, which had been refreshing their crews and carrying out essential repairs in Puerto Deseado had been ordered to leave by Spanish frigates: David Mackay, *In the Wake of Cook*, London, 1985, p. 41

[5] This paragraph crossed out in the MS journal.

[6] A less accurate method of rating chronometers than by equal altitudes. First apparent time is obtained either on shore or from a ship at anchor by observing the altitude of the Sun by sextant when it is some distance from the meridian. Then knowing the ship's latitude and the Sun's declination it is possible to calculate the Sun's hour angle at the time of the observation as noted by chronometer. If the altitude is observed after noon the hour angle expressed in time is the local time of the observation. If it is observed before noon it has to be subtracted from 24 hours. Then by applying the equation of time the local mean time of the observation can be obtained; the difference between it and the time by

itself, obviated the need to have an observation spot on shore, thus avoiding two serious inconveniences: having our people mingle with the native Patagonians, and the risk in having boats plying back and forth in waters beset by such strong and persistent currents.

So that same morning, altitudes were observed on board, although we later preferred to use those observed the following day for deducing the longitude and for adjusting the rates of our chronometers. The main yards and topgallant masts were duly struck, and all the boats were lowered. That morning, Don Juan Vernacci was commissioned to undertake a study of the tides, which he saw as of the greatest importance in this port.

With the turn of the tide, and abandoning in consequence the hope that the *Atrevida* might anchor before nightfall, I decided to put the day to use by examining the harbour, hoping at the same time to foster friendly relations, if possible, with the Patagonians. Don Antonio Pineda and Don Cayetano Valdés accompanied me in the pinnace. I also took two armed marines so that they could hunt as well as being responsible for our safety and for the same reason we ourselves were also armed. We also took as gifts some trinkets which might prove pleasing to the Patagonians.[1]

They had not yet appeared on the opposite shore when we arrived on board the brig. We hunted along the southern shores with only moderate success and then returned to the brig about noon, where we were informed that one or two natives had shown themselves in the vicinity a short time earlier. We went immediately in that direction and a Patagonian on horseback did indeed appear on a hillock, not far off, seemingly watching our movements. Our gestures of friendship, and the way I walked towards him without my gun, persuaded him to wait for me and eventually accept some trinkets. As the rest of the band, who were close by hidden behind a knoll, saw this they gradually all walked their horses forward, and dismounted. They then sent for the women who soon arrived and also dismounted. The whole tribe, at that time, consisted of some forty people, including ten women and twelve children,

---

chronometer gives the chronometer error on local mean time. If the same observations are repeated at the same place or anchorage several days later a further error of the chronometer on local mean time can be obtained. The difference between the two errors divided by the number of days between gives the daily rate of the chronometer. For an example of the working of observations to rate the chronometer by this method see Charles F. A. Shadwell, *Notes on the Management of Chronometers and the Measurement of Meridian Distances,* London, 1861, pp. 49-56 and R. C. Mayne *Practical Notes on Marine Surveying and Nautical Astronomy*, London, 1874, pp. 23-8.

[1] Malaspina had been looking forward to this encounter ; before setting out, he wrote to the distinguished scientist, Antonio de Ulloa, 'On the manners of the Patagonians, ... which are so robust and sociable ... we shall omit ... no effort for which chance, our encounters and our reach may give occasion. ... Penetration of those regions, and of the Patagonian lands in particular ... is an object of much relevance for the history of the propagation of the human race ...': Novo y Colson, *Viaje político-científico,* p. 7; F. del Pino Díaz, 'Los estudios etnográficos y etnológicos de la expedición Malaspina', *Revista de Indias,* 1982, p. 417. The account of the parley which follows should be understood in the context of debate about the juridical status of nomadic peoples, who, according to conventional orthodoxy, could have no rights of landed possession or, a fortiori, territorial sovereignty: E. Vattle, *Le Droit des gens ou Principes de la loi naturelle appliqués à la conduite et aux affaires des nations et des souverains,* Paris, 1758. Malaspina, however, reported that 'all the evidence tends to confirm that [the Patagonians] are free possessors of a country which is immense in relation to their numbers.' Pimentel, *La física,* p. 205.

Plate 14. Encounter with the Patagonians, 1789, by José del Pozo. Museo Naval, Madrid

three or four of them still at breast. Only two of the women were elderly, but exceedingly nimble despite this. Among the males, the *cacique* and one other were elderly. There were also five who seemed nearer the age of puberty than full manhood. In general they were all (including the women and children) very large and solidly built. Their height was not in proportion to their build but they were tall: the *cacique* Junchar who was carefully measured by Don Antonio Pineda and found to be six Burgos feet and ten inches in height, and almost twenty-three inches broad from shoulder to shoulder.[1]

At length we were all seated in a circle and as any distrust on either side had subsided, there began to manifest itself that desire, innate in mankind, to wish to know one's fellow man more closely. In this scene, composed of course more of people than of words, it was not long before the Patagonian women appropriated the principal role. Whether from curiosity or from a greater propensity for conversation, they soon took it upon themselves to answer our questions. More than one of the women made an effort to use Spanish words and, displaying the facility for language which travellers have long admired, they brought a certain charm to this side of the conversation, so that despite their dress and their rustic customs, this major characteristic of their sex began to shine through. We presented them with various glass ornaments, some ribbons and necklaces. In return they gave us a guanaco[2] fur and a guanaco bezoar[3] as well as a very small live guanaco, which could well have been the model for a stylish painting Commodore Byron had made of a similar animal.[4]

Our own questions were mainly directed to their language and customs.[5] Don Antonio Pineda and I agreed that we should work separately as far as the language was concerned; so that having made a small collection of words in one session we should try to go over them all at the next session, before learning any others. Finally, since it would be highly misleading to learn about the customs before we had the least idea of the language, we also agreed to leave that, for the most part, until subsequent visits, when Piloto Peña would be able to come with us. This we did, and as I have already mentioned, two women were particularly useful to us as they knew quite a number of Spanish words and had met Pilotos Tafor and Peña.

At the very outset, the natives asked that the two armed marines should withdraw.

---

[1] Height six Burgos *pies* and ten *pulgados* and breadth twenty-two *pulgados* and ten *lineas* in the MS journal. (one Burgos foot = approximately 11 inches). The supposedly 'gigantic' height of Patagonians had been the focus of European enquiry in the region ever since the Magellan expedition had transmitted the first reports. The quest for monstrous deformations of humanity, which had inspired early explorers nourished in the medieval tradition of mirabilia, was gratified by discoveries of peoples whose physiques appeared to be outside the normal range. See M. Hodgen, *Early Anthropology in the Sixteenth and Seventeenth Centuries,* Philadelphia, 1969.

[2] The journal of Francisco Javier de Viana, *Diario del viaje explorador de las corbetas Descubierta y Atrevida en los años 1789 a 1794,* Montevideo, 1958, p. 49, concentrates, in its account of the ecology of Patagonia, on the relationship between the native hunters and this quadruped, defending the nomadic way of life against charges of barbarism on grounds of its efficiency.

[3] A concretion with hard nucleus found in the stomach or intestines of certain animals; formerly much prized in pharmacology and as a protective amulet.

[4] There is no illustration of a guanaco in Hawkesworth, *Voyages.*

[5] Malaspina's and Pineda's Patagonian vocabulary is printed in María Dolores Higueras Rodríguez and Juan Pimentel Igea, eds, *La expedición Malaspina, 1789–94,* Tomo V: *Antropología y noticias etnográficas,* Madrid, 1993, pp. 43-5.

We immediately complied with their wishes, and resulting confidence seemed to reassure us both so much that when Valdés asked whether they would object if he shot a nearby bird, the men immediately agreed to him firing his gun. He did not do so, however, because the women and children looked as if they were afraid, or might even leave.

It was already half past two in the afternoon when we thought about separating, not without the expression on both sides of the warmest friendship, and the hope on our part that they would come on board the following day at our invitation, having promised them many gifts. They all rode away on horseback, and as the tide was in our favour we lost no time in returning to our ship, keeping a good look-out for any of the many birds that came within range which were likely to provide new specimens for our natural history collection.

Now that the *Atrevida* had anchored, our only concern was that every moment should be made use of from the following day onwards, so as to keep this visit both as short as possible and of the greatest usefulness. That night we agreed with Piloto Peña on a signal which would let us know when the natives were in sight, so we could meet them and so save the time which otherwise we would have spent in vain searching for them.

It would have been a serious mistake not to engage in fishing and hunting while in this port. Accordingly, we arranged each day that a boat from one of the two ships would go fishing, under the direction of the *guardiamarina*, and very soon the game, which was so abundant that even the most inexpert hunters would wax enthusiastic, provided daily sport for our officers in their leisure time. Our table was laden day after day with nothing but the fruit of our excursions, and our crew were able to enjoy in the seafood, the fish and the many tasty seabirds, a plentiful and varied supply of delicious food.[1]

We did not manage another session with the Patagonians until the afternoon of the 8th, when after midday, the brig signalled that the Patagonians were in sight. On hearing this, Pineda and I lost no time in going to meet them and the artist Don José del Pozo went with us to draw their portraits.[2] When we reached the brig, we found

---

[1] The generosity of nature in Patagonia was the key to Malaspina's search for what he called 'the philosophical truth of the failure of our highly expensive and unfortunate attempts to settle the Patagonian coast.': Higueras and Pimentel, *Antropología y noticias etnográficas*, p. 29. As he argued in a memoir titled 'Suelo de las costas de la tierra patagónica e islas Malvinas, algunas noticias de los patagones y demás habitantes de la costa hasta Chiloé', the prodigious benevolence of the sea was nature's 'recompense' for her hostility to agriculture: colonization had to be focused on the river-banks, where alone farming and therefore civilization would be possible. Malaspina was an enthusiastic proponent of recolonization, which he proposed to achieve not by coastal settlements, which he, like other visitors to the region, deemed useless, but by means of grants of agricultural land in the vicinity of Maldonado. He imagined a society of 'solid moral principles' in place of the 'pursuit of gain'. He was particularly anxious to avoid a recurrence of the rancher-culture of the pampa, where, he opined, 'the custom of bathing in blood of cattle-slaughtering' 'makes men forget every principle of religion and society.': Pimentel, *La física*, pp. 194-5.

[2] Pozo immortalized the scene of the conference with the Patagonians on 9 December in a form which continued a tradition already well established in European efforts to depict these peoples. They were idealized first by being shown with physiques and in poses modelled on classical statuary and contemporary European genre painting; the children resemble putti; the posture and even dress of some of the women suggest scenes of peasant life painted by Goya. Secondly, the level of material civilization is

that the *cacique* had been on board waiting for us for some time along with some members of his family and three or four women, with their unweaned children. The latter were all asleep, tucked up in the deck-house. The rest (so Peña told us) had been feasting on ship's biscuits and both raw and cooked vegetables, often at the same time smoking cigars and drinking wine. We watched while they smoked and drank once again, realizing that the use of brandy was not new to them and that they continued to enjoy it. They told us that the reason they had not returned immediately was that their herd of horses had become scattered and the young men had been sent to find them but had not succeeded until the third day after they had set out.

We had brought with us such trinkets as were likely to please them: some scissors and small knives were given to all, a large knife and a mirror was given specifically to the *cacique* and there were ornaments for the women. These so re-established our mutual friendship that it was easy after that to paint their portraits and a long and interesting conversation began. We confirmed the words learnt during the first session and many new words were added to our stock. It was possible then to acquire a clear idea of some of their customs, especially concerning their ties of kinship and their love for their parents and children. Finally, with Peña's help, we managed to gain some idea of their religion which led us, little by little, to talk about their present quarters, about three leagues distant from the beach. We proposed to go there the next day, if they could provide us with some horses.

A Patagonian girl aged about fourteen, whose good looks, great charm and exceptional loquacity had made us choose to portray her rather the other women, was drawn to our attention even more when the time came for them to go ashore. Peña had given them some ship's biscuits and vegetables to take back to their camp with them in such quantities that they had to put a portion of what they could manage to carry in their restricted clothing. But there was still some left and the young woman, who had asked for them to take to her parents, had nowhere to put them. The others advised her to take off her poncho[1] and use that as a bag. This poncho, which formed part of her dress, was worn under a guanaco skin which, together with other skins, covered her completely. To take off the poncho, then, she had to show her bare back. This prospect made her hesitate for some moments, and she only finally removed the poncho after the others persuaded her. This she then did, with such dexterity as to give a new lustre to her modesty.

---

flattered in a way reminiscent of John White's drawings of natives of Virginia at the time of the first attempted English colonization: one woman wields a pair of scissors, evoking the armillary sphere clutched by a child in one of White's works; others are well dressed and deliberately coiffured and one has a musket. Similar conventions had been applied in pictures of other 'noble' or supposedly civilizable 'savages' – especially North American Indians, South Sea Islanders and Easter Islanders – almost throughout the century, despite the signs of disillusionment which began, in the last quarter of the century, to be detectable in written accounts: P. J. Marshall and G. Williams, *The Great Map of Mankind: Perceptions of New Worlds in the Age of Enlightenment*, Cambridge, Mass., 1982, pp. 202-3 and F. Fernández-Armesto, *Millennium*, New York, 1995, pp. 481-3.

[1] The use of this garment of Peruvian origin is generally considered a sign of Araucanian influence – even perhaps imperialism – in late eighteenth-century Patagonia and of social changes as a result of trade and cultural contacts across the Andes.

9 December

The following day, the Patagonians arrived punctually, and told Peña that they had brought a horse so that I could go to their camp. On this occasion the officers of both corvettes accompanied me to the beach where the natives were. But whether it was our greater number, or my decision not to go to their camp, or (as Peña conjectured) their lingering suspicion that we might be Englishmen,[1] because of our dress and the fact that several of us had fair hair, we were mortified to see their strong reluctance to come on board. We made them a present of a very large amount of victuals, and when they left us at five o'clock, young Jujana's modesty was once again evident. She took advantage of a moment when no one was looking to mount her horse and seeing that she could not do so, even though her parents who had stayed behind were trying to help her, she decided to get on the horse in an unsafe and uncomfortable manner rather than be guilty of any immodest movement. Then, at two gun shots distance from us, she slipped off the horse to mount again with all the freedom so vital to these people.

This was to be our last meeting with the Patagonians, of whose customs and language, in so far as we could ascertain them, we shall, in the next chapter, provide a description, if not as detailed as we would like, at least as true and dispassionate as Europe would demand.[2]

Our desire to study the Patagonians closely did not cause us to overlook, in this location more than anywhere else, the need to attend continuously to many aspects of our present mission.[3] Since the morning of the 4th, Tenientes de Fragata Quintano and Salamanca had been detailed to explore the watering places along the coast immediately to the south, where Commodore Byron had replenished his stocks of water.[4] They found the water so scarce and so far from the beach that even taking on a moderate amount seemed to them an arduous task. They preferred to take on water at the village of Antigua.[5] While this was three leagues from the corvettes, it seemed more convenient since the launches could go directly there on the flood tide and return on the ebb, regardless of the prevailing wind. However, there were a number of facts that had to be considered: at slack water a launch would not be able to enter or leave the shallow inlet. Indeed, the actual scarcity of water would not allow more than about twelve casks to be filled between the two launches in twenty-four hours. The passage was fraught with danger because of eddies and whirlpools caused at

---

[1] British interloping in Patagonia had been feared at least since 1770, when Francisco Millán y Maraval had published a scaremongering work, *Descripción del Río de la Plata*. In 1774, the Mancunian Jesuit and former slaver, Thomas Falkner, published a call to Britain to occupy the region on the grounds that it was of great strategic importance and neglected by Spain: Thomas Falkner, *A Description of Patagonia and the Adjoining Parts of South America*, London, 1774. Malaspina claimed to find empirical evidence for his claim that Patagonia was immune from foreign interference in the reports of Don José de la Peña, which he received when he got to Valparaíso. According to Peña, the only Englishman on the coast was one young man 'apparently a fugitive' living with a Patagonian tribe; only one foreign naval vessel had been seen, although English, French and American-loyalist whalers were numerous.

[2] There is no such description in the next chapter, but possibly printed as 'Suelo de las costas de la tierra patagónica', in Higueras and Pimentel, *Antropología y noticias etnográficas*, pp. 23-32 and 43-5.

[3] This sentence crossed out in the MS journal.

[4] Hawkesworth, *Voyages*, I, p. 20.

[5] Presumably the site of the abandoned Spanish settlement.

times by the tide itself at the different bends and at other times by wind against tide. For safety therefore two launches would always go together and would never be overloaded. They would be accompanied by the yawl carrying an officer from either corvette whose sole task would be to direct operations and to assist whenever this might be necessary. With these precautions in force, the launches had on the 6th already made two successful journeys, under the intelligent direction of Tenientes de Navío Valdés and Tova, when a new development that very afternoon, diverted our boats elsewhere. As a result of a strong gust from NNW, combined with the ebb tide, the *Atrevida's* western cable parted, and even though their reaction was extremely fast and appropriate, they were unable to avoid their sheet anchor fouling our western cable or [the *Atrevida*] striking our port side, although without causing any major damage. We were very fortunate when weighing our anchor and recovering our entangled cable to do so before it chafed or parted. Then, by means of hawsers, the *Atrevida's* cable was returned to her without unbending it or dropping it to the bottom. This was done without using the launches, which were away at that time or out of commission. Finally at nine o'clock, when the tide turned, the other corvette was able to moor again, although far too close to some rocks. Since it was not possible to use buoys or buoy-ropes in a place with such strong tides, the *Atrevida's* anchor was not provided with this useful means of recovery. It was therefore necessary to use two launches to drag for it, a difficult manoeuvre, in view of the short period of slack water, which was the only time that a diver could go down to pass the buoy-rope. Our efforts failed on the morning of the 8th, but on the 9th we were lucky enough to find the anchor. A diver from the *Atrevida* passed the buoy-rope to it, and it was weighed that afternoon and rehoused safely on board.

For his survey of the harbour, which extended on the seaward side as far as Isla Reyes, Don Felipe Bauzá had displayed his usual energy and ability. On the morning of the seventh, having measured a base on the southern shore, he proceeded to various points along the same coast where useful bearings could be taken to link all the essential features of this harbour; he then crossed over by boat to the northern entrance point of the harbour; finally, either with transits or additional bearings taken with the theodolite, he had gradually extended his triangles inshore, proceeding eventually on the 10th and 11th with Teniente de Fragata Quintano and Piloto Peña as far up the inner harbour as the yawl would take them. They took bearings from one side of the harbour or the other, choosing vantage points on higher ground where they could see the many islands with which the channel is strewn. They also took soundings, guided by the excellent information that Piloto Peña had provided. Thus, with this trip another aspect of our visit, perhaps the most essential, had been concluded.

Our chronometers were compared by the method already mentioned of absolute altitudes, by which, on the morning of the 4th, we deduced the longitude of the anchorage to be 9°50′ by number 61, although when this was corrected by some of the officers, taking into account the [mean][1] rate compared with that determined in Montevideo between that ascertained here, the longitude was rather 9°44′. On this

[1] Illegible in the MS journal.

occasion all three chronometers on board the *Descubierta* showed quite different rates from those determined in Montevideo. Number 61 in particular varied from 53·78″ to 1′1·43″ fast, causing us to be very suspicious of its results, although it agreed daily with *Atrevida*'s number 10.

Given this ambiguity, it seemed prudent to seek clarification from astronomical observations although my first response to this idea had been negative because it entailed the setting up of a station on shore, with repeated trips by boat day and night, which we considered really dangerous. On the morning of the 5th we set up the observation tent on the southern shore opposite the corvettes. The large astronomical quadrant and chronometer 72 from this corvette were taken there; Don Dionisio Galiano set to work deducing the longitude from the meridian passage of the Moon compared to a very reliable star, since it was not possible to observe the satellites of Jupiter at that time, nor any occultations of the stars by the Moon.

To ensure the safety of the observatory as well as good order among the crew sent ashore each day, a small hut was erected near by, constantly manned by a corporal and two marines. A signal mast was erected to maintain communication with the corvettes day and night and one of the two *guardiamarinas* was in charge of the station from dusk to dawn.

As the peaceful nature of the Patagonians had by now been established and as they kept to the northern side of the harbour, there seemed to be little danger in setting up the station on the southern shore, particularly as everyone going ashore either for leisure or to hunt always went to that side.

Don Dionisio Galiano's efforts to take lunar observations on the first nights were fruitless, but he did succeed in determining the latitude as 47°45′33″ from meridian altitudes of different stars to the north and south. Variation by theodolite was 19°50′NE. On the morning of the 9th the officers of the two corvettes observed nine sets of lunar distances, arriving at a value of 9°22′ west of Montevideo, a longitude that was somewhat further to the east than that determined by the chronometers.

Every passing day in this harbour convinced me more and more of the possible consequences to our anchors or our boats, or even to the voyage itself, of remaining in this anchorage for much longer. Water was scarce and poor, there was no wood, and the fact that we needed the right combination of wind, light and tide to leave harbour meant that we could not set sail any day that we wanted. Bearing this in mind, I ruled out the idea of waiting for an opportune moment to make further observations of the meridian passage of the Moon. All instruments and observation equipment were brought back on board on the morning of the 9th, and we thought only of our readiness for our next undertaking. I left completion of our water to Puerto Egmont, where further astronomical observations would throw fresh light on the true state of the chronometers and the astronomical position of that harbour.

Pineda and Neé were very pleased with their work. From the moment we entered the harbour they had taken every opportunity to increase their respective scientific collections. The former, who was particularly addicted to examining stones, shells, quadrupeds and birds, had gathered such large number of specimens that he had more than enough to study over our next, somewhat protracted passage round Cabo de Hornos: Don Luis Neé, with his customary discernment, application and perseverence managed to collect many especially rare and interesting plants despite the arid appearance of the environment.[1]

This harbour also provided material for the elegant work of artist Don José del Pozo. The contours of the landscape provided attractive studies of perspective which were easily committed to paper by means of the *camera obscura*, and the Patagonians were worthy subjects for his paint brush. Furthermore, he also made sketches from life of the many species of waterfowl, in greater numbers and beauty in this port than in any other harbour along the coast of Patagonia.

The ability of our hunters could not be judged by the luck they had here. Guardiamarina Ali Ponzoni surprised a guanaco close to the watering place and brought it down with a single bullet. When taken on board it was found to weigh 195 pounds. One of the excellent hares to be had in this area weighed 18½ pounds. Several foxes were taken, but their stench soon obliged us to get rid of them. And on a trip to the inshore islands, Teniente de Navío Valdés came across fifteen sizeable guanacos isolated by the tide on one of them. It would have been easy to cut off their retreat, had those that came behind him been skilful enough to corner them, but the animals scrambled and half swam to the mainland.

As we were about to sail, and knowing from recent experience how much the brig's speed differed from our own, I decided that our next tasks should be carried out separately. In that way, we should not only save much time, but our passage would be more productive and our very safety would also be put on a more certain footing. I therefore gave instructions to the captain of the brig *Carmen* to accompany us until in sight of Puerto San Julián, from where the corvettes would sail across to Puerto Egmont in the Malvinas, while he was to continue along the coast and enter Río Santa Cruz and Río Gallegos. From the latter harbour, he was to set a course for Puerto del Año Nuevo on Isla de los Estados[2] if conditions made it prudent for him to do so. There he would either rendezvous with us, or would learn, by means of bottles, what news had to be communicated to his excellency the Viceroy in Buenos Aires.[3] I gave the captain the few necessities that he asked for, and advised him to be ready to sail on the morning of the 13th.

10 December

Thus, on the night of the 10th everything had been brought on board, with the exception of Bauzá and Quintano who had not yet returned from their trip to the inner harbour in the launch belonging to the brig *Carmen*.

---

[1] On his return to Spain Neé prepared a list giving the number of plants collected at various places visited during the voyage. At Puerto Deseado and Port Egmont in West Falkland, which he did not differentiate between, he collected 235 plants, but as he listed various types of grasses, mosses, algae and fungi separately, without giving the localities where they were collected, the number of plants collected at these two places is probably nearer 300: 'Plantas colectedas en la expedición alrededor del mundo', AMN, MS 1407. See also Engstrand, *Spanish Scientists*, p. 106.

[2] Cook's New Year Harbour on the north coast of Isla de los Estados [Staten Island].

[3] It appears that Malaspina intended to leave a message for Peña in a bottle at Isla de los Estados if he called there.

CHAPTER 2

From Puerto Deseado to Puerto Egmont

11 December

Quintano and Bauzá did not get back until midnight. They had been up river since early morning with the captain of the brig *Carmen* and Don Antonio Pineda, the former to take bearings with the theodolite and to complete the triangulation and the latter to explore the country and its inhabitants of whatever kind they might be. The north wind which had blown violently all afternoon and evening had persuaded us to hoist all boats. [12th] The same wind continued blowing quite strongly next day, when it veered to NE, making it impractical for us to unmoor. We spent that day rating our chronometers and drawing a chart of the coast that we had surveyed between Río Negro and Puerto Deseado to give to the captain of the brig *Carmen* and thus ensure its prompt and safe delivery to the Ministro de Marina in case we got separated from that vessel once again. I included a brief explanation with it, stating that time did not allow all the data compiled on board both the corvettes to be included.

At the same time I wrote to the Viceroy in Buenos Aires, reporting our encounter with the English vessel on the parallel of Cabo San Antonio and stating that we had not found the slightest trace here of new foreign establishments for fishing purposes since the action carried out by His Majesty's ship *Santa Elena* on Isla Reyes, about which Piloto Peña had thoroughly informed me.[1]

The chart, letters and a small box in which Don Antonio Pineda was sending twenty-two stuffed birds, almost all of them unknown species, could not be handed over until dawn on the 13th, at which time I had already given new instructions to the commanding officer of the *Atrevida* and the captain of the brig concerning our immediate destination.

13 December

Soon after sunrise, there were signs of a moderate wind from the south;[2] the tide was due to begin to ebb at about ten o'clock in the morning and since the *Atrevida* could not remain on her western anchor because of her closeness to the rocks, both corvettes took advantage of the rising tide to weigh this anchor. At ten our other

---

[1] The fishery that Malaspina was referring to was whaling: see p. 84, n. 4 above.

[2] To conform to AMN, MS 753, since AMN, MS 610 gives 'viento bonancible del S y NO' which is clearly erroneous.

anchor was up and down and we were ready to sail. The brig, which was already under way according to a previously agreed signal, now found it impossible to round some small islands between her and us, and immediately dropped anchor once more. A gentle breeze, which had now veered to the east and ESE, obliged this corvette to do likewise, having scarcely weighed the anchor a couple of fathoms. We anchored again with thirty fathoms of cable and the *Atrevida* did the same and like ourselves, spent the rest of the day with a gentle breeze from east and NE. The tide was running at three to four knots and the weather was very misty, becoming somewhat squally with distant thunder in the afternoon.

Some brief passing showers during the night convinced me that there would be a breath of land breeze at daybreak and since the ebb tide would cease at four-thirty in the morning, this seemed to me the most opportune time to accomplish our departure from these narrows, particularly since this should be avoided once the sea breeze set in at eleven o'clock, when it would be slack water again.

14 December

Westerly light airs set in at four o'clock, when both corvettes set sail with great haste, and by the turn of the tide we had passed the two small rocks in the middle of the fairway[1] with the *Atrevida* in close company. Neither the wind nor the tide were by now favourable for the brig to attempt a departure. It was not long before light winds from south and SSE sprang up making it impossible at times to hold our own against the tide; on two occasions the *Atrevida* was almost forced to anchor, as we were both being carried rapidly towards the northern shore.

At eight o'clock a moderate breeze set in from SSE from the second quadrant and we considered both corvettes to be clear of the coast now one-and-a-half leagues distant. Soon after, the weather turned misty and the wind dropped considerably. We were under easy sail to allow the *Atrevida* to catch us up. At eleven o'clock, with a moderate breeze blowing once more, and anticipating that by noon we should be able to take useful bearings of the nearby coast and also obtain satisfactory observations for latitude and longitude, we altered course to SW5°S. At a quarter past one in the afternoon our latitude was 47°37′50″, which, when referred to dead reckoning, placed the most easterly of Islas Reyes or Penguines bearing S20°E from us, the northern point of the harbour entrance S23°W and the most salient land to the north, N25°W, all by compass. In the afternoon variation, by means of several azimuths, was 20°4′NE and longitude by the two chronometers was 9°38′, somewhat different from that taken from the chart, probably because of the unreliability of our horizon.

The bearings indicated that we were falling off considerably to the north. We therefore thought it prudent to alter course again to the east and we hauled the wind to this tack until five o'clock in the evening, when the wind dropped completely, with the sea remaining calm. At that time the weather was clear and mild. At sunset Isla Reyes bore S20°W by compass, distant five to six leagues. Soon afterwards as light airs had set in from the north and almost immediately became a moderate breeze

---

[1] Rocas Dos Hermanas.

from NW, we took advantage of them under all sails, steering first SE and then SSE by compass.

My first idea was to fix the position of the entrance to Puerto San Julián at once with good observations and with suitable views to enable our national shipping to navigate with greater safety in these waters. But against this idea there were a number of considerations: the more pressing need to establish the true position in longitude of the Malvinas, the work already carried out along this part of the coast by Churruca and Cevallos,[1] the very frequency with which our ships called at this harbour, either seeking to set up establishments or fisheries or for the acquisition of salt, and finally our need to proceed as soon as possible to Tierra del Fuego, in accordance with the original intentions of His Majesty.

At dawn the weather was extremely fair and with the wind now blowing from the west we were making five to six knots, with soundings of fifty fathoms, black sand. We had lost sight of land, and having decided to make for the Malvinas (and perhaps anchor in Puerto Egmont) we steered SSE, at liberty to use any wind that would take us to the south. By two series of azimuths taken during the morning variation was 23½° and 22½°NE. At noon our latitude was 48°49' and longitude by both chronometers was 9°6' west of Montevideo and soundings were sixty to sixty-five fathoms, sand and hard mud. By this time the wind had stengthened from NW and NNW, moderate, which I immediately decided to make use of under all sails, signalling the *Atrevida* to stop taking soundings. At four o'clock it seemed advisable to set a direct course for Los Salvajes[2] proceeding SE true under a press of sail, anticipating being in sight of them by the following afternoon if the wind, sea, and skies continued in our favour, giving us a speed of some seven to eight knots.

In determining our landfall in the vicinity of Puerto Egmont, we thought it advisable to refer the longitude by our chronometers to the chart drawn by Tenientes de Navío Galiano and Belmonte,[3] whose meridian difference between Isla Rasa[4] and Cabo Blanco (adjusting its longitude according to our observations) was 3°25', almost identical with the reasonably accurate charts in use at present for navigating between Río de la Plata and the Malvinas. Various errors, which I shall refer to in more detail below, made the chart from Captain Cook's second voyage untrustworthy.[5] Variation by western amplitudes was 23°10'. Around ten o'clock in the evening the wind began to shift round to the west.

16 December

At dawn there was a fresh breeze from SSW. We were close-hauled under the

---

[1] Tenientes de Navío Cosme Damián Churruca and Ciriaco Cevallos y de Bustillo surveyed this coast in 1778-9 in the snows *Santa Eulalia* and *Santa Casilda*. Cevallos joined the *Atrevida* in Acapulco in April 1791.

[2] Steeple Jason and Grand Jason.

[3] Galiano and Alejandro Belmonte, who had previously worked under Vicente Tofiño during his hydrographic survey of the coasts of Spain, served together on Antonio de Córdoba y Lazo's surveying voyage to Estrecho de Magallanes in 1785-6 in the frigate *Santa María de la Cabeza*.

[4] Jason West Cay

[5] '*A Chart of the Southern Extremity of America*', James Cook, *A Voyage towards the South Pole and round the World*. 2 vols, London 1777, II, f.p. 198.

principal sails and staysails and only occasionally with the main topgallant sail because the heavy swell made this sail somewhat vulnerable during rolling. As the sun gathered strength the haze which obscured the horizon dissipated, but although the wind had abated considerably, the very rough seas from SW did not subside, causing us to fall to leeward, and be driven too far to the east. The *Atrevida* kept up well with us, but appeared to have lost the slight advantage in speed which she had shown at times during the previous afternoon and morning.

At noon our latitude by observation was 50° 19′ and longitude by chronometers 6° 44′ 13″; according to our charts Isla Rasa and Los Salvajes were bearing SSE distant sixteen or seventeen leagues. The wind decreased even more in the afternoon, shifting to NW, and we made the most of it by crowding on all sail, despite the extremely heavy swell, steering SEbyS by compass in case the wind should back to SW once more during the night.

The barometers that evening heralded a deterioration in the weather and since neither the late hour nor the skies encouraged us to close the land, we abandoned the EbyS course we had been steering by compass since eight o'clock. We took in one reef of the topsails, struck the topgallant yard, and hauled the wind to WSW under topsails. By now the sea had risen very high, the weather was gloomy with drizzle and the wind, already gusting from WNW and west, persuaded us to alter course to north and then at four o'clock in the morning to EbyS for the land, with a strong SSW wind blowing.

17 December

At six o'clock in the morning the weather was still very overcast. We believed we were close to El Salvaje[1] and Isla Rasa, but considered it dangerous to sail towards them with such unfavourable winds and skies. At the same time it was important not to sail beyond them without identifying them thoroughly so as to provide a safe landfall for making Puerto Egmont. With the wind at the time blowing from the SW, the wisest thing to do seemed to be haul the wind to WNW. We took in two reefs in the topsails and sent down the main topgallant mast; and persuaded that the wind, now settled from SW, would bring clearer weather, we did not hesitate to stand to the east[2] under easy sail.

At nine o'clock the weather did indeed begin to clear, and we immediately altered course to ESE by compass, applying to the most likely position of Los Salvajes, as has been indicated, the errors which might have crept in regarding our own position since the day before. A large number of birds of different species, especially ducks and Wandering Albatrosses[3] appeared to be coming from the east and ESE. We were still out of soundings, obtaining no bottom with 110 fathoms of line, and, under foresail and topsails, we were still making five to six knots.

Hour angles were taken and referred to the chronometers at noon, which gave the

---

[1] Presumably Steeple Jason, which would be a little closer than Grand Jason

[2] O = west in the MS journal, which is clearly an error for E, since Malaspina's track chart confirms that he altered course to the east and not to the west.

[3] While Malaspina may have encountered ducks it is more likely that he sighted Imperial or King Shags (*Phalacrocorax albiventer*) or one of several species of penguin, while he may have applied the Spanish for Wandering Albatrosses (*carnero*) to any species of the larger albatrosses.

longitude as 5°45′30″, which was some 30′ to the east of our position by dead reckoning; at the same time our latitude was 51°5′30″. As the weather had cleared, we were able to see isolated high land bearing true east at a distance of about eight or nine leagues, whose latitude, fortunately, was not in the least doubt. It was seen, however, that the latitudes both on our charts and that of Captain McBride's[1] were somewhat too high. It was not long before we altered course to NEbyE to look for Isla Rasa, which indeed we sighted at one thirty, and a little while later, when on its meridian, we observed its longitude by means of the chronometers.

The distance we still had to cover to reach Puerto Egmont (approximately some sixteen leagues) was too much for us to attempt during what remained of the day. So, having obtained a new position from our bearings of the many islands in the vicinity, we observed hour angles once again on the meridian of one of the Salvajes. In the afternoon and evening we observed all possible bearings and transits, with bases measured by log-line and we hauled the wind under topsails, shaking out the second reef since the wind was so moderate that it scarcely allowed us to make headway against the current which appeared to flow towards the nearby islands and then to the NE.

18 December

At dawn the wind changed to WNW and west and we found ourselves somewhat driven to leeward and some distance from where we had been at nightfall, despite having made short boards overnight in order to remain in the same area. Los Salvajes bore SW from us by compass, distant five leagues. The sea was high and there was a moderate gale from WSW, but the outlook was bright and fine. It therefore seemed prudent, with all sails set, to haul to the south, thereby succeeding in returning to the islands from where we had sailed close-hauled the previous evening and so obtaining immediate shelter from the heavy swell and the wind, which were causing us considerable inconvenience.[2] By seven o'clock we were able to bear away to ESE and east by compass, leaving the five islands, the Low Green Keys of McBride's chart,[3] two miles to our right, and then keeping very close to the coast which led towards the harbour. Hour angles were taken to determine longitude on the meridians of conspicuous points and, a mile off Punta Bluff,[4] soundings gave fourteen to nineteen fathoms, small shells and gravel.

In general, if one excepts the position of an occasional island in a labyrinth so full of them and the latitude, which we determined slightly better, we found Captain McBride's chart very accurate, while the chart we acquired in Montevideo, even though recommended as the best one of this area, was most inaccurate, at least for this part of the coast. The coast, or rather islands, which front this harbour, are generally steep, rising in the form of cliffs. There are some ledges to be seen towards the

---

[1] Malaspina held a copy of Hawkesworth, *Voyages*, which includes Macbride's chart of the Falkland Islands titled 'A Chart of Hawkins's Maidenland'.

[2] In the lee of Steeple Jason and Grand Jason (Los Salvajes).

[3] The 'Low Green Keys' depicted on McBride's chart comprise The Fridays, Flat Jason, Seal Rocks and North Fur Island; they are depicted as Las Llaves (The Keys) on Malaspina's chart.

[4] Now Elephant Point, the name Bluffs Point on McBride's chart, having been transferred at some time to the southern entrance point of Brett Harbour.

shallows which one leaves to the left when approaching from the west; and to avoid a sandbank,[1] which in conformity with the English and Spanish charts we have placed two miles south of a low island,[2] it is also necessary to hug the southern shore and Punta Bluff before going beyond this bearing.

The bottom of white sand recorded by Byron as a most useful anchorage,[3] is really a very good place for it has twelve to twenty fathoms of water and is very close to land, sheltered from all winds that could impede entry into the harbour. With its two bays at its entrance[4] and its inner reaches, we can justly call it [Port Egmont], along with Byron, one of the most beautiful harbours in the world.

As we neared the anchorage we signalled to the *Atrevida* that our priority was to water ship and it was our wish that she should anchor close to us. With topsails, jibs and main topmast staysails set, we continued to follow the coast at about a good musket shot distant in twelve, eleven or ten fathoms. At length, at twelve o'clock , coming in sight of the mouth of a brook and the remains of an old establishment, we dropped anchor close by in eight fathoms and moored successively to WSW and ESE, with the bower cable to the east, scarcely half a mile from the beach. La Vigía[5] bore N3°E from us, the north point of the entrance was in sight, bearing N10°W and the narrow channel leading to the watering place bore S80°W, all by compass.

The *Atrevida* anchored not far from us to the north, and to the south was a *sumaca* in the King's service[6] six days out of Puerto de la Soledad.[7] Her captain, a *pilotín* in the navy, had already left with an English harbour pilot and six horses to examine some coves and the coast where foreign ships could be sheltering and whose existence would be useful for our own fisheries and privateering.

[1] Malaspina appears to refer here to the shallows extending south from Sedge Island.

[2] Malaspina is somewhat confused here, which is hardly surprising since the low-lying Sedge Island and Wreck Islands, a group of three islets, little more than an above-water reef, are not always easy to see. He also appears to have used here the Spanish *isla rasa* (flat or low island) as a descriptive rather than as a proper name, which he had already used for the first island he encountered on making the Falklands. On a MS chart of the Falklands (AMN, Sig. XLVIII.D.(1)) Sedge and Wreck Islands are named Las 2 Hermanas (The Two Sisters), but on his return to the Falklands in 1794 Malaspina refers to them as Los Hermanos (The Brothers). On McBride's chart a below-water rock rather than a sandbank is charted 2 miles south of Sedge Island.

[3] Commodore Byron, who visited the Falklands in 1765 wrote 'here we brought to, and having sounded, we had forty fathom water, with a bottom of white sand': Hawkesworth, *Voyages*, I, p. 46.

[4] On the western side of the entrance to Port Egmont.

[5] Gull Hill, 254 metres in height, situated near the NW tip of Keppel Island; named Monte de la Vigia (Lookout Hill) on Malaspina's chart.

[6] Identified as the *San Juan Battista* by Malaspina in a letter to Valdés from Concepción, 26 February 1790: W. I. Morse, *Letters of Alexandro Malaspina (1790–1791)*. Published as Supplement to *The Chronicle*, No. 240. Boston, Mass. 1944, p. 15.

[7] Berkeley Sound on the NE coast of East Falkland Island.

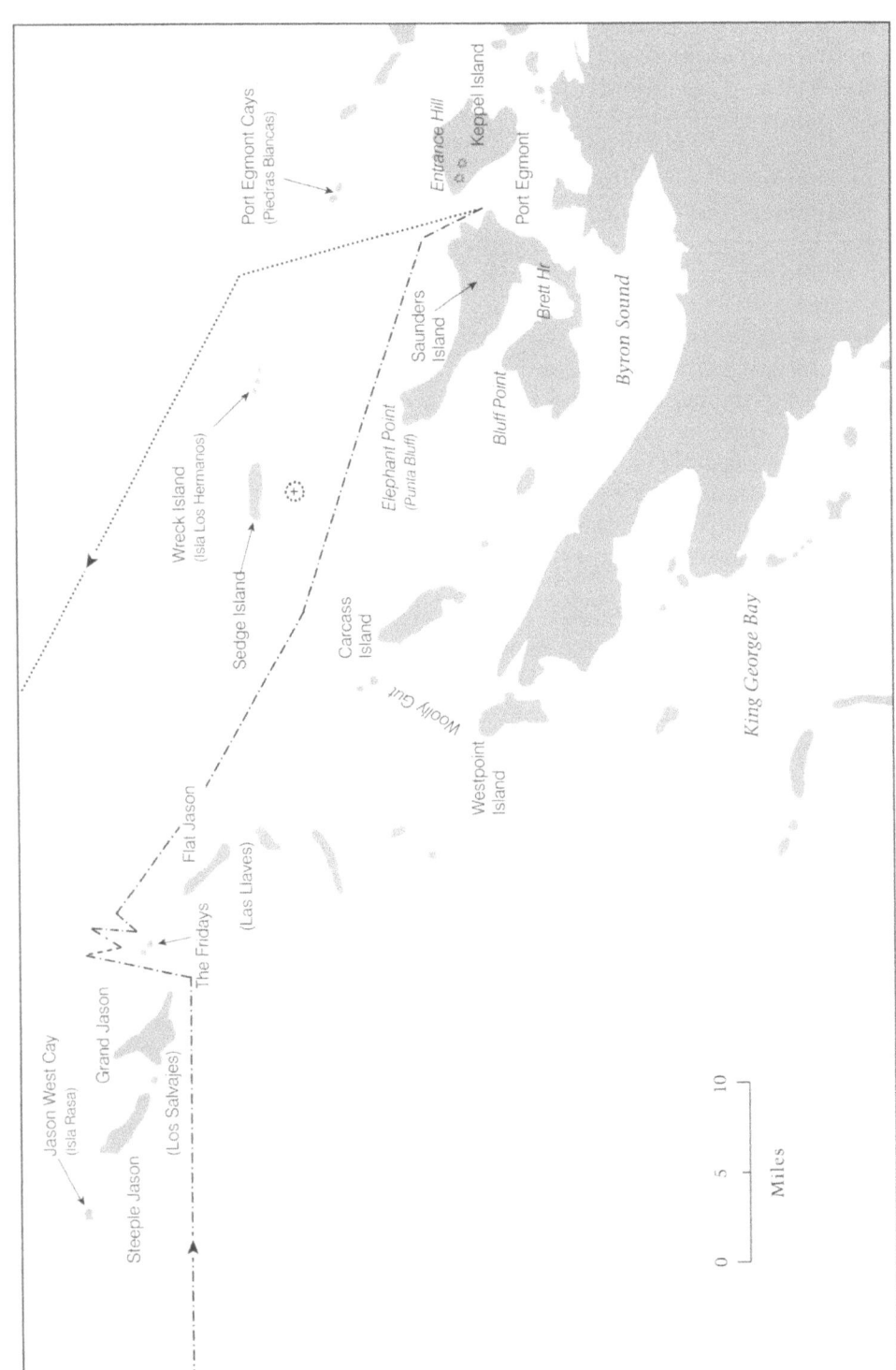

Fig. 4. West Falkland Island, December 1789

CHAPTER 3

At Puerto Egmont

18 December

In no time I set off in a boat with Teniente de Navío Valdés to examine the watering place, which I found highly satisfactory. It was a small stream that ran down from the nearby mountains,[1] rendering the ground fertile on either side in a most pleasant way and ending up in a sort of pond, close to the beach, where an abundant supply of water made it possible to fill up our water casks and carry them on board with the help of the tide. The spot was nicely sheltered by the remains of an old quay.[2] On the bottom of the pond were many dressed stones and, all round to a fair distance, the area was full of celery and scurvy grass.[3] The rocks along the shore were invitingly full of shellfish, while ducks' nests already with young, turned the site into a most attractive place.

We returned promptly on board and, having sent the pertinent orders to the *Atrevida*, from two o'clock in the afternoon the two launches were employed in ferrying the empty casks to the shore to fill them up. At the same time, those of us left on board proceeded to fit a new suit of sails. Meanwhile, in the afternoon, hunters and the naturalists separately searched the neighbouring countryside, with notable success to the satisfaction of all.

Considerable attention was paid to astronomy today. In such a high latitude, with so few hours of darkness, we were nevertheless pleased to have the opportunity to observe the emersion of Jupiter's second satellite. The evening, particularly after six o'clock, promised to remain clear and peaceful, as only moderate breezes were blowing at the time from SSW.

A small tent was speedily set up on a hillock near the watering place. The Ramsden astronomical quadrant was landed to measure meridian altitudes, together with chronometer number 72, in place of the astronomical clock, and all the achromatic telescopes of the *Descubierta,* accompanied by Galiano, Concha and Vernacci. A *pilotín* and a marine, also from the *Descubierta,* were detailed to look after the servicing and security of the observatory.

---

[1] Probably rising in Egmont Hill, 830 feet high. The highest point of Saunders Island, Mount Harston, is only 1421 feet high.

[2] This was the site of the first British settlement in the Falklands established in 1765 by Commodore Byron.

[3] Probably wild celery (*Apium australe*) and European scurvy grass (*Cochlearis officinalis*), the latter possibly introduced by the early settlers to combat scurvy.

No sooner had all the preliminary arrangements been completed and the chronometer compared from the shore with those on board both corvettes, than the sky began to cloud over and the astronomers began to doubt whether our intended observations would be successful. Jupiter's satellite disappeared from view and the measurement of several meridian altitudes was frustrated. Only by dint of the utmost vigilance was it possible to obtain the meridian altitude of Sirius, which turned out to be far from correct because the astronomical quadrant was not sufficiently steady. [19th] The meridian altitude of the Sun the following day, 51°21'45",[1] provided evidence of the same misalignment of the cross-wires previously noticed in Montevideo.[2] Our men worked at the watering place all that day and, by nightfall, that task had been completed on both corvettes. Pineda, Bauzá and Pozo, who had been out since daybreak at Cerro de la Vigía,[3] on the opposite island,[4] returned contented and tired shortly before nightfall, each with items related to his speciality. Astronomy had not been neglected since, apart from taking equal altitudes to check the rates of our chronometers, the officers from both corvettes had observed various distances between the Sun and the Moon. During the early part of the night Don Dionisio Galiano had observed the occultation of ξ Aquarii by the dark side of the Moon. Ever since the first few days out of Cádiz we had realized the difficulty of using the main capstan to good effect while it remained installed in the waist. We therefore took advantage of the peaceful waters of this harbour to transfer it to the quarter deck; and as the forge was set up on board for that purpose we also arranged for any other pieces of machinery requiring attention to be seen to. The night was cloudy and dark.

20 December

The dawn presented a most beautiful sight. As it was Sunday and we had completed our set tasks, most of our crews (duly supplied with some soap) went ashore to relax and launder their clothes. A boat was sent fishing, which was not very successful. Three different places were tried but only a few fish of Linnaeus's genus *Mulus* were brought back.[5] In this place a dead calm is needed for the fish to come up to the beach.

One of the *Atrevida*'s boats was sent to sound the bay. Don Felipe Bauzá took a number of bearings. He also marked with flags headlands in the direction in which we wanted to extend the base, which was to be measured using the masthead heights of both corvettes since the adjacent coastline did not appear to offer any place suitable for the purpose.

Today most of our time was occupied with astronomical observations. Taking altitudes with the quadrant, observing the meridian passages of the Sun and the Moon and working out a number of absolute altitudes to find the longitude by means of

---

[1] This is the resulting latitude not the meridian altitude. In 1843 Captain B. J. Sulivan obtained a latitude of 51°21'S near the ruined settlement, during his survey of Port Egmont.

[2] But not noted earlier.

[3] Gull Hill, see p. 99, n. 5 above.

[4] Keppel Island.

[5] A member of the genus Mullus – the mullet family. The locally named mullet (*Eleginops maclovinus*) is one of the more common Falkland Island fish, found in shallow water in creeks and estuaries.

their hourly movement, obtaining variation by theodolite and repeated observations of the distance between the Sun and the Moon were among the many tasks undertaken by Galiano and Vernacci as astronomers, with the assistance of the other officers. From these observations the following longitudes were calculated for the point where the portable observatory had been set up the day before.

			West of Montevideo
Referred to noon on the 20th	{	No 61	3 52 38
	{	No 72	3 51 25
Mean of 155 sets of lunar distances observed today and the day before		....	3 49 15
The mean given by the *Atrevida*'s three chronometers			3 59 30

The day was disturbed, even to the point of upsetting part of our astronomical work, by one of those incidents which, while inevitable when sailors are gathered together, are none the less painful and distressing. Some of the seamen landed from either the *Descubierta* or the *Atrevida* (it was impossible to find out exactly which) set light to a pile of peat and soon we saw the nearby bush catching fire, while the smoke rising from the flames hid even the nearest objects. At six o'clock appliances and tools were taken ashore from the *Descubierta* so that all crews could be employed checking the advance of the fire under the direction of the officers and of the boatswains who were ashore at the time. At around nine o'clock it seemed that they had succeeded; nevertheless, a party of about eight or nine men remained watching over the most suspicious area, in order to proceed immediately to wherever the smouldering fire might rekindle.

We were therefore hopeful that we would be able to observe some meridian altitudes to the south during that night which, seemed clear and peaceful, and thus confirm the latitude of the observatory. But in the middle of the night, with a strong breeze now blowing from WNW, the fire broke out again with such force that it became impossible to control it. It was therefore not without some risk and considerable effort that we managed to observed the meridian altitudes of β Reticuli and Volantis, which confirmed, when compared with the observation of the Sun, that the latitude of the observatory was 51°21′3″, with the same misalignment of the astronomical quadrant as at Montevideo.

[21 December]

At dawn there was a thick fog with a strong WNW breeze. The smoke was still masking most of the objects we needed to take bearings to. At one point we despaired of ever being able to measure a base using the corvettes's masthead heights as we had planned.[1] At last, the base was measured, but at eight o'clock in the morning a new difficulty arose when it became necessary to explain the difference of 3′ more in the

---

[1] By measuring the angle subtended from the water-line to the truck of the mast and knowing its height the distance between the two ships can be calculated. If at the same time horizonal angles are measured from the base of the two masts to flags on shore and vice-versa a system of triangulation can be built up. See also pp. 325-6.

measurement of the angle as observed by the *Atrevida*, when compared with the angle observed on board the *Descubierta*. Don Felipe Bauzá and I transferred to the *Atrevida* to remeasure with her officers the masthead height of the *Descubierta*, charging this vessel to repeat the observation when ordered to do so by signal. Again the same difference of 3′ or 4′ was found in the subtended angle on board the *Atrevida* in this second observation. Greater care could hardly have been taken in applying this method of masthead heights than we had done on both corvettes. The instruments, the accurate measurements, the mathematical calculations, the considerable expertise of the observers and taking full advantage of whatever favourable circumstances were at hand, did not guard us from the possibility of serious errors, something we have frequently found in the hydrographic survey of coastlines. This has led us on many occasions to prefer measuring bases by log-line or astronomical differences[1] to this method.

From first light we employed the launches of both corvettes gathering a good store of wild celery and scurvy grass, while the hunters, though not very expert, had been employed acquiring more agreeable and plentiful food. In the event their spoils did not reflect the abundance and size of birds in the area, among which bustards and snipe,[2] to use the names they are given in our Puerto de la Soledad, are particularly noteworthy.

The fog prevented Vernacci making any observations for equal altitudes, nor was he able to observe the Sun's meridian altitude. After noon, however, the weather cleared up and he was able to obtain absolute values[3] which, when accurately compared with the chronometers immediately afterwards, gave a new starting point for checking their rates. The dip of the needle was also observed and its declination[4] with different azimuths was also noted. Around four o'clock, the last of the instruments were brought on board. The ship's carpenters remained ashore with instructions to rehouse the portable observatory in its box.

Although the weather was not conducive to marine expeditions, I set out with Bauzá to take a few soundings and bearings inside the harbour. We did not get far with the soundings because of the heavy swell; but, as I had been wise enough to take with us in the pinnace the boatswain of the *sumaca*, who had been sailing in these parts for a number of years, we were able to record in detail a number of features, such as bearings to several conspicuous points, the situation of many reefs and the openings of several channels. By this time the peat fire had made rapid progress and the smoke was such a hindrance that the commanding officer and officers of the *Atrevida*, who had spent the afternoon on board this corvette, had to avail themselves of a compass and a lamp to be able to return to their own ship. Anyone sailing in these waters should take every precaution against the danger of fire which, besides being destructive in itself, immediately frightens off the game and its smoke obscures the marks most needed for safe navigation.

---

[1] Measuring a base by astronomical distances probably refers to obtaining the latitudes of two points roughly north/south from one another and measuring the true bearing between them from which their distance apart can be calculated.

[2] Possibly Upland Geese (*Chloëphaga picta leucoptera*) and Common Snipe (*Gallinago gallinago*).

[3] I.e. absolute altitudes.

[4] An alternative term for variation.

Before nightfall, we hoisted the boats after bringing back everything that had been taken ashore and we were ready to set sail with the next morning's tide. A volunteer sailor was sent to the *Descubierta* from the *sumaca* as a replacement for a farmer from Montevideo whom I had originally impressed and then transferred to the *Carmen* at Puerto Deseado so that he might return as soon as possible to his home and family, where he would be of more use than he was on board, Finally, I entrusted the boatswain of the *sumaca* with a letter for Buenos Aires, addressed to His Excellency the Ministro de Marina, in which I informed him about the progress of our mission and the principal results of our observations regarding the astronomical position of this harbour. The night was still and [22nd] dawn brought a variable WSW to SW wind, which we had every intention of using to set sail. Once the signal had been given to the *Atrevida*, we proceeded to unmoor. By six o'clock the small bower was being weighed; but by then there was quite a stiff SW wind and a considerable sea and it was clearly becoming very hard to get it up and down and, seeing the difficulty of clearing the small islands to the north of the harbour, we were forced to abandon the idea of getting under way and we remained riding to a single anchor, having hoisted the launch and taken down the main topgallant mast which we had swayed up to measure masthead heights.

When drawing up the plan of the harbour, we were unable to reconcile the length of the base, as measured between the two corvettes, with the distances and positions of our survey marks. We therefore took advantage of the need to remain in harbour for the day to send Don Felipe Bauzá ashore in the pinnace to measure a short base near the watering place as a check when completing the important plan of this harbour. This operation was successfully completed and the pinnace returned at five o'clock in the afternoon, by which time the wind had dropped considerably, the weather had turned misty and the smoke was fairly thick. We took advantage of this stillness to heave in the best bower cable which was almost completely paid out and night found us on some forty fathoms of cable.

The immersion of Jupiter's first satellite which was due at three o'clock the following morning had roused so great an interest among our astronomers that they wished to observe it from on shore. However, taking into consideration that it would be twilight, as the Sun at the time would be only eight degrees below the horizon, and my determination to sail at dawn if the wind was favourable, I begged them to give up the idea, particularly as it would be most inconvenient to land the instruments again and to spend the night on the beach without the slightest shelter. As it was, during the night some light airs from WSW and west brought with them a misty sky that would have made it quite impossible to make any observations. The *Atrevida*, perhaps because she had not paid out enough cable, dragged her anchor during the night and was carried by the tide before enough additional cable could be veered, eventually being brought up by our stern, some distance off the shore. Dawn had scarcely broken before a strong steady breeze sprang up from the north and increased in strength with the Sun. We therefore remained at anchor in the same position as the day before. Since the sea was not too high, we sent the pinnace to replace the drinking water we had consumed.

By noon, although the wind had shifted to NNW and, at times, to NW, we knew

Plate 16. Native Pansy or Violet, *Viola Maculata*. Engraving in

Plate 15. Native Pansy or Violet, *Viola Maculata*. Original

Plate 18. Scurvy Grass or Vinaigrette, *Oxalis enneaphylla*. Engraving in Cavanilles, *Icones ...*, 5, p. 6. ARJB

Plate 17. Scurvy Grass or Vinaigrette, *Oxalis enneaphylla*. Original drawing by José Guio (ARJB, Div. VI, No 12)

that we would not be able to make our departure that afternoon. The wind having settled down somewhat, a second trip was undertaken by the *Descubierta*'s pinnace to replenish the water supplies and we signalled the *Atrevida* to do the same. Don Antonio Pineda made good use of this second trip to collect a few stones and the skeleton of a penguin. He also supervised the collection of a fresh and abundant supply of celery and scurvy grass and at about three-thirty he was back on board. For the rest of that afternoon, as the wind had slackened and the idea of a further visit ashore had been discarded, the *guardiamarinas* from both corvettes were allowed to take the pinnaces out and race them beating into the wind. With a breeze that was fresh enough at times but would suddenly drop, both sides were naturally led to believe that they had got the better of the day, and so all were quite satisfied with their own victory.

Night fell with some haziness and a moderate NW breeze which, after fading away altogether by midnight with a slight drizzle, settled to a light NW breeze at two o'clock in the morning. In both corvettes we immediately hoisted our pinnaces, raised the main yards and brought the cable up and down. At three-thirty we got under way and by five o'clock, making the most of the light airs from the third and second quadrants, we were clear of the harbour and set a course to run between a low island[1] and the two small islands to the NE of the harbour entrance. By this time, not only had we taken all the bearings we could possibly take to fix the positions of many useful landmarks, but we had also drawn up a chart of the entire inner harbour with the outer bays and the many islands in the vicinity.

[1] The flat island comprises Sedge and Wreck Islands, subsequently named Las Hermanas (The Two Sisters) on Malaspina's survey (see p. 99, n. 2 above). The two small islands are Port Egmont Cays, named Islas de Piedras Blancas on Malaspina's survey.

CHAPTER 4

From Puerto Egmont to Puerto San Carlos de Chiloé[1]

24 December

We were only some two miles from the harbour entrance when the light airs fell away completely and the greater part of the adjacent islands disappeared in thick haze. A heavy swell from north and NW began to push us towards the shore west[2] of the harbour; we sounded nineteen fathoms, sand. Our situation being far from comfortable, our sole concern was to keep off the reefs around the islands east of the harbour[3] and to stay closer to the western shore so as to be able to anchor, should that be necessary. However, this unpleasant situation lasted only until seven-thirty in the morning when a moderate breeze from SW soon cleared the skies and the horizon. We crowded on sail and regained our intended track, passing the two small islands at a distance of two miles, until at about a quarter past ten we were able to haul to the wind in the fourth quadrant while having in sight the sandy shores of a low island[4] bearing WSW, distant one-and-a-half leagues.

At noon our latitude was 51°2′21″ and longitude by our chronometers was 4°2′ west of Montevideo, at which time the low island bore true south. Having placed it in the same longitude on our arrival was a satisfactory proof, both of the excellent rates of the chronometers and of the undoubted accuracy of our astronomical observations. There was still a big sea running from NW, but the wind had died down almost completely and the skies and horizon were clear. For this reason, because of the rolling and the rather inconvenient position of the stars,[5] we spent that afternoon observing distances between the Sun and the Moon. The mean of the 102 sets of observations gave a longitude of 4°16′, with 4°10′ being the longitude by number 61. From five o'clock in the evening we were becalmed and without steerage way. The land in the vicinity of Puerto Egmont remained visible, though somewhat misty, and at sunset the harbour entrance bore S36°E, Pan de Azúcar[6] S10°W and Los Salvajes, further to the west, S48°W, all by compass.

---

[1] Bahía de Ancud.

[2] Possibly a slip for east.

[3] The two small islands comprising Port Egmont Cays; Islas de Piedras Blancas on Malaspina's survey: AMN, Sig.XLVIII.D.(1)

[4] Isla Rasa in the MS journal, presumably the small islets comprising Sedge and Wreck Islands, see p. 99, n. 2 above.

[5] Presumably none of the tabulated stars were suitable for observing lunar distances that night.

[6] Elephant Jason, a conical shaped island about 200 metres in height.

25 December

Shortly before midnight a moderate breeze sprang up from the north and with this we set our course true west under full sail. At three o'clock in the morning we could see the summit of Pan de Azúcar as well as Los Salvajes. At four o'clock, the eastern-most of these was four leagues away true south and by half past five, despite the heavy haze, we could make out Isla Rasa the westernmost island in the archipelago, bearing SSW true. The wind's force was increasing rapidly although still from north and NW, but the outlook had become much more overcast and there was a much heavier sea. We were making nine to ten knots under the four principal sails and the fore top-mast staysail.

By nine o'clock it had been necessary to strike the yard and main topgallant mast, to furl the mizzen topsail and to take in two reefs of the main and fore topsails. We were still steering true west so as to close the coast and to continue surveying it in detail as we had been doing up to now. Our latitude by the Sun's meridian altitude was by then quite doubtful. Our longitude, which we had observed by chronometer number 61 at ten o'clock in the morning and with a less rough sea, was 6°47'32". There was some discrepancy between numbers 13 and 72 and when we compared them at noon they showed a significant difference.

The rest of that afternoon was rather gloomy and at about seven, after some showers, the wind shifted to WNW and at the same time dropped almost to a calm, consequently rolling made us most uncomfortable for the rest of the night, which we spent under topsails only, without steerage way.

26 December

At dawn it was very overcast. Then the wind, a gentle SSW breeze, freshened and gusted before finally settling into a moderate one from SW that cleared the skies and the horizon. First under the four principal sails, the topsails with two reefs and later with all reefs and staysails shaken out, we sailed close-hauled in the fourth quadrant in order to make as much westing as possible. But by ten o'clock we had to put about as the wind had veered to WSW. Our latitude at noon was 50°48' and longitude by number 61 was 8°58'30" west of Montevideo.

For some days variation, which we had observed routinely as often as possible, had shown considerable discrepancies. It varied between 23° and 25° although, in truth, these differences may have been caused by our severe rolling in the continuously high seas. At noon on that day we settled on 22°30', which we had determined in the morning by a number of azimuths taken in completely favourable conditions. The wind was variable the next afternoon and at about three o'clock it had steadied to WNW and NW, freshening, and we were able to take advantage of it under full sail, making between seven and eight knots as we steered west and a few degrees to the south.

After five o'clock in the afternoon, we took in two reefs of the topsails as the weather gave clear signs of worsening. Indeed, at around seven o'clock the wind backed, blowing in almost the opposite direction, from SSW and SW, compelling us to tack to port. A high sea developed from NW and the sky, quite heavily overcast, looked ominous. At nightfall the wind, still in the third quadrant, began to abate. We sailed close-hauled in the fourth quadrant and at nine o'clock we sounded seventy-three fathoms, black sand.

27 December

With a fine dawn we crowded on sail and by seven o'clock in the morning, with the wind then from WSW, we were able to alter course to south by compass, making five or six knots. At noon our latitude by observation was 51°11' and longitude 10°42'55"; Cabo Vírgenes[1] bore S34°W from us, twenty-eight leagues distant according to the position determined by Belmonte and Galiano. From ten o'clock it was very squally from WSW and west; so in the afternoon we proceeded with caution under reduced sail and, indeed, at one o'clock we were struck by a very strong and sudden gust from WSW with a lot of hail, forcing us to strike the topsails, keeping only the foresail hoisted. But an hour later the weather fell to its earlier calm and we could again carry a full set of sails. Then, at around three-thirty, there was another squall, although not so violent, which we withstood with only the foresail and topsails lowered; the wind then shifted to SSW and with it we immediately put about and steered to the west.

Later the wind veered to SW, a moderate gale, and the sea began to rise considerably. We took in two reefs of the topsails and at nine o'clock we changed tack and stood to the SSE close-hauled. This seemed necessary to me, even though it was taking us away from the coast north of Cabo Vírgenes, both because that part of the coast had been closely examined by Churruca and Cevallos and also because by sailing due north for a considerable time we would lose a considerable amount of latitude for the sake of gaining only a few minutes in longitude, without even the benefit of sighting the coast. Therefore, with a moderate gale now blowing from SW, a high sea and the promise of completely clear and fair skies, we stood to the SSE under our four principal sails at five to six knots and thus avoided a considerable amount of leeway while being agreeably reminded of the outstanding sailing qualities of both our corvettes.

28 December

At two o'clock in the morning the wind began to ease and the sea calmed down immediately. We crowded on sail and at eight o'clock the wind, now a moderate westerly, allowed us to set a course to the NW. However, this new course, which would not allow us to change tack, was taking us away from Cabo Vírgenes, which I was anxious to sight, more for the sake of comparing the corvette's chronometers than for the verification of its longitude, which I had no reason to doubt. At noon our latitude was 52°8', longitude 1°04' [east of Cabo Vírgenes],[2] so that the cape was bearing W20°S forty miles distant. Although these were quite reasonable ideas, the weather chose to take a hand in our plans during the afternoon. The wind shifted gradually to a moderate NW which allowed us, after three o'clock, to set our course

---

[1] The northern entrance point of the Atlantic entrance to Estrecho de Magallanes.

[2] 0°58' in MS journal which has been crossed through and amended to 1°04', but misread as 10°58' in the published journal; the bearing and distance of W20°S, 40 miles, shows that the *Descubierta* was just over one degree east of Cabo Vírgenes at the time.

true west and thus to sight Cabo Vírgenes at around five-thirty, bearing WbyS true. We immediately bore up to it and soon, with a good bearing, we were able to observe for longitude by number 61, which gave the longitude of the cape as 62°18′48″. The figure determined in the frigate *Cabeza*[1] by Belmonte and Galiano was 62°9′43″.[2]

By eight o'clock in the evening we were no more than three leagues from Cabo Vírgenes. As my plan was to close the coast of Tierra del Fuego at Cabo Espíritu Santo,[3] a signal was made to the *Atrevida* to come within hailing distance and I asked her to lead the way that night as she carried with her a copy of Sarmiento's[4] *Voyage* which warned of a shallow of only four fathoms located just off the mouth of the strait.[5] She was to pass between the four-fathom shallow and the coast and after a run of about ten leagues be off Cabo Espíritu Santo at dawn. The *Atrevida* crossed our bows and although there was only a moderate breeze, we took in two reefs in our topsails as our run was to be so short. At the time the wind was from the north, the horizon obscured by drizzle. The *Atrevida*, guided by the lead, set courses from SSE to SEbyE changing even to SE whenever necessary.

29 December

These courses were so well planned that at four o'clock in the morning we were able to sight land close to Cabo Espíritu Santo despite the heavy drizzle which was obscuring the horizon at the time. As we passed shortly afterwards close to the *Atrevida*, she informed us that during the night she had sounded fourteen fathoms and was now in over forty-five fathoms. It was not long before the wind shifted to NE, fair. As the land was by now somewhat clearer, we were able immediately to fix its true position, while keeping a keen eye on our steering, for the wind showed every sign of remaining strong from NE and the direction of the coast was inclining much more to the east than was shown on Cook's charts and on those in the *Viaje al Estrecho de Magallanes*.[6]

Guided by the Nodales's[7] *Derrotero*, even though it is rather confusing, we were

[1] The frigate's full name was *Santa María de la Cabeza*.
[2] Belmonte's and Galiano's longitude by lunar distances; their longitude by chronometers was 62°24′30″ west of Cádiz: Luis Rafael Martínez-Gañavate Ballesteros, ed., *La expedición Malaspina*, Tomo VI: *Trabajos astronómicos, geodésicos e hidrográficos*, Madrid, 1994, p. 142.
[3] The southern entrance point of the Atlantic entrance to Estrecho de Magallanes.
[4] Pedro Sarmiento de Gamboa was sent by the Viceroy of Peru in 1579-80 to Estrecho de Magallanes in an unsuccessful attempt to intercept Drake on his return to England. For an English translation of this voyage see Clements R. Markham, ed., *Narratives of the Voyages of Pedro Sarmiento de Gamboa to the Straits of Magellan*, Hakluyt Society, 1st ser., 91, London, 1895 and for a brief account see O. H. K. Spate, *The Spanish Lake*, Canberra, 1979, pp. 264-78.
[5] Banco Sarmiento, which extends 20 miles SE from Cabo Vírgenes, is unstable and is liable to vary in extent and depth.
[6] [Antonio de Córdoba], *Relación del último viaje al Estrecho de Magallanes de la fragata de S.M. Santa María de la Cabeza en los años de 1785 y 1786*, Madrid, 1788.
[7] Bartolomé and Gonzalo García de Nodal were sent in 1618 from Lisbon to examine the recently discovered Estrecho de Le Maire and the strait discovered by Magellan. Having passed through Estrecho de Le Maire, which they named Estrecho de San Vicente, they discovered an isolated group of islands which they named Islas de Diego Ramírez after the expedition's cosmographer. They then transited Estrecho de Magallanes from west to east in fifteen days before returning to Lisbon. Their account of the voyage, *Relacion del viaje … los capitanes Bartholomé García de Nodal y Gonzalo de Nodal … al descubrimiento del estrecho nuevo de S. Vicente, que hoy es nombrado de Maire, y reconocimi° del de Magallanes*, was

able to survey the coastline close to Cabo Espíritu Santo. Following Anson, we retained the name of Cabo de la Reina Catalina[1] for the southernmost point of the moderately high land adjacent to that cape. It was then our first and principal objective to determine the extremities of Canal de San Sebastián[2] and to this end we followed the low-lying coast at a distance of two or three leagues. However, as the weather was again turning rather hazy and the breeze was now strong from ENE, making it most inadvisable to approach too close to these shores, we were finally left in some doubt about the southern extremity of that channel whose latitude, by our observations of longitude and successive positions at noon, differed considerably from that of the Nodales; not, to be sure, because we lacked sufficient reliable information to verify it, but because it became impossible to match the latitude with that given by them. There was general agreement about the outline, particularly on their note that the higher and snow-covered lands began at the southern end of Canal de San Sebastián and at Cabo Peñas.[3] At noon our latitude was 53°23' and our longitude 11°14'30" west of Montevideo, the latter being deduced from the mean of the observations made during the morning and afternoon, which, in conjunction with our dead reckoning, did not reveal any effects of the currents.

The wind, which had already shifted to ENE, increased in the afternoon. At four o'clock the land was getting darker and, as we were no more than two-and-a-half leagues from it, we identified it to be close to Cabo Peñas. Soundings continued to be between thirty-nine and thirty-five fathoms, mud. We therefore altered course to north and not much later, as the wind shifted to the east and ESE, we sailed close-hauled in the first quadrant with the soundings increasing to forty-four, forty-two and forty fathoms, gravel and small shell. The land had become entirely shrouded in haze and the wind was losing its strength to the point that by ten o'clock it had almost fallen to a calm. The coast could no longer be seen and at times it was difficult to keep in company with the *Atrevida*. The NE breeze fell to a complete calm towards ten o'clock but by midnight there was a settled south and SSE moderate breeze. Dawn brought fine weather.

30 December

At four o'clock in the morning, we saw the coast some five leagues away and since at sunrise the weather appeared very pleasant, it seemed a most inviting prospect to begin surveying and plotting the land right away, all the more so because our tasks for the day included returning to Cabo Peñas where our previous day's bases had been

published in Madrid in 1821. For an English translation see Sir Clement Markham, *Early Spanish Voyages to Magellan Strait*, London, Hakluyt Society, 1911, pp. 169-272 and for a brief account see O. H. K. Spate, *Monopolists and Freebooters,* London, 1983, pp. 25-6.

[1] Q. Katherine's Foreland, on chart in George Anson, *A Voyage round the World*, London, 1748, f.p. 95; charted 20 miles SSE of Cabo Espíritu Santo by Malaspina; now Cabo Nombre.

[2] An opening on the east coast of Tierra del Fuego, shown on old charts about 65 miles south of Cabo Vírgenes, which was said by the Nodales to be the mouth of a channel leading to Estrecho de Magallanes. The existence of this channel was disproved in June 1830 by Robert FitzRoy, *Narrative of the Surveying Voyages of His Majesty's Ships Adventure and Beagle*, 3 vols and an appendix, London, 1839, I, p. 458.

[3] Situated about 90 miles SSE of Cabo Vírgenes.

completed. In fact, at this point, the coast seems to rise and become snow-covered, but not so steeply that one cannot find near the sea several valleys and plains in which the vegetation can display all its greenery and beauty. The snow or ice was generally only to be seen on the sheerest mountain peaks, towards the southern parts and as it was sprinkled, so to speak, in low mounds glinting in the sunshine, it suggested a pleasant contrast between the two most opposing seasons. This made us conjecture that the summer was well advanced, an assumption which was wholly confirmed by the generally mild weather we had enjoyed all along the Patagonian coast and by the calm and pleasant weather we were experiencing today. As we crossed the meridians of various landmarks on the coast, longitudes were observed with number 61 and all our work was then joined up by means of short bases. In order not to interfere with this, the task of sounding had been entrusted to the *Atrevida* which reported, at eight o'clock and again at ten, depths of forty and thirty-seven fathoms, when we were about two leagues distant from the shore.

Captain Cook's astronomical work had begun at Cabo Santa Inéz,[1] and that navigator assures us that from that point to Cabo San Juan[2] on Isla de los Estados and to Isla Recalada[3] west of Cabo Negro[4] all his longitudes had been deduced by chronometer and were related to his determination, based on many sets of lunar distances, of the longitude of Cabo de Hornos [Cape Horn] as 61°30′ from Cádiz.[5] Therefore, whatever error there might have been in such determination, and the captain himself suspected it to be as much as a quarter of a degree, it would have to manifest itself at Cabo Santa Inéz in the same way that it would at any other place.[6] With this in mind, as soon as we were in the right position, we signalled the *Atrevida* to observe longitudes at the same time that we would observe them ourselves by number 61 whose rate, virtually unchanged though subject to a correction for minor alterations, we were checking to our satisfaction by means of daily comparisons with our other chronometers. The resulting longitude for Cabo Santa Inéz was 60°47′, equal, but for 2′ more, to that of the English navigator.[7] Yet our linking of subsequent longitudes did not agree so well with his. Thus, our longitudes for Cabo San Diego at the [NW] entrance to Estrecho de Maire and for Cabo San Juan on the eastern end of

[1] Cabo Santa Inés, situated about 30 miles SE of Cabo Peñas.

[2] The easternmost cape of Isla de Los Estados; Cape St John on Cook's chart.

[3] Situated about 4 miles off the SW side of Isla Desolación, on the south side of the western entrance to Estrecho de Magallanes; Land-fall Isle on Cook's chart.

[4] Cape Noir on Cook's chart.

[5] In calculating his longitude for Cabo de Hornos Cook took the mean of the longitudes obtained on his first and second voyages, placing the cape in 67°46′ west of Greenwich. Malaspina appears to have converted this to the Cádiz meridian by subracting 6°16′, although the accepted difference in longitude between the two observatories at the time was 6°11′. The correct longitude of the cape is in fact 67°16′ west of Greenwich, very close to that obtained by Cook on his second voyage.

[6] Malaspina is somewhat confused with regard to Cook's astronomical activities. Cook only passed Cabo Santa Inéz during his first voyage when his longitudes depended solely on lunar distances. It was on his second voyage that his longitudes from Cabo Negro to Cabo San Juan were additionally determined by chronometer.

[7] Malaspina appears to have taken Cook's longitude for Cabo Santa Iñez from where the cape appears to be placed on Cook's 'A Chart of the Southern Extremity of America' since Cook neither mentions this cape in his first and second voyage journals, nor is it named on the above chart.

Isla de los Estados, both showed that the longitudes determined by the English captain on his second voyage were affected by an error of approximately 21′ which placed them too far to the west.

At noon we were in latitude 54°10′ and longitude 10°19′ west of Montevideo, variation by different azimuths 25°19′NE. We took bearings of Cabo Santa Inéz to the WNW, distant four leagues, and Cabo San Vicente at the [entrance to the][1] Strait was bearing SEbyE twenty leagues distant. We continued under full sail until three in the afternoon. Then, as the wind was increasing steadily from WNW and the skies and horizon were becoming overcast, we furled the small sails and followed the coast, under all topsails and the foresail, so as not to lose sight of it nor to miss either the longitude of its salient points nor any views of such points as might serve as a guide for future landfalls. Among these, Los Tres Hermanos and Pan de Azúcar[2] are the most deserving of mention. The position which Frézier[3] accorded them seemed to us to be mistaken although it had merited the praise of Lord Anson,[4] but the position on Captain Cook's chart is perfectly correct.[5]

We were unable to fetch Cabo San Vicente until six o'clock in the evening, when it was half a league distant. Up to that point the currents had not affected us in any way, as was indicated by the agreement of our various corrected base lines, our many observed longitudes and the exact agreement of our latitudes by dead reckoning with the corresponding ones shown on the English captain's chart of that coast. By four o'clock that afternoon, I had abandoned the idea of passing through Estrecho de Maire and anchoring in Bahía del Buen Suceso.[6] I had been led to this decision by the following considerations. First, a survey of the strait could be considered as superfluous in view of the chart that the English captain had published with such precision. Second, as the corvettes were not suffering any defects and nothing of interest in the field of natural history could be achieved without a stay of at least five or six days in the area, to enter the bay for only two or three days would be a waste of precious time. Third, we had reached the entrance to the strait at six o'clock in the evening, the time when the ebb tide begins to run, and we were only two days short of the full Moon, when the flood starts at one or two in the afternoon. Finally, the appearance of the land, the temperatures we were experiencing, and the weather that we had enjoyed so far showed how advanced the southern hemisphere summer was this year, so that we could expect weather that was pleasant and suitable for carrying on our survey of the western coast between Cabo Victoria[7] and Chiloé; but we could

---

[1] Illegible in the MS journal, but *entrada del* in AMN, MS 753; Cabo Vicente is situated about 5 miles west of Cabo San Diego.

[2] Cerros Tres Hermanos and Monte Campana, a prominent mountain which resembles a large bell.

[3] Amédée-François Frézier, *Relation du Voyage de la Mer du Sud aux côtes du Chily et du Pérou fait pendant les années 1712, 1713 et 1714*, Paris, 1716.

[4] 'Frezier has given us a very correct prospect of the part of *Terra del Fuego*, which borders on the Streights': Williams, *Anson's Voyage round the World*, p. 82.

[5] 'A Chart of the S.E. part of Terra del Fuego including Strait le Maire and part of Staten-Land': Hawkesworth, *Voyages*, II, f.p. 39 on which they are named '3 Brothers' and 'sugarloaf'.

[6] Bahía Buen Suceso on the NW side of Estrecho de la Maire, which Cook surveyed in January 1769.

[7] The northern entrance point of the Pacific entrance to Estrecho de Magallanes.

also expect adverse conditions as the season advanced and drew closer to the autumn. On the other hand, it appeared to me that to set a course eastward of Isla de los Estados would be so much more pleasant and reasonable. I was swayed towards this not only by the detailed description Captain Cook had already made of this island, but also by my belief that it was of great consequence for the safety of our nation's navigation to ascertain fully the longitude of Cabo San Juan and to take views of the adjacent land in order to identify a good landfall and the possibility of finding suitable landmarks on those shores.

The crossing from Cabo San Diego to Isla de los Estados thus deprived us of the little light remaining that day, so that it was already nine o'clock that evening when we were on the meridian of Cabo San Antonio at a distance of two leagues. By that time the wind had swung round to NW, strong and gusty; then, as we were closing the island, a strong southerly current took us by surprise and began to drive us towards the coast so rapidly that by ten o'clock we were barely two miles from it. We had to luff hard round to NE and maintain quite a disproportionate amount of sail which brought us nonetheless, at eleven o'clock, very close to the meridian of Islas del Año Nuevo[1] and finally allowed us by midnight to run true east under regular sail.

As far as we were able to see with the darkness and the speed at which we were following the island, Captain Cook's description of the outline and trend of its shoreline could not have been more accurate. The islands that form the anchorage of Año Nuevo are recognizable at a good distance jutting out to the north and, as they are low-lying, while all the land on Isla de los Estados is high and craggy, this anchorage can be considered as the easiest to identify for anyone who looks for it. This was the harbour that had been chosen as our rendezvous with the captain of the brig *Carmen* in the last set of instructions given him at Puerto Deseado. However, we could not believe that the brig had been able to reach the harbour with the weather we had encountered and in any case we had nothing special to communicate to the Viceroy at Buenos Aires; so, pressed on by the wind as we were, we passed the harbour entrance at about midnight, and concerned ourselves solely with being close to Cabo San Juan by dawn, as I considered its longitude and appearance from the sea to be particularly important information for our country's navigation.

31 December

Thus, at three o'clock in the morning, it was already easy for us to take some coastal views. Later we were also able to obtain hour angles at two different times and bearings, both of which confirmed the longitude of Cabo San Juan as 7°25' west of Montevideo, using only number 61, as both 13 and 72 differed considerably, one to the west the other to the east, although they gave together a mean quite consistent with the results observed by 61. This longitude, as already mentioned, differed 21' to the west from the longitude that had been determined by Captain Cook and, as in

---

[1] A group of five islets on the north side of Isla de los Estados, named New Year Islands by Cook on 1st January 1775.

the *Carta del Magallanes*,[1] Cook's error, when related to the longitude of Cabo Vír-genes, gave a figure of 33′ in the opposite direction, thus making the coast from Cabo Espíritu Santo to Cabo Santa Inéz lie almost north and south. The resulting error in that chart for Cabo San Juan was 47′ further west than had been determined by our observation; therefore, for the safety of our navigation henceforth, it seemed best to refer the entire chart to the longitude east or west of Cabo San Juan after determin-ing its position in accordance with our own observations, always assuming the differ-ence between meridians as determined by the chronometers of the English navigator to be correct.

From daybreak, the wind, sometimes clear and sometimes squally, had shifted round to west, SW and SSW. We sailed close hauled but occasionally we had to fall off in order to counter the strong current which, with the wind abating, was setting us to the south and bringing us far too close to Cabo San Juan. But at eleven o'clock a squall from SW brought in its wake strong winds and the sea began to rise, forcing us to take in two reefs in the topsails and to steer SSE under these and the foresail, all the more so because our position and the appearance of the skies and the horizon threatened imminent bad weather.

At noon, we were in latitude 55°4′, longitude 6°50′, when the northernmost point of Isla de los Estados bore N79°W, Cabo San Juan bore N52°W and the southernmost point of the island bore S80°W, placing us about six leagues off Cabo San Juan.

To round Cabo de Hornos [Cape Horn] there was a choice of two routes, both of them well supported by previous experience. One of these was to keep close to the land, the other to make for higher latitudes in search not so much of favourable winds, but of more sea room for tacking freely whenever the wind blew from the third or fourth quadrants. Our sailing practice supported firmly this second course, mainly on the basis of usage. Captain Cook recommended this same route, although stating also how well he had done following the first route.[2] In deciding whether to favour the outer course or a coastwise navigation, there were two considerations we could not afford to ignore: first, that it was of the greatest importance for the service of His Majesty that we should reach the western coast of America as soon as possible and second, that Captain Cook's survey of the entire coast between Isla de los Estados and Cabo Pilares[3] could be regarded as complete and, indeed, that it was impossible to undertake such a task except by running from west to east, as had been done by that navigator. Furthermore, adding to these considerations the observations I had made during my earlier voyage to these waters, particularly regarding the effect of currents, I concluded that it was best to chance the whims of the winds, naturally variable at this time of year, in order to make progress to the west as far as possible. I could thus reach a position where, by tacking to the north with winds from west and

---

[1] '*Carta esférica de la parte sur de la América meridional en la qual se ha colocado el Estrecho de Magallanes*', in Córdoba, *Relación del último viaje*; the chart extends from 47° to 66°S and from 50° to 75° west of Cádiz.

[2] Cook makes no such recommendation, stating 'But, in my opinion, different circumstances may at one time render it eligible to pass through the Streight, and to keep to the eastward of Staten Land at another.': Hawkesworth, *Voyages*, II, p. 68.

[3] Cabo Pilar, the southern entrance point of the Pacific entrance to Estrecho de Magallanes.

WNW, I could pass to windward of Cabo Pilares and CaboVictoria, and then to be free to choose whether or not to approach the coast, according to the strength of the onshore winds.

The prevailing wind at the time offered us few choices that would not be considered imprudent. All that afternoon it blew continuously strong and gusty from the SW so that by nightfall the seas were excessively high and we had to reduce our sails to only the foresail and a close-reefed topsail, which we would strike or hoist as necessary to keep company with the *Atrevida*. Even so, we had at times to ease the [main] sheet and topsail sheet a little.

1 [to 11] January 1790

Only next morning did the wind begin to die down and the corvettes, which had proved their outstanding sea-keeping qualities while on a steady tack, showed good speed at midday, making little leeway, when latitude by observation was 56°34′ and longitude 6°20′ west of Montevideo. At dawn the wind shifted to the west and by noon we were making the best use of it under the four principal sails and with two reefs in the topsails, the sea much abated. There were few birds to be seen and when next morning moderate breezes set in from SSW, we altered course to the west, under a press of sail, and finally to SSW.

We hauled off to the starboard tack under full sail and by noon on the 2nd this had brought us to latitude 57°38′ and longitude 7°13′. Cabo de Hornos on the *Carta del Magallanes* therefore bore N51°W fifty-three leagues distant, all our longitudes being adjusted to our measurement of that of Cabo San Juan.

From then on, the rounding of Cabo de Hornos became as pleasurable as if in the tropics rather than the arduous passage anticipated both in this vessel and by [earlier] navigators. The sea was constantly smooth, the winds variable from NNW to south, generally moderate and occasionally accompanied by mist or fine drizzle, or some short-lived hail. We were able to take the Sun's meridian altitude and hour angles for longitude every day and on the 4th and 5th we were able to observe azimuths, which gave the variation as 26°30′. Meanwhile fifty-six sets of lunar distances observed on the morning of the 8th were close to the chronometer readings, which allowed us to hope that they were keeping a uniform rate.

On the 6th, the highest latitude of 60°43′ was observed. There were considerable differences in our [observed] latitudes compared with our dead reckoning, sometimes to the south and sometimes to the north. At first there were no important differences in longitude, but from the 8th to the 12th we experienced some strong easterly currents between the parallels of 57° and 59°. Considerable discrepancies developed between our chronometers and we could even see that the error was becoming progressively greater. Consequently we concerned ourselves with the best way to investigate this and once we had found that the daily comparisons could indicate with high probability and great precision the changes in the rate of each chronometer, unless all three of them were erring in the same direction, we drew up a table in which the various differences each day since the departure from Montevideo would provide an equation whereby the longitudes given by the chronometers could be reduced to a single result.

At noon on the 11th, we observed the following longitudes:

Uncorrected longitudes	Longitudes corrected by the corresponding equation
N° 61 . . .19° 30' 7"	19° 31' 11" . . . .19° 30' 7"
N° 72 . . .19 7 1	19 31 11 . . . .19 31 11
N° 13 . . .20 15 2	19 36 31 . . . .19 29 27
Longitude by lunar distances observed on the 8th	19° 20' 12"

The latitude was 57°52'0" by observation with the dangers off Cabo Deseado[1] bearing N3°E ninety-seven leagues. Our sick list at the time consisted only of a marine gunner with his customary rheumatic pains and a second carpenter and an able seaman with colds. All the remaining crew were enjoying good health and we were most pleased to note that this was also the case on board the *Atrevida*.

12 January

At three o'clock the following morning we sighted a ship to the south about two leagues distant. From the set of her sails we could see that she was bound for the South Sea and this of course gave rise to the hope that she might have sailed directly from a Spanish port and, consequently, with much more recent news than any we had. The moment we unfurled our ensigns she acknowledged them by hoisting the national merchant flag and as soon as we fired a gun from this corvette to summon her, she crowded on sail to approach us. But as the wind just died away at that instant she could not close us until noon, although we took the precaution of lying-to from seven o'clock on. When she was close and since it continued calm, Teniente de Navío Valdés went in the pinnace to identify her, taking with him our latitude and longitude at noon, namely 58°3' and 20°19' west of Montevideo, together with a copy of the *Carta del Magallanes* and with instructions to offer her whatever assistance was within our power.

Meeting the *Santa María Magdalena,* a merchant vessel from Cádiz, under her Captain and Master Martín Antonio de Ichurriaga, was a very pleasant occasion for us. She was bound for the ports of Valparaíso and Arica, after 112 days at sea, having sighted Isla de los Estados on December 27th. They informed Valdés that her crew numbered forty-four men all of whom were in good health and they did not require anything to complete their voyage. She was using a chart (quite an accurate one) by Piloto Moraleda of Lima.[2] The continued good health of our beloved monarchs and

---

[1] Cabo Deseado is the NW tip of Isla Desolación, situated on the south side of the Pacific entrance to Estrecho de Magallanes. It is a prominent headland, off which there are some rocky islets. Magellan's Cabo Deseado, the northern tip of the island, is now called Cabo Pilar.

[2] José de Moraleda had charted a great deal of the coastline of Chile in the three years preceding Malaspina's arrival, especially the island of Chiloé and Archipiélago de los Chonos, extending his surveys later to Callao. In 1831 Bauzá presented Captain Beaufort, Hydrographer of the Navy, with a chart by Moraleda, which he considered was tolerably correct and also a copy of Moraleda's two-volume manuscript 'Reconocimientos de Chiloé por Dⁿ José de Moraleda'. These two volumes were issued to Captain Robert FitzRoy for his South American surveys in the *Beagle* (1831–6). They are now held in the Public Record Office, Adm7/842 and 843. See also Hugo O'Donnell y Duque de Estrada, *El viaje a Chiloé de José de Moraleda (1787–1790)*, Madrid, 1990.

the flourishing state of our merchant marine formed almost our only topic of conversation; which was far too agreeable to be shared with others of less importance.

At three o'clock in the afternoon, after the pinnace was hoisted, we tried to take advantage of various light airs to resume our course, but we were not able to make any way for the whole night as we were virtually without steerage way apart from when we encountered a fleeting squall from the east.

13 January
In the morning a gentle breeze sprang up from the fourth quarter when we immediately altered course to SW and WSW with all sails set. We hailed the *Atrevida* to inform her of the previous day's news. To our great satisfaction we learned at the same time that the longitude obtained at noon the previous day by their chronometers, which agreed with each other, agreed to the minute with our longitude, referred to number 61 corrected by the equation mentioned above.

At noon our latitude was 58°6′ and longitude 20°28′ west of Montevideo. The merchant vessel was about three leagues from us to the SE. The wind soon backed from the fourth to the third quadrant and, with good prospects and a calm sea, we immediately altered course to NNW under full sail while the merchant vessel stood to the south away from the course followed by the corvettes and disappeared from sight at about six o'clock in the evening. At nine o'clock, however, we had to tack again to SWbyW as the wind had veered back to the fourth quadrant, with even lighter breezes than in the morning, shifting round to NNE with a most pleasing prospect so that, by hoisting full sail and steering WNW by compass, [14th] by noon we were in latitude 57°49′, longitude 21°56′. Various azimuths observed the following afternoon gave the variation as 28°25′ NE. The wind remained moderate in the first quadrant all evening but during the night it veered quickly through SE, south and SSW, finally falling to a dead calm. [15th] As a result Francisco Viana and Antonio Pineda went in the pinnace, the following morning, to the *Atrevida* and we were thus able to report to each other the results of our activities, not only in terms of the astronomical observations but also with regard to the detailed examination of the coasts already surveyed which kept us busy all day on both corvettes. Latitude at noon 57°7′ longitude by the three chronometers in agreement 22°36′, variation by two azimuths 27°40′. With regard to this last matter, we had compared the results of the two azimuth compasses, one an English instrument by Gilbert and the other by the instrument-maker Martínez of La Carraca. Much to our disquiet we saw that the latter differed considerably from the true value and the knowledge that the two instruments from the same makers in the *Atrevida* had sometimes agreed only made us take even more care and attention in our examination, but with no better result; so we delayed looking for the causes of this difference until the first port where circumstances would permit similar observations in greater detail.

The following afternoon we took advantage of the moderate westerly breeze to alter course to the north and as the following night (after a couple of hours of calm) brought gentle breezes from NW and NNW, we hauled our wind under full sail on the starboard tack until the next day [16th] when, with the breeze blowing fresh from NE and ENE, we were able to tack again to NNW.

With our latitude 56°48' by observation, longitude 22°48' from Montevideo, Los Evangelistas,[1] at the entrance to Estrecho de Magallanes, bore N24°E, 190 leagues distant on the chart by Belmonte and Galiano, which on this point agrees with the one by Churruca and Cevallos.[2]

The next two days' runs were very favourable. Throughout the 17th we were able to steer NNW by compass with fresh breezes first from east and SE, and then from SW and west. But a number of showers made our position at noon on the 17th doubtful. The following night we had to steer SW for a few hours with winds from the fourth quadrant, which, however, later backed to WSW and provided us with more favourable tacks until noon [18th], when our latitude was 53°16' and our longitude (according to the three chronometers) 22°28'. The wind did not take long to turn stormy, with squalls, heavy seas, and showers. We continued to steer NNW under the four principal sails and two reefs in the topsails until nine o'clock that night, when with the wind veering to the fourth quadrant, we altered course to the third quadrant.

19 January

At ten o'clock next morning we were totally becalmed and our position at noon was latitude 52°35', longitude 22°43' west of Montevideo, with Cabo Victoria bearing E5°N, forty-nine to fifty leagues distant.

At this point, having entered the Pacific I set out a plan of action which, even if modified according to circumstances, would still encompass reasonably well the objects of our expedition. These were (as expressed in our instructions)[3] to establish the longitudinal limits of the west coast of Patagonia from Cabo Victoria to Chiloé for the greater safety of the national shipping in these waters; not to endanger our vessels unnecessarily; to combine our tasks with those that had been carried out on other occasions at substantial cost to the treasury and which deserved public confidence; and finally to reach Chiloé at a suitable season for the more detailed and scientific survey of the coast further to the north, a task which should, according to the plan drawn up and approved, occupy the whole of the present year of 1790. I was well aware that various stretches of this coast had been surveyed on other occasions, though since not all the results were equally reliable, particularly with regard to longitudes, they could not yet be trusted by a navigator. At the same time I knew very well that the prevailing winds on this coast are stormy westerlies accompanied by high seas and, furthermore, in the coming months of February and March it would not be surprising if we were again to encounter those northerlies which, with good cause, are much feared along those coasts.

Cabo Victoria and Los Evangelistas at the entrance to Estrecho de Magallanes could already be considered to have been fixed with the greatest hydrographic precision which depended on reliable bearings taken by English sailors adjusted to the latitude and longitude of Cabo Pilares as determined by Churruca and Cevallos. The

---

[1] A group of rocky islands in the western entrance to Estrecho de Magallanes, named by the Nodales in 1619 in honour of the Four Evangelists; Narborough's Islands of Direction.

[2] During the surveying voyage in 1789 of the snows *Eulalia* and *Casilda* in 1789 under the command of Brigadier Don Antonio de Córdoba.

[3] Malaspina appears to be referring to the 'Plan for a Scientific and Political Voyage Around the World' that he submitted on 10 September 1788 to Antonio Valdés y Bazan, Ministro de Marina, which was approved on 14 October 1788; see pp. 311-15 below. No formal instructions were given to Malaspina.

survey carried out in 1765 by Piloto Machado and that recently made by Alférez de Fragata and Piloto Moraleda had been tied in to the position of Chiloé and the coasts to the south as far as approximately 49°30′. Therefore, the only stretch that could be considered as insufficiently well known was between 52° and 49°, where Sarmiento's account, although written by a seaman who was outstanding in his day, still retained many doubtful positions, not only in terms of latitude but also longitude, later clarified by the accurate work of Galiano and Belmonte.

In view of this and with good weather still allowing us to devote fifteen or twenty days to this survey without detriment to the other objects of our mission, I determined to make a landfall south of Sarmiento's Cabo Santa Lucía[1] so that everything in sight to the south would tie in, at least in terms of direction, to Cabo Victoria and from there to follow the coast northward as far as circumstances and prudence would allow. In that way the passage to Chiloé would not be slowed down since in the vicinity of the coast the prevailing winds are deflected somewhat to the south. Thus in a few days of good weather, I might solve a problem of great interest to our national shipping which is still subjected to a long and difficult passage, given the well founded suspicions of His Excellency Ulloa about the trend of the coast of Archipiélago de Chonos.[2]

The presence of seals and numerous birds, particularly Wandering Albatrosses, certainly confirmed in me the idea I had formed from seeing them on my earlier voyage, that they were in no way signs that land was near. On various occasions I have seen them fifty leagues from the coast and only eight leagues from coasts in depths of more than 120 fathoms. These albatrosses are truly enormous, as described by Mr Banks. We could have wished for a few hours of those spells of calm weather which, on his voyage, allowed him to hunt and kill these creatures with the aid of boats.[3]

Since remaining on the same parallel as the entrance to the strait would probably make our situation precarious and expose us to the currents which naturally set strongly to the east, my first intention today was to make some latitude to the north as soon as possible. Therefore when NNW and NW winds got up after midday, we steered to the west until half past four when, with a wind from SbyW, we tacked again to NbyW, still keeping the topsails double-reefed. The weather continued to be overcast with squalls and with very high seas from SW and NW.

Shortly after midnight, during a strong squall with heavy showers, the wind shifted to south and SSW, settling finally into SW with the skies and horizon clearing. This was too favourable an opportunity of closing the coast for us to miss. We signalled our impending change of course to the *Atrevida* and immediately bore up to true east with foresail and topsails set, making six to seven knots which we could not increase by setting more sail because of the excessive rolling. We signalled the *Atrevida* to clear away her anchors. [20th] The flocks of birds and pods of seals increased and by noon

---

[1] The western extremity of Isla Diego de Almagro.

[2] For Antonio de Ulloa see p. xciv, n. 1 above.

[3] 'As the weather was frequently calm, Mr Banks went out in a small boat to shoot birds, among which were some albatrosses and sheerwaters. The albatrosses were observed to be larger than those which had been taken northward of the Streight; one of them measured ten feet two inches from the tip of one wing to that of the other, when they were extended.': Hawkesworth, *Voyages*, II, pp. 66–7, 26 January 1769.

we were in latitude 51°57′ and longitude 20°47′ which put us twenty to twenty-two leagues from the coast according to the *Carta del Magallanes*. The weather remained fine, but in the afternoon the wind, now from WNW, increased greatly in strength with high seas, squalls and frequent showers which made it unwise to stand in for the land. Thus at four o'clock, when we found ourselves seventeen leagues offshore, we put about to SW being certain that had there been anything to be seen to the west of Cabo Victoria it would have been in sight. It was a rainy night with squalls and high seas. We hauled our wind to the third quadrant under full sails until eleven o'clock [21st] , when we tacked to NNW with winds from west and WbyS. Around noon the weather had improved and the outlook was favourable. We were in latitude 51°17′, longitude 20°46′, about twenty-two leagues from Cabo Santiago[1] bearing E21°N. It would have been useless to run along this parallel during the afternoon because it led to Canal de San Sebastián[2] and its narrow entrance would obscure our view of land, and to make matters worse we could not be fully confident of its latitudes. On the other hand, it seemed best not to wait for a latitude observation in case, close to the coast, a NW wind should blow up, but rather to proceed to the parallel where we intended to begin our next survey.

I therefore decided to close the coast to the north of Cabo Santiago and to do so early in the morning so that surveying for a whole day might produce the greatest results in the shortest time. Thus we sailed as far north as possible that evening with a moderate breeze from SW and WSW and quite a smooth sea. Having informed the *Atrevida* by signal that we intended to stand in for the land during the night, we altered course to east at eight o'clock. By midnight we had covered about six leagues when, the weather having turned hazy, we altered course to the north under topsails, promising ourselves that we would sight land at daybreak.

22 January
In fact, at half past three in the morning we could already discern on our beam some mountain ranges and at the same time the *Atrevida,* which was some distance astern confirmed this by signal. The land we had in sight stretched from NE to the ESE, high and rather steep and very similar to the land we had seen in the eastern part of Tierra del Fuego. About six leagues from the coast we obtained no bottom with 120 fathoms of line. The number of birds in sight was not excessive and the weather promised to meet our wishes. We certainly believed (as did the *Atrevida* which was within hailing distance) that the coast that was in sight was the land which, according to Sarmiento, lay between Cabo Santiago and Cabo Tres Morros.[3] He had drawn in a channel which might lead WSW, neither end of which joined the adjacent land, which ran from NbyW to SbyE, true. We immediately crowded on sail and as the wind had steadied to a moderate breeze from the west, we expected to extend our survey to beyond Cabo Corso.[4]

At a quarter to six we observed longitudes by chronometers which, once adjusted

---

[1] The SW extremity of Isla Duque de York.
[2] Probably Canal Concepción, which leads east of Cabo Santiago.
[3] Cabo Tres Picos, the NW extremity of Isla Madre de Dios.
[4] Cabo Primero, the southern extremity of Peninsula Corso, the SW extremity of Isla Mornington.

by the equation, gave our longitude 20°9′30″ west of Montevideo; variation observed with different azimuths 21°28′. Other hour angles measured at ten-thirty gave the longitude as 20°10′48″ and at noon our latitude was 51°4′30″.

Much as we tried, we made very little progress along the coast during the morning as the wind became much more moderate and by eleven o'clock it had swung round to WNW, fresh, obliging us to alter course to SW. Nonetheless, we had been fortunate enough to sight the whole of Canal de la Trinidad[1] at eleven o'clock and also the extreme northern part of Cabo Corso. There were no longer any doubts about its latitude and longitude and we saw with relief that we had reached the same point from Chiloé as the hydrographic work of Pilotos Machado and Moraleda.[2]

The channel in question was about six leagues wide at its mouth and our bearings placed Cabo Corso in latitude 49°27′30″ and longitude 69°48′ west of Cádiz. It was, therefore, further west than *Carta del Magallanes* indicated but much further east than was suspected by His Excellency Ulloa.

At a great distance behind the coast could be seen other extremely high mountain ranges the elevation, position and direction of which, together with the fact that they were covered in snow, left us in no doubt that they were part of the Andes. Everything in view appeared arid and (as far as one could tell from such a distance) seemed to consist of dark granite.

After noon, we were able to observe longitudes by lunar distances. The mean of sixty-eight sets which agreed well with each other gave a result 12′ east of the longitude by chronometers. The weather soon thickened, with rain, squalls and high seas which obliged us to steer WSW and west by compass, and with winds from the NW and NNW under only the foresail and double-reefed topsails. It even became difficult to keep company with the *Atrevida* because of the gloomy weather, especially at night. It was difficult for us to take observations for latitude and longitude at [23rd] noon the following day. By now the high seas were causing us considerable leeway. Later with the wind less stormy from the fourth quadrant, but always overcast with rain, we were able to proceed from four o'clock on with double-reefed topsails. At the same time we were accompanied by a number of seals and quite a few birds. They became even more numerous the next morning [24th] when the wind, after constant rain, shifted to the west and the skies cleared a little allowing us to steer NNW under a press of sail. At noon our latitude by observation was 50°56′, 20′ further south than by dead reckoning and 50′ further south than our latitude on the 22nd. Thus, with winds which were sometimes moderate, sometimes calm and sometimes fresh from the west, we sailed that day and the next morning [25th] steadily NNW to north under full sail. However, from ten o'clock in the morning the wind began to increase again and even shifted to the fourth quadrant. At noon our latitude was 49°32′ and longitude by the three chronometers was 22°17′. We were thus about thirty-five leagues from Cabo Corso.

Since our survey had reached this latitude on the 22nd, we tried steering to the north to reach the 49th parallel so that (without abandoning our progress towards our

[1] Golfo Trinidad, entered between Cabo Tres Picos and Cabo Primero.
[2] I.e. the southern limit of their survey.

destination of Chiloé) it would be easy to link another considerable stretch of coast with our previous [survey]; but we had hardly started to come closer [to the coast] that afternoon when, with winds from west and WNW, the weather closed in with squalls and heavy rain and the sea rose considerably.

This third attempt left me in no doubt that in these latitudes the proximity of the coast not only produced thick weather but also winds from NW, with which it would be difficult (if they shifted to the west, as might be expected) to stand offshore without making considerable loss of latitude, which would delay our arrival in Chiloé. The coast surveyed by Piloto Machado in 1769 extended to latitude 49°22′ and even if it were not easy to reconcile his position for Cabo Corso with our own, it must be believed that Alférez de Fragata Moraleda (whose accuracy was known to me) would have clarified any doubts on this score in his last expedition to these coasts during 1787 and 1788. However, if that survey was not adequate for maritime safety, it would have to be undertaken in greater detail and with smaller vessels from Chiloé, in which case any further delay along these coasts would only be a waste of time.

With these thoughts in mind and since the weather was now completely overcast with squalls and showers, at six o'clock in the evening, after lowering the topgallant yard, we tacked to the SW and sailed under the four principal sails with the topsails double reefed. A sudden change of wind took us aback at ten o'clock at night but the wind soon returned to the fourth quadrant and we tacked again to the third.

26 January

At dawn it was overcast with rain and a strengthening wind accompanied by a very high sea which obliged us to keep as much sail as possible to make some westing without going too far off course. Despite this, the latitude of 49°55′ which we were able to observe with some uncertainty at noon showed that the currents had increased our latitude to the south from 12′ to 15′ compared with dead reckoning. Our longitude was 23°00′ west of Montevideo.

In the afternoon and early evening, far from abating the weather became worse and strengthened greatly from the fourth quadrant. At five o'clock, after the wind backed to west, we altered course to the north, but at nine o'clock we had to haul off to port again. From then on, thick cloud, rain and squalls increased to such an extent that by eleven we had to run, at some risk, under only the foresail and the double-reefed topsail. The wind and the rain continued unabated until four in the morning.

27 January

But by that time the wind had backed to west and WSW and the skies and the horizon cleared progressively so that we were able to carry a press of sail steering NWbyN by compass, making four to five knots despite an extremely high sea from the NW, which sometimes made us pitch excessively. Our latitude at noon was 49°32′ and longitude 23°26′.

Thus for a period of five days we had not been able to make more than half a degree's progress northward in spite of the fact that we had put our topmasts at risk more than once and that our ships were undoubtedly of the best quality both for sea-keeping and for tacking to windward. Although we had made good about thirty leagues to the west, this had been necessary to find more suitable weather for our passage and particularly to avoid the foul weather which had until then made it impossible for us to see even the closest objects and had raised exceedingly high seas.

For this reason, we decided to abandon these coasts and sail directly to Chiloé in order to examine the earlier surveys by Machado and Moraleda and plan our forthcoming activities so as to avoid any useless duplication of tasks or expenses while ensuring that our country would not lack the detailed and reliable information which it could rightly demand of us. The coming months of February and March would certainly not be wasted if we conducted a full examination of the geographical, political and natural situation of our colonies from Chiloé to Valparaíso, working in such a way as to be beyond reach of the northerlies and so make good use of the time allotted to our voyage.

28 January

Convinced by our recent experiences that approaching the coast, unless necessary for surveying it, would only cause considerable delay to our passage, we altered course to NNW keeping our advantageous longitude so that, despite the moderate breezes that sprang up from west and WNW the following night, we were able to continue on a favourable course and by noon we were in latitude 47°41' and longitude 23°57', finding ourselves with breezes, outlook and seas which for the first time proclaimed the idea of a Pacific Ocean.

We had missed no opportunity to observe variation which this evening we calculated to be 19°2' and 18°50'NE by western amplitude. The night was extremely pleasant with fair weather. We stood to the north under full sail and next morning [29th], when the *Atrevida* was within hailing distance, she told us that the longitude by their chronometer 71 was the same as by our chronometers, but a little further to the east. The results of number 10 came closer to the mean of their distances from the 22nd and therefore to ours but, as mentioned, they were 6' east of the mean of our chronometers.

The health of both our crews was at the time extremely pleasing to us: no one was sick in the *Atrevida* and on board this corvette the sick list included only the master gunner, already recovering from the symptoms of an inflammatory fever accompanied by considerable bleeding, and a marine gunner with a weak constitution who had been suffering from his habitual rheumatism since Puerto Egmont. Three or four sailors who had come down with the beginnings of chills and fevers, or perhaps overheated blood, had recovered in only a few days and, even more surprisingly, despite the cold and the rain, there had not been the least sign of any venereal disease in all that time.

The daily pint of wine was suspended from this date and we issued a ration of sauerkraut three times a week, advising the *Atrevida* by signal of both these changes for her to copy. At the same time, the evening *sopas en aceite*[1] were replaced by gazpacho[2] and these, together with good ventilation below decks and cleanliness, were the main elements in our method of maintaining good health until now.

[1] Sops (probably ship's biscuits) soaked in olive oil.
[2] See p. 5, n. 4 above.

At noon our latitude was 45°52′, longitude 21°56′ and variation 18°35′. Punta Quillán[1] on Isla Chiloé (according to the position charted by His Excellency Ulloa) bore N27°E[2] eighty-four leagues distant and Islas de San Fernando[3] and Inchin[4] almost due east forty-five leagues away. We altered course to NNE by compass with a moderate breeze from the west which later eased more and more as evening fell and even veered a little to a northerly direction. The weather promised to remain extremely pleasant and the difference in the rigging showed clearly that the *Atrevida* had achieved a fair advantage in speed over this corvette.

Azimuths obtained during the day's run gave the variation as 18°35′ and by amplitude next morning 17°55′.

From early that evening the weather had begun to close in with a thick haze and the wind veered from west to NW. Nevertheless we made the most of it with all sails set and a speed of six knots. From eight o'clock in the morning, when soundings were being taken both corvettes noticed the sea had taken on the colour of shallow water.

30 January

As the day grew on, the haze thickened and then degenerated into a constant drizzle which made it impossible to take any observations for latitude or longitude. At noon we had to rely on dead reckoning which put us in latitude 44°25′ and longitude 19°53′. As a result we deduced that we were on the same parallel as Isla Guafo and (according to the chart by His Excellancy Ulloa) Punta Quillán at the southern tip of Isla Chiloé bore N10°E forty-nine leagues.

In the evening the weather closed in considerably and the wind veered to NWbyN which obliged us to steer NEbyE true and to haul off to port. At the same time, looking at the longitude of the anchorage at Chiloé on a chart which had been sent to me from the Archivo de Indias in Madrid, I discovered that the longitude assigned to it and which could therefore be taken as that of Punta Quillán, given that the trend of the coast was north-south, was one degree further west than on Ulloa's chart. Thus, if this difference really existed and the chronometers had an error to the east, as seemed to be indicated by the distances, we could not be more than a few leagues from Isla Guafo and even from the mainland [of Chiloé].

If so it would have been unwise to continue on the same tack, especially as the gloomy weather was not clearing and the wind was driving us on a course that would take us close to land. At six o'clock in the evening we sounded and finding ninety fathoms, rock, we went about and steered west and WSW to take us away from the coast which by then we could not doubt must be very close.

But as soon as night fell, the weather began to close in with violent gusts which slowly backed from NW to west with quite a lot of rain and finally, at half past eight,

---

[1] Cabo Quilán, the SW extremity of Isla Chiloé.

[2] The published journal has E27°N (i.e. 063°) which clearly should be N27°E (027°), not only the more convential method of expressing a bearing, but the bearing calculated from the given position, taking into account variation, is indeed N27°E.

[3] Possibly Isla Tenquehuén in the southern part of Bahía Darwin.

[4] Islas Inchin, close SW of Isla Tenquehuén. In 1769 Piloto Francisco Machado examined the area from Islas Inchin to Isla Campana, 140 miles to the south.

backed to a moderate gale from WSW while the sea rose considerably and the mist began to clear. We immediately wore ship and heading first NNW then NW, taking in two reefs in the topsails and finally sailing with foresail and topsail, but keeping the fore-topsail clewed up to be able to manoeuvre as required. I was in no doubt that the land must be very close; therefore, apart from the stated precautions with the sails, I ordered that the ship should be steered with the greatest care and that both watches should be on deck ready for any manoeuvre which might require both speed and safe handling.

Indeed, it had just turned midnight when, with the aid of the Moon, we sighted land four leagues to starboard. It was still a cloudy night with a moderate gale from SW and a very high sea. The land in sight must undoubtedly be part of Isla Chiloé as it ran from NNE to NEbyE by compass forming a kind of bay without much elevation and then fell away from sight towards the north and south so that we actually believed that it could be the vicinity of Punta Quillán.

A signal was immediately made to the *Atrevida* to crowd on sail and, steering NWbyW under the four principal sails, with the topsails double-reefed, we were soon clear of the coast and, despite the high seas and strong wind, were out of danger. At half past one the coast was sighted bearing true east about five leagues distant. From that time on, we attempted to keep a very precise dead reckoning in order to tie this stretch of coast to our observations the following day. By half past two we had lost sight of the coast although the Moon was shining and the sky was already beautifully clear.

At dawn the same stretch of coast bore ESE from us by compass about twelve leagues distant. We bore up to true north and soon afterwards to NE running under a full press of sail with moderate breezes from SW and with seas not so rough as on the previous night.

Our latitude at noon was 42°39′ and our longitude 19°4′; from this fact and our dead reckoning, the land seen the night before had to be in latitude 43°33′ and longitude [blank in the original] west of Montevideo.

Shortly afterwards land was sighted again about nine or ten leagues away, which confirmed to us the great extent to which it stood out from the north-south direction of the island, extending as it did towards the west at the southern part or towards Punta Quillán.

The land now in sight was much higher than that seen the previous night, and sloping down from its mid-point towards the north, it really presented an appearance that was as attractive for its leafy woodlands as it was fearsome for the steepness of its coasts which, irregular though they were, showed not the slightest sign of any harbour. The irregularities were those which run down from the heights of Cucao[1] towards the southernmost tip of the island, so that we could hope to reach Puerto de San Carlos[2] next morning, especially since the winds seemed to have settled down into the SW with excellent prospects.

We steered NEbyE by compass until three o'clock in the afternoon in order to reach a position four or five leagues off the coast, which we later continued to survey

---

[1] Alturas de Cucao; prominent heights attaining elevations from 600 to 900 metres near Cabo Quilán.
[2] Now Bahía de Ancud.

for the rest of the afternoon steering to the north. At quarter past five, our longitude by observation was 18°22′, which confirmed our noon observations and, therefore, those of the previous night.

By seven-thirty we were a good distance from the coast and considerably to the south[1] so that we need have no fear of falling to leeward under the influence of the currents, and we tacked for the offing with only the topsails set. At midnight, seeing that the weather had greatly improved, we bore up to the east once more under full sail so that by dawn we were only three or four leagues from the coast, although the amount of haze meant that we could not make out its shape clearly.

The chart we had received from the Archivo de Indias, as I have already mentioned, was no use to us on this occasion but rather threw us into doubt and confusion. To the south of the true Punta de Cocotuya[2] there is an bay which is not very deep but is encumbered with many small rocky islets, with some hills that look a little like breasts in the high land in its extreme south.[3] The distances on the chart were evidently erroneous since the mainland around Punta Capitana[4] could not possibly be made out, nor could we see any other entrance with extremities that lay almost north and south; and so we mistook Punta de Cocotuya for Punta Capitana and ran into the bay which lay before us.[5] We signalled the *Atrevida* to take a sounding but no bottom was obtained. Soon afterwards, within hailing distance, we discovered that they felt the same uncertainty as we did.

At nine o'clock we were no more than two leagues from the head of the bay and although many signs seemed to indicate that there should be an anchorage there,[6] we could neither confirm nor dismiss this possibility because of the wind which had by then died down almost completely. As a result, the sea was setting us hard into this bay so we decided to tack to the west with a gentle breeze from NWbyN, hoping that its latitude and a more favourable wind would allow us to discover what the truth was.

1 February

At half past eleven, after a passing shower, the wind veered to a moderate WNW breeze. We immediately hauled round to north and after the haze obscuring the sky had cleared a little at noon we were able to obtain our latitude 41°55′ by observation, while the *Atrevida* signalled that her observation was 41°56′. This was the latitude of the anchorage according to the chart we had and this seemed to be confirmed by the observations made during the last war by His Majesty's ships under the command of Jefe de Esquadra Vacaro.[7] Thus, as I thought it extremely dangerous to fall to leeward

---

[1] This appears to be a slip for north since Malaspina was steering to the north.

[2] Punta Caucaguapi.

[3] Bahía Cocotué, situated about 6 miles south of the NW extremity of Isla Chiloé, with Tetas de Teguaco (Tetas de Cocotue on Malaspina's survey), twin hills, standing near the southern part of the bay.

[4] Punta Capitanes situated on the mainland in 41°09′S, 73°56′W and about 35 miles north of the NW tip of Isla Chiloé.

[5] Malaspina was attempting to round Península Lacui to enter Bahía de Ancud, but having mistaken Punta Caucaguapi (Punta de Cocotuya) for Punta Capitanes he was some 30 miles farther south and so entered Bahía Cocotué instead.

[6] I.e. the anchorage in Bahía de Ancud.

[7] Lorenzo Vacaro.

and as no previous identification had been made to our complete satisfaction, I decided to bear away again with a moderate wind and overcast skies.

The *Atrevida* then signalled that she wanted to speak to us and so we shortened sail to wait for her, but as the wind was rapidly dying down and since, with land now in sight, I considered it important to remove all doubts about the location of the anchorage, I signalled back that I was going ahead to reconnoitre the land and we crowded on sail while at the same time lowering the pinnace.

Eventually, at a distance of about one league from the head of the bay, we confirmed that there was no opening whatsoever. The soundings which at noon had been thirty fathoms were now twenty-two fathoms, shell and gravel. After recalling the pinnace which we had earlier dispatched with a *pilotin* to examine a bend behind which we suspected that there might be an opening, we hove-to until the *Atrevida* closed us and Don Felipe Bauzá went on board her to compare our charts with hers.

At about half past two, the pinnace returned with some other charts of the anchorage which the *Atrevida* had on board, but these did little to clarify its true position and direction. We hoisted the pinnace immediately and under all sails hauled our wind to the third quadrant using the light airs from NW which picked up strength by five o'clock and gave us the chance, once we had got some distance from the coast, to alter course to NbyE. The *Atrevida* had made considerable progress to windward, so we signalled to her to go on ahead and find the anchorage and she was able to ascertain (as she reported by signals) that the anchorage was on the other side of the point we had sighted, which was actually Punta de Cocotuya.

At half past six, with the coast about half a mile on our beam, the slight wind proved insufficient for us to steer to the north, while Punta de Cocotuya, which we could not clear, bore NNE about two miles from us. The *Atrevida*, which we had ordered to come about, did so straight away; but not seeing the signal to close which we had hoisted at sunset, probably because of the overcast skies, she remained to windward about two miles from us.

During the night the wind veered from west to NW. We hauled the wind under full sail to the third quadrant and then we put about to NNE with the coast about four leagues distant, and shortened sail as the lie of the land seemed to indicate that the current was setting us considerably to the north.

From three o'clock in the morning the weather turned ugly and the wind increased from the NW with the sea becoming quite rough. By means of a rocket we had ascertained the position of the *Atrevida* which was well to windward. However, when we decided to go about to the starboard tack she apparently did not see the signal to go about and so next morning she was nowhere to be seen. The extremely overcast weather and constant downpours which accompanied the now tempestuous winds from NW made matters worse.

2 February

We took in two reefs in the topsails and sent down the topgallant yards before continuing to make an offing with the foresail and the single topsail lowered, the only sails that the wind and the sea would allow, and even then with great risk of damage.

No observation was possible today nor did we see the *Atrevida* although we knew

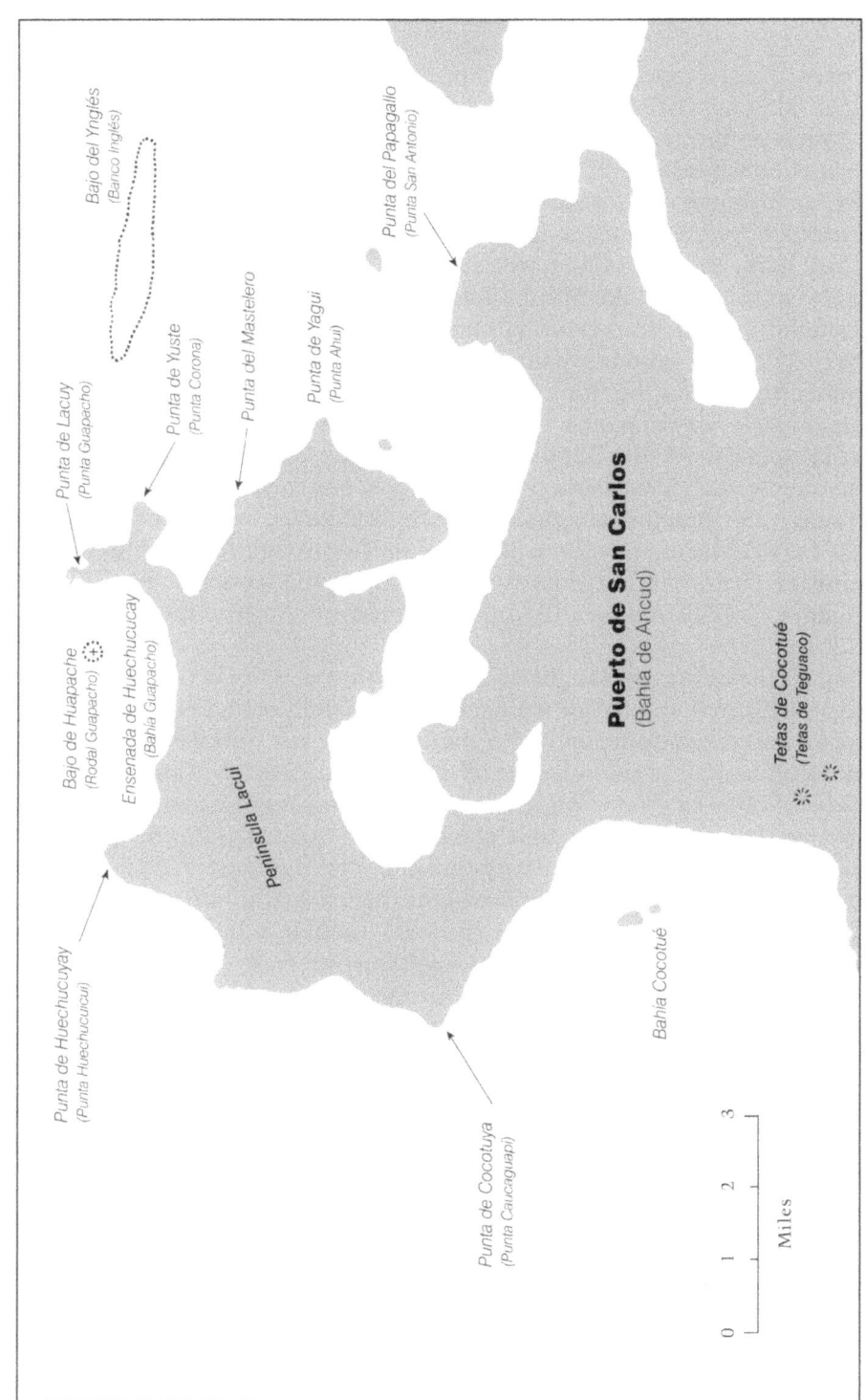

Fig. 5. Puerto de San Carlos, February 1790

that our separation would not be long since we would meet again in Puerto de San Carlos.

The overcast skies continued with rain, squalls and high seas until midnight the next night, although from eight o'clock with westerly and WSW winds we had altered course to the fourth quadrant. Later, as the weather began to look much brighter and the wind shifted to SSW, south and SSE we altered course to NEbyE to close the coast, under the foresail and a single reef in the topsails. Helped by the bright moonlight, we managed to sight the coast at a distance of five leagues by about two o'clock in the morning. There were high sierras among which we could clearly see two lower hills shaped like breasts. They stood out to the right of a large bay and so we thought (imagining an error in longitude) that we were at the entrance to the anchorage.[1] [3rd] We sounded without obtaining bottom and so remained in the offing under topsails, waiting for daylight to retrace our steps with more success.

At five o'clock next morning we were only two leagues from the coast but the sight of the open bay and the fact that some of the points we saw to the north seemed to be the ones we had identified on the first day soon persuaded us that this was not the anchorage. We therefore concluded that the anchorage must lie to the north and that what we were seeing now were the Tetas de Cucao which we knew were also a feature on the coast.[2] At that time we were closing the land and the falling wind shifted suddenly to NE and then to north, obliging us to steer to the west once again under full sail.

Making for the offing, we caught sight of the *Atrevida* to the WNW at a distance of three leagues. From then on, our main aim was to rejoin her and with this sole purpose we tried to take advantage of the variable light airs from the first and fourth quadrants. Our latitude at noon was 42°5' and longitude 18°12' [from Montevideo]. At the time of observing hour angles we had taken bearings particularly of Punta de Cocotuya, the latitude and longitude of which we had already fixed on the 1st. In consequence we found today that the headland's longitude was 17°53'24", a difference of less than one minute from our original determination of 17°54'20". It was a very rainy afternoon and the wind was extremely variable and light, so that we could not rejoin the *Atrevida* before nightfall. Then the wind began to increase from NW and so we kept under way for the rest of the night under only a foresail and with the topsails lowered, manoeuvring as necessary to rejoin the *Atrevida*.

4 February

At dawn it was overcast with some rain, a strong wind from NW with violent squalls and a very high sea. We could not make out the land and the weather made it inadvisable to search for it, so we took in two reefs in the topsails.

At eight o'clock, however, we felt the wind suddenly back to the south. We immediately took advantage of it even though the overcast skies made us very fearful that it would be short-lived. We started to crowd on sail and steer to the east. At ten o'clock we were already heading NE under full sail and at eleven the skies and the horizon

---

[1] The bay was once again Bahía Cocotué, with Tetas de Teguaco the two hills shaped like paps.

[2] Malaspina appears to be confusing this feature, now called Tetas de Metalqui, with Tetas de Teguaco, some 21 miles farther north.

cleared and we could see the two stretches of coast between which we had found ourselves embayed the day before. We immediately set a course to close the coast and at noon the latitude by observation was 42°00′ and with a strong wind still from south we altered course to close Punta de Cocotuya from which we were about six leagues distant at this time.

At half past two it bore east of us about four miles distant and as some of the charts showed shoals in this vicinity, we steered true north until we were on the parallel of Punta de Huechucuyay.[1] Eventually we bore up to close this particular point and the succeeding ones to the extent that the various hazards surrounding them allowed. The wind favoured us on this occasion, the weather remaining clear and the breeze fresh, shifting slightly to the west as we altered course to the south. But the ebb tide was very strong, particularly as we rounded Punta de Lacuy and Punta de Yuste,[2] causing us considerable delay, with the result that we only reached Punta de Yagui[3] at nightfall. By that time the wind had died down completely and, with the tide continuing to ebb, we dropped anchor in six fathoms, sandy mud, about four or five cables from the battery of Yagui, from where they had already hailed us by megaphone as we passed.

From mid-afternoon on we had noticed gunpowder smoke or cannon fire on a hill-top close to [Punta de] Cocotuya. The battery at Yagui responded and we realized that it was undoubtedly the approach of unknown ships that had set them off.[4] Hence we hoisted our ensigns and these were acknowledged by the battery and also by the town fort which we could see close to Punta del Papagallo.[5]

Although it was difficult to believe that a chart submitted to His Majesty could contain as many errors as we had just discovered, the one that was intended to serve as our guide had proved that our doubts were not unfounded. The loss of four days of our time which those errors had caused seemed a small price for freeing navigation of a chart which might have brought about such fatal consequences.

---

[1] Punta Huechucuicui, the NW etremity of Península Lacui.
[2] Punta Guapacho and Punta Corona, the NE points of Península Lacui.
[3] Punta Ahui, the easternmost point of Península Lacui.
[4] Malaspina was convinced that 'any intent by any European power against the Pacific coast of America in the latitudes of Patagonia is an imaginary danger which should not for a moment concern our defence policy': Pimental, *Descripciones y reflexiones políticas*, p. 55. See also p. 90, n. 1 above.
[5] Punta San Antonio on which Ciudad de Ancud (Pueblo de San Carlos) stands, situated about 2 miles SSE of Punta Ahui.

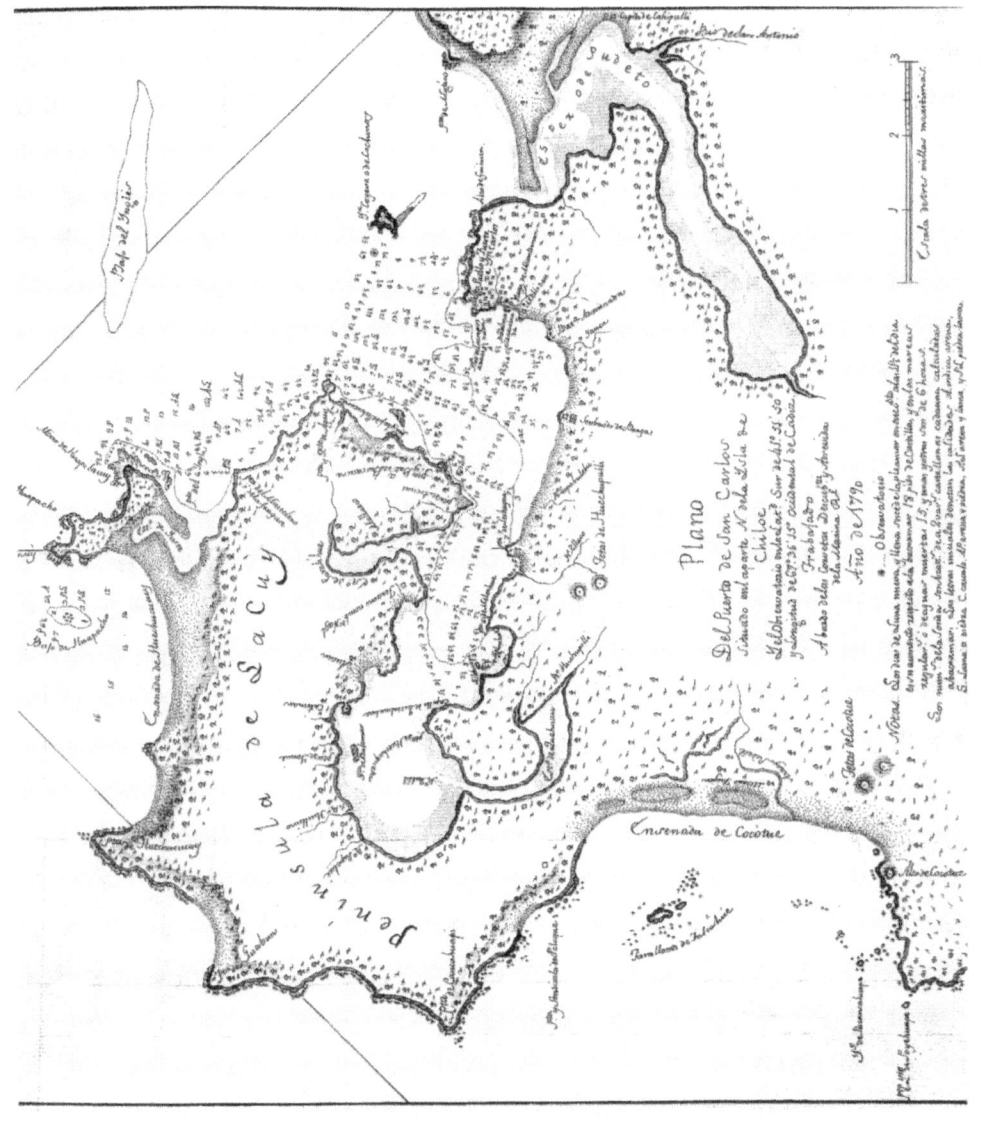

19. *Plano del Puerto de San Carlos situado en la parte N dela Isla Chiloé ... trabajado abordo delas Corbetas Descub.ᵃ y Atrevida ... Año de 1790. Museo Naval, Madrid*

CHAPTER 5

At Puerto San Carlos de Chiloé

Just after nightfall, the senior sergeant of the garrison Don Antonio de Mata,[1] royal army engineer Don M. Olaguer,[2] and senior *piloto* of the royal navy Don José Moraleda, breveted *alférez de fragata*,[3] came on board by piragua[4] from the settlement. The first two, as well as presenting the compliments of the acting governor Coronel Francisco Garós,[5] carried out the health inspection. The third, who had been employed in surveying the coasts of this island for the past three years, was to hand over all his plans to us by order of His Excellency the Viceroy of Lima. The Viceroy had already in the previous November sent to the port an atlas of maps of these shores drawn up by Don Lázaro Ribera[6] to be handed over to me on my arrival together with a chart of the coastline of Peru from Chiloé to Lima, surveyed by Piloto Moraleda.

This gentleman [Moraleda] had rendered considerable service to the crown, and indeed to humanity itself, working with uncommon steadfastness and intelligence on plans of the ports and on establishing most precisely the astronomical positions of the

---

[1] For maps attributed to Mata see P. Torres Lanzas, *Catálogo de mapas y planos, virreinato del Perú (Perú y Chile)*, Madrid, 1985, pp. 83-6.

[2] Manuel Olaguer Felíu was born in 1759 and posted to Valdivia in 1788; he accompanied Francisco Garós to Chiloé in December 1788, returning to Valdivia in 1790. He took charge of the repopulation of Osorno, which had been made uninhabitable by Indian hostility, in 1796.

[3] For Moraleda, see p. 119, n. 2 above.

[4] A term originally applied to a long narrow canoe-like craft made from a single tree trunk, but later, as in this instance, to a locally built two-masted sailing barge.

[5] Francisco Garós replaced Francisco de Hurtado y Pino in charge of Castro, the chief town of Chiloé in 1788, when the salute fired in his honour caused part of the fort to collapse. Manuel Felíu Olaguer, born in 1759 and posted to Valdivia in 1788, accompanied Francisco Garós to Chiloé in December 1788 and returned to Valdivia in 1790.

[6] Lázaro Ribera had published his finding in 1782, before becoming Governor of Moxos in Peru and joining the commission on boundaries with Portuguese territories in Montevideo in 1784. His *Descripciones exactas e historia fiel de los indios, animales y plantas de la provincia de Moxos en el Perú*, together with another manuscript called *Libro de las maderas* and *Relación de Gobierno*, dated 1792, were given by Ribera in person to Godoy in Aranjuéz in the summer of 1794. His subsequent career as Governor of Paraguay from 1795 to 1807 revealed him as a reformer somewhat in Malaspina's mould – interested in economic liberalization and peaceful compromise with hostile Indians – albeit distinctly authoritarian. M. Palau and B. Sáiz, eds, *Moxos. Descripciones exactas e historia fiel de los indios, animales y plantas de la provincia de Moxos en el virreinato del Perú por Lázaro de Ribera, 1786-1794*, Madrid, 1989.

entire coast, having adopted the longitudes of Father Feuillée and M. Frézier[1] and observing for himself all the latitudes with standard instruments. Lately he had been entrusted with the single-handed hydrographic survey of this island, in a wretched, poorly equipped piragua, but he had nevertheless completed it, partly by land and partly by sea, so that the survey of this section of the coast, including Isla Guafo, could now be regarded as properly completed. At nine o'clock they returned ashore, charged by me and by the captain of the *Atrevida* with offering to the Governor the most sincere expression of our gratitude.

The night was almost completely calm and very bright and pleasant and at nine o'clock the next morning [5th] with the first of the flood, we weighed anchor and were towed into port under staysails, anchoring at half past ten in the vicinity of an excellent stream and a thick wood where we could take on all the water and wood we needed.

Earlier, at eight o'clock, I had landed with the astronomical instruments, some from this corvette and others from the *Atrevida*, and accompanied by Don Dionisio Galiano, I went to the town, with the double intention of calling on the Governor and of setting up the observatory so that our chronometers could be rated once again, and reliable observations taken for latitude and longitude. We found a house that was very suitable for the purpose, and the astronomical clock was immediately set up. We left a *pilotín* and a marine there to guard the instruments.

When I arrived back on board I found that Teniente de Navío Valdés had moored the corvette north and south, with the bower cable laid out to the north, towards the entrance to the harbour. The *Atrevida* was some three cables from us and in this position the observatory flagpole bore S81°E, two miles distant from us. The land we could see presented a pleasant appearance and the port, on account of its position and safety, looked worthy of serious attention.

All officers went ashore in the afternoon, and the same night Galiano, Concha and Vernacci observed various meridian altitudes to north and south with the astronomical quadrant to determine the latitude of the observatory. This was found to be 41°51′50″. They intended to observe the occultation of a star by the Moon at three o'clock in the morning. Such is the thankless nature of astronomical observations, however, that the task was rendered impossible by some cloud which interposed itself just three minutes before the event.

6 February

As the next day dawned with fairly clear skies, all the officers wasted no time in going

---

[1] Louis Feuillée (1660-1732) was a Capuchin, who, thanks to prodigious mathematical skill, became a pupil and protégé of Cassini, whom he accompanied to the Levant as a chart-maker. After official cartographical expeditions to French islands in the West Indies he obtained a royal commission to undertake a mission of botanical, zoological and cartographic research, 'travaillant', as he said, 'à la perfection de l'astronomie, de la géographie et de la navigation,' to Chile and Peru from 1707 to 1711. He became embroiled in a controversy with Frézier, whose expedition began in the year Feuillée's ended, over corrections suggested by the latter to some of his readings of co-ordinates on the Chilean coast. See L. Feuillée, *Journal des observations physiques, mathématiques et botaniques faites par ordre du roi sur les côtes orientales de l'Amérique méridionale et aux Indes occidentales*, 3 vols, Paris, 1714-25.

to the observatory to measure lunar distances which would contribute to the determination of longitude. At the time of the observations, the sky and weather were not as favourable as I could have wished. The Moon was visible for only a few brief intervals, and despite our vigilance we only managed fifty sets of distances which agreed pretty well, and determined the longitude to be 17°14′19″ west of Montevideo.

The day was made particularly interesting to us by the official visit paid to the Governor by the *cacique* and a number of warriors of the Wiliche Indians[1] who had recently arrived in the town. For some years now Governors of Valdivia had made successful efforts to be on good terms with the Juncos and the Wiliches in order to open up, with notable advantage to the crown, a means of communication by land between Valdivia and Chiloé which might also in time be extended to Buenos Aires. They had therefore been surveying the regions of the interior, of which a more precise account will be given when we have been able to obtain one.[2]

The *cacique*'s speech to the Governor was lengthy and dignified.[3] It was interpreted by a sergeant from Valdivia who for eleven years had lived almost entirely among these Indians. He then replied in the same language saying that the Governor had said that His Majesty had entrusted him to look after them here and in the city of Valdivia.[4] The speech had been preceded by each of the Indians embracing the Governor and by a second circuit during which they shook hands with every one of us

[1] Malaspina's attitude to the Wiliche (also spelt Viliche or Huiliche) Indians should be considered against the background of their existing reputation as among the most 'barbarous, troublesome and menacing' of the South American cone – to quote the botanical explorer, Hipólito Ruíz, who preceded Malaspina in the region by a few years. Stories of their cannibal rites and hostility to cities were part of a barbaric stereotype, which makes all the more remarkable Malaspina's willingness to see them as potential partners in empire-building. See R. Evans Schultes et al., eds, *The Journals of Hipólito Ruíz, Spanish Botanist in Chile 1777–88,* Portland, 1998, pp. 208-9.

[2] Attempted peaceful assimilation gradually replaced war as a means of extending the Spanish frontier in America in the second half of the eighteenth century. Of the native groups Malaspina encountered on Chiloé, the Chilotas and, less securely, the Pehuenche, had already accepted Spanish sovereignty, while the Wiliche remained unabsorbed. (See J. Hidalgo et al., eds, *Culturas de Chile: Prehistoria desde sus orígenes hasta los albores de la conquista,* Santiago, 1989, p. 10). The situation of the Wiliche appeared, from Malaspina's point of view, paradoxical. They were agriculturists and therefore on the usual scale more 'civilized' than their neighbours, who, nevertheless, by virtue of their hostility to the Wiliche had established themselves as the supposedly natural allies of Spain.

[3] The use of the term *magestuosa* (dignified) reveals Malaspina's positive view of the civilized status of native American peoples and a concession to their claim to possess sovereignty in their own right. See A. Pagden, *Lords of All the World: Ideologies of Empire in Spain, France and Britain, 1500–1800,* New Haven, 1995. Malaspina's perception of the Indian representatives as spokesmen of an independent nation was consistent with the practical policy, embraced by the governor, of treating with them on equal terms. Catiguala's speech was recorded, partly in the words of the original language, in Viana's account of the meeting: Novo y Colson, *Viaje politico-cientifico,* p. 581.

[4] The main purpose of the policy to which Malaspina alludes was the re-settlement of the town of Osorno, which had been destroyed by the Mapuche in the early seventeenth century, for which it was necessary to open a land route between Valdivia and Chiloé, which officials dubbed 'the colony's outer defence'. Such a road would have to cross Wiliche territory. Despite the failure of earlier efforts, the project made some progress in the years before and after Malaspina's visit. Between 1777 and 1787 a string of missions was opened from Valdivia to Río Bueno but in 1792, after an apparent entente encouraged the Spaniards to attempt to extend the road, the Río Bueno station was destroyed in a Wiliche raid. The

Plate 20. Catiguala, chief of the Wiliche Huiliches Indians and his son, 1790, by José del Pozo.
Museo Naval, Madrid

present, saying the word *comzà*, which means friend. There followed a brief speech by another two Indians, the one apologizing for the absence of the *cacique* of Río Bueno who had a son gravely ill, the other announcing that he was very pleased to have made this journey to the town to which he had been invited by Cacique Catiguala. Their music, which was a passable imitation of the sound of horns, was produced with two very long canes which ended in a much wider opening. This was closed with a tree leaf, leaving just a small outlet for the breath, which was blown in at the other end from the side. Finally, they sat down, drank brandy and bade us farewell.[1]

The next night Galiano, Concha and Vernacci, under the most favourable conditions and to their complete satisfaction, observed the immersion of the first satellite of Jupiter, which when compared with the Paris and London almanacs, gave the longitude of the observatory as $17°35'30''$ from Montevideo, the latter assumed to be $50°5'45''$ west of Cádiz, according to our own observations.[2]

Another sixteen sets of lunar distances taken the following day by the Galiano,

Spaniards responded bloodily and re-imposed peace the following year, when normal relations – resolution of differences at meetings called *parlamentos* – were resumed and meetings were held at Osorno and Quilcahuín: R. Molina Otárola, 'Los mecanismos del despojo del territorio Mapuche-huilliche de Osorno,' in M. Orellana Muermann and J. G. Muñoz Correa, *Comunidades indígenas en su entorno*, Santiago, 1962, pp. 23-44. For the generally peaceful purposes of Indian policy in this period see F. Silva Vargas, *Tierras y pueblos de indios en el reino de Chile: esquema histórico-jurídico*, Santiago, 1962.

[1] The following paragraph in the MS journal, marked *no va* (not to be included) in the margin, has been omitted.

[2] Several paragraphs which follow in the MS journal, marked *no va* (not to be included) in the margin, have been omitted.

Concha and Vernacci gave a mean of 17°15'49". Variation from many azimuth observations taken by theodolite was 17°30'. Gilbert's azimuth compass agreed pretty well with this result.

From the morning of the 6th, we occupied ourselves on both corvettes in diligently providing ourselves with water and wood. The marines with their sergeants were also detailed to cut wood once every three days and, in order to avoid any sort of unruliness and to enure that the work did not slacken, an officer from one or other of the corvettes was present every day. At the same time, Don Felipe Bauzá had taken advantage of low tide to measure a base on the beach at the far end of the first harbour.[1] He was able to extend this to 1,450 English feet and had extended the triangles to Punta de Cocotuya, measuring another base in Ensenada de Huechucucuy[2] to locate Bajo de Huapache.[3] Pineda and Neé had started their excursions and made notable progress in natural history, especially in botany. Only fishing, which we had undertaken from the first day, failed to match either our needs or our hopes.[4]

The astronomers were no more fortunate on the night of the 8th when an untimely though short-lived mist made it impossible to observe another immersion of the first satellite of Jupiter, which was visible in Europe.

As the skies continued to be almost cloudless and the winds fair from SW we applied ourselves with diligence, in both corvettes, to determining the rate of our chronometers by means of equal altitudes with the astronomical clock compared later at noon with the chronometers by means of pistol shots. At the same time Tova, Valdés and Quintano went on an excursion to Castro,[5] to gain political and physical knowledge of the interior. Don Antonio Pineda and Guardiamarina Ali Ponzoni set off towards Chacao[6] and Don Felipe Bauzá went in the same direction with Piloto Sanchez to continue his triangulation.

Hunters from both corvettes made themselves busy collecting every possible variety of bird for the study of natural history.[7] The *Atrevida's* forge was taken ashore close

[1] Possibly at the western end of the harbour.

[2] Bahía Guapacho, on the north side of Península Lacui, west of Punta Guapacho.

[3] Rodal Guapacho, a dangerous reef, usually marked by breakers, extending 1¾ miles west of Punta Guapacho.

[4] Malaspina expected that the rewards of whaling and cod fishery would alone justify Spanish imperialism (without, in his opinion, need for permanent establishments) in the South American cone: Pimentel, *Descripciones y reflexiones políticas*, p. 60-2.

[5] Ciudad de Castro, the provincial capital of the Province of Chiloé, founded in 1567, and situated on the east side of the island, about 40 miles south of Puerto de San Carlos. Despite the limited local economic opportunities it had been continuously occupied to prevent pirates from using the excellent harbour.

[6] A town on the north coast of Isla Chiloé about 12 miles east of Puerto de San Carlos.

[7] The questionnaires with which the members of the expedition interrogated natives concentrated almost entirely on ethnographic questions in Chiloé, but began with a question about whether there were any high mountains and, if so, the maximum habitable altitude. The ethnographic compilation was made during a three-day excursion as far as Castro. At Osorno, the independent Wiliches maintained a political system based on chieftaincies inherited in the male line. Women were abducted for marriage and their deaths were unmarked by rites of passage. According to observations recorded by Espinosa, they did not keep the solar calendar, despite their reliance on agriculture: Novo y Colson, *Viaje político-científico*, p. 604: Higueras and Pimentel, *Antropología y noticias etnográficas*, p. 52. He and Viana agreed that the Wiliches believed the world was ruled by an evil force: Viana, *Diario*, p. 84. This

Plate 21. View of Santiago de Chile, by José de Pozo. Museo Naval, Madrid

to the watering place and was used to make various items of galley equipment or utensils which we needed. By the 11th the various parties and officers, detached for the tasks mentioned above, had already returned on board. On the previous day we were visited for almost the whole day by the forty-four Wiliches I previously mentioned. This encouraged us to learn their customs as thoroughly as possible, to acquire some of their costumes and weapons, and to make a portrait of the chief and a small son of his. This was done by Don José Pozo, with his characteristic accuracy and speed; see plate 20.

12 February

We expected therefore that on the next day, the 12th, after carrying out two more observations of the satellites of Jupiter and taking equal altitudes, we would be able to complete the rating of the chronometers and continue with our tasks en route for Valdivia, to whose Governor Coronel José Pustela I had sent a letter by piragua which was proceeding there with some articles of trade. I wrote that our arrival at that port would be on the 15th or 16th and that our stay would be very short, because what was taking me there was collecting objects related to natural history rather than maritime objectives. Consequently I requested him to start collecting for the Royal Government in Madrid before our arrival.

But the imminence of the new Moon caused the winds to veer to the north and NW that very night, stormy and followed by rain. As a result our plans were to a large

---

was seriously heretical from a Christian point of view and strictly incompatible with the claim, common among sympathetic observers, that pagan Indians were privy to a partial revelation of divine truths. It could, however, be considered an observation in the tradition of commentators who exculpated Indian pagans as victims of diabolic delusion.

extent frustrated, since it was not until the 15th that we could obtain equal altitudes and so complete the rating of the chronometers. That same afternoon all the astronomical instruments were brought on board. Don Manuel Novales and Don Felipe Bauzá, who had set off to visit the inner harbour early that morning to extend the triangles to Ensenada del Chasco Doble,[1] returned at nightfall. We were therefore in every respect ready to make our departure the next day.

Despite this, we could not ignore the fact that there was to be that very night, visible here and in Europe, an emersion of the third satellite of Jupiter the observation of which would do much to confirm our longitudes. For this purpose officers from both corvettes landed before midnight on the nearby beach with all the achromatics[2] and chronometer 71 from the *Atrevida*. The heavy night dew which constantly misted up the telescopes and the mistaken idea that the emersion would take place on the apparent upper part[3] of the planet led to our being several seconds late in observing the phenomenon, but we were nevertheless able to determine it as follows:

With 20″ of the maximum correction of the assumed lag in determining the phase according to the light perceived at the satellite	By Don Juan Vernacci .... 17[4] 47 31 time recorded on board by chronometer 71[5] & Berthoud's N° 10 By me using a telescope of lesser power .... 17°4 47′ 36″

Immediately after the observation we had been careful to compare chronometer 71 by means of pistol shots with those on board both corvettes. Special comparison of this chronometer made before and after it was taken ashore, as well as the above-mentioned precaution, assured us that its rate had been uniform and that in this respect the observation would be as reliable as that obtained with an astronomical clock. The longitude according to Vernacci's observation resulted from the mean of the times calculated in the *Connaissance des Temps* and in *Almanak Nautico*[6] to be 17°34′25″ [from Montevideo].

While the weather and the circumstances had so far combined to give us the very flattering hope that our tasks would be regarded as worthwhile and useful to the nation, on the other hand the conduct of the marines and seamen in this port contributed to our dismay. This port could not have been more suitable for me to adopt (on the basis of personal experience rather than tradition or whim) a policy of relaxing somewhat our military discipline as not being in any case the most appropriate to this particular mission and thus of introducing the fairest possible regime in the many ports we were due to visit. The population of Chiloé, which has virtually no contact with the mother country, contains almost nobody who is Spanish born.[7] This adds to

---

[1] A nearby bay or bight not identified on Malaspina's survey.

[2] Types of telescopes.

[3] Viewed through the telescopes Jupiter would be inverted and thus the actual immersion was by the lower part of the planet.

[4] 12 or 12° in original MS, which is clearly in error.

[5] Since chronometer 71 was on shore at the time presumably the comparison was by pistol shots as noted below.

[6] A precursor of *Almanaque náutico y efemérides astronómicas*: see p. 9, n. 2.

[7] According to the census of 1784 Isla Chiloé had 15,076 Spaniards and 11,627 Indians. The imperial effort these figures imply had been stimulated in the 1760s by fears of an English attempt to settle the island. See G. Guarda, *Flandes indiano: las fortificaciones del Reino de Chile 1541–1826,* Santiago, 1990, p. 34.

the importance of those who are, particularly in relation to marriage. The local women combine a grasping nature with a passion for vice and loose-living common to all the provinces of Peru, while the men are sunk in a perpetual idleness[1] which is reinforced, as one would expect, by the continual use of strong drink. Inevitably, at such a distance from any city, there is much slackness in the government and public administration. So the sailors were sure to find here precisely those inducements which help to form their erroneous notion of happiness.[2]

We needed to send boats repeatedly to the settlement for our astronomical, chart-making and natural science purposes as well as for the thousand other urgent necessities that a large vessel has. This obliged us to fill the settlement with our servants, sailors and marines, yet at the same time we became aware that many of them were missing at the time prescribed for their return to the ship. Some stayed ashore when expressly forbidden to do so; more than a few indulged in drink.

I lost no time in offering the local community and the shore patrols two *pesos fuertes* for each individual caught and handed over and I asked the Governor and staff officers of the garrison to participate wholeheartedly in these measures. In both corvettes any man who had remained a single night on shore was punished with some severity. At the same time a marine from the *Atrevida* who had been left on shore to guard the forge and the cattle near the watering place, though well aware that it would soon be time to return on board, was lured away by a local farmer who, being rather old and tired, saw him as a companion for his work on the land and as a husband for one of his daughters. The marine made off at night, taking with him many of the utensils for the forge and all the blacksmiths' clothes, as well as some other men's clothes which they had washed near the landing site.

Our suspicions were confirmed at daybreak with reports from the sergeant of the Yaquí shore battery.[3] Alférez de Navío Viana immediately went ashore in search of

---

[1] Cf. Viana on the natives of Chiloé: 'their dominant characteristic is laziness': Higueras and Pimentel, *Antropología y noticias etnográficas*, pp. 61-2.

[2] While blaming his men's indiscipline on strong drink, negligent policing and meretricious women, Malaspina displays some of his characteristic prejudices: the conviction that colonies grow ungovernable in proportion to their inaccessibility from 'home'; his conviction that seamen are close in condition to 'natural man'; his assumption that the frontier can only be civilized by more intensive colonization; and his enlightened notion of happiness, which he equates in his report home on conditions in Chiloé with 'rectitude, good faith, disinterest, true equality ... and uniform compassion.': Pimentel, *Descripciones y reflexiones políticas*, p. 68. His solution to the problem of maintaining a Spanish presence was based on what he claimed was an English model of fortified outposts in India. Fortifications should be few, modest and lightly manned; Spanish officials should function as visiting arbitrators, earning native loyalty by using their objective status to resolve local disputes; and commerce with the Wiliches should be stimulated by establishing fairs and providing a modest level of naval protection. Religious missions should be withdrawn on the grounds that the Wiliches were evidently 'deaf' to preaching. This programme is in line with his constant themes: the superiority of commerce over tribute as a means of exploiting the empire and the need to reduce the crown's cost in order to make the empire profitable. Ibid., pp. 69-72. Malaspina's idea of happiness was essentially political: he usually spoke of 'public happiness' ('felicidad pública') - a stage of social collaboration in what he understood as a process of socialization in the course of which egotism would wither. See J. Vericat, 'Fuentes del pensamiento socio-político de Malaspina,' in Palau and Orozco, *Malaspina '92*, pp. 13-18.

[3] Possibly in the fort on Punta Ahui.

the man with a party of marines. They first apprehended the local farmer and sent him on board, then they searched his house thoroughly. By chance at this time a peasant arrived at the house on horseback; he was arrested by Viana, made to talk, and soon confessed that the marine and the stolen articles were at his own house, about two leagues away. Don Francisco Viana went straight there with two marines and succeeded in finding the deserter still asleep and all the stolen goods stowed away. By noon the marine and his farmer accomplice were on board the *Atrevida* and all the stolen clothing and utensils restored to their owners.

There was no need for an inquiry to investigate the crime. Even so, summary proceedings were completed in the afternoon and in these it was again confirmed that the peasant had been an accomplice and that one of his sons had helped to carry away the numerous stolen items. The next morning, the marines from both corvettes were ordered to fall in with the *guardiamarinas* at their head when the marine was made to run the gauntlet three times. The peasant was forced to watch the punishment; the fact that he had a large family to support stopped us from handing him over to the justice of San Carlos to be punished according to his crime.

Despite our diligent searches, we could not prevent the loss of five seamen from the *Atrevida* and three from this ship, either led astray or fearful of their due punishment. I had no option but to write officially to the Governor, sending him a list of the deserters with a clear statement of their personal details, and specifying how to apprehend them and have them sent safely to Lima, where I offered to make good any expenses that might arise from this transfer. I put on record how important it was for the reputation and safety of the nation's shipping in these waters that governors and ships' captains should work together to root out this age-old transgression of the seamen.

The Governor had asked me in an official letter dated the 10th whether I could carry 20,000 planks of larch in the corvettes to Valdivia which, by order of His Excellency the Viceroy of Peru, had to be taken to that town as the balance of 50,000. I replied that since we had replenished our water and wood there was no stowage room left and that our departure from the port was imminent. In particular the necessity of making good use of the season for our hydrographic operations along this coast would not allow me to tie myself to entering any particular port or to be encumbered with loading or unloading and so perhaps missing an opportune moment. However, since this was a place of the finest timber, very inexpensive and very suitable for different uses, I arranged to procure for the corvettes a quantity that would not inconvenience us. We each loaded two hundred and fifty planks of larch and also ordered a number of oars for the boats, as well as sixteen, rough-hewn, for use by the corvettes themselves when becalmed in the waters we were to navigate.[1] However, bad weather only allowed eight of the large ones to reach this ship.

16 February

Since two o'clock in the morning we had been getting ready to weigh anchor and prepare to sail to the entrance to the harbour on the morning tide, there to ride at

---

[1] Usually referred to as sweeps.

anchor and wait for a favourable wind, which naturally would not spring up until midday. Indeed, when dawn broke we were up and down; at first light we weighed anchor and, towed by the launch and pinnace, we made slow progress despite the tide. Later with the tide already on the turn, we told the *Atrevida* at half past six not to get under way and we dropped anchor in six fathoms fairly near Bajo de la Ensenada.[1]

All morning there was either a dead calm or light airs from north and NNW, outlook cloudy, but at one-thirty in the afternoon, the weather cleared, and the wind shifted to WSW, with a moderate breeze which we tried to take advantage of immediately, especially since we could count on an ebb stream until nightfall. Nevertheless, the wind soon shifted again to the west, becoming calm, so that we had to anchor once more just off the Yaquí battery, with the sole advantage of having reached the position that we had wanted to reach in the morning.

Some hours of the afternoon remained, and Pineda, Valdés and Quintano made use of them to our great advantage by setting out in the pinnace to hunt, with the particular intent of killing a species of amphibious cat that we frequently saw running over the rocks and sands by the sea. They did indeed succeed in shooting one with a shotgun. Don Antonio Pineda later recognized it as being almost identical with what Count Buffon calls a Canadian otter. They also brought back a number of birds, many shells and other species of shellfish, all very useful acquisitions for natural history.

17 February

The night was dead calm, and the outlook the following morning decidedly cloudy with the usual north and NW light airs. We sent the pinnace to a nearby stream to take on water and sand, remaining in the same position as the day before. At one in the afternoon, however, almost at the turn of a favourable tide, a promising fresh wind sprang up from SW. We immediately got under way (having hoisted the pinnace) and set a course to pass Punta del Mastelero.[2] But the wind did not take long to shift to the west, slackening to a gentle breeze so that we had to come about and seek our former anchorage. We soon reached Punta de Yaquí and again picked up a fresh breeze from SW which left the skies increasingly clear and which seemed set to continue. We went about again on the port tack under full sail, which brought us once more near Punta del Mastelero, but again we met a westerly wind which obliged us to alter course towards Los Farallones[3] and Doña Sebastiana.[4]

The tide was driving us strongly towards the open sea; but it was now half way towards the turn and we were well aware that with the beginning of the flood stream after sunset, with the wind falling to a dead calm and night coming on, we would either be swept towards one of the many hazards that surround the entrance to this harbour or be obliged to anchor to our considerable disadvantage. It therefore

---

[1] Not shown on Malaspina's survey, but possibly the shoal area extending west from Punta San Antonio.

[2] Situated 1¾ miles NW of Punta Ahui.

[3] Farallones de Carelmapu, situated about 10 miles NE of Punta Huechucuicui.

[4] Situated about 6 miles ENE of Punta Huechucuicui.

seemed to me more prudent to take advantage of the little wind that remained to return to an anchorage to the east of Punta de Yaquí rather than the one to the west of it, so as to see, if the calm continued, whether a piragua could bring us the oars which were still lacking and which I had been promised by the following dawn.

The tide held us back considerably. We passed Punta de Yaquí at a gun-shot's distance in six or seven fathoms of water; later we sailed close-hauled as much as possible, finally at three thirty we let go the best bower in eight fathoms, with the same bearings that we had on anchoring on the first night. The *Atrevida* anchored close to us, in nine fathoms, but she dragged her anchor and was soon in twenty fathoms. The night was calm, like the previous ones. When the ebb tide ceased at the beginning of the night, the *Atrevida* worked herself into a shallower anchorage, first with a warp to this ship and afterwards towed by the two pinnaces. We both rode almost up and down.

18 February

The morning dawned very fair. The pinnace was sent to the watering place near Yaquí and as soon as it returned we hoisted it. At nine a gentle SE breeze set in and we immediately signalled the *Atrevida* to set sail, as we were ourselves getting under way; but since the tide was against us and would be running strongly until half past one in the afternoon, we delayed weighing anchor until half past ten, hoping that the wind would be strong enough to keep us close to Punta del Mastelero until the flood tide was well established and we could take full advantage of it. The breeze had remained fresh for more than two hours and at that moment seemed likely to strengthen so we lost no time in weighing. Indeed at first our bearings showed that we were making considerable headway, but after half an hour we found ourselves at the mercy of some light winds from the east and NE. With these we could make no headway although we had unfurled all small sails and the tide was carrying us towards Isla Cochinos.[1] We held out as long as possible against dropping anchor, in the hope that soon the sea breeze would start blowing from SW, but at length, a little before noon, we were obliged to do so almost in the middle of the harbour mouth, in ten fathoms, sand. As this position was not at all promising, particularly in view of the great strength of the tide, at nightfall we ran a long warp to the west with our boats, which first we heaved on and later the *Atrevida*, bringing ourselves to our former positions four cables SE of Punta de Yaquí, where we passed the night which was absolutely calm.

19 February

For three days, deceived by seemingly favourable winds, we had been continually working the anchors. The alternative was either to remain where we were and wait for a steady wind or to expose ourselves to a racing tide in calm weather, surrounded by dangers. Both possibilities seemed equally hazardous. Since I had to send the pinnace to the town that morning to look for some items, I sent Don Felipe Bauzá in it

---

[1] Isla Cochinas, situated 2¼ miles ESE of Punta Ahui; a wooded island of 4 hectares, also known as Caicué, north of Chiloé in Bahía de Ancud.

to enquire of Piloto Moraleda how many hours of sea breeze and what strength of tide we could expect with the present outlook and whether southerly winds that would certainly favour us were frequent or rare. The *Atrevida*'s pinnace was sent to replenish water for that corvette.

Since the skies were clear and we had an unimpeded horizon at eight in the morning for taking absolute altitudes with the sextant, and since our longitude was established from bearings referred to our plan of the harbour, we made sure both on this day and the day before to use this useful method of verifying the rates of our chronometers and the accuracy of our comparisons. We obtained the following results, which we could assume to be satisfactory, taking into consideration the slight errors in the measurement of mean altitudes with the sextant.

Date	Longitude by bearings	N° 61 Uncorrected	Corrected	N° 72 Uncorrected	Corrected	N° 13 Uncorrected[1]	Corrected
18	1′4″ West	1′31″	1′15″	6′10″	1′16″	4′29″	0′43″
19	1′9″	0′30″	0′15″	4′42″	0′51″	4′44″	0′51″

From which it can be deduced that Number 61 was keeping its rate, while number 72 and number 13 had altered considerably.

In order to consolidate still further our calculations of longitude in the future, signals were pre-arranged to allow the corvettes to compare their chronometers with each other from time to time and so determine the corrections to rates with a much greater degree of probability, particularly those which depended on the alteration in the three chronometers' rates on the same day and by any amount in the same direction.

At last, that same morning the piragua hove in sight with our large and small oars. The full complement of sixteen large oars was made up for both vessels. We also received twelve small oars that needed shaping; the person who had paid for them was reimbursed.

Very light airs from the fourth and first quadrants prevailed during the morning, with an outlook that was extremely pleasant and mild. At half past eleven a fresh sea breeze came in from WSW. We immediately recalled the pinnaces and when ours had returned with some effort because of the contrary wind and tide, Don Felipe Bauzá told me that Piloto Moraleda had assured him that winds from SW, inside the harbour, immediately turned westerly once Punta del Mastelero had been rounded. They then fell calm toward nightfall, leaving a vessel exposed either to the vagaries of a contrary tide or the risks of a very poor anchorage. Moraleda himself had already presented us, for our subsequent navigation, with a Mercator's chart from south of Valdivia to Chiloé, another of the entire island and the adjacent mainland and another of the anchorage at Juan Fernández, all compiled by him with great diligence and uncommon accuracy.

---

[1] *Dist.* in original MS but *Directa* (direct, i.e. uncorrected) in AMN, MS 753, which seems much more likely.

BOOK THREE

FROM PUERTO SAN CARLOS DE CHILOE
TO COQUIMBO

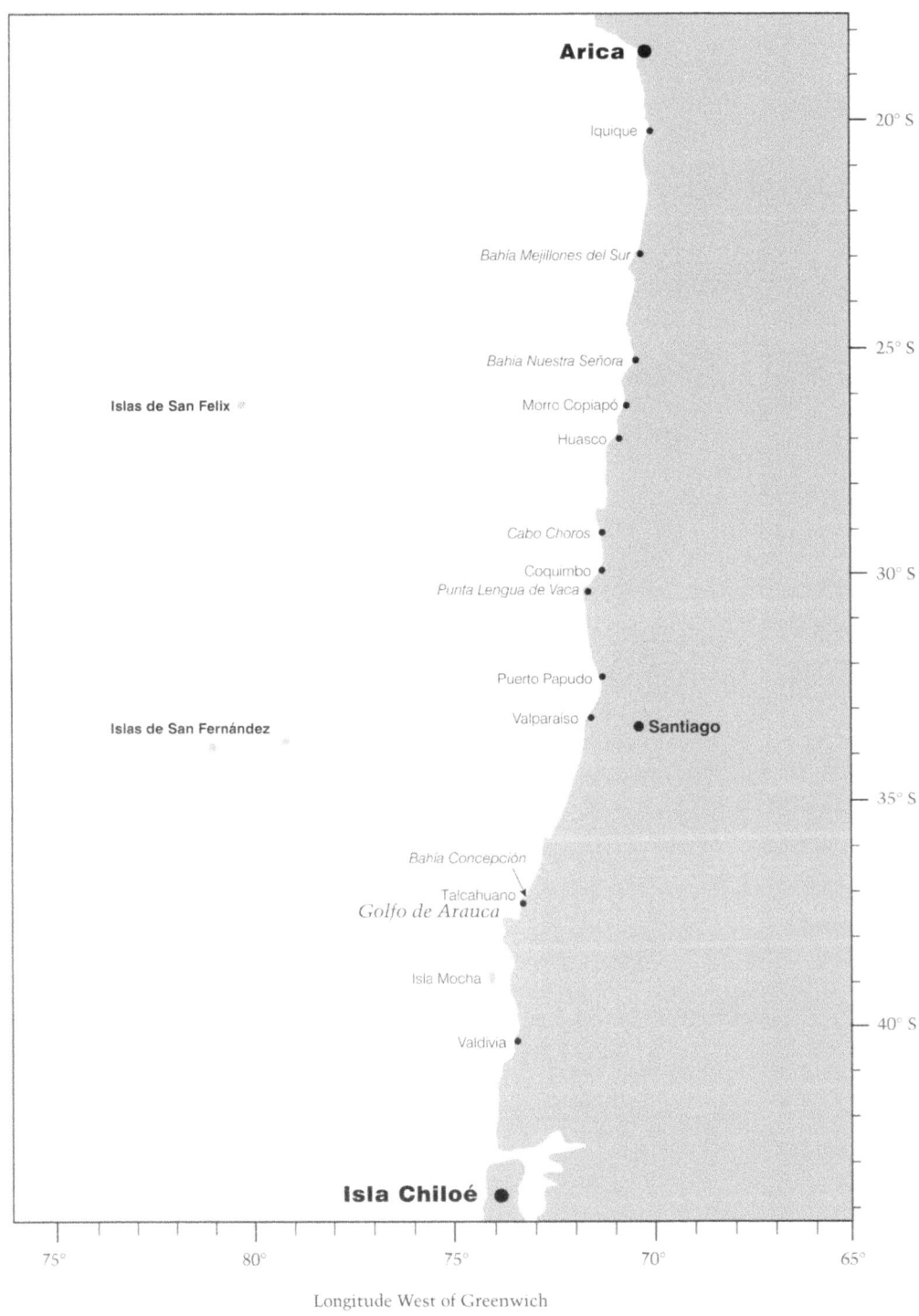

Arica ●

Iquique ● — 20° S

Bahía Mejillones del Sur ●

Bahía Nuestra Señora ● — 25° S
Morro Copiapó ●
Huasco ●

Cabo Choros ●
Coquimbo ● — 30° S
Punta Lengua de Vaca ●

Puerto Papudo ●
Valparaíso ● ● Santiago

— 35° S

Bahía Concepción
Talcahuano ●
Golfo de Arauca

Isla Mocha — 40° S

Valdivia ●

Islas de San Felix

Islas de San Fernández

Isla Chiloé ●

75°   80°   75°   70°   65°

Longitude West of Greenwich

Fig. 6. West coast of South America from Isla Chiloé to Arica, February to May 1790

148

CHAPTER 1

From Puerto San Carlos de Chiloé to Talcahuano

[19 February]

Despite the aforementioned advice from Piloto Moraleda, since the weather outlook was exceedingly favourable and the tide to our advantage until eight-thirty in the evening, I felt compelled to get under way and head out once more and, if necessary, to come about in the stretch of sea between Los Farallones and Punta de Cocotuya. In any case, since the weather seemed fine, there would be no inconvenience in dropping anchor anywhere and continuing on our way with the sea breeze the following morning.

So at two o'clock in the afternoon we hoisted the boats and set sail, passing close to Punta del Mastelero under a press of canvas. Thus we were able to exploit the favourable tide at its greatest strength. Far from backing to the west, the wind continued set fair from SWbyW. We continued to close-haul this wind all the time. At six o'clock, when it fell almost completely, Punta Capitana bore N7°W, Punta de Huechucucuy S25°E, and Farallón Mayor de Carelmapú[1] S62°E, by compass in each case.

The land breeze dropped completely and at nightfall the horizon was hidden by dense mist. With only light airs from the north and NE, we set a course to the west and WNW to keep as clear as possible from Punta Capitana and from the land immediately adjacent to the port. Thus at dawn we could already consider ourselves to be clear of the coast, although the land around Punta Capitana, the only part we could see, was very indistinct because of the great amount of mist and fine drizzle.

20 February

The wind began to blow strongly from north and NbyE, and we hauled this wind on the starboard tack, keeping to our course as much as possible. As somewhat heavier seas set in after that, we remained under mainsails and topsails, as the tack we were on out of necessity only served to take us away from the coast.

Since the previous afternoon we had once again been finding that the *Atrevida* was sailing considerably faster than this corvette. As a result, during this morning we were obliged to adjust our ballast. And indeed, after we had shifted sixty *quintales* in guns and pig iron from the stern to the hold, we immediately increased our rate of sailing, but still not enough to compete with the *Atrevida*.

At noon, we began comparing the rates of the chronometers on board the two

---

[1] Farallón Grande, the outermost and highest island of Farallones de Carelmapú.

ships by means of signals. This was done with pistol shots, signalling the time by our 61 for comparison with the *Atrevida's* number 10, and having deduced the rate of the latter from our notebooks, we obtained the following results, having already applied the equation to our numbers 13 and 72.

$$N^\circ\ 61\ \ldots.1^\circ4'37''\ \text{west of San Carlos}$$
$$N^\circ\ 72\ \ldots.1\ \ 4\ \ 28$$
$$N^\circ\ 13\ \ldots.1\ \ 4\ \ 18$$
$$\textit{Atrevida}\ \ldots\ldots\ N^\circ\ 10\ \ldots.1\ \ 3\ \ 47$$

Observed Latitude 41°20'30"

This comparison (as mentioned earlier) had both to confirm the correctness of our daily tally of the differences in the rates of the chronometers and to derive from all six chronometers the same or a very similar longitude, so that the geographical positions which we based on it might be more reliable. It was consequently with great satisfaction that we saw that those longitudes which were derived from number 10 could be regarded as highly reliable. Number 61 continued at its even rate and differed little. However, number 13 was gaining a little, while 72 varied quite considerably first in one direction and then the other.

Today we maintained an offshore course all afternoon with the wind remaining moderate from the fourth quadrant, with considerable low cloud or rather mist. But at nine o'clock in the evening a gentle breeze began to blow from SSW and south with the prospect of fine weather. We immediately altered course to the north, as in this case it seemed preferable to keep clear of the coast, to make even more sure of this favourable wind, and in the morning to seek and follow the coast from a lower parallel, less exposed to the northerlies.

21 February

At dawn, which was very fair but, as is usual on these coasts, rather misty, our look-outs mistook the changing shapes of the haze to the east for the appearance of land, which they thought they saw stretching from ENE to NNE by compass. So insistent were they, that despite the trend of the coast given by Moraleda for this area, I was forced to believe either that he was somewhat mistaken or that we were the victims of some strange effect of the current. We altered course to NNE, and only the speed we were making until around noon and our observations of latitude and longitude were finally sufficient to rid us of this delusion. Our latitude was 40°14' and our longitude 1°1'45" west of San Carlos.

With this information, we could have no doubt that we were out of sight of land and indeed some considerable distance from it and that we could easily pass Puerto de Valdivia.[1] I did not wish to call there, because making a plan of its inner port would require the sacrifice of a good deal of time and the accurate work done by Ingeniero Garlan and Piloto Moraleda had in any case rendered it virtually unnecessary. Moreover Puerto de Chiloé had impressed upon me the inconvenience, in terms of time and of discipline, that such ports along this coast could occasion us,

---

[1] Malaspina's proposals for streamlining the empire included abandoning this outpost entirely: Pimentel, *La física,* p. 218.

particularly those that were exposed to northerly winds. However, I certainly was concerned, in the interests of shipping, to locate this port in its correct latitude and longitude and to take some views of the land immediately adjacent to the entrance.

Accordingly, at noon we luffed round to true east, under a press of canvas, and making a good seven or eight knots we sighted land at around two o'clock in the afternoon. It was this land on the parallel we were on which formed the surroundings of Río Bueno;[1] it was high, precipitous, and arid rather than luxuriant. At four-thirty in the afternoon we altered course to close the coast and having observed hour angles, being some three leagues from the coast, we bore away so that we could extend our survey of it as far to the north as possible. By sunset we were in full sight of the port. A little earlier *Atrevida* had sounded without obtaining bottom and sixty-two sets of distances between Sun and Moon gave a mean longitude that differed from the chronometers by 7' 57" to the west.

Having completed our survey of this part of the coast to our entire satisfaction, we proceeded under full sail from nightfall to pass in sight of [Isla] Mocha. The intervening coast had been surveyed by Moraleda. Closing it was always dangerous because it was inhabited by the Junco Indians who were enemies of the Crown and also because there was no sheltered port along its length. As Isla Mocha was the most suitable landfall for those making for Puerto de Concepción it was important to our shipping to fix its position astronomically and obtain views of it from different bearings. These considerations obliged us to prefer the outer channel to the inner one between this island and the mainland.

22 February

At sunrise we sighted Mocha to the NNE at a distance of six or seven leagues. At the same time the haze greatly hindered our recognition of the coast and at times even obscured part of the island itself. The wind at this time was blowing strongly from south and SSE and from the appearance of the sky and horizon we could expect a steady trade wind. We therefore altered course, making for the island, and when we were only about two leagues from its southern extremity we set a course rather more to the west, to steer clear of the ledge which extends for about a league SW from that extremity. At the same time we shortened sail to obtain good observations for latitude in sight of the island. We started measuring bases at about nine o'clock but were partly frustrated by a rather fierce squall from the SE at ten o'clock, which forced us to take in sail. However, between eleven o'clock in the morning and noon we were able to measure a good base, standing off from the shore to the NE. By means of this, with the north end of the island a league away from us, we were able to obtain its latitude and longitude and thus position it to our entire satisfaction, while at the same time connecting it with the nearby coast.

The passage between this island and the coast has been much frequented. The bottom is visible in depths of eighteen to twenty fathoms and there is an anchorage sheltered from the northerlies on the eastern side of the island, near its southern end.

---

[1] Río Bueno enters the sea through a ravine at the head of Ensenada Dehui about 33 miles SW of Valdivia.

Nevertheless, only absolute necessity or the danger of foundering would bring large vessels to this place.

Thus by noon we had completed our hydrographic tasks with respect to Isla Mocha and had carried them out under easy sail. We then continued in the direction of the coast and in the course of the afternoon and evening we were fortunate enough to complete its description and situation as far as Punta de Rumena.[1] The sixty-two sets of lunar distances taken that day gave a longitude of 8′14″ further to the west than that obtained from the chronometers. By the latter at noon on the previous day we were in longitude 9′8″ west of the meridian of San Carlos.

We had not been able to sight Isla Santa María,[2] whose position we thought it important to determine for those who sailed in these latitudes and wished to make their landfall at Concepción. Thus soon after nightfall we remained under half-reefed topsails although the extremely strong wind and the very rough sea caused us to roll excessively.

At one o'clock in the morning, having already sailed the right distance to bring us some two leagues south of the island, we luffed round to the east. And then at three o'clock, considering ourselves well ahead, we hove to under the main topsail and the flying jib to await the first light of day and the beginning of our daily work.

23 February

Indeed, at four o'clock we could clearly see Tetas de Bíobío[3] and the intervening island,[4] and by five o'clock we had already begun to measure a base, the ends of which we could determine exactly and which we would be able to connect up with points on shore. The wind had by then dropped considerably. The sea was smooth and the day was very clear and sunny. There were a large number of whales, in pleasing variety, spouting water high into the air in several places.

We closed the coast and passed very near to the entrance to Puerto de San Vicente[5] and once we sighted the shore we bore away to leave it to starboard at a distance of a cable and a half. At eight-thirty that evening we were close to the northern point of Quiriquina.[6] The large rocks that surround it conceal no hidden dangers and it was possible to pass close to them. Immediately beyond them we luffed, still with a gentle breeze from SSE, which until then had gusted somewhat, and we began to tack towards the anchorage of Talcahuano.[7]

The coastline was clear of dangers in both directions and so invited us to extend our tacks as far as possible, but we knew that when one approached too close to the land the wind lost much of its force. Moreover it varied considerably in the vicinity

[1] Situated about 2 miles SW of Punta Lavapié, the SW entrance point of Golfo de Arauco.
[2] Situated about 6 miles NNE of Punta Lavapié in the entrance to Golfo de Arauco. A focus of anxiety concerning foreign interlopers because the island has plenty of good anchorages, water and wood.
[3] Tetas del Bío Bío, two prominent hills, situated 10 miles SSW of Punta Tumbes, the northern extremity of Península Tumbes, which forms the western side of Bahía Concepción.
[4] Presumably Isla Santa María.
[5] A bay open to NW winds between Tetas del Bío Bío and Península Tumbes.
[6] A large island in the entrance to Bahía Concepción.
[7] Founded in 1764 in the SW corner of Bahía Concepción, Talcahuano was regarded as the best harbour in Chile, although historically Valparaíso was preferred as being closer to Santiago.

of Quiriquina. Finally, quite close to the anchorage, it sometimes fell completely calm and at other times blew contrary to the tack that would be most helpful to us. Nevertheless, the advantageous set up of the running rigging allowed us to sail scarcely five points off the wind, and by two-thirty in the afternoon we were able to anchor at a distance of three to four cables to the east of Castillo de Gálvez[1] in seven fathoms, sandy ooze. The *Atrevida* anchored close to us and both corvettes were soon afterwards moored north and south with their bower cables to the north.

---

[1] Situated on the SE side of Península Tumbes about one mile north of Talcahuano.

CHAPTER 2

At Talcahuano and Bahía Concepción

[23 February]

In the port we found the merchant ship *Hércules* of the Lima trade about to set sail for Valparaíso; she had taken on half a cargo of wheat, intending to take on the other half in that port.

The commandant of the Talcahuano detachment immediately came on board in one of the local barges. He gave me a letter from the Gobernador Intendente of the city of Concepción,[1] Brigadier Don Francisco Mata Linares,[2] in which he advised me that he had gone to Chillán[3] on a visit of inspection, and knowing of our arrival in these waters by His Majesty's advance orders he had charged two highly energetic and reliable people to supply us promptly, in his absence, with every assistance we might need.

Without delay we received letters which had been sent to Buenos Aires and, not finding us there, had been forwarded to Concepción. Among them was a letter from His Excellency Señor Bailío Frey Antonio Valdés enclosing the circular instructions issued by the ministry of his most Christian Majesty,[4] ordering that whatever assistance we might need should be extended to us while we were in his colonies. He also informed me of the favour that His Majesty had shown us on the occasion of his coronation by promoting to the next rank most of the officers taking part in this

---

[1] Concepción, was founded in 1550 in the SE corner of Bahía Concepción. It reached the peak of its prosperity in 1751, when the population was said to be about 20,000. In March 1685, by order of the Governor of Chile, Don José de Garro, the natives of Isla Mocha were transferred to the mainland and resettled on the north bank of Río Bío Bío, 6 miles from Concepción, thereby giving the name La Mocha to their new home. After a period of near-abandonment, caused by a tsunami in 1751 and a series of destructive earthquakes, the main settlement moved from Concepción to La Mocha in 1764. This explains Malaspina's later reference to 'La Mocha or Nueva Concepción' (see p. 155 below).

[2] Francisco de Mata Linares was a radical strategic thinker, governor of Concepción from 1789, who had proposed the abandonment of Valdivia. For his part Malaspina came to see Chile as epitomizing the past mistakes and current problems of the Spanish empire in America – 'without question, the country which ... has cost most blood and treasure and has yielded fewest advantages.' Ingenuously he claimed that prospecting territory could have been peacefully acquired if the natives had been left in peace: Pimentel, *La física*, p. 219.

[3] An inland town about 45 miles ENE of Concepción. The old town was a short distance from the present town, which replaced it after the earthquake of 1835.

[4] I.e. the King of France.

expedition,[1] which also served to show those who, because of their lack of seniority, had not shared in these favours that His Majesty would remember them when the time was right.

It is not easy to judge whether on this occasion our feelings of rejoicing were greater than our sense of gratitude and our desire to respond to the royal generosity with whatever it was in our power to do.

Since the Gobernador Intendente was absent, the Coronel of the company of dragoons in the city of Concepción, Don Pedro Quijada,[2] who was at the time in command of the settlement, sent on board various types of refreshment all of the best quality and flavour. [24th] He also wished to entertain the commanding officer of the *Atrevida* and a number of officers from both corvettes for lunch on the following day and these officers went to the town to present their own compliments and mine.

The inhabitants of La Mocha or Nueva Concepción had scarcely had time to draw breath after the dreadful devastation of a smallpox epidemic which had almost instantaneously struck down some 2,500 people in the city and surrounding region, without respect for sex, age or status. Until now this province had always been spared this dreadful scourge when it had ravaged the other provinces of Chile, so its inhabitants had neglected to inoculate themselves. This invaluable safeguard was immediately brought to the city by a mulatto from Santiago, sent by the President.[3] But many had already lost their lives, and the features of the fair sex, really noteworthy in this province, had to a great extent fallen victim to this plague.

25 February

Today there was a moderate northerly breeze but with no sign of bad weather and as the same weather continued the next day, we took advantage of it by taking soundings in the bay, particularly to locate precisely the shoals between Talcahuano and Quiriquina. Two theodolites set up on El Morrito de Talcahuano and Batería de Gálvez were used to fix the position of the soundings until five in the evening, by the method already adopted by Don Vicente Tofiño[4] of signals from the sounding launch which enabled bearings to be taken simultaneously from the two observation points.

26 February

We were in an area where, four years earlier, members of Comte de la Pérouse's expedition had deployed all their activity and intelligence. Considering this encouraged us to follow in their footsteps, while wondering whether this would be possible. However, as it turned out, our work had to differ considerably in accuracy to that undertaken by the *Boussole* and the *Astrolabe*. How closely we agreed with their results only time will tell.

In San Carlos de Chiloé, we had determined the absolute longitude by means of

[1] These promotions included that of Malaspina to capitán de navío.
[2] Quijada was a particular protégé of President O'Higgins and became governor of Chiloé in 1794.
[3] See p. 170, n. 3 below.
[4] See p. 327.

observations of the satellites of Jupiter, enabling the chronometers to be rerated, although there had only been a short interval since they were last checked. The season was still favourable, inviting us to undertake many useful astronomical tasks in these latitudes and, given the position of Valparaíso and its sky, we were inclined to prefer working there rather than in this port. The natural history of these provinces was also such that we could barely give it more than a rapid and so to speak imperfect examination, when considerable money and time had been spent on it by Señor Pavón and Señor Gálvez for his Catholic Majesty and by Monsieur Dombey for his most Christian Majesty.[1] Thus only the hydrographic task remained for us and we devoted ourselves especially to this without dropping the other tasks which we could not omit without being accused of negligence.

On this day equal altitudes were taken with the large astronomical quadrant and referred initially to our chronometer 72, taken ashore for this purpose, and then by means of pistol shots to the remaining ones on board the corvettes. Early in the night some meridian altitudes of stars were observed to obtain our latitude, after our large astronomical quadrant had been erected in the observation tent in a small square next to the Casa del Cura,[2] in the same place where Monsieur Dagelet had taken his observations.[3] Mist prevented the observation of an occultation of a star by the Moon as it had at the same time on the previous night and the two following ones and the same thing occurred for two occultations of the satellites of Jupiter which would have provided fresh results for the longitude of these coasts.

The Gobernador Intendente Don Francisco Mata had already returned to La Mocha[4] from Chillán, where he had been on official business. I went immediately to call on him with the commanding officer of the *Atrevida* and a number of other officers. Afterwards we visited the bishop, Coronel Pedro Quijada, Teniente Coronel Juan Zapatero[5] in command of the artillery, and some other leading citizens who had not been able to come on board on the previous days.

We discussed our immediate requirements with the Governor, who effectively offered us every assistance in his power. This was the speedy supply of twenty casks of good wine, the curbing of desertion and help with an overland expedition to Santiago, including visits to some volcanoes and mines. This was be undertaken by Don Antonio Pineda and Teniente de Navío Valdés with, as their guide, the priest Don Juan de Ubera who had formerly been a senior surgeon in the royal navy.[6]

---

[1] In 1777 a botanical expedition under the Spanish botanists Hipólito Ruíz and José Antonio Pavón, accompanied by the French botanist Joseph Dombey was sent from Spain to examine the plant life of the viceroyalty of Peru and parts of Chile. They were accompanied by the artist José Gálvez: see Ruíz and Pavón, *Systema vegetabilium florae peruvianae et chilensis*, 3 vols, Madrid, 1798-1802. Two further volumes were prepared but were not published until the late 1950s. Dombey (1742-94?) studied at Montpellier and had previously served at the Real Jardín Botánico.

[2] Literally the priest's house.

[3] Joseph Lepaute Dagelet, La Pérouse's astronomer.

[4] I.e. Concepción.

[5] Juan Zapatero and his son José were engineers with special expertise in gunnery.

[6] For Pineda's expedition see A. Galera, ed., *La ilustración española y el conocimiento del Nuevo Mundo: las ciencias naturales en la expedición Malaspina (1789–94): la labor científica de Antonio de Pineda*, Madrid, 1988 and M. Muñoz Garmendia, ed., *La expedición Malaspina*, Tomo III: *Diario y trabajos botánicos de Luis Neé*, Madrid, 1993.

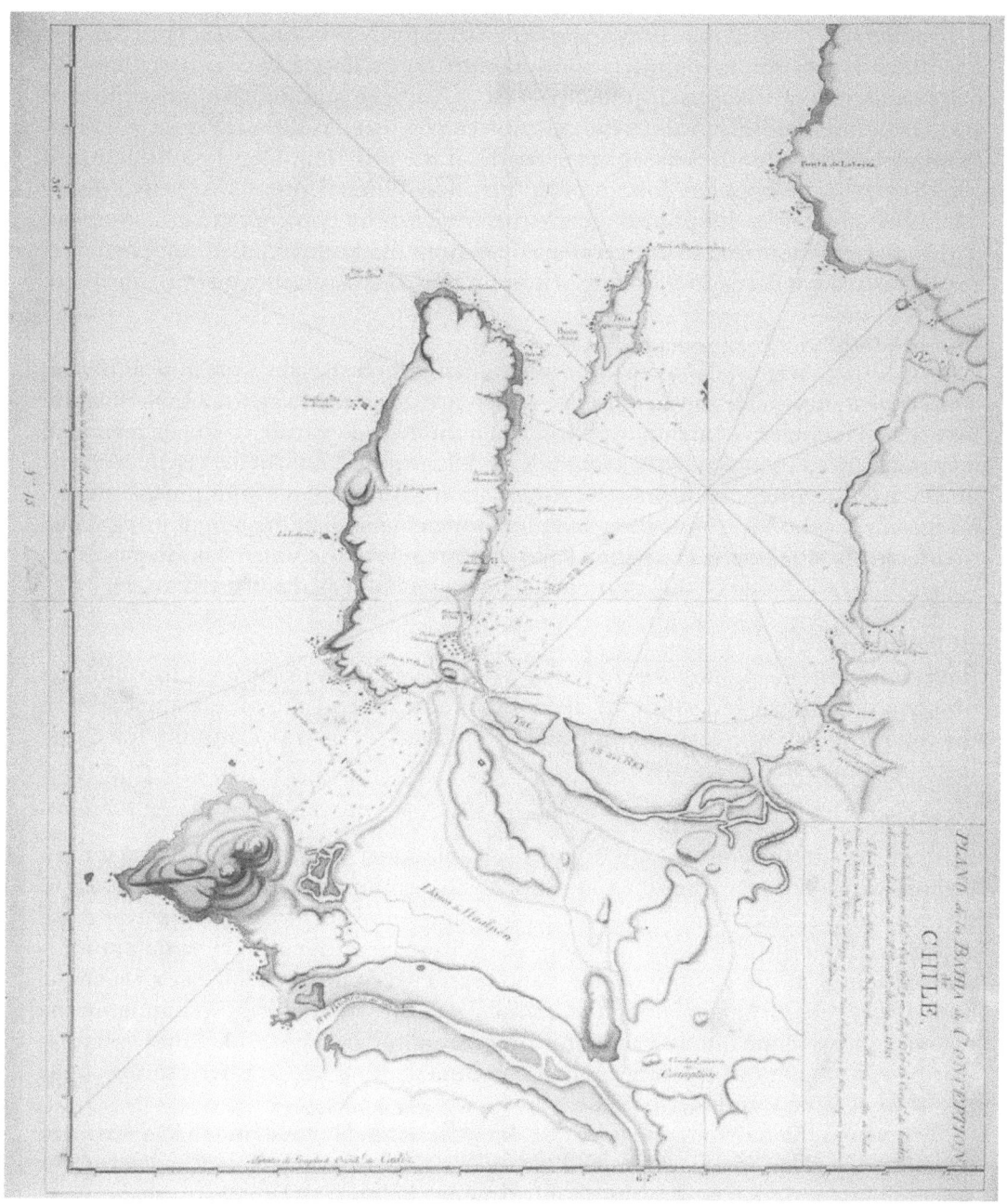

Plate 22. *Plano de la Bahía de Conception de Chile*; manuscript copy of Malaspina's 1789 survey obtained in South America 1814–15 by Captain Philip Pipon, HMS *Tagus;* UKHO, 144. Reproduced by permission of the Controller of Her Majesty's Stationery Office and the UK Hydrographic Office.

In addition I decided the *Atrevida* should go to Valparaíso immediately with the best instruments and our astronomers to work on the catalogue of southern stars, to observe a series of triangles, if possible, as far as Santiago and to make other physical and astronomical observations as the occasion might arise. At the same time, after discussing our budget with the Capitán General of the province, Don José Bustamante was to oversee the baking of four months' provision of bread for each corvette. Meanwhile I would stay in this port to receive the wine for the two corvettes, to complete in the port and its surrounds any geodetic operations that were required, and then proceed to Islas Juan Fernández, before rejoining the *Atrevida* in the port of Valparaíso.

27 February

The next day, there was a north wind with a fairly high sea and a great deal of rain. This made it impossible for the *Hércules*, which had got under way, to sail and it would have left us completely unemployed had the imminent departure of the general mail boat not spurred us into activity. I sent His Excellency the Ministro de Marina a report of the main occurrences during our voyage from Montevideo to this port. I advised his Excellency the Viceroy of Peru and the Capitán General of the kingdom[1] of Chile of our arrival and intended activities. I also sent a report to the Viceroy in Buenos Aires concerning our survey of the coast of Patagonia which he had entrusted to me.

28 February

It was not until the night of the 28th that we observed appropriate meridian altitudes of stars to the north and south for deducing the latitude of the observatory; and on the following day we were able to observe a second set of equal altitudes for rating the chronometers.

1 March

Early in the afternoon (since the day was very pleasant) Quintano and Bauzá set out by boat with Piloto Sánchez for Quiriquina. Their intention was to take bearings to the extremity of Talcahuano[2] and to various points on the island, stay there over night and at first light the following morning take advantage of the calm by taking numerous soundings in the vicinity of the island, particularly in Boca Chica or Canal del Sur.[3] They were fortunate enough to complete all this satisfactorily, sounding on the seaward side with the pinnace and returning the following afternoon. [2nd] Guardiamarina Ali Ponzoni in the other pinnace sounded along the southern shore of the bay as far as Penco[4] and Cerro Verde.[5]

By seven in the morning the corvette *Atrevida*, assisted by our launch and favoured by extremely clear, fine weather, had set sail for Valparaíso. On board, as well as our astronomers and the astronomical instruments belonging to the *Atrevida*, were

---

[1] For the use of the term *reino* or *reyno* = kingdom see p. lxxxvi, n. 1 above.
[2] Probably to the northern extremity of Península Tumbes.
[3] The channel between Isla Quiriquina and Península Tumbes.
[4] A small town situated in the SE corner of Bahía Concepción on the former site of Ciudad de Concepción which was moved to its present site after being frequently devastated by earthquakes and tsunamis: see p. 154, n. 1 above.
[5] A distinctive hill situated about a quarter of a mile north of Penco.

Teniente de Navío Don Cayetano Valdés and Teniente de Fragata Don Juan Vernacci with the tent, astronomical clock, astronomical quadrant, large achromatic telescope and water level,[1] all belonging to the *Descubierta*. Vernacci was assigned to the astronomical tasks, the main objective of which I have already mentioned. It had been necessary to change Valdés's route to the city of Santiago, not only because this was too long a time to devote to a scientific expedition (and the rains, moreover, had come very early this year, making the going somewhat difficult and dangerous) but also because it was important that he should stay in Santiago for some time so that he could make a compilation of the considerable information both old and new that would be there. To that end I had given him the order issued by His Majesty that the President should facilitate our examination of all the papers of the expelled Jesuits.

By nine o'clock the *Atrevida* was clear of the port. The wind was very favourable on this day and the next, so that after completing the survey of the intervening coast she should be able to reach port at the same time as the *Hércules*, which had left the previous day. And so, in order to be able to sail at the earliest opportunity, we strove to make the best possible use of our time and we set about the tasks in the following way. [3rd] Don Francisco Viana, accompanied by a pilot from Talcahuano, went to look for a shoal which a merchant ship was reported to have struck some years earlier and which proved by chance to be the very one we had surveyed on the first day. Quintano and Salamanca set off in a boat to navigate as far up Río Aldarién[2] as they could go. I went with Bauzá to examine Puerto San Vicente and to complete the geodetic operations which still remained to be done. At the same time, six seamen went to Quiriquina to spend the rest of our stay in port making charcoal.

By chance, moored in Puerto San Vicente was a snow of the merchant marine whose launch was very useful to us in sounding the port. I therefore left this task to Piloto Sánchez, while Bauzá and I set about measuring a base at the inner end of the harbour. From the two ends of this base all the necessary bearings were taken to enable an accurate plan to be made. We then rode on horseback to the more southerly of the Tetas de Bíobío and from its summit we took by theodolite all the bearings needed to link up the numerous points we could see along the coast. We then rode down to the mouth of the Bíobío[3] and from a rock which formed its northern entrance, further bearings were taken to the outer and inner points of the river. These bearings had to link up with the chain of triangles which I intended to observe the following day along the river as far as Plaza Mayor in La Mocha or the city of Concepción. At this stage night overtook us and we spent it in a hut not far from the river.

4 March

Day had scarcely dawned when, despite the threat of a north wind and unsuitable weather, we began our geodetic work. We measured a base to fix the position of a nearby, isolated hillock known as Cerrillo de los Conejos,[4] and since we could see

[1] A common instrument at the time though few have survived, they were more robust and easier to make than the hand ground tubes for spirit levels.
[2] Río Andalién which enters Bahía Concepción 1½ miles SW of Penco.
[3] Río Bío Bío which enters Golfo de Arauco a short distance south of Tetas del Bío Bío.
[4] Literally hillock of rabbits.

the Tetas and other points that had been well fixed, we were able to tie in their position so that we could take useful bearings from there to various points on the Bíobío and in particular to Alto de Chepe[1] which rises steeply from the river. This height, where we went immediately to take bearings, afforded views of the main landmarks in the town and on the other side of the river. Having sighted these again from Cerrito Gavilán[2] which we fixed by means of a base measured along the Camino Real, we were able to complete our work around midday and to get ready to start next day on the survey of the harbour at Coliumo,[3] about twelve leagues from La Mocha and north of Tomé[4] and Ensenada de Talcahuano.

5 March

The Gobernador Intendente wanted to offer us, from his own house, all the assistance necessary to render the journey both comfortable and speedy. The weather was very favourable and so we reached Tomé before noon and at two in the afternoon we were able to start measuring a base in Puerto del Coliumo. By four, the angles at both extremities had been measured and since we had the opportunity of talking to a fisherman who had fished no other shores than these for the last thirty years, this lucky encounter even supplied us with sealskin rafts which we intended to use for our sounding. The bottom and other features of this harbour are such as to invite its use as a port of call by any ships that fall off to leeward of Puerto de San Vicente in winter and need shelter or supplies. Foodstuffs are plentiful in this neighbourhood, the watering place is very close and in the near distance can be seen mountains covered with apparently excellent building timber.

We were able to reach Tomé before nightfall and obtained the necessary bearings from its prominent headland, while the generous hospitality no less than the useful information offered by the deputy mayor Don Juan Ferrer contributed to making the remaining hours of our day exceedingly agreeable. Ferrer is at present the owner of most of the surrounding woodland. He combines unusual energy with uncommon perception and, since he is much loved by all the inhabitants of the district, he is undoubtedly the most suitable agent for any construction work which might need to be carried out in that harbour or in Coliumo, an enterprise which I think would profit both the entrepreneur and the country.

6 March

After taking further bearings the following day at Batería de Penco and examining in the surrounding countryside the grim evidence of the destructive earthquake of 1742, I was able to reach the city of Concepción and, that same night, to return on board after a four-day absence.

7 March

Both Don Manuel Novales, who had taken command of the corvette, and all the other officers had done everything in their power to ensure the speed and precision

---

[1] A height about one mile west of Concepción.
[2] A hill NE of Concepción.
[3] Bahía Coliumo, a present day holiday resort 13 miles NNE of Talcahuano.
[4] Situated in the NE corner of Bahía Concepción.

of our operations. They had watered ship, worked daily on the examination of the tides and variation, taken delivery of some wine and prepared the ship's casks to take the remainder the moment it arrived. The excursion of officers Quintano and Salamanca, though rather laborious since, for two nights running, the tides had prevented them from returning on board, had produced exact information on the course and depth of the Aldarién as far as the upstream side of the city of Concepción.

So it was easy for me to decide on the 10th for our departure, not least because for several days now the weather had been threatening to deteriorate again and it would not be surprising to see the northerlies and the seas beginning to establish their dominion over this truly hazardous port.[1]

Together we took leave of the Gobernador Intendente and the other military chiefs of the settlement; and as the Governor's help and attention to us had been unfailing, we expressed our most sincere gratitude and spoke of the solid advantage that the service had gained from these courtesies. The fine friendship of Coronel Pedro Quijada and Teniente Coronel Juan Zapatero will always retain a prominent place in our memories, together with a keen desire to return it.

8 March

Next day, at first light, the members of the crew who had been fishing and making charcoal for the last few days on Quiriquina returned in our launch. They had not managed to secure either commodity in any quantity. The rest of the wine for both corvettes was delivered on board; and while we intended to check the rates of the chronometers by absolute altitudes taken on board, it proved impossible by this method to achieve the accuracy we required, because of misty horizons for most of the time and the rolling and other movements of the ship. [9th] Thus on the morning of the 9th I went ashore myself with a highly reliable sextant and chronometer 72, which I wanted to refer to the other two chronometers by signals. Two sets of hour angles were taken to our complete satisfaction, one before and the other after the comparison signals, and we obtained the following results by amalgamating the latter observations with the former and applying, as always, the equation of correction according to the method already adopted.

	Date		Longitude by 61 east of San Carlos	Indicated by 72	Indicated by 13	Mean
Equal	February 26		0° 38′ 42″	0° 38′ 55″	0° 38′ 53″	0° 38′ 51″
Altitudes	March	1	0 38 21	0 38 15	0 38 8	0 38 15
Absolute						
altitudes		9	0 38 7	0 37 44	0 38 16	0 38 2
Mean						0° 38′ 22″

The latitude of Talcahuano observatory by different meridian altitudes taken to the north and to the south by Don Dionisio Galiano (as entered in the astronomical log) was 36°42′28″, variation obtained by theodolite 15°29′ and the establishment of the tide on the days of the full and new Moon at ten-forty in the morning with a maximum rise of 5½ to 6 feet.

[1] Strong northerly winds raise a swell in the bay, which may interrupt boatwork.

Our observations agreed very closely with those taken by Father Feuillée[1] who had obtained a latitude of 36°42′53″and a longitude of 75°32′30″ west of Paris for the town of Penco. These observations, transposed onto our plan at the Talcahuano observatory, gave the following results:

Latitude South		Longitude west of Paris	
Descubierta	*Feuillée*	*Descubierta*	*Feuillée*
36° 42′ 28″	36° 42′	75° 39′ 8″	75° 40′ 00″

Consequently it was all the more surprising that Father Feuillée, having repeated his observations using the satellites of Jupiter, and having obtained the corresponding values by Monsieurs Maraldi and Cassini[2] in the Paris Royal Observatory,[3] still failed to attain preference [for his longitude] in the list of longitudes published in the *Connaissance des temps*, where it was replaced by another which 32′ less and which was marked with the asterisk indicating the source to be a fellow of the Académie des Sciences, although we had no idea, nor could we trace, who that academician was.

The fact that the rates of the chronometers had been checked and found to be uniform and that the consolidated results were identical with those made in Chiloé and repeated in Valparaíso, indicated to us that we should not distrust our calculations, which we could also refer to the observations made by Monsieur Dagelet, the astronomer on board the ships of Comte de la Pérouse.

From almost the first days of our arrival in this port there had been desertions from both ships, which we found surprising, not only because I had made very clear the dire consequences they were exposing themselves to, but also because the people guilty of this crime were precisely the ones who had seemed most trustworthy and who made the greatest sacrifices. Two marine gunners from the *Atrevida* and one from the *Descubierta*, a marine, and four able seamen from the latter ship made up the already significant number of deserters, attracted no doubt by the deceptive lure of vices in the midst of idleness and by a social standing that they could never know in Europe. As soon as the desertions began I offered thirty *pesos fuertes* for each deserter from our ships who was brought back and, as I had already noted in the case of the merchant ship *Hércules*, how inclined the merchant seamen were to unruliness. I also offered ten *pesos fuertes* for every deserter from a merchant ship who was apprehended. The Gobernador Intendente, for his part, had already taken all measures likely to help in this aspect of discipline and good order. Indeed, although no one from our ships' companies was apprehended as a consequence, at least it was possible to arrest two merchant marine deserters and for each of these I immediately handed over the promised reward of ten *pesos fuertes*.

[1] See p. 136, n. 1 above.
[2] Jean-Dominique Maraldi (1709-88) and César-François Cassini de Thury (1714-84), noted French astronomers. Maraldi edited the *Connaissance des temps* in continuation of the work of his uncle, Jacopo Filippo Maraldi (1665-1729).
[3] If the occultation had been observed simultaneously in Paris and Penco the difference between the Paris time of the occulation and the local time at Penco would give the longitude of Penco west of Paris.

Today I had the satisfaction of having the Governor and several of the leading citizens of Concepción to lunch on board. We dressed ship,[1] obtained some fresh water by distillation and when the guests returned ashore our officers accompanied them a considerable part of the way on horseback.

---

[1] I.e. decorated the ship with flags and bunting.

CHAPTER 3

From Talcahuano to Valparaíso
and visit by *Descubierta* to Islas de Juan Fernández[1]

10 March

Before dawn the next day we weighed the north anchor and hauled the shoreward one up and down. Don Antonio Pineda had already returned on board, having spent the seven previous days on a scientific excursion to the various frontier *presidios*, accompanied by Don Juan de Benavente, acting *capitán* of the frontier dragoons. We now only needed a breath of favourable wind to set sail.

The light airs from the north which prevailed even after midday, with thick haze, had by now almost convinced us that it would be impossible to depart today, but finally at four in the afternoon a light breeze sprang up from the south and we saw that it was gradually extending and increasing offshore. We were able to fill our sails quite quickly and by sunset we judged that we had cleared the harbour. I had chosen to pass to the east of the outer shoal, both to be able to proceed more quickly without being forced to be constantly sounding, and because in the short time that remained of the day I intended to survey and take views of the coast as far as Puerto del Coliumo, which we were approaching.

Eventually, however, we saw that we were likely to be becalmed, with a rather high sea from the south setting us hard onto the rocks. With night drawing on we abandoned the idea of approaching the entrance to the port and we hauled the wind under full sail. Variation at sunset was 13°46′NE. At six o'clock in the evening, in latitude 36°29′30″ and longitude 42′0″ east of the meridian of San Carlos de Chiloé, we made our departure. In this position the NW point of Talcahuano[2] bore S24°W, the SW point of Quiriquina S9°E, and Punta de la Herradura N50°E.[3]

At nightfall the wind shifted to SSW and south and increased considerably in strength with a very heavy sea, accompanied by constant and violent rolling. The skies were clear, and under courses, topsails, main topmast staysails and jib we steered WNW5°W by compass, having accepted the dead reckoning distance and

---

[1] Archipiélago Juan Fernández, which consists of Isla Robinson Crusoe (formerly Isla Más a Tierra),
[2] Probably to the NW extremity of Peninsula Tumbes.
[3] Possibly Punta Mela 26 miles NNE of Talcahuano.

course between Isla Quiriquina and the inner island of Juan Fernández[1] as determined by His Excellency Don Antonio de Ulloa, referred to the position we had determined for Talcahuano. The scrupulous attention to detail to be seen in all his works or directions left not the slightest doubt that his determinations should be preferred to Moraleda's, although we assumed the trend of the coast between Concepción and Valparaíso to be almost north-south, giving a slightly greater distance. The dead reckoning gave a distance of 115 leagues from the said island to Valparaíso.

When the wind began to freshen considerably and we had been obliged to furl the topgallant sails and some of the staysails, we noticed that the corvette had an excessive propensity to come up into the wind which not only caused us considerable loss of speed, but also exposed us to the danger of shipping a green sea or the awful consequences of being taken aback. So at first light we began shifting weight from the bows to the stern, moving one hundred pigs of iron, which much improved our steering.

11 March

As we cleared the coast the wind veered to south, became less gusty and moderated, the sky giving every indication of windy weather. At ten o'clock in the morning, we were able to crowd on all sail once more and, while maintaining a speed of seven or eight knots, we were able to observe at noon a latitude of 35°15′S and longitude by the three chronometers 1°53′ west of Chiloé. The afternoon was pleasant and the night hazy, with bands of heavy cloud constantly gathering in the second quadrant and the wind blowing with more or less powerful gusts. From nightfall we ran under studding sails on the same course as the day before.

12 March

At dawn we noticed a few common types of birds and on the basis of the distance we had covered we looked for a break in the clouds which might disclose the island, for the sky was still overcast with thick misty clouds even after the Sun had risen. Indeed it was not long before we sighted the island right ahead. At six-thirty in the morning it was about eight leagues distant from us and despite the haze its shape was easily discernible as it rose to a considerable height. We bore away somewhat to bring ourselves on the same parallel as its northern extremity and at nine-thirty we hauled the wind to WSW and began to measure bases to fix its SE coast, and the islet which Anson calls Islote de los Conejos.[2]

Little by little the weather became clearer, and the wind gained strength somewhat, so that by eleven-thirty we were obliged to heave-to scarcely a league from the island, in order not to lose the horizon to the north at a time when we needed to take the meridian altitude, which gave us a latitude of 33°49′45″. The three chronometers gave our longitude as 5°0′53″ west of Chiloé; so on this passage there was little or no difference in latitude by dead reckoning and 24′ difference in longitude, with the dead reckoning being less, or further east, than the true one.

[1] Isla Robinson Crusoe.
[2] Isla Santa Clara, Anson's Goat Island, but literally Island of Rabbits. Anson makes no mention of rabbits in his account.

However, almost all this error had been noticed at noon on the previous day and it was probably more the result of the variable strength of the wind and the force of the waves, with the steering at this time very unreliable, than of any unknown current in these waters.

After the Sun had crossed the meridian, having obtained no bottom with one hundred fathoms of line, we immediately filled all sail and continued measuring bases steering to the west which would have taken us two or three miles to the south of the islet; in this way we could (if the weather permitted) reach and then round Isla de Más Afuera.[1] On the way back we would survey the northern end of the island that we now had in view.[2] I considered it useless, not to say dangerous, to seek out the anchorage where, to judge by the detailed reports from Anson, His Excellency Ulloa and all our navigators, as well as the precision with which we could determine its latitude and longitude from seaward, we would merely lose several precious days, not to mention an anchor, as commonly occurred in an anchorage which afforded so little shelter.

At one-forty in the afternoon we observed longitudes by chronometers, the mean being 5°9′18″. By this time we had satisfactorily fixed certain points to the north on the large island and after correcting the bearings we had taken of the southern extremity of the islet, we obtained for that point a longitude of 5°11′14″. Its latitude was 33°29′, which we deduced from the distance at which we had passed it, since it was impossible to combine the result of the bases which were too far away with the short distance by which we undoubtedly were from that point. We were able to take several azimuth readings during the afternoon as well as an amplitude, in order to obtain variation. There was some discrepancy in the results, the first being 13°37′ and the second 14°32′NE.

The sea had calmed completely. It appeared that birds did not frequent these coasts. Only seals and whales were occasionally bold enough to show their faces. The skies continued to alert us of the imminence of the new Moon, and this was the more perceptible as our course took us nearer to the torrid zones. What at first was a falling wind from the south veered to SW. The sky became very overcast with dense cloud. Later on light airs with some moments of dead calm so delayed our passage during the night, even though we did not overlook the slightest puff of wind or the smallest of sails, [13th] that at first light on the following day we were still nine leagues from Isla de Más Afuera. From five o'clock in the morning, however, we could see it very clearly, and at six we could make out the left-hand tangent bearing S70°W and the right-hand tangent bearing S76°W. At the same time all the high peaks of the inner island were clearly visible from the quarter-deck well above the horizon.

The wind was still light and variable and since it was from the west it compelled us to abandon our original intention of circumnavigating Isla de Más Afuera – an enterprise of little importance since the English Captain Carteret had surveyed it carefully

[1] Isla Alejandro Selkirk.
[2] Isla Robinson Crusoe.

at the cost of innumerable hardships and perils.[1] It was therefore necessary to steer to the north and thus begin to measure a base to determine how far we were from Isla de Más Afuera, which, referred to the successive observations of latitude and longitude, would accurately determine these two coordinates for this island. At six-thirty we took bearings and steered in that direction under full sail. However, as the wind continued to blow from the same direction as at dawn, we had scarcely managed to complete an acceptable base of two leagues by ten-thirty. Fortunately we were able to take satisfactory altitudes of the Sun to obtain longitude. Consequently at eleven o'clock we bore up and, as we also succeeded in taking the meridian altitude of the Sun at noon, our position was now well established by these observations and that of both islands could easily be inferred. Our latitude was 33°31'9", mean longitude 6°7'34" and, having completed the triangle to the first island, the position of the northern extremity of Isla de Más Afuera at ten-thirty was 28·2 miles from the corvette, in latitude 33°37'33", longitude 6°44'33".

We succeeded in lassoing one of the many seals that surrounded the corvette when the wind was falling and it immediately fell victim to our enthusiasm for natural history.

After midday the weather brightened considerably and, as the wind backed again to SW and freshened appreciably, we felt confident that sailing at a reasonable rate we would be able to repeat our observations at the inner island. Variation by azimuth was 13°30' and by western and eastern amplitudes no greater than 13°0'.

Our longitude at 4.48 in the afternoon was 5°44'5" and our estimated latitude at the same time was 33°30'9". The difference between this longitude and that determined the day before at Islote de los Conejos[2] referred to the same time, determined the latitude of its southern extremity as 33°44'49". What we had deduced the previous day by the meridian altitude observed in its vicinity was 33°45'20" so that by rectifying accordingly the supposed difference in longitude, the position of the NE extremity of that island was established with all certainty as latitude 33°37'39" and longitude 5°6'30". The distance from one island to the other was seventy-nine miles.[3]

At sunset the northern headland of the inner island bore S80°E at a distance of six leagues, and the centre of Isla de Más Afuera bore S 65°W, by compass in each case. The elevation to which the latter rose left me in no doubt that it could be seen from the extreme heights of the island before us.

His Excellency Ulloa's very detailed information on the anchorage,[4] and the chart of the whole of this island made any further survey of it pointless. As the few remaining days of summer had to be regarded as precious, we decided to take advantage of that night by setting a course ENE5°E under full sail, with a moderate breeze and a fair outlook.

---

[1] Carteret visited Isla Alejandro Selkirk (Isla Más Afuera) in May 1767. Malaspina is referring to 'A View of the N.W. Side of Mas-a-Fuera', which also contains soundings: Hawkesworth, *Voyages*, I, f.p. 553.

[2] Isla Santa Clara.

[3] They are in fact 90 miles apart.

[4] Bahía Cumberland on the north side of Isla Robinson Crusoe.

14 March

At dawn the most northerly extremity of the inner island was in sight bearing S52°W, six or seven leagues distant. The haze caused us to lose sight of it towards nine o'clock. The weather outlook remained pleasant, the wind from SSE had strengthened steadily, and the sea was now beginning to rise from the same direction. At noon our latitude was 33°24′ and longitude 3°52′15″ by the mean of the three chronometers which agreed closely with each other.

As we increased our distance from the island, the sea and the wind regained the same strength we had experienced during our outward passage. We were struck by some heavy seas which forced us to reduce sail but the weather nevertheless remained fine, except for a few hours early in the night when the sky grew very overcast and the horizon thick with dense cloud. [15th] After steering somewhat to the south, we found ourselves at noon in latitude 33°38′ and longitude 0°56′ west of Chiloé.

It seemed likely that the trend of the coast between Concepción and Valparaíso was correct as shown on Moraleda's new chart, who had surveyed these parts on several occasions. Since it virtually ran almost north and south, whatever errors there might have been in Father Feuillée's longitudes, we had to assume that difference in longitudes in this stretch was the same as we had observed between Juan Fernández and Concepción. Early in the afternoon our lookouts thought they could see land through the haze and so insistent were they that for a time we bore up to NE5°N, later returning to ENE5°N by compass once we realized that their assumption had been incorrect.

That afternoon variation by amplitude was 14°8′30″. The outlook remained clear and the sea had fallen almost completely calm. By midnight we had made good the supposed difference in longitude. For that reason we lay-to heading west [16th] until two-thirty in the morning, at which time we bore up to NEbyE. At this time the SSW wind dropped and was soon replaced by a gentle little breeze from NNW which compelled us to sail close-hauled, since, as we had not been expecting it, we were being set too far to the south for our intention of closing the coast and identifying it at a slightly higher latitude than the anchorage.[1] It remained overcast until ten in the morning, by which time the wind had backed to WNW and freshened, giving us the opportunity of sailing with a quartering wind, seeing that we were already on the parallel of Valparaíso, as well as carrying out our observations, according to which at noon we were in latitude 33°00′ and longitude 1°23′30″ east of San Carlos.

We could now see a fairly wide sweep of the horizon, yet the land was still not in sight. From this we could now infer without a shadow of doubt that the longitudes assigned to Concepción and Valparaíso in *Connaissance des temps* were quite wrong and in the opposite direction. Also that Moraleda had followed these so slavishly, despite what he had seen, that he plotted them quite erroneously on his chart. Soon after midday the wind backed to north and the eastern horizon was much obscured by haze. This made it practically impossible to sight land and avoid falling off to leeward of the port before nightfall. Land was finally sighted very indistinctly at about two o'clock in the afternoon. It was quite high, about seven leagues distant, and as we

---

[1] I.e. off Valparaíso.

closed it the outline looked like the area around Punta Coroumilla.[1] At four our longitude by observation was 1°41′49″ and at this time Punta Coroumilla bore true east, four to four-and-half leagues distant.

Just before sunset, as we were close to the point, we went about with our ensigns hoisted so that the Valparaíso lookouts could identify us. We then sailed some four leagues under our principal sails in the fourth quadrant with a moderate breeze from NNE. Finally, we hove to under topsails on the same tack until two-thirty in the morning, at which time we sailed close-hauled once again to ENE and east, to be a little to the north and very close to the harbour entrance by dawn.

17 March

Our position at daybreak was very satisfactory. We were only three leagues from the harbour entrance. Despite the hazy weather we had a clear view of the coast on either hand as we headed east by compass and the wind from NNE made it likely that we could enter harbour on the same tack. Contrary to all our expectations the wind which was already falling, changed to variable light airs from north and NE, which obliged us to change tack three times, the last time being when we were already in the harbour. But the corvette's excellent speed and our resolve to take advantage of every breath of wind finally allowed us enter the bay at one-thirty in the afternoon and soon after to anchor there in seven fathoms of water, sand and gravel, too close to Punta de la Pratería Vieja where the poor holding ground exposes ships' cables to considerable risk in that locality.

---

[1] A rugged rocky point about 9 miles SW of Valparaíso.

CHAPTER 4

At Valparaíso and visit to Santiago

[17 March]

The corvette *Atrevida* had been anchored in the port since the 11th. Her commanding officer later told me officially that calms and the inaccurate plotting of the coastline on Moraleda's chart had made his passage slower than we had expected. The mist and the heavy sea from SW caused him much inconvenience, not only making it impossible for him to keep the coast in sight, but almost forcing him on two occasions to drop a kedge because he had been set too far towards the land. In spite of this he had managed to plot with full confidence the coast from Quiriquina to the mouth of Río Itata and then as far as Ensenada del Cerro, the vicinity of Morro de Topocalma, the bluff itself, Bajos de Rapel, Playas de Cartagena, and finally the section between Puntas Coroumilla and Los Angeles.[1] All these were exceedingly interesting operations which redounded all the more to the reputation of the commanding officer of the *Atrevida* because of the great difficulties he had to overcome with regard to the weather.

The observatory had already been well set up in the northern corner of Castillo del Rosario,[2] for which purpose the Governor, Coronel Don José Salvador, had supplied as much aid as might contribute to its success. Galiano, Concha and Vernacci were lodged in a room which had previously been an armoury and a *pilotín* and a marine were attached to them for service in the observatory. The astronomical clock had been set in motion from the first days and the astronomical quadrant mounted. Only storm clouds and mists hindered our projected catalogue of right ascensions and declinations of southern stars which fortune might place within the reach of our operations.

Bustamante had also been informed of the safe arrival in Santiago of Teniente de Navío Valdés, and of the kindness of the Capitán General and Presidente of the kingdom, Ambrosio Higgins,[3] who intended to leave the

---

[1] Río Itata, 24 miles NNE of Talcahuano with Bahía Cobquecura (Ensenada del Zorro), about 15 miles north, Punta Topocalma about 70 miles SSW of Valparaíso with Bajo Rappel, Bahía Cartagena and Punta Curaumilla between Punta Topocalma and Punta Angeles, the western entrance point of Bahía Valparaíso.

[2] Abandoned by 1775: Guarda, *Flandes indiano*, p. 185.

[3] Ambrosio O'Higgins, born in Ireland but educated in Spain, settled in Santiago and entered the service of the crown in 1760. He became garrison commander of Concepción in 1786 and Capitán General of Chile in 1788. In 1796 he was made Viceroy of Peru and was ennobled as Marqués de Osorno. At the time of Malaspina's arrival, he already had an established reputation for hospitality to

spa[1] where he was at the time and go immediately to the capital to ensure personally the provision of whatever assistance our task required.

With regard to the baking of some ship's biscuit, which I had entrusted Bustamante to oversee in Valparaíso, for the sake of the price and the convenience, which I had been told could be obtained there, Bustamante told me that although they had scoured the entire neighbourhood they were only able to obtain two hundred pounds per day and in view of this small quantity he had given up ordering the biscuit. This was a good decision since we had no need of supplies until we reached Lima and the funds in the Royal Treasury in Santiago were not so plentiful as to support unnecessary expenditure.

A few days earlier two ships in the Lima trade, the *Valdiviano* and the *San Miguel,* had set sail for Lima with a load of wheat and the *Aquila* and the *Hércules*, at present anchored in the bay, were preparing to follow them. Their captains and supercargoes came on board immediately to present their respects. This was the occasion I had been waiting for to cut short some of the unruliness in the merchant navy by dealing with the two deserters who had been handed over to me in Talcahuano. [18th] The following morning, with the marines under arms and officers despatched to both merchant ships to bring on board their crews and either the *pilotos* or the boatswains, fifty lashes were ordered to be delivered at the hands of the merchant seamen to one of the deserters, since there was some doubt whether the other had in fact deserted. All the others were warned at the same time not to get involved in unruliness and to be obedient and punctual in their duty on board our ships if they did not wish to be severely punished.

The day was hazy. Both vessels were employed in improving our moorings. For that purpose our corvette set up both her fore and aft mooring lines, with one hundred fathoms of the bower cable to the north and the small bower cable to an anchor to the south, backed by a kedge fixed onshore. Such precautions are necessary, although often of little use in a place which has been given the name of a port only out of sheer necessity. The *Atrevida* had her southern cable secured ashore and to the north an anchor as well as a kedge, the intention being to drop a second anchor if so required by the wind. The ships were in nine or ten fathoms of water, the outer anchors in twenty to twenty-two fathoms.

We were not very pleased with the news which Bustamante had obtained about the idea, which was also laid down in the instructions, of observing a series of triangles as far as Santiago;[2] although the delightful capital[3] of a flourishing kingdom like

---

and interest in scientific missions, from his time as garrison commander in Concepción: Schultes, *The Journals of Hipólito Ruíz*, p. 198. In a letter of 20 June, 1786, addressed to the Marqués de Sonora, the minister responsible for Spain's New World possessions, O'Higgins had proposed an expedition similar to that on which Malaspina was now engaged, as part of a strategic programme for defending the empire in the light of La Pérouse's voyage: A. Galera Gómez, *Alejandro Malaspina: en busca del paso del Pacífico,* Madrid, 1990, p. 10. In his report home, 'Examen político del país comprendido entre Chiloé y Coquimbo', Malaspina regarded him as an ally in the cause of boosting the role of agriculture in the economy and reducing military expense by adopting a 'commercial' instead of a 'conquistador' mentality towards the colony: Pimentel, *Descripciones y reflexiones políticas*, pp. 77-91, at p. 80.

[1] Due to volcanic activity, Chile is a country rich in hot springs and spas, with a number within 100 miles of Santiago.

[2] See p. 158 above and Higueras, *Diario por Bustamante*, p. 135.

[3] Founded in 1541, with 34,000 inhabitants, according to Ruíz, at the time of Malaspina's visit: Schultes, *The Journals of Hipólito Ruíz*, p. 250.

Chile did deserve to be fixed with precision while it was so close. Nevertheless neither of us lost hope of managing the task. Since we had decided to go to the capital for various official purposes, foremost among them the close study of the political state of the kingdom[1] and in order to pay our compliments to and receive orders from the Capitán General, we left the decision about whether that enterprise would be feasible until we had carried out an examination of the terrain.

19 March

Leaving Tenientes de Navío Tova and Novales to take charge of the corvettes and Dionisio Galiano of the observatory, I prepared to set out for Santiago with Bustamante early the next morning. On that day, which was fine and calm, I had managed to see the results of our chronometers and those of the *Atrevida* and of two observations of satellites of Jupiter. These had been obtained at the beginning of the night and might perhaps have corresponding observations in some observatory in Europe. The result, compared with the tables,[2] gave the longitude of our observatory as:

By the emersion of the first (completely reliable) satellite			
Longitude West of Paris	74°	8′	30″
By that of the second (fairly reliable)	74	11	00
By the *Atrevida's* chronometers	74	8	30
By the *Descubierta's* three chronometers			
(adjusted by the corresponding equation)	74	9	50

These results supported the determinations we had made since we left Chiloé and gave our work the consistency that we wished to attain. There still existed the difference in longitude, which we had noted earlied in Talcahuano, given by our chronometers and by those of the *Atrevida*. It was decided to adhere to our longitude which depended principally on chronometer 61, so that the detail of the conspicuous points of the coastline which depended on it would not have to be modified until the true longitude of one point or another was determined by observations of satellites of Jupiter which corresponded to ours.[3]

The road from Valparaíso to Santiago, which is almost entirely rocky and very tortuous, traverses three rows of mountains which increase considerably in height as they approach the foot of the *cordillera*. The first plain is quite extensive and partially used for pasture or for cultivation. The hamlet of Casablanca[4] makes the second one

---

[1] Chile's continuing role as a military frontier made it a particularly testing terrain for Malaspina's political project of displacing a 'conquistador' mentality with a commercial one. In practice, however, most of the new research undertaken by the expedition in Chile was economic and commercial: questionnaires were prepared and circulated on food prices, the balance of trade with Peru and Spain, and the nature of the exploitable resources of the interior: AMN, MS 337, ff. 120-82 and Pimentel, *La física*, p. 228.

[2] I.e. from one of the nautical almanacs held on board the *Descubierta*, presumably *Connaissance des temps* since Malaspina gives the resulting longitude west of Paris.

[3] Since *Connaissance des temps* gave predicted values for the occulations, a better value for longitude would be obtained if they were simultaneously observed in Paris or in another European observatory.

[4] About 25 miles SE of Valparaíso.

Plate 23. View of Valparaíso, by José de Pozo. Museo Naval, Madrid

more agreeable and useful, apart from the two small valleys of Viñilla and Puanghi[1]. The third plain is the fine valley which is watered by the Mapocho[2] and where, on the side of the *cordillera*, Ciudad de Santiago is situated.

The whole road is no longer than twenty-four to twenty-six leagues, although the local inhabitants believe it to be thirty-six. It runs ESE to SEbyE, but turns several times towards the east. The individuals who have a small-holding along this route are accustomed to offering lodging to travellers in that spirit of hospitality that has become so delightfully ingrained in the inhabitants of this region. The land, the fertility of which cannot easily be described, contributes much to the traveller's enjoyment of that abundance that he so often longs for in vain.

We left Valparaíso at five o'clock in the morning, rested for several hours in the afternoon in Viñilla, in the *hacienda* of the Señores Azagra, and spent the night in Puanghi, in the *hacienda* belonging to Don Xavier Bustamante. [21st] We arrived in Santiago at noon the following day.

21 March

The Capitán General had already reached the capital the previous day; his energetic kindness not only to us but also in all things related to our commission will ever remain impressed upon our hearts, difficult as it is for us to describe it with the same expansiveness.

Don Cayetano Valdés had already got to know in great detail not only everything held in the Archivo de Temporalidades[3] but also much information specific to this kingdom which could facilitate the gathering of more material useful either for our work or for gaining a more exact knowledge of the interesting country that we were observing at the time.

I naturally doubted very much if we could carry out the proposed scheme of triangulation without great sacrifice of time and money. Bustamante and Valdés were of the same opinion but they agreed that we should on no account neglect the precise determination of the geographical position of the capital. So I unhesitatingly took all the necessary steps so that we could be certain (if the occasion allowed it) of using the most reliable methods to carry out operations which at first sight seemed impossible. I sent orders to the ship for Quintano, Viana, Vernacci and Bauzá to come with all speed to the capital, bringing with them a good collection of astronomical and geodetic instruments, several flags and chronometer 105 by Arnold[4] which belonged to the commanding officer of the *Atrevida*. Piloto Sánchez was charged with the safe transportation of the instruments and so that we could better coordinate our successive tasks we were to measure hour angles and the latitude in Casablanca. With these measures I would be able to carry out all the operations that could be undertaken during our short delay in Santiago and also, if the triangulation could be observed, to

[1] Puangue in the modern department of Melipilla.
[2] Río Mapocho rises in the Andes, passes through Santiago, and eventually empties into Río Maipó.
[3] The archives kept by the Jesuits of records, transactions and documents relative to their secular properties and revenues.
[4] Being a pocket chronometer, Arnold 105 could be transported more easily and safely than the larger box chronometers.

do so with two teams, each having first marked the stations with tall poles and signal flags.

The unavoidable irregularity in eating and sleeping times which naturally resulted from these hurried operations and the necessity of sharing living quarters for the better coordination of our work, persuaded me (in agreement with Bustamante and Valdés of course) to decline the generous offers made by the Capitán General and some of the leading citizens to accommodate us in their houses. We rented a residence towards the centre of the city which had a patio large enough to allow a view of most of the sky. The way we ate and lived conformed with how we lived on board.

So on the morning of the 26th, after the officers arrived as quickly and efficiently as could be desired, bringing the instruments we had requested from on board, we were able to start our operations. The next day, using 105, we deduced the longitude of the Santiago observatory, which was situated 1°6′51″ east of Valparaíso. Next, using a base and the angles measured at the extremities and summits of the nearby Santa Lucía, San Cristóbal, and Renca we undertook the mapping of the city and the whole valley. Don Juan Vernacci was mainly occupied with the latitude, the rate of the chronometer and the variation by theodolite, while Valdés, Quintano and Viana summarized a considerable quantity of papers which were as interesting for navigation as for the current political state of this kingdom.

On Maundy Thursday [1 April] we completed our geodetic operations. We observed the emersions of the first satellite of Jupiter during the nights of 2 and 4 April,[1] and set out once more on the morning of the 7th for Valparaíso, where we arrived the following day.

8 April

The following data applied to our observatory in Santiago:

Latitude by six stars, three to the south and three to the north . . . . .S 33° 26′ 16″

Longitude	Emersion on the 2nd (unreliable) .73	16	30
	Emersion of the 4th (very reliable) 73	13	00
	Chronometer 105 (Valparaíso assumed to be in 73°8′30″) . .73	1	39

Variation by theodolite . . . . .NE 13 20

With the arrival of the mail from Buenos Aires during the time we were in Santiago I received the glad news of the brig *Carmen*'s successful return to Montevideo. His Excellency the Viceroy of Buenos Aires was so kind as to send me a summary of the journal of Piloto Don José de la Peña, whose experiences after he had parted company with us in Puerto Deseado were of great interest and value to our expedition. He had anchored in Puerto San Julián and then in Río Santa Cruz. From this latter position he had sent a team by land to survey Río Gallegos.[2] In each of those ports he had managed to maintain continuous close communication with the Patagonians, which allowed him to make certain that there was no foreign settlement in those parts, except that there was a young Englishman, seemingly a fugitive, living

---

[1] Malaspina clearly regards these observations as astronomical and not geodetic.

[2] Río Gallegos rises near the present Chilean/Argentine border and flows east to enter the South Atlantic 44 miles NNW of Cabo Vírgenes.

among one of the tribes of Patagonians and that years ago the English had tried in vain to settle on the Río de Santa Cruz. On his return he surveyed Bajo Bellaco[1] and saw many English, French and American whalers. He reached Puerto de Montevideo after undergoing numerous dangers, but having suffered only slight damage to his mainmast.

The Viceroy advised me at the same time of the measures taken so that Don Tadeo Haenke, the botanist appointed to our expedition by order of His Majesty, who had been unable to join us either in Cádiz or in Montevideo, could travel to Valparaíso, where I had arranged in advance that he should join us between the 15 March and the 15 April. This worthy gentleman had not delayed in reaching Santiago to join us. In the Royal Order he had been asked to deliver, His Majesty authorized the salary he was to receive and expressed how much he valued the services of someone who, in addition to a truly remarkable ability in various sciences, could invoke the recommendations of Conte de Graneri, the Ambassador of the King of Sardinia in Madrid, of adviser Born and the botanist Jacquin of Vienna.[2]

Don Tadeo Haenke, after the extreme mortification of arriving at Isla de León only two hours after the corvettes had left Bahía de Cádiz, had to travel to Montevideo on board a merchant ship, in which he had the misfortune of being shipwrecked near Punta de las Carretas,[3] a little distance from the port, solely because of the inexperience of the captain or *piloto*. The corvettes had sailed eight days earlier and so he was obliged to present himself to the Viceroy requesting the necessary orders and assistance to continue his journey in search of us. He set off on 24 February through the pampas and the *cordillera* and finally joined us after eight months wandering, all because of an unavoidable delay of a few hours in his arrival in Cádiz.

These incidents, which were really annoying for the person who endured them, proved very useful for the progress of our botanical work and for the greater glory of our expedition, since Haenke had used all his time examining the surrounds of the Río de la Plata, taking a particular interest in Las Conchas and the Paraná, areas where Neé had not penetrated for lack of time. Later, in the pampas, the Sierras de Mendoza and finally in the peaks of the *cordillera*, he had gathered close to 1,400 plants which were either new or had been poorly described, as well as making several investigations in other branches of natural history. He was duly signed on according to the terms prescribed in the Royal Order and thus from 1 April he became a member of the ship's company.

I had been forced to abandon the idea of a programme of triangulation, both because Holy Week and Easter occurred while we were in Valparaíso, which prevented the men from doing any work, and because of it was impossible to obtain any reliable information about the higher mountains where we could site suitable stations and signals. To find this for ourselves would have meant a further sacrifice of time and might have exposed most of us to an unnecessary health risk.

---

[1] Roca Bellaco, a dark rock which dries 20 feet in the form of an upturned boat, situated 10¾ miles SE of Cabo Guardián and about midway between Puerto San Julian and Puerto Deseado.

[2] Ignatz de Born (1741-1791) and Nikolaus Joseph Jacquin (1727-1817), both Viennese naturalists.

[3] Punta Carretas, a name no longer in use, situated 5 miles NE of Punta Brava.

We had also planned to study the elevation of the land in Santiago, and even of the nearby mountains, by using Señor Magallanes's barometer. Unfortunately, it broke as it was being brought to the capital and consequently the observations which were made on the coast at Valparaíso were worthless. At any rate, there remains no doubt that the terrain in that area is elevated considerably above sea level. The mercury in the barometer demonstrated this clearly, and the road itself confirmed it.

9 April

On our return to Valparaíso I found that the officers left in command, Tova and Novales, had carried out the duties entrusted to them with great efficiency. Both ships had completed their watering, the port had been meticulously sounded, all available supplies of coal, of which we previously had absolutely none, had been stored. The *Descubierta*'s boom-irons,[1] which had been slightly damaged on the afternoon of Maundy Thursday when excessively strong gusting winds in the area broke the *Atrevida*'s southern mooring line and she became entangled with the *Descubierta*, had been repaired.

Don Dionisio Galiano and Don Juan de la Concha had fully displayed their ability and energy in fixing the latitude of the observatory. They also took advantage of the few nights that were clear enough to add to the catalogue of star declinations with the same astronomical precision. Some 150 stars had now been newly identified and many more in Abbé Lacaille's catalogue[2] had been confirmed. With regard to the longitude relative to our observations in Santiago, the corresponding observations made from the observatory had been unsatisfactory. However, the officers mentioned, like all those who were not in Santiago, had not neglected to observe the longitude by lunar distances. Thus 361 sets of extremely consistent observations gave the longitude as 73°53'22" west of Paris. In the remaining days that we stayed in the port of Valparaíso, most of which were clear, Don Juan Vernacci, who had joined the two officers already mentioned, worked constantly with them on the same task. Their work, which is so useful for astronomy and even more so for navigation, will forever give them a distinguished place among those who have dedicated their time and health to the public good. Two fully satisfactory observations of the first and second satellites of Jupiter taken on the night of the 11th corroborated the result that had been deduced from the observations of 19 March.

One matter more than any other that soured and disrupted our activities in this port was the constant unruliness of both sailors and marines. For my part it was impossible to restrain them even by force. The lack of assistance available locally and particularly the small number of marines that we had in the two ships who could not be much trusted in any case, obliged me not to compromise respect for authority by revealing the weakness of its resources. At first, a few sailors deserted. Some of them, armed with knives, challenged one of the marines from the *Atrevida*, as a result

---

[1] Metal rings fitted on the yardarms through which the studding-sail booms traverse.

[2] Nicolas-Louis de La Caille (1713-62), a French astronomer who visited the Cape of Good Hope in 1750-52 where he compiled a catalogue of nearly 2,000 southern stars and measured an arc of the meridian. Malaspina was referring to La Caille's *Caelum australe stelliferum* published in Paris in 1763.

of which one of the sailors was mortally wounded.[1] Then the two divers and a corporal from the *Descubierta* deserted. On the night of the 12th two marines from the *Atrevida*, one of whom was on sentry duty, had the audacity to make off in one of the boats when the officer on watch took a moment's rest. The efforts of a sergeant and four marines who were immediately sent to capture them proved unsuccessful.

In accordance with the system adopted in Talcahuano whereby we decided to give a little leeway to those who had been diligent in their duties by overlooking the first few faults of the good men and never to confuse the one kind with the other, I tolerated this incorrigible unruliness with as much patience as annoyance. Those I was able to apprehend I punished only with the penalty appropriate to their misdemeanours. To my satisfaction I learnt in a letter from the Gobernador Intendente of Concepción that two seamen who had deserted from this corvette had been captured and that, according to our agreement, they would be sent to Lima. However, as the Governor, in order to encourage further action on the part of the inhabitants, asked not only for the reward of thirty *pesos fuertes* in its entirety but also for what it had cost to keep them, I had to devise a method of pursuing future deserters and put it forward in the reply I sent him, including at the same time the 715 reals that were owing for the two seamen.

I passed on a copy of this communication to the President in Santiago, requesting him to take it as my offer to those who either declared or handed over some of the deserters. The four main points were, first, that only those who could reach Lima before the end of August should be sent there; second, in calculating the rewards for each apprehension His Majesty would not be held accountable for the life and safety of the captives until they had been detained; consequently, the captor would be paid only half and not the full amount of the reward promised, the other half being paid when the deserter arrived at his destination; third, for the time they remained in the custody of the relevant authority they should be kept on half rations or be made to earn their keep by working for public authorities or for private employers with the proper degree of responsibility; fourth and finally, for those who might be detained too late to be transferred to Lima, they should be sentenced to hard labour or to armed service depending on the circumstances of their desertion and at the rate of thirty *pesos fuertes* per deserter so that anyone, whether in Chiloé, Valdivia, Concepción or Valparaíso, who captured a deserter during the following three years could have access to this fund, it being sufficient grounds for claiming the reward that they could establish the sentence that had been imposed. Teniente de Navío José Robredo was given responsibility for substantiating the case against the marine who had attacked the seaman from the *Atrevida,* whose life was still in serious danger. The marine's defence was assigned to Teniente de Fragata Secundino Salamanca.

Gathering supplies of firewood and water would be extremely difficult in the ports between here and Lima. We had completed taking on water, bringing it in barrels from the fountain in the square. When we found an individual who was selling a considerable supply of firewood, it seemed appropriate to buy as much as the ships could

[1] Premature! Malaspina later referred to this seaman's life as being still in serious danger.

carry, which amounted to 200 quintals for this corvette and 230 for the *Atrevida*. We also acquired several quintals of fine oakum to stuff birds with, along with some grease so that we could paint the ships' sides. By the 13th we could finally feel completely ready to set sail. At that time twenty-one of the crew of the *Atrevida* were missing, as well as fifteen from this corvette, all of whom had deserted in Chiloé, Talcahuano, and Valparaíso, except for the marine of the *Atrevida* who had died, and the seaman who had been injured recently.

I considered the survey of the coast as far as Lima to be of supreme importance for Spanish shipping, and knew that the fog which was very common at this time of year[1] might delay our journey for some considerable time and above all could subject us to major mistakes, particularly at the entrance to ports where it was very easy to fall to leeward given the constant northerly current.[2] With this in mind, and with no doubt that in Valparaíso, the inhabitants of which are almost all seafolk, we would find a coastal pilot familiar with the next section of the coast, I had charged Bustamante to make some inquiries about this. He was in fact able to find someone who fitted my requirements and when I arrived in Valparaíso he presented him to me, together with excellent references. This person was Don Domingo Velázquez, who had sailed these coasts for a long time in command of merchant ships and was perhaps without equal in his knowledge of the entire coastline from Chiloé to Acapulco. He agreed to act as pilot on board this corvette for the duration of the voyage; so he was signed on and accorded the salary to which he was entitled under Peruvian regulations and assigned satisfactory quarters.

13 April

Don Antonio Pineda did not return to the ship until after midday on the 13th. He had previously gone to study the rocks in the area close to the city of Santiago and had recently made a trip to the San Pedro Nolasco mines and the adjacent volcano. His physical observations and his acquisitions in the field of natural history were a great credit to his energy and ability, just as they had been in the other places he had previously investigated.

As the peace and calm of our forthcoming passage to Lima might allow us to sort out quite a lot of the work we had done up to then, I thought it appropriate to gather together all the documents concerned with the plans arising from the work of the two ships. For this purpose I ordered Teniente de Navío Galiano to transfer from the *Atrevida* to this corvette bringing with him all their notebooks and observations. I replaced him on that ship with Teniente de Fragata Francisco Viana. To allocate accommodation properly, I was obliged to send Guardiamarina Ali Ponzoni to the *Atrevida* as well, since the naturalist Haenke and the pilot Velázquez had also been assigned to the *Descubierta*.

14 April

Finally, we spent the entire night of the 13th taking observations of the stars and

---

[1] Fog may be encountered off the coast of Chile, particularly in April, when upwelling brings very cold water to the surface.

[2] The Peru Current flows to the north along the west coast of South America throughout the year at an average rate of between ½ and 1 knots.

returned on board the ship early the following morning with the tent, the astronomical quadrant and the astronomical clock, the only things that had been left on shore. We cast off the mooring line from the shore, leaving a single hawser in place and prepared to set sail at the first breath of favourable wind. The calculations of the chronometers had been completed at noon on the 12th, and chronometer 72 had been mounted in gimbals since the 7th to see if this would be as advantageous as it had been for number 71 of the *Atrevida*.[1]

---

[1] *Sobre esferas* (in spheres) in original MS, but clearly some sort of crude gimbals since Arnold No 71, which is now held in the Museo Naval in Madrid, is still housed in its original box without gimbals which were not favoured by Arnold at the time.

CHAPTER 5

From Valparaíso to Coquimbo

Instead of the usual hours of calm at the beginning of the morning being followed by a southerly wind, which customarily sets in at about noon,[1] a thick fog set in from the north and prevented us from seeing even the closest objects and as there was absolutely no wind until four o'clock in the afternoon we were almost obliged to abandon any hope of setting sail that day. But when the fog was finally dispersed by some light breezes from the south, it seemed that the help of a tow, with the assistance of the current, together with the short distance we would have to run to consider ourselves out of danger, would allow us to get under way at the intended time without the risk that the light northerly breezes so common during April would keep us in port for several more days. I hailed the commanding officer of the *Atrevida* to inform him of my intention. So both ships hauled in their northern cables and recovered the hawsers secured ashore. At four-thirty we weighed anchor and, towed by the launch, we tried with our topsails, topgallants and staysails set to catch the favourable southerly winds, scarcely perceptible though they were. At first progress was slow and we were being driven towards the rocks at the entrance to the port, but later the situation began to improve. By nightfall we had made two miles to the north of Punta de la Batería, with the wind still giving us steerage way.

The *Atrevida* followed close astern, with equal success. We hoisted the launch and at six o'clock in the evening fixed our point of departure by bearings in latitude 33°0′ and longitude 2°7′39″ west of San Carlos de Chiloé.

Our first intention was to anchor in Puerto Papudo,[2] ten to twelve leagues from the port of Valparaíso, having been persuaded that it would afford us some shelter. Moreover I had been led to believe this because the port had been frequented by French ships at the beginning of the century when they flocked in such numbers to the coasts of Peru and Chile.[3] There were still many *bodegas* to be found among the ruins, although it was always illicit trade more than anything else that had drawn them

---

[1] Off the coast of Chile, north of about 40°S, the sea breeze commonly sets in during the mid-forenoon. The land breeze is usually weaker and blows as a light offshore wind from around midnight and fades soon after dawn.

[2] Anchorage can be obtained in Puerto Papudo, but is exposed to winds between west and north.

[3] During the War of the Spanish Succession(1702-13), in which Britain was faced by France as well as Spain, French vessels were allowed to enter South American ports to trade, which they did in large numbers.

to this sparsely inhabited place. The President, whom I had informed of my intention, had been kind enough to order in advance the *subdelegados* of the region[1] to
assist us. In the end, however, we had to abandon the idea, not only because of the
lack of time, the risk of further desertions, the prevailing calm and the murky appearance of the horizon, but also because the previous year Ingeniero Pedro Rico,[2] by
order of the Government, had drawn up a chart of that roadstead, as well as that of
the following one, Pichichanque or Rada del Gobernador.[3] The engineer in fact
recorded this second port as sheltered, but Pilot Velázquez insisted that there was neither shelter nor even water in it.

Most of the night, calm, without steerage way, and enveloped in dense fog.

15 April

At dawn a favourable wind set in from SSE and SE, under which, though it was no
more than a gentle breeze, we steered NNE and NE in order to close the coast,
which at that time was shrouded in fog and hard to distinguish, even though we were
never more than two or three leagues from it. There was no doubt whatsoever that
the currents or the moderate swell from the south had carried us considerably
towards the north, so that without the assistance of the pilot it would have been difficult to identify the coast clearly since the latitudes in the three charts of Ulloa,
Moraleda, and Hervé[4] that we had in our hands were very misleading.

By eight o'clock in the morning we were able to get a clear sight of the little port
of Quintero[5] bearing SEbyE; our surveys from Valparaíso had reached this far, consequently we were not too far from Papudo, nor from Ligua,[6] and, as the coast could be
seen quite clearly since we were not far from it, we began to run bases under full sail
towards NbyW, with a slight breeze from SSE and south. At nine-thirty at just less
than two leagues from the coast we sounded ninety-two fathoms, coarse white sand.
At twelve o'clock the mouth of Puerto Papudo bore east two leagues distant. The
coast we had seen until then appeared to be low-lying, with a few long and wide
beaches, especially towards Ligua and there was the occasional small island between
Papudo and Puerto del Quintero with a few reefs visible near the latter. To the north
of Papudo stood an isolated mountain which might possibly be used as a landmark to
identify the port.[7]

In one locality, where the coast ran in a north/south direction, we were in no

---

[1] The President's regional representatives.

[2] Pedro Rico Ortiz was made *ayudante de ingeniero* in 1776 and given the rank of *capitán* in 1785,
when he was posted to Valparaíso. He arrived in 1787 and made a tour to the north of the country with
O'Higgins the following year. Despite requests to return home owing to ill health, he was retained to
work on several projects, including the Santiago-Valparaíso road in 1792. He died in July of that year:
Guarda, *Flandes indiano*, pp. 251-2.

[3] Puerto Pichidangui, situated 21 miles north of Puerto Papudo, is an indifferent anchorage exposed
to seas raised by NW winds.

[4] Teniente de Navío Juan Hervé, senior pilot in the South Sea. For MS sailing directions for the west
coast of North and South America by Hervé and Espinosa see UKHO, Misc. Papers, Vol. 58.

[5] Bahía Quintero, situtated about 17 miles NNE of Valparaíso, provides anchorage during the
southerly winds of summer.

[6] Caleta Ligua is situated about 8 miles north of Puerto Papudo; it is subject to heavy surf at all times.

[7] Possibly the conical-shaped Monte Papudo, 1 mile SW of Puerto Papudo.

doubt about the effect of the current. One can imagine how unhappy we were at not being able to obtain our latitude because of the continuous foggy weather. The lack of observations for latitude caused us concern about the accuracy of the work we had already carried out. We watched for any clear spells, however short, to take two altitudes of the Sun, but it was only possible to take one at two-fifteen in the afternoon, from which, without the latitude, we could only work out an extremely unreliable longitude. We signalled the *Atrevida* to sound at one-thirty when she obtained a depth of eighty-five fathoms. The wind had begun to strengthen and was maintaining its south to SSW direction. We therefore thought it better to maintain our course since we were aware, from an old tradition, that the weather on this coast was usually overcast and hazy and so we continued measuring bases until nightfall, when the most northerly piece of land, apparently just past Silla del Gobernador,[1] bore N3°W six leagues distant.

The night was still, but continued to be hazy with patches of fog. We steered NNW until midnight and then stood offshore and lay to under topsails. At four o'clock we got under way once more under the same sails in order to close the coast at dawn; but at five-thirty we discovered we were too close so we stood offshore until six o'clock.

16 April

Early in the night we saw some fires on land. At one o'clock we obtained no bottom with the entire 120 fathoms of line, but at five o'clock we eventually obtained a depth of eighty fathoms, silt and pebbles. We then hove-to for several hours with the weather still remaining misty, but at dawn we were quite close to land. With a light wind setting in from SSE, we were able, before sunrise, to start measuring bases a short distance from the coast, steering NNW and NbyW under full sail. The coast we had in sight was that which extends from Silla del Gobernador past Punta del Negro to the vicinity of the town of Conchalí,[2] which we could see. From our position the land appeared to be quite high, to the extent that we could relate it to that seen the previous afternoon, from which the error caused by the currents was seen to be no more than three leagues. The impossibility of taking observations on the second day was to cause us even more concern than it had on the previous one, not only because the errors caused by the currents, swell, and a thousand other unpredictable causes (unavoidable in the course of any voyage) were increasing, but also because we now had behind us a considerable section of coast which had not been tied into any observation. Indeed we were on the look-out for clear spells which, for short periods, we were fortunate to get in greater number than on the previous day. We observed several altitudes of the Sun at different intervals, some close to noon, and the results promised us a value not far from the true latitude. When the sky finally cleared in the afternoon we were able observe two sets of satisfactory observations at three o'clock and at four o'clock and these, when combined with those taken in the morning, increased the likely accuracy of our search for the latitude while they also they gave the true longitude according to the useful method devised by Don Dionisio Galiano.

---

[1] Cerro La Silla del Gobernador, also known as Cerro Santa Inés, a saddle-shaped hill about 20 miles north of Puerto Papudo.
[2] Situated in Bahía Conchalí, 37 miles north of Puerto Papudo.

For this purpose we adopted the last two sets observed in the afternoon, one by Galiano and the other by me, and the altitude on which we had agreed at 11.39 that morning. The two results were in agreement and gave the latitude at noon as 31°36′16″ which was 19′ north of our dead reckoning. The longitude deduced was 12′ west of Valparaíso and the meridian altitude of τ Navis[1] at a quarter past six in the evening gave the latitude as 31°30′3″ which, in addition to confirming the latitude at noon, also served to confirm our longitudes, from which it had been obtained.

The irregular rates of chronometers 61 and 13 and the consistent rate of 72 obliged us to adopt the latter as our master, although it would have been easy to adjust the others by means of the daily equation. Its results, when compared with dead reckoning longitudes and that observed on the afternoon of the 15th, permitted us to correct that noon's latitude. Thus we were finally able to infer from this happy combination that the errors had been proportional and that the position calculated at noon on the previous two days was remarkably close to the true one.

At three o'clock in the afternoon we obtained no bottom with 110 fathoms of line and as soon as suitable light winds began to set in from SW and south, with fair prospects, we steered NNW and followed the coast at a constant distance from it. We tried to measure new bases so as to tie it in as accurately as possible. Variation by azimuths was 14°NE. By sunset we were almost on the same parallel as the southern end of Altos de Chuapa[2] and the furthest land to the north was six leagues distant from us. Having made good this distance, early in the night we ran with a fresh breeze from south and under easy sail we stood offshore and lay to under topsails with the intention of combining the work of the last two days, as we had done on previous days.

On the following night while we were lying to, the effect of the current to the north became clear from the difficulty we had in turning the corvette to windward. We naturally surmised that most of the errors we would discover in the latitude of the following day would be attributable to the hours spent lying to. At dawn we found ourselves somewhat further to north than we had expected. The nearest places to us were Río de Chuapa[3] and Quebrada de Limarí,[4] which was a fairly prominent landmark along this part of the coast, which was only about four miles distant.

17 April

The breeze, which had remained fresh all night, began to abate in the morning; the weather was clear and everything was conducive to continuing with the task of charting the coast. No bottom was obtained with 110 fathoms of line. The directions of our bases were N5°E and then NNW5°N with which, and maintaining a speed of three knots under full sail, the coast still remained at a constant distance from us. Our latitude at noon was 30°39′ and our longitude by number 72 was 18′ west of Valparaíso.

Quebrada de Limarí bore SE, true, three to four leagues distant, and Lengua de

[1] τ Argo Navis, now τ Puppis.
[2] Altos de Choapa, in the vicinity of Río Choapa, 16 miles north of Bahía Conchalí.
[3] Río Choapa.
[4] The gorge, through which Río Limarí enters the sea, 50 miles SSW of Coquimbo.

Vaca[1] bearing NNE about six leagues distant appeared to be farthest land in sight. In coastal navigation, this is the name that has quite properly been given to the low headland that protrudes into the sea, from which point the large bay that Frézier named Tongoy[2] extends to the north towards Puerto de Coquimbo.

The error in latitude during these 24 hours was 9' to 10' to the north and in longitude we found no difference between dead reckoning and observation.

After midday the trade wind began to blow with considerable force. Steering north and NNE5°E by compass at five to six knots, we took advantage of the fine weather to observe azimuths and hour angles. The first gave a variation of 13°NE and the second at four-thirty in the afternoon by No 72 gave our longitude as 9' west of Valparaíso. At that time Punta Lengua de Vaca bore some two leagues NE and so, according to our chronometers, it was almost on the same meridian as our observatory in Valparaíso. At about five o'clock, close to that point, we sounded seventy-one fathoms, white sand, and found that the current was setting us considerably to the north. Since four in the afternoon we had been sailing under foresail and topsail since it was impossible to make the port that same afternoon. We rounded Punta Lengua de Vaca under topsail alone and altered course to NE and NE5°E to enter the bay that Frézier named Tongoy, so as to fix more accurately the positions of the nearby points. We made use of the little light that remained before nightfall to measure a new base in the direction NbyE and shortly after prayers we were off the foot of Cerro Guanaquero, which stands at the northern end of Bahía de Tongoy. We obtained no bottom with one hundred fathoms of line. By then the trade wind, which before sunset had given us a speed of nine knots under only our topsails, had abated considerably. We had already sighted the neighbourhood of the port, from which we were some five leagues distant at six-thirty.

To avoid any possibility of falling to leeward during what was already becoming quite a long night, it seemed best to remain in our present position and stay so close to land that our bearings would themselves indicate how much we were being set. To this end we stood off and hove-to under the fore topsail braced aback until nine o'clock in the morning, when, being too close to [Cerro] Guanaquero, we tacked and stood off again. Since the wind was abating too quickly and we were being set towards the coast, we hoisted the same sails. At two o'clock in the afternoon, being a little distance off the coast, we stood off and lay to once more.

18 April

At four o'clock we realized that the current had already set us over two leagues to the north, but keeping the coast well in sight with a moderate breeze, when we obtained no bottom with 120 fathoms of line, we steered NE under topsails to close the anchorage.

Later that afternoon our progress became easier, the wind having got up considerably, although without the sky or horizon becoming clear. By dusk we had passed Punta Lengua de Vaca, off which we sounded seventy-one fathoms, white sand, and from where the entrance to Puerto de Coquimbo was in sight some six leagues

---

[1] Punta Lengua de Vaca (literally Cow's Tongue Point), 23 miles SW of Coquimbo.
[2] Bahía Tongoy.

distant,[1] so that if we lay to that coming night we would have no difficulty in making it with the first breath of wind the following day.

We drifted so far during the hours we lay to, mainly because of the current setting us to the north, that at dawn on the 18th we were scarcely a league from the entrance to the little port of La Herradura.[2] The southerly wind seemed to have freshened somewhat and did not appear likely ease after sunrise so that we had no need to fear either a calm or a land breeze. This, combined with the natural desire to survey closely the vicinity of an important port, led us to close the coast to the distance of barely a mile, with all the more reason since the *Atrevida* had sounded eighty fathoms at seven-thirty. Thus we set course to the east towards Punta del Lobo[3] from which, by eight o'clock, we were no more than a mile distant as planned.

But at that moment and in that position, contrary to all our expectations, some light airs began to blow from NE. These light airs and a high sea from the south rendered steering impossible however much effort we put into trimming the sails and so we found ourselves to be at the mercy of a counter current and being rapidly swept towards the nearby shore which at nine-thirty was no more than four cables off. Neither the pinnace we had lowered earlier nor the launch, which we then lowered, could over-come the power of the waves to turn us in the right direction to benefit from the light sea breeze which was beginning to make itself felt. Finally, with the launch at our bows and the pinnace at the stern hauling in opposite directions, we succeeded in turning towards the land, heading NNE, setting all sails and being towed at the same time by the boats. The *Atrevida,* which had managed to keep clear of the danger by making use of her sweeps, although at the cost of breaking most of them, at once sent her launch to assist us. The launch stayed with us for a while helping at the bows until the sea breeze began to increase in strength and we sent it back to the *Atrevida.* We also hoisted our own launch, and left only the pinnace in the water to tow us with the help of its sails.

By midday we had already passed the mouth of La Herradura and since the breeze remained light from SWbyW it seemed to me both foolhardy and unprofitable to enter the port through the channel between the inner islet and the mainland.[4] We set a course to leave the outer Pájaro Niño[5] to starboard and before one o'clock we passed it at a pistol shot's distance. Then after leaving the extremely steep rocks to starboard, we suc-ceeded in reaching the anchorage. It was my intention to keep under way on the port tack under topsails to the position where we were to drop the northern anchor. How-ever, the anchor-stopper parted, while we were clearing away the anchor, forcing us to anchor before we had reached a suitable spot so that we were forced to take in the sails.

The position of the port and the usual fair weather it enjoys persuaded me to spare the best bower cables. So I used instead the old small bower cable and moored to the south with two hawsers firmly attached to a single rock called La Tortuga near the

---

[1] Coquimbo is situated in the SW corner of Bahía Coquimbo.

[2] Bahía Herradura de Guayacán, about 3 miles south of the entrance to Bahía Coquimbo.

[3] Not identified.

[4] I.e. through Paso Interior, between Islotes Pájaros Niños, the southernmost of Islotes Pájaros, and Península de Coquimbo, which is only suitable for small vessels.

[5] Islotes Pájaros de Afuera, two rocky islets off the western entrance point of Bahía Coquimbo, the northernmost of Islotes Pájaros.

beach. Everything was carried out immediately. We were warped into a suitable position and the corvette was moored a cable from the shore in a depth of five fathoms at low tide, with the entrance to the port bearing N8°W and Torre de Santo Domingo de Coquimbo bearing NE. The *Atrevida* anchored about a cable and a half to the north, and was secured in almost the same manner as this corvette.

An assistant to the Subdelegado of the district soon came on board to present the respects of the Subdelegado, who followed, accompanied by a number of the city's leading citizens. While they were on board we learned of the kindness of the President of Chile who had sent an express message to advise them of our visit and ordered them to assist our somewhat hurried operations as efficiently as possible.

The arrival of the Subdelegado and Capitán José Antonio Corvera of the Army was most opportune in enabling us to set up our observatory without delay, which was all the more important since we knew that clear days in this place were very few. On the shore near the corvettes were some well appointed warehouses, belonging to a resident of Coquimbo, which were immediately put at our disposal for setting up the observatory.

19 April

That night, under excellent conditions, fully reliable observations were made of the emersions of the second and first satellites of Jupiter using *Atrevida's* chronometer 71 and comparing it by signal with the chronometers of both ships. The following morning, after some of the crew from both ships had emptied and cleaned the best of the three warehouses, the astronomical clock and chronometer 71 were installed there to obtain their rates by equal altitudes. The previous night's observations on the beach had confirmed that our concerns about 72's rate was well founded. The rates of the chronometers had been established on the 12th in Valparaíso, but equal altitudes on the 13th had indicated to us that all three had suffered some alteration during the previous twenty-four hours, although it was much less in 72 than in 61 and 13. The alteration may have been caused by our farewell gun fired on the morning of the 12th. The next check on its rate at the Coquimbo Observatory confirmed these conclusions and we were able, without being thought remiss, to determine the following results:

	Corrected in time	Longitude east of Valparaíso	Longitude east of Chiloé
N°72	– 3·78″	15′ 17″	2° 28′ 07″
N°61	+29·40	15 47	2 25 37
N°13	– 18·22	16 15	2 26 5
N°10 of the *Atrevida*		16 54	2 26 48
Emersion of the second satellite of Jupiter			2 24 05
Two emersions of the first satellite observed after the 20th and corrected according to the Errors of the Tables[1]			2 26 00
Further emersion observed on the night of the 20th		16′ 30″	2 26 00

[1] Presumably these emersions were observed simultaneously in an European observatory enabling the actual rather than the predicted values to be computed subsequently.

While the various tasks we had to perform in the port were progressing, I set off at nine o'clock in the morning with most of the officers to the city of La Serena,[1] not far from Coquimbo, riding horses which the chiefs of the settlement[2] had been kind enough to provide. For much of the way the road runs along the beach, which is very pleasant at low water although rather inconvenient when the tide is in. The road then goes inland to avoid the marshy area lying between the sea and the higher ground where the city stands. According to the natives the distance is three leagues, but it can be covered easily in forty-five or fifty minutes.

The position of the city could not be more beautiful or convenient. The view of the coast, the abundance of crystal-clear water, the surrounding plains which can all be irrigated, without risk of flooding, from a constantly flowing river,[3] rendering the fields fertile while providing several mill-races for water-mills and ore-crushers, the rich mines not being far off, the excellent port, the sea abounding in fish, the cheap and delicious foods and an agreeably mild and constant climate all year round – all these form one of those marvellous combinations of nature which seem more like a poetic fiction than a reality to those who limit their inquiries to the examination of less fortunate parts of the world.

The work in the mines and the fertility of the fields combine to make the city seem deserted. Both banks of the river are settled as far as the *cordillera*, up to which exploration and the exploitation of mines have actually reached, although they are some forty or fifty leagues from the sea. Thus the population of Coquimbo[4] can be said to consist of some fifteen or twenty miners, or rather merchants who provide for the miners, six or eight wealthy families descended from conquistadores, some employees of the King, and a large number of monks of the orders of St Francis, St Dominic, Sisters of Mercy, St Augustine and St John of God; the Augustinians occupy the house of the expelled Jesuits.

Subdelegado Don José Antonio Corvera had prepared a large and delicious meal for us, but as the day remained clear, which is uncommon in those parts, we thought it best that almost all of us should return to the observatory immediately to measure lunar distances to the Sun. This we did, between two and three in the afternoon, with a set of eighty distances observed in very good conditions, enabling us to determine a mean longitude for the observatory of 2°24′50″ east of Chiloé.

An equal number of observations in the same conditions gave 2°15′ at noon the next day. Accordingly, the average of 160 sets gave a longitude of 73°56′ west of Paris, only 15′ greater than that indicated by the observations of the first satellite of Jupiter.

[1] Situated on the eastern shore of Bahía Coquimbo about five miles NNE of Coquimbo. At the time of Malaspina's visit La Serena had a population of about 5,000 and was the principal city in the area, having been founded in 1543, while Coquimbo was merely the port or landing place for La Serena and a settlement for the native Changos.

[2] I.e. the settlement of the native Changos.

[3] Río Elqui,

[4] From the context Malaspina clearly means La Serena, since the rest of the description relates to this city and in the next paragraph he refers to the need to return to the observatory, which had been set up in Coquimbo.

Bauzá and Maqueda, having measured an excellent base at the end of the harbour and taken bearings with the theodolite on the points and on a ridge nearby, had now returned on board, as had Guardiamarina Ali Ponzoni who had sounded much of the port in the *Atrevida*'s launch. Consequently we were able to have lunch at half-past three with the pleasure of seeing our work considerably advanced, a pleasure much increased by the success of Galiano and Vernacci in making a satisfactory calculation of the latitude of the observatory by means of meridian altitudes of stars to the south and north, the result being 29°56′40″.

On the following day, which was, as the previous day had been, extraordinarily clear, Don Felipe Bauzá took the *Descubierta*'s launch out of the main port to that of Herradura, where he drew a plan, having measured a base, sounded the port in detail and extended his triangulation to the furthest points in sight. Guardiamarina Murphy sounded the channels between the mainland and the two islets known as the Pájaros Niños. One of our boats went fishing, which was not very successful, and at noon the rest of the officers occupied themselves measuring lunar distances. Eighty sets gave a longitude of 2°15′; accordingly, the mean of 160 sets observed during these two days gave a longitude of 2°20′ east of San Carlos de Chiloé.

Both launches returned soon after nightfall. The latitude previously determined was confirmed with new observations and as the noon comparisons with the astronomical clock also now gave us the rates of the chronometers, we could consider all our operations to have been completed in the space of forty-eight hours.

As soon as we had arrived Pineda, Haenke and Neé had begun their natural history excursions, the last-named botanizing and the others examining some nearby deposits of petrified shells which generally resembled those on Isla de León.[1] However, this area which was so rich in mines of every description and particularly the mercury mine at Punitaqui only some thirty leagues from the port,[2] demanded as detailed an examination as possible with respect to their yield and the means of exploiting them, matters as important to the Royal Exchequer as to the promotion of science and national prosperity. The Punitaqui mine was being worked at His Majesty's expense and was therefore under the charge of an administrative superintendent, Don Miguel José Lastarria,[3] who had met us on the beach with the other gentlemen on the afternoon of our arrival and, at my request, had agreed to accompany Pineda and Haenke on the planned excursion. They were to leave on this day and meet at the place of the menstruating stone mentioned by Frézier;[4] they would

---

[1] The island at the eastern end of Bahía de Cádiz.

[2] Actually about 60 miles south of Coquimbo.

[3] Miguel de Lastarria (1758-1815), a graduate of the University of Lima and professor of mathematics in Chile, who was appointed to supervise the mercury mines of Coquimbo in 1788.

[4] It is not clear from the English edition of Frézier's account (*A Voyage to the South-Sea, and along the Coasts of Chili and Peru, in the Years 1712, 1713 and 1714*, London, 1717) what feature Malaspina was referring to, since on p. 133 Frézier describes a feature in La Serena 'as it were from an Ampitheatre, appears a curious Landskip, form'd by the Town, the Plain which reaches down to the Sea, the Bay and the Mouth' and in describing the countryside around La Serena (p. 135) that 'ten Leagues to the Southward of the Town there is a blackish Stone, from which flows a Spring only once a Month, at an Opening like unto that humane Part, whose regular Flowing it imitates, and that Water leaves a white Track on the Stone' and also six leagues east of the town 'There is a gray Stone of the Colour of Lead Ore, as

go by way of the goldmines at Andacollo and then make for Punitaqui, which they were particularly instructed to examine. On their return by a different route they were to inspect some copper mines and they were to be on board by noon on the 28th at the latest, by which date our stay in this port would coincide with our astronomical aims since that very night there was to be an eclipse of the Moon which we would have been very remiss not to observe.

Teniente de Navío Quintano had taken it on himself to accompany the naturalists on this excursion. I was all the more willing for him to do this because, while he was actively seeing that everything went on efficiently, he could study with astuteness and accuracy the interior of the country and above all the financial management of its mines, both the private ones and those in royal ownership. Supplied with all necessities, therefore, they set out at noon this day, having previously been provided by Teniente Coronel Tomás Shee, a man well acquainted with those parts, with all the information which might make their journey faster and more useful.

Nothing in particular happened on the following days. Don Felipe Bauzá took bearings from the tower of San Francisco in the city to tie in the position of the port more accurately; the sounding of the port was finished and the tides had been carefully observed. Various men went hunting, which, because of the abundance of partridges in these parts,[1] served both the table and our natural history studies; nor was fishing ignored. Several mornings were given over to training the combined marines from both ships, sending them ashore with the sergeants and the *guardiamarinas*, so that Teniente de Fragata Secundino Salamanca could instruct them in musket drill, marching and manoeuvres. Finally, on the morning of the 27th, they performed firearms drill, firing six cartridges each, three while on manoeuvre and three at fixed targets. By these means, which served to divert them from idleness and with the excellent example of the officers which manifested itself in this port even more than in previous ones in a dedicated concentration on their work, we were able to keep all the men calm and united and so to treat them correspondingly with all possible kindness.

At the time of my departure from Santiago, the President and Capitán General had spoken to me of the important measures he had taken when visiting Coquimbo to encourage fishing, particularly of conger eels, which acquired considerable value when salted and did not require a large investment of funds. I considered these measures to be so sensible that I offered to contribute to them as far as I could and no doubt even if I had not committed myself, I would have been urged to do so by the zeal, knowledge and courtesy of Teniente Coronel Tomás Shee who, in full agreement with Subdelegado Don Victor Ibañez, had undertaken to build a fishing launch.

From the first days of our stay in the city, the chief carpenter of this ship, Juan del Río Miranda, had been placed at the disposal of the city council and Don Tomás Shee. His arrival was so timely that the launch could be built in accordance with better principles and he was also able to show them various types of mills and presses of which they had no knowledge whatsoever, much to the detriment of the rural

---

smooth as a Table, on which there is exactly drawn a Buckler and a Head-piece, both red, the Colour sinking deep into the Stone, which has been purposely broken in some Places, to see it'.
[1] Probably one of several species of quail found in South America.

economy. The launch was to have eleven oars and two masts and to be an exact copy of those used on the Cantabrian coast. The town council then asked me for certain ship's stores that they particularly needed to make it seaworthy. I was happy to agree to this request and ordered that they should be given from our ship's stores all the canvas, needles, sailmakers' palms, bolt ropes and thread necessary for their sails, some tar and grease and a boat's compass. As for repayment of their value from the royal treasury, after this had been determined by both paymasters, I asked the council to have the sum paid to the order of the Capitán General of the kingdom in Santiago, whom I told, with the appropriate documents, that it could be used as part of the funds for the apprehension of deserters, asking him to be so good as to send word to me in Lima of its receipt or subsequent distribution.

28 April

Pineda, Haenke and Quintano had returned the previous night. Excellent progress had been made in botany, mineralogy and knowledge of the interior as a result of this excursion, during which they had also had the good fortune to meet, at the Punitaqui mine, the engineer and superintendent of the mine who, at His Majesty's orders, were on their way to inspect the mine and who had sailed with us to Buenos Aires. Our officers were most grateful for the generous attentions of the royal appointee as director of the said mine, Don José Lastarria. They were indebted to him for abundant provisions and lodgings which were clean and as comfortable as was possible for an excursion of this kind.

We had not seen the Sun for several days and the constantly overcast skies had not only caused us to miss an immersion of the first satellite [of Jupiter] on the night of the 27th, but had also interrupted the rating of the chronometers.

On this day the clarity of the sky was of the greatest importance, as, apart from the approach of the time of departure, when we would have to complete rating the chronometers, the following night, if clear, would allow us observations of the greatest importance during the eclipse of the Moon, the occultation of α Libræ[1] and the immersion of the fourth satellite of Jupiter.

The day, although cloudy, did allow us to take satisfactory equal altitudes to determine the exact rate of the astronomical clock and the chronometers and with this we had to be satisfied; but as darkness approached and the sky remained covered in a thick haze, we became the more eager to complete the work that we had previously thought to be impossible or only achievable in part. Our astronomers and anyone who could get hold of a telescope, were ready to carry out the observations and any calculations necessary to make them more reliable and significant. Only the haze in the east seemed likely to hamper all these measures.

Contrary to all our expectation, the moment of the eclipse, or complete darkness, could be determined during a very fortunate break in the clouds during the early hours of the night by Galiano, Concha, Vernacci and Valdés. The haze continued until nearly seven o'clock, when a light breeze from SW providentially cleared the

---

[1] Zubenelgenubi.

sky, so that the observations of the end of the eclipse were numerous and highly satisfactory. The eclipse ended at 8h 56m 20s (apparent time). At 9.12.34 the occultation of the fourth satellite of Jupiter took place and finally, between midnight and two in the morning, the immersion and emersion of two stars in the constellation Libra[1] behind the Moon were observed. The rest of the night was most beautiful and peaceful, and the day dawned with the prospects of perfect weather.

29 April

We would have made our departure then if we had not known that the lack of equal altitudes on this day for calculating the rate of the astronomical clock exactly would detract from the desired reliability of our previous night's observations, as astronomers would see them. Accordingly we put off our departure until the following day and with the embarkation of the many instruments, the observations of variation and inclination and of the tides, the delay of twenty-four hours was not wasted. When the equal altitudes had been taken, the astronomical quadrant and clock were packed in their boxes and before prayers everything had been brought on board.

One of those unpleasant incidents which overwhelm the minds of our seamen, just as the waves are driven by the wind, was now to cause me great uneasiness, particularly as it greatly affected the King's service and the safety and speed of our commission. At the evening muster three marines were missing from the six who had been given leave the previous afternoon to wash their clothes at a small lagoon near by. A marine gunner had been missing for two days and at nightfall a seaman absented himself, after being chosen by me the previous afternoon to take a compass to a nearby ridge where Don Felipe Bauzá was going to take some transits. During the same night three seamen from the boat in which we went ashore to make observations also abandoned their posts and finally, a sudden apoplectic seizure threatened to cause me the loss of the best seaman I had. I considered the loss of the gunner and the three marines neither surprising nor of great importance; but that of the five seamen, which, apart from being so unexpected, involved some of the better, happier and quieter men, was a heavy blow to me personally especially as it meant too great a reduction in the number of men available for service.

From early in the morning Don Fernando Quintano, with his usual energy and knowledge of the country, had been riding around to search all the huts in the district, in case any of the fugitives had stayed in one of them to sleep off exhaustion or an excess of wine. It was all in vain, and the news he brought when he returned at half past two in the afternoon served only to convince us that they really had deserted. The marines had stolen two mules which would allow them to get away more quickly; the seamen were still on foot when last seen. With two seamen who had deserted from the *Atrevida* and with the greater number who went missing in Valparaíso, both crews were now reduced to approximately half of their marines and seamen, making working the ship and particularly anchor-work, difficult and rather dangerous in the future, if we had to sail as close inshore as we had done until now.

---

[1] One was the relatively bright Zubenelgenubi (see p. 191, n. 1 above) and the other a nearby star of lesser magnitude.

I gave the Subdelegado the names and descriptions of the deserters, with a copy of the same terms for their capture as those issued in Valparaíso, Talcahuano and Chiloé, and, giving up all hope of apprehending them, I concentrated on ensuring that none of the remaining crew could desert and that we could make our departure the following day.

Almost all communication with the shore was cut off and orders were given that, for the future, the few boats which had to go ashore, even on the most deserted beaches, would always carry an officer or *guardiamarina* on board, as well as two armed marines who were to fire on anyone who tried to escape. The water, firewood, stores and provisions that we had on board would relieve us of any necessity in subsequent ports, so that, by cutting all communication with the shore we would ensure that none of those who still remained could desert. At the same time all strangers were sent ashore, since any one of them was now suspect and finally, at nightfall, three of the five boats were hoisted, leaving us ready to set sail at the first favourable opportunity.

The breeze, which had been moderate from the WSW since midday, with signs of very clear, calm weather, dropped at nightfall, as usual, but the sky remained clear all night.

At about nine o'clock the man previously mentioned as suffering from apoplexy died. He was Seaman Gunner Francisco García, who had sailed with me during his first years and had then reached the rank of senior gunner in the King's service. He was a useful carpenter. He had come from Cartagena de Levante[1] with the sole object of taking part in the voyage and, as well as having a lively nature, he was well spoken and good mannered, qualities rare among seamen. It may be imagined how much I, no less than the other officers, would feel the unhappy loss of such a man.

30 April

At three o'clock in the morning the *Atrevida* was instructed to unmoor and bring her cable up and down. The same was done in this corvette soon afterwards and the pinnace was sent to let go the hawsers ashore, with the above mentioned precautions. At sunrise the weather was clear, but calm; we cast loose the fore topsail and hoisted the ensigns. As three young natives appeared on the beach, apparently asking to be taken on board, I sent Don Manuel Novales to fetch them in the pinnace and indeed they did wish to join our crew, although completely ignorant of seamanship. I had asked the Mayor of La Serena to send me two other seamen who had been imprisoned for minor offences. He agreed to my request, but at half past nine they had not yet arrived.

---

[1] To distinguish it from Cartagena de Indias in present day Colombia.

BOOK FOUR

FROM COQUIMBO TO CALLAO

CHAPTER 1

From Coquimbo to Callao
and visit by *Descubierta* to Islas de San Félix

[30 April]

At half-past nine in the morning when a moderate breeze set in from WNW, both corvettes got under way so as to make their departure from the port, passing to the north of the outer Pájaro Niño.[1]

As a great many highly satisfactory observations had been made in this port to determine its longitude, I thought it best to relate the next stretch of the coast directly to this meridian, particularly as it was my intention to use the chronometers for determining the respective differences in longitude until reaching Lima, rather than other astronomical methods which would now take too long and be inappropriate given the present state of our work and equipment. We therefore abandoned the meridian of Chiloé and adopted that of our observatory at Coquimbo to determine the exact state of the chronometers.[2]

When we last rated the chronometers, we found Number 72 to be as accurate as it had been on the passage from Valparaíso, thus confirming not only the determinations previously made with it, but also a strange irregularity in its movement, which was found to be 58″·16 in Valparaíso and was 1′3″·45 when we reached this port, this being the average of all the observations, which were, however, very different from each other. Number 13 no longer gave any hope of even moderate accuracy. It varied considerably from one day to the next, although it was kept absolutely still and at a very uniform temperature and handled with all possible care. Chronometers 71 and 105 from the *Atrevida* continued at a uniform rate. We had noticed some alteration in number 10, a chronometer which until now had been outstandingly accurate.

Until noon the breeze was very light from NW, although it showed signs of soon backing to the west and freshening a little. For this reason we made short tacks and stayed close to the southern shore, rather than the northern. At noon the outer Pájaro Niño bore WSW half a league distant and we stood to the north, intending to clear the harbour mouth on the next tack. Indeed, having tacked to starboard shortly afterwards, by one thirty-six in the afternoon it bore true south, two cables distant. As this

---

[1] Islotes Pájaros de Afuera, the northernmost of Islotes Pájaros off the western side of Península de Coquimbo, not to be confused with Islotes Pájaros about 23 miles NNW.

[2] Malaspina appears to refer here to the errors of his chronometers rather than their rates.

was an excellent position for correcting the rates of our chronometers over the last twenty-four hours, we took the opportunity to observe hour angles. The results of these, when compared with our plan, confirmed the excellence of number 72 and showed a new gain of 6 seconds in number 61. As for number 13, which had stopped during the morning because of an accidental failure to wind it , we were able to set it going again by chronometer 72, in the certainty that it was correct.

No sooner had both corvettes passed abeam of Pájaro Niño than the breeze began to slacken and die. We continued on the same tack until nightfall, nonetheless, and made good progress, being now clear of all danger, even though the calm seemed likely to last for some time. At sunset the harbour entrance bore N43°E by compass, three leagues distant, and the coast was clearly visible as far as Lengua de Vaca. Various observations by azimuth gave the variation as 12°59′NE. Don Felipe Bauzá took advantage of our present position to take fresh bearings of Cerro Guanaquero and Punta Lengua de Vaca, to confirm the positions determined while approaching Coquimbo from the south by tying them into points, already well established, in the vicinity of the port. We were pleased to find that they agreed beyond our expectation, particularly after we had incorporated the bearings taken the following morning, when we still had both points in sight.

It was calm all night, except from two to four in the morning, when there was a slight breeze from NNE and NE, which we used to make good a course of NWbyN under full sail. The weather continued hazy and at dawn it was calm once more, but the land surrounding the harbour mouth was visible five leagues distant and almost east and west. We had made some attempts to gain ground to the west, as it is commonly said by sailors on these coasts that the currents set very strongly towards Pájaros Niños and that the channel between them and the mainland is full of reefs.[1]

1 May

It was now ten o'clock in the morning and there was still no sign of a breeze to give us any hope of continuing our survey. With the misty and almost motionless cloud that covered the sky Don Dionisio Galiano and I had only managed to observe together some altitudes of the Sun which, with the reliable latitude provided by the bearings, could determine the longitude by chronometers. At about half-past ten, however, a moderate sea breeze got up at last from SSE, which soon cleared the mist and allowed us to steer first NNW and then north, under full sail. Our latitude at noon was 29°53′30″. The three chronometers agreed in giving a longitude of 21′49″ west of the Coquimbo observatory and our dead reckoning, although somewhat affected by the many hours of calm, showed a difference to the south rather than to the north, not at all unusual on these coasts at the beginning of winter.

It was very important to the method adopted for our tasks to make an accurate survey of Islas de Pájaros,[2] which lie at some distance from Punta de Choros.[3] We had already taken a bearing of the westernmost of them from a height near the port in

---

[1] Malaspina should have had up-to-date knowledge of this channel following Guardiamarina Murphy's survey.

[2] Islotes Pájaros, see p. 197, n. 1 above.

[3] Cabo Choros.

transit with the outer Pájaro Niño. We made directly for them, therefore, while measuring bases with the log to link the stretch of coast now in sight again with that adjacent to the harbour and Punta de Teatinos.[1] With the fresh breeze blowing at the time we were making five knots and the sky was very clear. At this speed we had almost reached the former by four o'clock and we could see Isla Choros[2] bearing NbyE. Observations were made for longitude; the latitude depended upon the dead reckoning being kept in great detail and everything was tied in to the bases measured since noon. The channels between the two larger of the Islas de Pájaros and Isla Choros looked very clear.[3] A reef ran out towards the coast from the Pájaro closest inshore; its extremity was also visible, revealing the channel to be apparently free of dangers.[4]

By sunset we had passed the Pájaros, when Isla Choros bore N5°E three or four leagues distant. To the NNE, in the far distance, we could see a point ending in a promontory, joined to the mainland by a strip of low ground. This was possibly Isla del Cañizal, mentioned by Frézier.[5] We shortened sail and steered a somewhat more westerly course, so as to avoid coming too close inshore, in case of a calm, to make sure of having the breeze for a little longer, which would allow us a choice of the better course of action. Far from dropping, however, the trade wind freshened to such an extent that very soon we had covered the distance previously fixed upon for our first course, and after having made another five leagues to the NE and sighted land, we hove to, first heading to seaward and then towards the land.

Either because of the proximity of the coast, or because of the regularity of the winds in these parts, the breeze began to slacken from two o'clock in the morning, and at five the sky and shore were shrouded in haze.

2 May

At dawn the coast was in sight and no more than three leagues abeam, but the haze was so thick and the breeze, still from the south, so light, that we could not start on our hydrographic work. We were able to take bearings of the farthest point of land visible to the south, however, and calculate its distance from us as five to six leagues and as we were able to take hour angles at eight o'clock, while the latitude was referred to that of noon the following day, we could consider that we had accurately determined the position of the southern extremity, which could be linked to the previous day's work. Fortunately the sea breeze, which began to set in at that time, dispelling the mist and giving us a speed of five to six knots under full sail, promised us some compensation for the considerable sacrifices to our intended course that we had been obliged to make during most of these excessively long nights.

---

[1] The northern entrance point of Bahía Coquimbo.

[2] Malaspina may have been mistaken since Isla Choros is situated 17 miles north of Islotes Pájaros, and possibly not visible from the vicinity of these islets.

[3] Malaspina was mistaken, the dangerous Arrecife Toro lies 5 miles south of Isla Choros.

[4] The channel between Islotes Pájaros and the mainland is indeed clear of dangers.

[5] Isla Chañaral, situated 15 miles NNW of Cabo Choros; according to Frézier, 'The next Morning we found ourselves four leagues N.W. and by N. of the Island of Channaral, join'd to the Continent by a Bank of Sand, which the Sea covers with a North Wind: It is four leagues from the Island of Choros, and 16 from Point *Tortuga*, this Island is almost plain, and very small.': Frézier, *A Voyage to the South-Sea*, p. 138.

At first we thought ourselves to be off Bahía del Huasco,[1] according to the reports we had of its latitude and position, but as we were sailing barely a league offshore, we were soon disabused, finding the latitude given to be mistaken. Our own latitude at noon was 28°31'40" and our longitude calculated as 29" west of Coquimbo. We soon sighted the detached rocks which surround the islet[2] lying to the south of Huasco and the nearby hill which, viewed from the west, ends in two points shaped like breasts[3] and were able to identify this anchorage in every detail. It was then half-past one and we had kept up a fair speed, so that the latitude of the anchorage was in fact 28°26', rather than that of 28°33' shown on the charts.

The beauty of the afternoon and the fresh sea breeze lent themselves to a detailed survey of the following stretch of coast running towards Totoral.[4] We therefore sailed along it at a distance of one to two leagues offshore, the entire stretch being of uniform height and having that arid and desolate appearance usually to be found in mining areas. In these parts there are rich copper mines and some of gold, the products of which are carried to Coquimbo by land, as are those of the very productive mines at Copiapó.[5]

Darkness put an end to our work before six o'clock. The extremity of land visible to the north was then some six leagues off and ended in a fairly low point, which was no doubt the one at Totoral. The longitudes observed during the afternoon were 3'36" to the east [of Coquimbo]. Variation differed greatly from what had been observed by azimuth and western amplitude on the afternoon of the 1st, which had been 13°15', while the former, by the same methods, did not exceed 11°45'. We assumed this to be incorrect, and continued calculating our dead reckoning on a basis of 13°.

That morning we had mounted chronometer 61 in gimbals. The fluctuations in its rate since Coquimbo had been compensated, so that it now gave longitudes equal to those of 72, the uniform rate of which daily increased our high opinion of it.

At sunset it seemed that the breeze might soon abate. Accordingly, we altered course so as to stand somewhat further offshore and continued under full sail, making five or six knots. After nightfall, however, convinced that the light sea breeze would continue until early the following morning, as it had on the night before, we continued under foresail and main topsail alone until we had made six leagues on a heading of NbyW. After that we steered NbyE for two leagues and finally hove-to heading towards the land.

With shifts in the wind matching those of the previous day, we were left in no doubt that the breeze would drop completely towards dawn and that the haze would prevent us from carrying on our survey if we were any distance offshore. With this in mind it seemed better to keep on an easterly course from two until four o'clock in the morning and, when the coast was sighted a short distance off, to lie to heading to seaward during the remaining two hours of darkness.

---

[1] Puerto Huasco.
[2] Islote Blanco.
[3] Tetas de Huasco.
[4] Situated 37 miles NNE of Huasco.
[5] Situated about 100 miles NNE of Huasco and about 40 miles inland.

3 May

This we did and indeed, although at dawn the breeze was failing and the sky overcast, we were able to make a detailed survey of all the coast in sight. The furthest visible points to the south were no more than five leagues away and if an isolated ridge bearing N5°E at about five leagues distant was Morro Copiapó,[1] these were certainly not the points that we had sighted the previous afternoon to the north. Consequently the current must have carried us a considerable distance northward. The land we could see to the south would therefore be in the vicinity of Bahía Salada[2] and an island not far from the hill would be that shown on the charts as lying off the mouth of Río de Copiapó.[3] An isolated rock lay one league to the east. The rest of the coast appeared to be clear and the shore mostly sandy.

We began measuring bases on a course parallel to the coast, but we soon had to leave off when the light southerly airs still prevailing were followed by a flat calm. According to our bearings, the current was setting us fairly strongly to the north.[4] The light airs, although weak, had finally settled in from NE. The appearance of the weather gave no sign of anything but continuing calms. We naturally took the precaution of observing many altitudes of the Sun to calculate our latitude in case we could not get a meridian altitude, but in the end we were fortunate enough to get a reliable altitude two or three minutes before noon and another as many minutes after. The resulting latitude was 27°14′45″, 10′ further north than by dead reckoning. Our longitude was 12′30″ east [of Coquimbo], showing a daily difference of 8′ to 10′ west of our dead reckoning, as it had on the previous day.

The calm persisted and as we were consequently at the mercy of the currents, we could, if it continued, fall off to leeward of the Puertos de Copiapó,[5] where I had intended to anchor, not so much for a detailed survey, but to undertake the transfers of men and other steps necessary so that the corvettes could complete separately the surveys that remained to be carried out before reaching Lima.

This plan would be at once more expeditious for our mission and now all the more necessary, given the poor state of health of both crews, many men having reported sick since Chiloé. One of the corvettes was to survey Islas de San Félix,[6] return to Cabo San Juan, on the coast of Peru in latitude 15°30′,[7] and make for Lima, while verifying Moraleda's chart of that coast. She was to arrive there some time before the other to attend to the preparation of provisions, the scheduling of tasks and the acquisition of preliminary information for putting in order the results of the recent season's work. It would fall to the other to follow the coast from Copiapó to the above-mentioned latitude of Cabo San Juan. As this corvette's work would clearly

---

[1] Situated about 80 miles north of Huasco.

[2] An open bay about 30 miles south of Morro Copiapó

[3] Situated 14 miles south of Morro Copiapó.

[4] The Peru current whose usual rate is between half and one knot.

[5] Possibly Bahía Copiapó, an open roadstead 12 miles south of Morro Copiapó, but more likely the reasonably sheltered Puertos Caldera and Calderilla, 6 miles NE of Morro Copiapó, which are still the outlet for the mineral regions of Copiapó.

[6] Isla San Félix and Isla San Ambrosio, two small volcanic islands 12 miles apart, lie some 400 miles off the South American mainland in approx. 25°15′S.

[7] Punta San Juan, about 235 miles SE of Callao.

take longer and be more complicated, her crew would have to be augmented with seamen from the corvette bound for Islas San Félix.

I told the commanding officer of the *Atrevida* of my plans a few hours before leaving Coquimbo and when I said that I would undertake the coastal work with pleasure, he, scorning both the need for rest and the strong attractions of Lima, insisted that the *Descubierta* should arrive there first and we agreed on the other measures necessary for this course of action.

Since the advantage of making Islas San Félix on the same tack might be reduced by making further northing, and since the Sun had passed the meridian without the least sign of wind, it was decided that we should spend the afternoon in ensuring that both ships were agreed on what had to be done for this separation. We lowered the launch and, after signalling to the *Atrevida* it was to come alongside, since the light airs at the time made this possible, Don Felipe Bauzá and Don Domingo Velázquez were transferred to the other corvette with six seamen and two marines. The instructions now given to the commanding officer were as follows:

The need to shorten our future passage times and the present condition of both crews require the corvettes to separate again and during that separation it will be necessary for you to take the principal part.

From Puerto de Copiapó, or from within sight of it, this corvette will cross to the Islas San Félix and, after rating the chronometers while in sight of them, will return to the coast of Peru at Cabo San Juan, latitude 15°30′, and then continue surveying the coast as far as Lima.

It will be the duty of the *Atrevida*, therefore, to plot the entire coast between Copiapó and Cabo San Juan, but this work must be restricted to the following limited number of objectives:

1st
It will not be necessary for you to enter any port other than Arica, which must be accurately charted. Its hydrographic position is to be determined by sextants and chronometers, rather than by any other method. Clearly, in this case, lunar distances must be observed with greater frequency.

2nd
You will endeavour to pass by day and, as closely as is prudent, the general anchorages of Nuestra Señora, Mejillones, Iquique, Ilo, Quilca[1] and finally Cabo San Juan. If the winds allow and without unduly extending the voyage, you may wish to anchor in the Bahías de Iquique and Quilca, the latter situated not far from Arequipa,[2] for a better examination from within [the bays] of the shape and, in particular, the bottom, of both anchorages.

3rd
In general the survey of the coast shall be conducted by the method followed so

---

[1] Bahía Nuestra Señora, Bahía Mejillones del Sur, Iquique, Puerto Ilo and Caleta Quilca, between Morro Copiapó and Punta San Juan.
[2] A major Peruvian city about 30 miles inland midway between Ilo and Punta San Juan.

far of lying-to for part of the night, taking care, at dawn, to be north of those points of the coast which were in sight on the said course on the previous afternoon. It will be necessary to take precautions against the currents, which appear to be still setting to the north, as shown by the experience of the previous days.

4th

To allow the same amount of time for the survey and description of the coast as has been so far possible in this corvette, I have ordered that Alférez de Fragata Felipe Bauzá and the pilot Domingo Velázquez should transfer to your command. Both have been extremely useful to me and I recommend them to you highly.

5th

So as not to upset the accommodation arrangements, you will instruct Alférez de Navío Martín de Olavide and Guardiamarina Ali Ponzoni to transfer to this corvette. To speed the transfer both may leave behind whatever part of their personal effects they may not require, as Don Felipe Bauzá will do on board this ship.

6th

To reinforce your crew I am sending at the same time four good seamen, two marines and two Indians recruited in Coquimbo, making eight people in all. All will be on secondment until we join company once again in Lima.

7th

You will endeavour to avoid all communication with land until reaching Lima and at all times, either when measuring bases, or taking bearings, or if for any other reason it is necessary to send a boat ashore, you will send it under the orders of an armed officer and two marines who are to fire upon the first person who leaves his station and as this method, the only means of avoiding desertion, makes the use of boats highly inconvenient, you will prevent any person, of any rank, from going ashore, unless there is an urgent cause for the purposes of naval service.

8th

Lastly, you will endeavour to combine all of the above-mentioned objectives so as not to delay your arrival at Callao beyond the end of May or the middle of June.

May God grant Your Excellency long life, etcetera, etcetera.

At three in the afternoon the launch returned with Olavide and Guardiamarina Ali Ponzoni. The commanding officer of the *Atrevida* , in his formal reply, confirmed his acceptance of the mission entrusted to him and meanwhile he was sending me a marine from his crew, accused of misconduct, as it would be easier to keep him in custody and take him more quickly to a place where a court martial could be held. He requested that I should signal to him if I wished him to leave immediately. I made the signal and, with a moderate sea breeze then setting in from the south, clearing the sky and horizon, the *Atrevida* got under way under full sail, making for the coast to the north of Morro Copiapó, where she began to measure bases.

A short time before I had taken the precaution of comparing our chronometers by means of signals with those of the *Atrevida*. We each sent the times of the comparisons, the adjustment and the rate of the most reliable chronometers to the other ship. These served to confirm the rates determined in Coquimbo.

When we had made our farewells to the *Atrevida* and hoisted the launch, we put to sea under full sail, close hauled and steering WbyS by compass. The sea breeze was weak and at times very light. The sky was now clear and we were able to observe longitudes and azimuths. These confirmed the latitudes and longitudes we had observed at noon the day before and in giving a variation of 13°44′NE, confirmed our suspicions that the magnetic observations of the previous day were affected by some error. As a point of departure we determined our latitude as 27°14′44″ and our longitude 12′22″ east of Coquimbo. The northern extremity of Morro Copiapó then bore N38°E and the southern extremity N59°E by compass, the headland being three leagues distant.

As night fell the *Atrevida*, already some three or four leagues off, was lost to sight. The sky was becoming overcast and the breeze freshening from south to SSW with slight gusts, leading us to furl the topgallants and staysails, but by midnight, the breeze having moderated, we were able to continue under full sail once more, although the sky remained dark and hazy.

4 May

At dawn land was no longer in sight. The wind had dropped almost completely and was shifting between east and north, with the sky still hazy. Our course was that previously decided upon and although we made very little headway all morning, by midday we were already some twenty leagues offshore. We were unable to obtain a meridian altitude of the Sun, but observed double altitudes instead, the first of which obtained only twenty minutes before noon, from which a latitude of 27°8′30″ was obtained as well as our longitude 53′22″ west of Coquimbo, which was 12′ to the west of our dead reckoning position.

In choosing the most suitable route to Islas de San Félix I had decided to make use of the position determined by His Excellency Ulloa, whose investigations into the exact positions of features included in his chart, as I well knew, had to be admired as much for their skill as for their attention to detail. Their latitude on his chart was 26°0′ and the longitude 8°52′ west of Coquimbo. Other charts varied considerably, as did the accounts that I had gathered from various navigators who said they had seen them; furthermore, some described them as small and consisting of no more than sand, others that they had distict ridges and were visible at seven or eight leagues in clear weather. It was even more difficult to make all this agree with what was added to these accounts. These placed them supposedly on the meridian of Islas de Juan Fernández and about a hundred leagues off the coast, which could only be so if they lay, as Ulloa believed, approximately on the meridian of Masafuera.[1] Mariners from Guayaquil, who sighted them now and again while on the landward

[1] Isla Alejandro Selkirk, formerly Isla Más Afuera.

tack making for Lima,[1] referred them to the meridian of Nasca,[2] where they usually made their landfall, with an error to the west in their dead reckoning. From this error it could also be concluded that these islands were not as far west as had been assumed and consequently I had to take the precaution of running along the parallel of 26° as soon as I could be sure of not falling off to a lower latitude, with winds from SSW which were fairly common near the coast. In view of this it was decided to alter course for the parallel of 26°00' when 5° to the west of Coquimbo, then to remain on this latitude until reaching a point fifty leagues beyond the meridian of Masafuera and, if they were not found, to continue the search to the east for one hundred leagues on the parallel of 25°40', on which they might be situated in view of an error in latitude arising from the incorrect use of the variation that we had observed in these waters.

The light airs from NNE and NNW (an indication of northerly gales at higher latitudes) abated completely in the evening, giving way to a flat calm which left us without steerage way until nine o'clock at night, when a favourable breeze got up from SSW and south, enabling us to resume our course under full sail. [5th] At noon our latitude was 26°52' and our longitude 1°58' west of Coquimbo.

6 and 7 May

As we got further offshore the weather became very fine and the breeze, while falling, backed further and further to south and SE, allowing us good progress on our course. The effect of a current setting to WNW was clearly perceptible in our daily observations[3] and as we saw a seal and some pelicans[4] on the afternoon of the 7th, our latitude then being 26°25', we altered course that night to NWbyN, [8th] and thus, by the following morning, we had reached latitude 26°01' and longitude 5°51' west of Coquimbo. Variation, according to an uninterrupted series of azimuth and amplitude observations, taken in very good conditions, remained at 12° to 13°NE.

We had naturally made use of the waning Moon, now in its last quarter, to observe lunar distances, considering them as an added confirmation of the position we were proposing for Islas de San Félix by use of the chronometers, in the accuracy of which we now had complete faith, particularly since both of them had been mounted in gimbals. So as not to waste the excellent opportunity, thanks to the calm conditions, we set about taking the necessary observations from the morning of the 5th, continuing on the 6th, 7th and 8th, with sextants by the best English makers with the greatest possible accuracy in both observations and calculation.

In such circumstances, our surprise may be imagined when we found the results of these lunar distances to be very different from those of the chronometers. The previous observations were compared to the later ones for confirmation. They agreed with

---

[1] The coastal route from Guayaquil to Callao is subject to the north-flowing Peru Current and contrary light southerly winds. These can be overcome by making an offing into the Trade Winds, reaching the coast by the westerlies and then running north with a fair wind and current. In so doing mariners might sight Isla San Félix or Isla San Ambrosio.

[2] Cabo Nazca, an easily identified dark brown bluff.

[3] With increasing distance from the coast the Peru Current becomes more westerly.

[4] The Peruvian or Chilean Pelican (*Pelecanus thagus*) which may be encountered in large numbers in the coastal waters of Chile and Peru, but it does not stray out of sight of land.

the *Connaissance des temps* and the *Almanak náutico* in the same determinations of the lunar positions. The lunar distances and the chronometers had agreed almost exactly in Coquimbo. Our calculations of the position of that meridian relative to that of Paris could not have been supported by more reliable information and finally, our consistency in observing such a large number of distances and the very agreement between the daily differences of longitude, calculated according to these distances and those indicated by the chronometers, appeared to require the same trust in one set of results as in the other.

Days on which we observed	Number of series	Mean longitude by lunar distances from Paris	By chronometer at the same time	Difference between lunar distances and chronometers
5	49	74° 53′ 21″	75° 41′ 00″ E	47′ 39″
6	143	76 47 57	77 34 20	46 23
7	88	78 6 10	78 55 15	49 05
8	16	78 48 22	79 34 38	46 16

Therefore the difference in longitude from the 5th to the 8th was:

By Lunar distances	3° 55′ 1″
By the chronometers	3 53 38
Difference	1′ 23″

The sky, generally hazy or cloudy near dawn, did not allow us to observe distances to stars west of the Moon, as we would have liked to have been able to do. This might perhaps have cast some light on the true causes of such a strange difference, by no means attributable to the sextants, which had been checked in the greatest detail and were, by chance, all made by the best English instrument makers: Ramsden, Dollond, Nairne, Stancliffe, Wright and Troughton.[1]

That afternoon we saw another seal and a few pelicans. During the following night we made some eight leagues from our position at nightfall; at one o'clock we hove-to heading SW, with a gentle breeze from SSE.

9 May

At dawn we altered course again, flattering ourselves with the hope of sighting the islands on this heading, both because of the light breeze and clear weather which had set in and the large number of birds that we saw coming from the west.

Our latitude at noon was 26°6′ and our longitude 7°24′30″. At three o'clock in the afternoon an island of medium height[2] was indeed sighted from the mastheads, bearing WSW 5°N by compass and we altered course for it at once. At half-past four, when it could already be seen clearly from the deck, we observed azimuths and hour angles and continued towards it under a press of sail until sunset, at which time it was still some eight leagues distant and we could not make out any other island nearby.

This persuaded me to steer a course during the night that would put us at dawn

---

[1] Jesse Ramsden (1735-1800), Peter Dollond (1730-1820), Edward Nairne (1726-1806), John Stancliffe (fl. 1770-1810), Thomas Wright (1711-1786) and Edward Troughton (1753-1836).
[2] Isla San Ambrosio, the easternmost of Islas de San Félix.

south and west of the island we had in sight so as to keep us to windward of the others if they lay further south, while surveying them during the early hours of the day, which were unsuitable for astronomical observations.

At nightfall, accordingly, we hauled our wind to SSW under the principal sails with a fresh SE breeze and, as the early part of the night was fairly clear, we soon sighted the island once again, after which it remained in sight. At two o'clock we obtained no bottom with one hundred fathoms of line. From then onwards we continued under topsails, [10th] and at dawn, although the sky was misty and overcast, we could see the large island clearly, three leagues distant, bearing N17°NW and two smaller islands bearing NW2°N and NW2°W,[1] which we were now sure must be all or most of Islas de San Félix. In this situation, with the weather now threatening to prevent any observations, we decided not to divide our efforts and to remain at some distance from the smaller islands in a position where neither ship-handling nor other tasks would distract us from waiting for a break in the cloud to observe our latitude and longitude. Fortunately our course of NbyE, which we were now running as a base under topsails to allow us to close the islets to the east of the large island to about two miles, gave us a transit line on the small islands and then of the two extremities of the large island with the three small islets,[2] thus avoiding any risk of a significant error in the bases, which was very much to be feared, given the constant westerly set that we had experienced every day since leaving the coast.

At half-past ten, having finished observing our transit lines, we hove-to very close to the large island and to the south of the other small islands to wait for noon in that position.

At dawn the breeze had got up from SE and east, with frequent gusts, a fairly choppy sea and a constant cloud cover, mixed with fine drizzle and mist at times. Later, when we were lying-to and very close to the northern side of the large island, it was frequently obscured, as were the others, by a dense cloud of vapour which covered them completely. Despite this, we had been able to observe two or three altitudes of the Sun at varying intervals, which gave us some comfort. At last, shortly before noon, we had the satisfaction of seeing the mist clear a little, enabling us to obtain a good observation of the meridian altitude of the Sun. After that the weather soon closed in again, obliging us to remain in approximately the same position as at noon, but at half-past one there were signs that there would soon be some breaks in the cloud and I decided to use them to survey the large island from closer inshore.

We approached it and, two cables off one of the islets, which we saw to be perforated right through,[3] we began to run along the coast at a distance of three or four cables. From this position we were able to observe hour angles and azimuths, while

[1] Isla San Félix, with Isla Gonzáles off its SE end.

[2] It is difficult to follow Malaspina's actions, but it seems that he approached close enough to Isla San Ambrosio to align its SW extremity with Isla San Félix and Isla Gonzales and then as he rounded the western side of Isla San Ambrosio to align its southern side with the three rocky islets off its eastern extremity.

[3] Roca Bass, the inner of three rocky islands off the eastern extremity of Isla San Ambrosio, which was described by Captain H. W. Bruce, who visited the islands in December 1837 in HMS *Imogene* as 'Through the West part of the "Bass" is a remarkable fissure, leaving a cavity through at the water-line, and apparently 20 feet high, shaped like a triangle': Alexander G. Findlay, *A Directory for the Navigation of the Pacific Ocean*, 2 vols, London, 1851, II, p. 795.

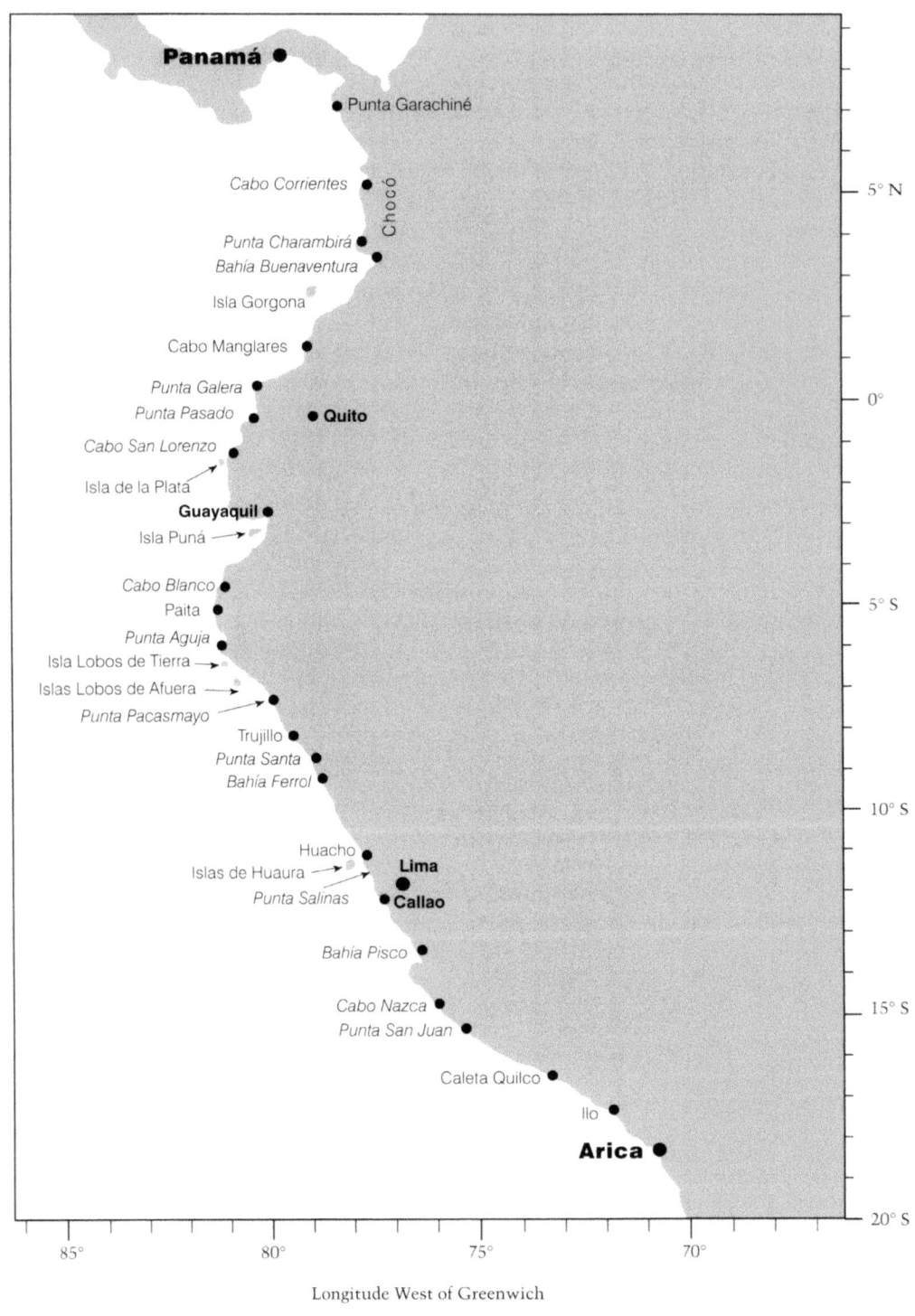

Fig. 7. West Coast of South America from Arica to Panamá, May to November 1790

Don José del Pozo was able to take a view from life of this and the three smaller islands,[1] and of a remarkable rock, standing alone some way to the north of them which looked very like a lateen rigged vessel.[2]

At three o'clock, having surveyed this, the only sheltered side of the large island, we made for the others,[3] with the intention of surveying them from a closer distance in what was left of the day. However, as the breeze then gave signs of falling calm and night was drawing on, I did not wish to take the risk of anchoring with the corvette's complement in its present reduced state and I decided that my survey was enough for the purposes of navigation, particularly as the structure and position of the smaller islands, which could now be seen in some detail, showed them to be as inaccessible to the mariner as the one which we had just examined from close inshore. At about half-past three, we altered course to NW and measured a base to fix the position of the rock which looked like a vessel and finally, at four o'clock, we set a course for the mainland under full sail.

This small archipelago consists of one large island, three of medium size,[4] and the ship rock, as well as three isolated hummock-like islets near the eastern end of the large island and another lying at a similar distance off its western extremity.[5] All appear equally bleak and steep-to and, as far as one can see, the two kinds of strata of which they are formed appear to contain a high proportion of ferrous particles. These strata slope somewhat towards the horizontal and the black or ferruginous ones are much thicker than those that are red, or of some cornelian-like material. The large island is certainly inaccessible on all sides, as from sea level to at least two-thirds of the way up it is everywhere so steep that the upper slopes lean outwards rather than inwards. However, on the summit we could make out traces of vegetation, although scanty. On a sort of tableland which occupies most of the higher end we saw a few green bushes, two or three feet high and some grass, but no sign of water or any shelter which might attract the mariner to these parts. The sea birds and seals, which we had thought to find in great numbers in such a safe sanctuary as this, seemed to have been driven away by the bleak appearance of these islands.[6] Neither on the entire northern side of the large island nor on the islets to the east did we see any sign of birds or seals. Some ten or fifteen petrels[7] did flutter around the corvettes, leaving us when we approached closer to the islets and when we later made our departure.

[1] AMN, MS 1723 (16).

[2] This view has not survived. The location of this rock is also uncertain as there is no such rock north of the three small islands. However, it may have been Roca Catedral de Peterborough, an isolated rock situated about 1½ miles NNW of the NW extremity of Isla San Félix, which was described by Colnett in 1793 as '… a remarkable small rock … which, in most points of view, shews itself, like a ship under sail.', James Colnett, *A Voyage to the South Atlantic and round Cape Horn into the Pacific Ocean*, London, 1798, p. 37.

[3] Isla San Félix and Isla Gonzáles.

[4] Malaspina must have imagined that Isla Félix comprised two islands.

[5] Roca Cónica, off the SW extremity of Isla San Ambrosio.

[6] From time to time these volcanic islands probably do support large breeding colonies of sea birds, but following the Chilean earthquake of 1922 gases of volcanic origin destroyed or drove away the greater part of the bird population: R. C. Murphy, *Oceanic Birds of South America*, 2 vols, New York, 1936, I, p. 259-60. It could be that the dense cloud of vapour seen by Malaspina was composed of such gases.

[7] Probably White-bellied Storm Petrels (*Fregetta grallaria*), specimens of which were collected off the islands in 1935: Murphy, *Oceanic Birds*, II, p. 761.

No doubt remains in our minds that not only the channel between the large and the smaller islands,[1] but also those between the smaller islands[2] are navigable throughout, as is the larger one formed by these and the rock that we called the 'canario' because of its resemblance to the craft of this name sailed by the Valencians in the Mediterranean.[3]

The latitude of the western extremity of Isla San Ambrosio was determined by observation as 26°20′15″, and its longitude west of Coquimbo:

By chronometers . . . . .8° 27′ 51″
By 296 sets [of lunar distances] 7 43 25

Variation remains somewhat doubtful, as either by chance or from the influence of the many ferruginous particles in the body of the large island, according to both an azimuth observed close to the island and a western amplitude observed at a distance of four leagues we obtained a variation of only 8°5′NE, while on the previous afternoon, at a distance of twelve leagues, we had observed a variation of 11°30′, proportionate to those we had carried from the coast. Finally, the differences in longitude and latitude between our observations of the previous afternoon and those of today were combined with the time of sighting the said island from the quarter-deck of the corvette in clear weather. The height of this island above sea level was calculated as 190 to 200 *toesas*,[4] and consequently it was easily seen from the quarter-deck at a distance of fourteen leagues.

According to the plan agreed with Don José Bustamante y Guerra, the next task of this corvette was to concentrate on the hydrographic survey of the coast from Cabo San Juan to Lima. Having taken the approximate latitude of that cape as 15°30′ and its longitude as 4° east of San Ambrosio, according to the meridian differences between Tetas de Biobío[5] and Punta Nasca,[6] noted on my previous voyage with pocket chronometer 71, it was clear that to steer NNE would take us directly towards this stretch of coast. However, I set our course a little more to the east as a precaution against the westerly currents, which, if combined with calms and light airs, could delay us considerably. Accordingly, the course of NEbyN, true, chosen at the start, had to be altered much further to the east the next day, as we seemed to have exactly those conditions of westerly currents and very light breezes, falling to calm, mentioned earlier.

11 May

At dawn the following day there was no sign of land. The sky was continually overcast, and the breeze, from ESE to east, varied constantly in strength and was excessively light at times. Don Dionisio Galiano, having taken some altitudes of the Sun as a precaution, decided to use them immediately to calculate our latitude, not so much from necessity, for he had taken the Sun's altitude only five minutes before it was on

---

[1] I.e. between Isla San Félix and Isla San Ambrosio.
[2] Malaspina is wrong; there is no navigable channel between Isla San Félix and Isla Gonzáles.
[3] I.e. between Isla San Félix and Roca Catedral de Peterborough. The *canario* is a lateen-rigged boat used in the Mediterranean and Islas Canarias.
[4] The *toesa* is the French *toise* of 6·3946 feet; the heights are therefore 1215 to 1280 feet.
[5] Two prominent hills marking the southern approach to Bahía Concepción: see p. 152, n. 3 above.
[6] Cabo Nazca: see p. 205, n. 2 above.

the meridian, but rather because we were now drawing very close to the equinox so that the examination of the accuracy of his method could be tested more rigorously. He had the satisfaction of achieving the following results at noon today:

	Latitude
By the Sun's altitude taken three minutes before noon	.24° 55′ 55″
By double altitudes, the second taken twenty minutes before noon	.24 55 25

Our longitude the same day by chronometers was 7°51′15″ west of Coquimbo.

These observations were further confirmation of the existence of a westerly current in these waters, the progress of which, since we had left the coast, had been as follows:

Day's run	Differences by dead reckoning	Ditto by chronometers	Difference to the west by chronometers
3 to 4	0° 53′ 0″ W	1° 5′ 22″	12′ 22″
4 to 5	56 48	1 5 13	8 25
5 to 6	1 37 00	1 50 8	13 8
6 to 7	1 6 36	1 18 27	11 51
7 to 8	33 5	0 43 43	10 38
8 to 9	1 16 50	1 33 46	16 56
9 to 10	42 35	1 4 13	21 38
10 to 11	0 58 5 E	0 40 22	17 43

Total error from dead reckoning over this period   1° 52′ 41″ to the west.

With this information we now altered course decisively and continued close-hauled steering from NEbyN to NEbyE by compass, as the wind allowed, this now having settled in from east to SE, moderate with haze.

12, 13 and 14 May

On the following days we were unable to make any observations, except during a few brief breaks in the cloud, for which we had to wait sometimes for up to an hour. Our latitude at noon on the 14th, calculated from the double altitudes of the Sun was 18°42′37″ and our longitude 3°9′ west of Coquimbo. Until now the sky had remained very overcast with frequent showers, with the same drizzle which often falls in Lima. The differences to the west [between dead reckoning and the chronometers] appeared to have ceased. We only found the set to continue to the north, having caused us an error of 45′ in that direction in the last three days.[1]

On the 14th we repaired with wooldings[2] some damage to the mainmast, which had several splits and cracks extending from the sheave of the topping lift to three feet below the cap. We attributed this to the brittleness of the timber rather than the particularly rough seas in which we were sailing.

From noon on this day, having almost made the distance in longitude that we had proposed, we altered course to NNE, intending, now that the winds appeared

[1] Nearing the mainland coast the Peru Current had lost its westerly component.
[2] Pieces of rope wound round a mast or a yard to support it where it has been fished.

favourable, to approach the coast further south and east, thus avoiding the effects of the full Moon on the part of the coast that we had decided to survey ourselves,[1] while taking precautions against any occurrence which might prevent the *Atrevida* from doing so.

16 May

We then continued steering true north, carrying a press of sail and, as the day dawned somewhat less hazy and the weather began to clear as the morning wore on, we were able to confirm our position with good observations and at last, at eleven in the morning, preceded by the sighting of several seals and petrels, we sighted land in the first and fourth quadrants. The land rose fairly high in the interior, but the shore consisted of nothing but sand dunes. The coast receded considerably to the north, so that at noon we could see land some eight leagues distant to the east, although we were at least five leagues from that to the north. A fully reliable latitude 16°29' by observation; our longitude, by the three chronometers in agreement, was 3°2'30" west of Coquimbo. We had no doubt that we were a considerable distance to windward of Morro de Acarí[2] and, therefore, having reached a reasonable distance offshore, we continued our hydrographic work from noon, combining the measurement of bases by log with frequent observations for longitude, as was our habit.

At nightfall the furthest point of land visible was some six leagues distant, bearing from west to NWbyN by compass; the land abeam was some three leagues distant, and azimuths and amplitudes had given a consistent variation of 10°15'NE. During the following night we adjusted our courses and distance run so as to approach, from the east, the points of land visible to the west. It was therefore necessary to lie-to from one o'clock. We got under way again at five and before sunrise we had already begun measuring bases.

17 May

The land now in sight was no different from what we had already surveyed: it extended in almost the same direction and terminated in the west in regular promontories beyond which the land rose abruptly like a step, concealing the next range to the north. The coastal strip, like the nearby high land, seemed bleak and almost a desert. We saw a few coves, which appeared to offer very little shelter, and as the breeze continued moderate from ESE to SE, with fairly clear skies, we made use of it under full sail, steering NW and NWbyW, following the direction of the coast. We remained in some doubt as to the true position of Morro de Acarí, but a satisfactory latitude at noon of 15°30' by observation, and the configuration of the coast, compared with Moraleda's chart, left us in no doubt of being off Punta de Peñas.[3] Our longitude was 3°58' [west of Coquimbo].

The favourable breeze picked up somewhat during the afternoon, allowing us to reach [Punta] Nasca, abeam of which we spent the night. The survey of this coast

---

[1] Around the time of the full Moon the angular distance between the Sun and the Moon would be too great to measure by sextant.

[2] Cerro Acarí, a prominent isolated hill 4 miles NE of Punta San Juan.

[3] Possibly Punta Penotes in 15°27'S.

was becoming increasingly important, since we regarded it as the landfall for those making directly for Lima from the south. Accordingly we made little distance that night, during which the breeze continued fresh from SE as we lay-to first on one tack and then on another.

18 May

At dawn a thick haze hid the trend of the coast from us almost entirely, although we were only four leagues offshore. We approached it immediately and at half-past six, having sighted Los Infernillos[1] about four miles distant, we bore away and made for Morro Quemado,[2] which was already in view. As the Sun rose, the haze cleared completely and, since the breeze remained favourable and a fairly strong northerly current had set in again, we could continue our survey with both speed and accuracy, particularly as regards Morro Quemado, from which we were one league distant at noon. Accordingly, we took various views of it and obtained two observations for longitude on its meridian at noon, so close to it that we were sure that future mariners could not attribute to us any omissions or inaccuracy. The determination of the longitude of this important point, which I had also obtained with the same Arnold 71 in the *Astrea,* when referred our longitude of Concepción in Chile and also our lunar distances observed today and yesterday agreed exactly with the chronometers.

Our latitude was 14°19′30″ and our longitude 5°2′ [west of Coquimbo]. The trade wind as usual favoured us rather more in the afternoon and we were able to survey at close range not only the small islands running north from Morro Quemado, but also the other three headlands as far as that of Lechuza, Islas San Gallán and the other islets that follow, but even with the help of the current we could not sight Islas Chincha, as we had hoped.[3] Variation, observed in very favourable conditions by azimuths and amplitudes, remained constant between 9° and 10°.

After nightfall we made only six leagues, and we then lay-to first on one tack and then on another, intending, according to our plan, to make use of the whole of the next day in a thorough survey of Islas Chincha and the adjacent coast to the north. [19th] However, as the day dawned extremely calm and overcast, with haze, a mistake in the calculations of the longitude observations made in the morning concealed from us the considerable westerly set of the current.[4] We sailed on until four in the afternoon, deceived by the appearance of the coast, when at last, making for the coast with a rather more settled breeze from SE, although not enough to dissipate the haze, we realized our mistake and had to abandon the survey of a short stretch of coast because of these circumstances. At noon our latitude had been 13°21′ and our longitude 5°36′ [west of Coquimbo], 17′ further west than by dead reckoning.

The flat calm that we had during the night, with a fairly heavy swell, obliged us to manoeuvre frequently under topsails and topgallant sails, in an attempt to use every breath of air to remain in a favourable position.

---

[1] Islotes Infiernillos, islets and rocks off Punta Doña María about 180 miles SE of Callao.

[2] A headland capped with reddish soil, about 25 miles NW of Islotes Infiernillos.

[3] Cerro Lechuza, Isla San Gallán and Islas Chincha are all in the vicinity of Bahía Pisco; the later island was formerly one of the principal guano islands off the coast of Peru.

[4] Off this stretch of coast, the Peru Current runs NW, parallel to the coast.

20 May

At dawn on the 20th we did indeed find ourselves close inshore and having sounded seventy-six fathoms, mud and small pebbles, we altered course, with the few light airs that we had, to survey Isla Asia,[1] which we thought we had in sight. We remained in the same position nearly all morning, finding this at noon to be latitude 12°56' and longitude 5°28'30" [west of Coquimbo]. The weather continued hazy and we would have been barely two leagues offshore. Finally, at about one in the afternoon, the trade wind began to freshen somewhat. We quickly continued the survey of the coast and by nightfall we were only four leagues off Punta Chilca,[2] having managed to sight Río and Pueblo de Mala[3] during the afternoon.

Having made only five leagues under easy sail during the early hours of the night, remaining in sight of the same point and observing the meridian altitude of α Centauri, we lay-to first on one tack and then on another, intending to survey Ensenada de Pachacamac and Morro Solar[4] next morning and then to make for port.

21 May

Even now the current continued to upset our plans. At dawn we found that we had been set to the north of Morro Solar and were between this headland and Piedra Oradada,[5] so that, choosing to pass north of Isla San Lorenzo[6] because of the uncertain duration of the breeze, we had to steer WbyN until we cleared Los Palominos.[7] We then bore away and passed a musket shot's length from Isla San Lorenzo, under shortened sail, to enter the bay and on the same tack, at about half-past ten, we reached the anchorage among the merchant ships in Callao.

---

[1] This island, situated about 55 miles SSE of Callao and white with guano, is one of the most easily identified features on this part of the coast.

[2] A prominent headland 33 miles SSE of Callao; also known as Punta Lobos.

[3] Mala, on the left bank of the river of the same name, is now part of the district of Lima.

[4] About 21 and 12 miles, respectively, SE of Callao.

[5] Islotes Horadada about 5 miles SE of Callao.

[6] An island over one thousand feet in height, 4½ miles long in a NW/SE direction and one mile wide, which protects Bahía del Callao from the prevailing southerly winds. The island is situated close west of La Punta, a peninsula forming the southern side of the bay from which it is separated by El Boquerón, a narrow channel leading directly into Bahía del Callao from south, which Malaspina elected not to use, passing west and then north of Isla San Lorenzo.

[7] Islas de Palominos, off the SE extremity of Isla San Lorenzo.

CHAPTER 2

At Callao and Visit to Lima

The proposed plan, as approved by His Majesty, involved a stay in Lima long enough to allow us to replenish our stock of provisions, to overhaul the vessels and their equipment, to carefully examine a country of such importance to the monarchy, and above all, to put in order the great quantity of hydrographic information that we had gathered and which was now piling up in great confusion and to some extent becoming muddled in our minds. To this was added the fact that strong rainy winds, entirely unfavourable for our work, were now prevailing on the coast to the north,[1] consequently, there was no route we could take, without the sacrifice of a considerable amount of time, to make the coast between Guayaquil and Acapulco, where the trade winds do not set in until December.

For these reasons, I had sent a request from Chile to His Excellency the Viceroy, begging that he would be kind enough to intervene on our behalf with the Buena Muerte fathers[2] and ask them to let us establish our quarters in their house in the township of La Magdalena,[3] while the corvettes were laid up in Callao. La Magdalena, an Indian village, like many others which add to the charms of the beautiful Rimac Valley, lies only two nautical miles [SW][4] of the city. The fertility of its soil, the salubrity of its air and water and its very separation from the noisy bustle of Lima, mean that it is much frequented by the sick and convalescents, for whom the air of the city is well known to be harmful, or even fatal.

A number of reasons had led me to set up our quarters at some distance from Callao and Lima, so as to bring all of us together for the many tasks to be undertaken and combine this with that natural freedom which alone forms the basis for rest and recreation. Apart from its notorious fevers, Callao would keep us too close to the crews, whose disorderly habits would cause us frequent vexation, while they, on the other hand, would be put out by our constant presence since we would find it impossible to overlook any misconduct. Furthermore the officers, who would need to

---

[1] The rainiest months at Guayaquil, Malaspina's next intended stop to the north, are from January to April: *South America Pilot, Vol. 3*, Taunton, 1987, p. 47.

[2] This order, also known as the Agonizantes or Camilos, was set up in Peru in 1712, and grew thereafter to occupy an important place in the life of the Viceroyalty. When Malaspina arrived, it had two houses, 36 professed members, 22 lay brothers, 13 postulants, one oblate and 16 slaves: *Mercurio peruano*, 3 February, 1791, pp. 97-8 and R. Vargas Ugarte, *Historia de la Iglesia en en el Perú*, Burgos, 1961, IV, 25-36.

[3] Occupying the district of Lima now known as Pueblo Libre.

[4] Blank in the original.

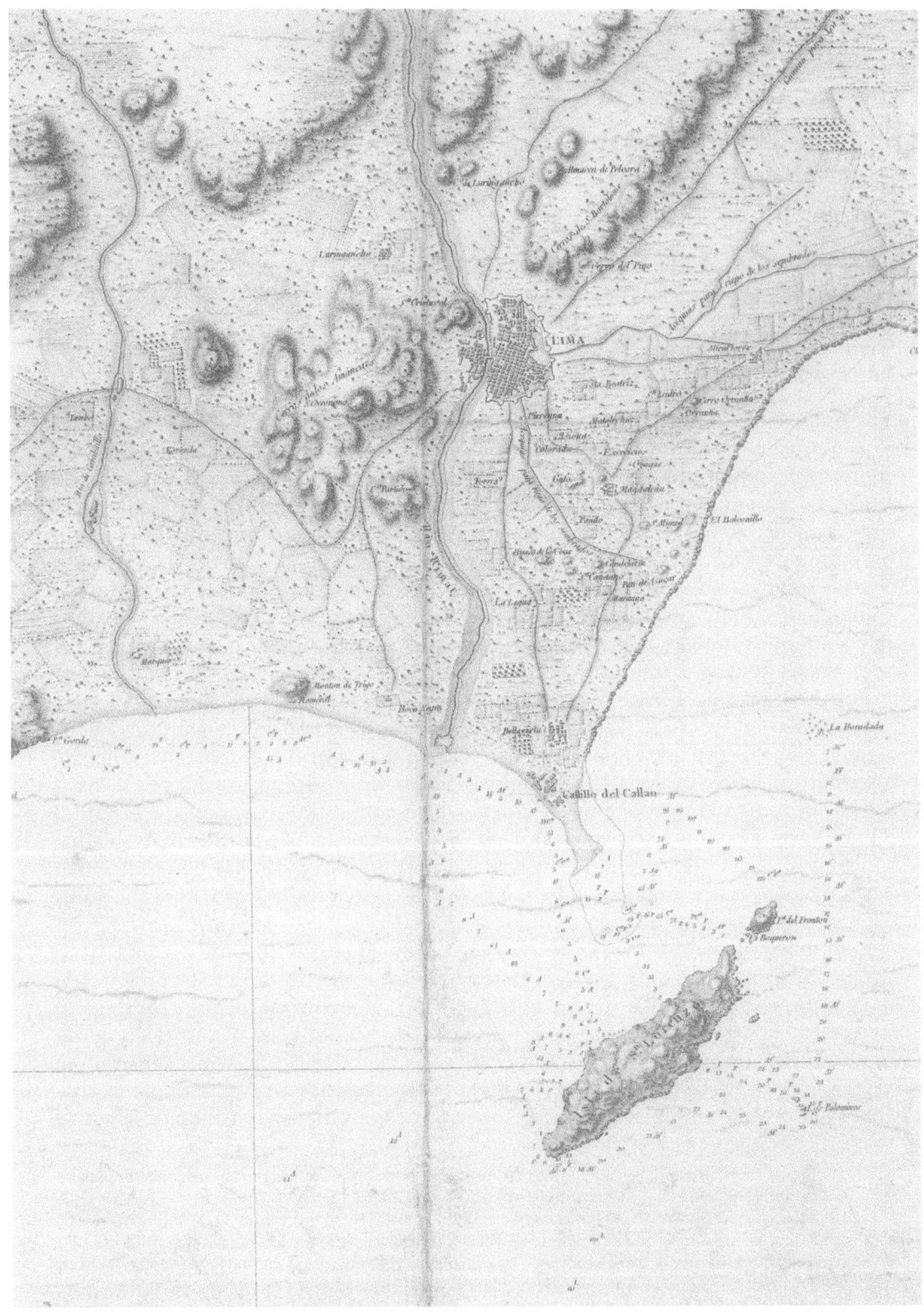

Plate 24. *Plano del Fondeadero del Callao de Lima … Construido por los Comandantes y Oficiales de las Corbetas Descubierta y Atrevida en 1790 y publicado en la Dirección Hidrografica año 1811* (detail); UKHO, 126. Reproduced by permission of the Controller of Her Majesty's Stationery Office and the UK Hydrographic Office.

work almost every day at various tasks, would find it difficult, at some distance from Lima, to spend the few free hours they had for well-deserved, honest recreation on the social and, no doubt, enlightened attractions which Lima could offer them if it were near by.[1] It would not be suitable to set up our quarters in the city itself, not only because of the distractions[2] and the greater difficulty of meeting together, but also because curiosity and natural idleness would continually attract large numbers of people to our place of work to the detriment of our intention to use our time to the best advantage. We could avoid all these problems at La Magdalena, where, away from an urban environment, we would have the benefit of a clearer sky for our astronomical observations, a much healthier climate and, in particular, a degree of rural freedom essential to that tranquillity which we needed to devote our unhindered attention to the many aspects of our present mission.

Accordingly, as soon as the corvette anchored in Callao, I sent Teniente de Navío Cayetano Valdés ashore and for the same purpose went ashore myself in the afternoon with some other officers, Pineda and Haenke. The proposed plan was approved by His Excellency and as the Buena Muerte fathers gave us full use of their house, we were able to start work on setting up our quarters the following morning, particularly the observatory. All the astronomical and geodetic instruments, the natural history collections and most of the books and charts were moved to our new premises at once. They were very soon followed by some of the officers, who, while taking care that everything should be kept in good order and properly installed, kept an eye open for any favourable moment for astronomical observations, so difficult to obtain because of the constant cloud cover in these parts. On the afternoon of the 24th the astronomical clock and quadrant were set up. Apart from a splendid introductory banquet, we evaded the generous offers of our hosts to provide for our sustenance (even for our spiritual needs) and from that date we could consider our quarters to be well established, at least as far as the officers and servants were concerned.

We could not yet adopt a constant and uniform method for our tasks, because the *Atrevida,* with whose officers we were to work, had not yet arrived and we were also missing Don Felipe Bauzá, who was in charge of the supply of the charts and maps, and because the coming days were precisely those chosen by the city of Lima for the ceremonial entry of the new Viceroy of Peru, Don Francisco Gil y Lemus.[3] This general, combined a fine character and admirable personal talents and qualities with the fact that he happened to have served in the Navy. He was to show us that same affection borne of being a member of our own service so that he knew personally some of

---

[1] The population was 52,627 by contemporary estimate: J. Hipólito Unanue, *Guía política, eclesiástica y militar del virreinato del Perú para el año de 1793,* Lima, 1792, p. 2.

[2] As well as theatres and bullrings, the city had six cafes with billiard rooms and various boulevards, including the Descalzos and the Aguas in Rimac.

[3] The entry took place on 17 May, 1790: R. Vargas Ugarte, *Historia general del Perú,* Lima, 1966, IV, pp. 99-100. Francisco Gil de Taboada y Lemus, a *teniente general de real armada,* served as viceroy until 1796, having been Inspector General de la Marina. He was personally well known to Malaspina as a member of the generation of naval officers whose formation took place in the 1780s in the Curso General de Estudios Mayores and the project for a maritime atlas of Spain, led by Vicente Tofiño. He was closely identified with reform and was the principal patron of the enlightened press, represented in Lima by the *Mercurio peruano.*

us who had served under his command and others under the command of his fellow captains in the *Esquadra Grande*. All this was in itself a major cause of our present rejoicing and was lent even more weight by our natural curiosity prompted by the truly magnificent preparations to be seen all about us.

His Excellency Ulloa,[1] in the account of his voyages, has described these pageants in detail, with such accuracy and elegance that it would be churlish to recount them again, particularly since, even if all the ceremonies proper to the occasion were restored to their past splendour from which they had declined considerably in recent years, his narrative would in no way be inadequate, except in so far as the population is now much larger and the acclamations now arise from a greater love for our august Sovereign who is so nobly represented at such a distance from the throne.

28 May

The ceremonies had hardly concluded when a new and exceptional cause for rejoicing was added by the arrival of the *Atrevida* which anchored near the *Descubierta* in the night of the 28th and was immediately moored and unrigged. Don José Bustamante, with remarkable energy and success, had surveyed and plotted in detail the entire coast from Morro de Copiapó, where we had parted, to Morro de Acarí or rather [Punta de la] Nasca. He had anchored off Arica and carefully determined its position astronomically with the chronometers and sextant and finally, keeping the coast in sight and enjoying better weather and no current between Nasca and Isla de San Lorenzo, he had collected new information with which to confirm or correct the results of the *Descubierta's* work.

From the moment of our arrival we had already undertaken in the *Descubierta* a comparison between the chronometers to confirm the rates determined in Coquimbo, preferring absolute altitudes taken with the sextant to other astronomical methods, not only because of the reliability and ease of this method, but also because, lying at anchor, we had a good horizon during precisely those hours of the afternoon when the Sun was more likely to be visible than in the morning. With the arrival of the *Atrevida* we repeated these operations for greater confidence by comparing our chronometers with theirs. Thus, by 3rd June, our fears in this regard were allayed, as we had now calculated the longitude of the anchorage according to each of the chronometers and rated them with a fairly satisfactory interval. Much to our surprise we found that these rates had all increased, even that of number 10 in which, however, the difference was small enough to be able to bring the longitude into line with that calculated by the emersion of the first satellite [of Jupiter], when it was observed a few days later. It therefore seemed wisest to adopt the change noted at Arica for number 10, and for chronometers 61 and 72, to adopt an average equation which would be used for a daily correction of longitude calculated with both chronometers, since there was uncertainty as to the interval during which the gain was observed.

The achievement of this essential aim was greatly helped by the clear skies that we enjoyed in La Magdalena during the first days of June. They were so clear, in fact,

---

[1] See p. xciv, n. 1 above.

that we were able to regulate the astronomical clock most satisfactorily by equal altitudes and, at nightfall on the 5th, to observe the emersion of the first satellite of Jupiter in good conditions, which, according to Don Dionisio Galiano, gave the longitude of the observatory as 79°28′30″ west of Paris although, as it could also be observed at the same time in Europe, we had the promise of even simpler and more reliable figures by which to determine it. We did not obtain the latitude until the night of the 14th, by which time we had already observed variation and inclination, so that neither the accuracy of our work nor our nightly rest were to be disrupted by the constantly overcast sky of the three winter months that followed. The longitudes by the chronometers, deduced from that of Coquimbo and referred to the observatory by means of His Excellency Ulloa's plan, were as follows:

	Atrevida			*Descubierta*	
West	Number 10	Number 71	Number 105	Number 61	Number 72
of Paris	79°30′23″	79°29′53″	79°31′53″	79°38′45″	79°39′9″

As for the observations previously made in Lima, these did in truth differ somewhat from those of Dr Durand,[1] quoted by Father Feuillée whose pupil he was, but they were very close to the results of Their Excellencies Juan[2] and Ulloa and to those of Dr Peralta,[3] whose longitudes, referring both to Lima cathedral, were as follows:

By the observations of Their Excellencies Juan and Ulloa ......79° 23′ 50″
By Dr Peralta 79 20 00
By Dr Alexandre Durand 79 9 30
By our observations at La Magdalena 79 26 30

It seemed better to keep the chronometers on board, leaving those by Berthoud stopped and keeping the others going, so that we could not only observe any alterations during this period but would also have a further reason for taking good care of them. It would then be easy to undertake any maritime excursion that we might plan for the near future.

The 1st of June had been the day on which we were to implement the arrangements for maintaining discipline in this port. Don José Bustamante and all the officers of the *Atrevida* had established themselves in a fine country house belonging to

---

[1] Of French birth, Alexandre Durand studied astronomy with Louis Feuillée (see p. 136, n. 1 above) during the latter's stay in Lima for a few months in 1709. Early in 1710 he made some observations of one of the satellites of Jupiter and made a calculation of the longitude of Lima relative to Paris: M. Moreyra Paz Soldán, 'Peralta astrónomo', *Estudios históricos,* Lima, 1995, iii, p. 536.

[2] Jorge Juan (1713–73) accompanied Ulloa on the La Condamine-Bouguer expedition (p. 263, n. 1 below) and collaborated in the publication of their extensive scientific findings. In 1752 he was appointed director of the Compañía de Guardias Marinas of Cádiz. Under his leadership the corps became Spain's elite school of mathematics and science. He founded the Observatorio de Cádiz.

[3] Pedro de Peralta y Barnuevo (1664–1743) included astronomy among a polymathic range of interests. He used lunar eclipses to make a calculation of the longitude of Lima relative to Paris in December 1713 and March 1717. As Professor of Mathematics and rector of the University of Lima, he was known as 'Dr Ocean' to his pupils. He published an enormous amount of heroic verse and rhetorical history, in which he combined creole patriotism with intense loyalty for and pride in the Spanish monarchy: Moreyra, 'Peralta astrónomo'.

Conde de San Carlos,[1] a short distance from La Magdalena, as the houses in the village itself were very exposed and uncomfortable. Every officer soon obtained a horse, which now made our visits to Callao, our almost daily trips to Lima and our occasional excursions as easy and frequent as they were healthy and enjoyable. The compilation of the information gathered, which was being carried out jointly in both houses, with the various topics carefully allocated between them, required the frequent resolution of one doubtful point or another. This was very easily resolved, as someone could be sent on horseback in his informal dress, to ask and settle the questions.

Many objectives needed to be considered in adapting a mode of discipline to our current circumstances. The commissioned officers had to be provided with the decorum and proper use of time that were essential for the performance of the mission that they had undertaken. The lower ranks should be permitted as little fraternization as possible, be gently dissuaded from the vices known to abound in Callao, made to attend regularly to their duties, prevented from deserting and, if possible, kept in good health despite the many risks which surrounded them. To accomplish these aims, the marines and marine gunners of both corvettes took up quarters at La Magdalena on the 1st [of June], while the detachments on board were to be of four men in each corvette under the command of a sergeant, master gunner or senior corporal. A single commissioned officer, taken alternately from either corvette, including the *guardiamarinas*, would be in charge of the anchor watch on board both ships, which were lying very close to each other; but the harbour, supply and incident logs were to be kept separately. The *pilotínes*, artists and the *Descubierta's* barber[2] also went to La Magdalena to carry out the mounting and other preparations for the natural history collection. The petty officers were allowed to live either on board or ashore, as long as their conduct was not disorderly, but the boatswain and two quartermasters were to take it in turns to sleep on board with a third of the crew and would be in charge of the launch, pinnace and jolly boat, which were beached at a convenient spot, so that the other two boats would be in use elsewhere as little as possible. During working hours nobody was to be absent without leave, the length of which was to be determined by the officers of the watch but was not to exceed three days and anyone who interfered with this equitable arrangement was to be severely punished.

Even if we had not known beforehand of the unruliness of the seamen in Callao, I could not have concealed from myself the fact that such measures would be of little or no effect if not attached to much more powerful instruments than advice or punishment. These are ineffective when pleasure, example, climate, idleness and ease of subsistence are all conducive to vice and difficult to apply when desertion is as easy as crime. Self-interest seemed, therefore, to be the only suitable weapon on this occasion. It was necessary to make some payments to the seamen, which did not need to be paid for until Manila, so that they would at least have money for wine and because

---

[1] Joaquín Manuel de Azcona y Buega, Conde de San Carlos, presided over the commercial court of Lisbon from 1778-82 and was elevated to the nobility in November 1781: M. de Mendiburu, *Diccionario histórico-biográfico del Perú*, Lima, 1878, x, p. 23. He was well advanced in years at this time and died shortly after Malaspina's departure: J. de Atienza, *Títulos nobiliarios hispanoamericanos*, Madrid, 1947, p. 488.

[2] In the navies of this period the ship's barber was also the sickbay attendant and the blood-letter (either with leeches or phlebotomy) in an age which believed devoutly in the benefits of blood-letting.

Plate 25. View of Lima, 1790, by Fernando Brambila. Museo Naval, Madrid

the few who had not yielded to the temptation of desertion or disorderly conduct on the coast of Chile deserved to be rewarded. This suggested to me the most appropriate means of uniting self-interest with the preservation of as relaxed a discipline as the proper service of the King would allow. Accordingly it was decided that during the months of the corvettes' stay in Callao, apart from the daily allowance of two reales, each seaman who presented himself for work, would be given four reales a day as part of his payment, so that, having to appear at daily musters, those who were absent would not get away with their faults or vices and would also forfeit a corresponding part of their pay to the royal treasury. Monetary punishment of this kind would also be applied to those who had committed a punishable offence but would rather wipe the slate clean by working than idle away their time in confinement. The marines, *pilotínes* and petty officers were paid at the beginning of each month, for the sake of greater propriety, but the right was reserved to withhold it as necessary from those who did not behave themselves. Finally, the sick were taken to the very well-run hospital at Bella Vista[1] but put in a separate ward under the supervision of our surgeons and under the immediate care of the *Atrevida*'s barber who was to be given a small monthly gratuity on this occasion, in recognition of his singular skill and compassion.

The first tasks on board ship were to strip the corvettes of everything but ballast and guns and put into storage all the provisions and supplies, while starting on the repair of the casks, sails and rigging, the last two being done on board to minimize any distractions for the crew. No caulking appeared to be necessary as the copper sheathing was still in very good condition, and not the slightest leak had been found in the timbers near the black strakes[2] or in the daily washing between decks.

As supplies of this quantity required some departure from the usual methods of the royal treasury, the Viceroy agreed that here, as had been done in Buenos Aires, we should be given what seemed to be the necessary amount of money in cash, leaving to the pursers the supervision of the daily distribution, while the retailer would only ratify the monthly certifications, which would from now on serve as data for all debits to be charged to my account. In this way we could provide promptly and economically for all the expenses arising from our work and start to embark the necessary provisions without being subject to the terms of contracts made in earlier times, when war was imminent, which would now be considered exorbitant and unreasonable. On the other hand, as we still had on board reasonable amounts of dried vegetables, salt pork, wine and oil, all of the best quality, we had only to take on a good quantity of bread, more dried vegetables and salt pork to have at least a year and a half's supply of provisions. As we wanted to clear the bread-rooms and fearing that the bread left over from Buenos Aires would rot in the humidity if taken ashore, we decided to take it over to the other corvette for a fortnight so that we could inspect the store-room in which it had been kept.

[1] The hospital at Bella Vista was built by the viceroy, Manuel de Amat, with funds confiscated from the expelled Jesuits; it was intended for naval personnel and merchant seamen. It ceased to function during the wars of independence between 1821 and 1824: S. Clavijo y Clavijo, *La trayectoria hospitalaria de la Armada Española,* Madrid, 1944, pp. 260-2.

[2] The strakes at and above the waterline of both corvettes were painted black.

Señora de Lima en 1793

Plate 26. Woman of Lima, by Felipe Bauzá. Museo de América, Madrid.

Meanwhile, we had started our scientific work and were carrying it out with all the energy required by our eagerness to ensure that it met His Majesty's wishes for the successful operation of our national shipping. The chronometers were checked again and their rates adjusted according to more reliable data, particularly for the periods elapsed between Buenos Aires and Chiloé, with the resulting alteration in the position of Puerto Deseado. The meteorological logs, tables of variation and tides, the astronomical log and the tables of daily variations in the rates of the chronometers were all put in order again by the officers responsible for them. Observations of latitude and longitude, serving as a basis for the hydrographic work, were checked and summarized. Quintano and Vernacci started work on the sailing directions, and Don Dionisio Galiano, in a well-argued preface which he provided to the astronomical work, described in great detail all the scientific and astronomical instruments that we had used and the methods by which, until then, we had applied to hydrography the most fundamental principles of astronomy. The naturalists were not idle either: indeed, they found everywhere new objects of interest in all branches of natural history and took advantage of the consistent climate to roam freely through the pleasant Rimac valley.[1] Don Antonio Pineda, who was in charge of all branches of natural history except botany, was not in fact able to leave La Magdalena but in the middle of June Neé and Haenke began making long excursions of great significance, the former making his way towards the Quebradas de Canta[2] and the latter going through Tarma[3] to the other side of the *cordillera* to Guanuco,[4] where the river, flowing eastward, meets Río de la Magdalena[5] and becomes navigable.

Don Luis Neé was given only thirty days absence, while Don Tadeo Haenke's absence was increased to fifty days. They were accompanied by Señores Tafalla and Pulgar,[6] botanists who were paid by the government from Lima, with two dragoons[7] as guides who had some fluency in the Indian dialect. Obviously, these measures needed at all times either the authority or the influence of His Excellency the Viceroy. Both of these were always forthcoming whenever we asked and indeed His Excellency granted Don Cayetano Valdés, who was in charge of examining the

---

[1] The river on which Lima stands. It rises at an altitude of 4,750 metres and flows for 80 miles before reaching the sea in the northern part of Bahía de Callao.

[2] A rich agricultural valley north of Lima, watered by Río Chillón.

[3] The chief town of the homonymous province east of Lima, enclosing vast stretches of Andean mountain country and drainage lands of the Amazon basin: A. de Alcedo, *Diccionario geográfico-histórico de las Indias occidentales o América,* Madrid, 1786-9.

[4] Properly Huánuco, a province NE of Lima, including a great extent of Andean and Amazonian territory.

[5] Linked to the Amazon by the somewhat dangerous Canal del Casiquiari, which crosses Colombia and flows into the Caribbean.

[6] Juan Tafalla, (d. 1812) was a Spanish botanist who arrived in Peru in 1778 as part of the Ruíz-Pavón expedition, accompanied by Fracisco Pulgar as the official artist. He published studies in *Mercurio peruano,* including one on cocoa and contributed to the collection of the Real Jardín Botánico of Lima under Fr Francisco Antonio González Luna, from its foundation in 1791. In 1796 he was appointed to the first Chair of Botany at the University of Lima: Mendiburu, *Diccionario,* x, p. 276 and J. B. Lastes, *Historia de la medicina peruana,* Lima, 1951, ii, pp. 266-73.

[7] The Regimiento de Dragones de Lima was formed in 1773 and stationed in the capital. On military conditions in the Viceroyalty see L. G. Campbell, Jr, 'The Military Reform in the Viceroyalty of Peru, 1762-1800', University of Florida Ph. D. thesis, 1970.

Archivo de Temporalidades,[1] all the information necessary for our work to cover as wide a field as possible, and we saw that the proposed plan[2] could be carried out in as many branches of knowledge as the length of our stay in these parts would allow.

The complement of both corvettes, very much depleted among both the marines and seamen since we had entered the Pacific Ocean, required me to give very serious attention to finding replacements, since there was reason to fear that the forthcoming visits to the ports of Guayaquil, Panamá and Acapulco would give rise to more desertions and unruliness. For the marines and marine gunners I immediately asked the Viceroy to allow me to make up our deficiencies with any volunteers from the permanent regiment of Lima[3] or the artillery regiment who were willing to transfer to our flag. I required them to be robust, of good conduct, and men of the veterans' regiments of Soria and Extremadura who would have completed their service in these regiments at the time they returned to Spain. I left it to Don Cayetano Valdés to decide, after a most careful examination, which among the many who volunteered were or were not acceptable. As for the seamen, knowing how unreliable men of this unfortunate class are and finding it possible to use the reduced numbers we still had to carry out the daily tasks and at the same time save public money, I left it until the last month of our stay in Callao to make up the complement, either from the numerous men who had come from Europe this year aboard merchant vessels or from those who were to arrive in the royal navy frigate *Liebre*.[4] According to the latest mails she was bound for the South Sea, particularly for the port of Callao, having left Cádiz about the middle of May.

The policing of the port, very much neglected since naval ships had stopped frequenting it, was something else we had to take care of while there was no other vessel to take on the responsibility. Accordingly, the captains of all merchant vessels were given precise instructions about anchors, mooring berths and the number of seamen on board, particularly at night, to ensure compliance with regulations. The officer of the guard was to patrol exactly as prescribed by the said instructions. He was also to examine the state of departing ships, so that they should not carelessly overlook those precautions that are indispensable in even the most limited commercial operation and lastly, he was to use moderation, integrity and prudence in cutting short all those daily complaints which the captains as well as the seamen of merchant vessels tend to bring up out of their confused and self-contradictory notions. Although all these precautions had just been enforced, they did not prevent a merchant vessel[5] catching fire on the night of 7 June because of the carelessness of a petty officer and a few seamen who had gone ashore without leave. This put the corvettes at risk, particularly the *Descubierta*, whose buoy line to her northern anchor remained for some time caught around the rudder of the ship. Guardiamarina Jacobo Murphy and the two

---

[1] See p. 56, n. 1 above.

[2] Malaspina's original plan for the voyage: see pp. 311–15 below.

[3] The Regimiento de Infantería de Lima was formed in 1787 and staffed by veterans. The permanent garrison also included a veteran artillery company.

[4] The *Liebre* was being sent to Callao as part of Spain's defensive measures following the Nootka Sound incident.

[5] The *Nuestra Señora del Socorro*.

boatswains of the corvettes, with both launches, showed remarkable energy and intelligence on this occasion, as they managed to tow the burning ship to a beach[1] and run her aground there, after failing in their attempts to put out the fire or to save the spars or some part of her equipment. On the merchant vessels recently arrived from Europe, the Intendente de Marina on Isla de Leon[2] had sent us various items which had not been ready at the time of our departure. Among these we had the pleasure of finding an equatorial instrument by Dollond, a number of natural history books collected in Paris and two boxes of glassware containing mostly the instruments for the air-quality experiment. In them were two of great importance, the eudiometers made by Volta and Fontana. However we could not but be surprised and dismayed, considering our intentions on this voyage, by the absence not only of the excellent collection of physics instruments collected in Paris at the orders of His Excellency Señor Conde de Fernán-Núñez[3] which had already been embarked at Rouen when we left Cádiz, but also of the electrical conductors and some other equipment due to be sent at the first opportunity from the dockyard at La Carraca to Lima. Meanwhile, we took possession of everything that had arrived and Don Antonio Pineda was soon examining the quality of the air to be inhaled from the atmosphere in various places around Lima.

These were in essence the activities that occupied both crews during the month of June. At the same time various orders from His Excellency the Viceroy concerning purely naval matters were carried out. Don Arcadio Pineda was entrusted with the legal examination of the circumstances of the loss of the vessel *San Pablo* which had been transferred to His Majesty's service by Don Juan Miguel de Castañeda[4] in order to identify some English ships that had been sighted off the coast of Arequipa,[5] and which had sunk near Valdivia under the command of a breveted alférez de fragata.[6] Don Francisco Javier Viana inspected the stock of rigging, which had either been received beforehand or was now arriving at the royal warehouses under a contract with some Chilean manufacturers. That officer and Don Juan Vernacci examined the condition of the merchant vessel *Mexicana* of the Cádiz trade which was soon to sail for Europe despite the severe winter weather and Don Mañuel Novales instituted summary proceedings over the accidental or malicious burning of the merchant vessel already mentioned.

---

[1] The location of this beach, stated to be in an inlet, is unclear.

[2] Don Joaquín de Rubalcava.

[3] The Spanish ambassador in Paris.

[4] In 1788 Juan Miguel de Castañeda y Amuzquivar offered his ship *San Pablo* to the government when an American whaler was reported in the Chiloé area. Viceroy Teodro de Croix accepted and the ship was fitted with 34 guns. She put to sea again in June 1789, but went aground on the rocky outcrops of San Pedro, near Valdivia, early in 1790: J. Ortiz Sotelo and A. Castañeda Martos, *Diccionario biográfico-marítimo peruano*, Lima, 1993, pp. 155-6 and T. de Croix, 'Relación que hace el excmo Señor Don Teodoro de Croix, Virrey de estos Reynos del Perú y Chile, a su suceso el excmo Señor Fr. Don Francisco Gil de Lemos desde 4 de abril de 1784 hasta 25 de marzo de 1790', *Memorias de los virreyes que han gobernado el Perú durante el tiempo del colonizaje español*, Lima, 1859, pp. 254-9, 263-5.

[5] The region to the north of Arica in present day Peru.

[6] At the time of the shipwreck, the *San Pedro* was under the command of Alférez de Fragata Antonio Cásulo, of the Callao naval station. He was investigated and apparently cleared of charges, continuing with his rank and duties until his death at an uncertain date, soon after 1792.

At this time almost all of those who had deserted at Chiloé had arrived in Lima, sent on by the various Governors. Among those from Chile were a seaman who had remained in Coquimbo, together with another and a corporal from the *Descubierta* who had deserted in Valparaíso. They were treated much less severely than they had a right to expect. Those who failed to report for muster on board were only obliged to serve their time wearing fetters or a chain as punishment for their arrest, which did not really preclude all means of escaping again and only the corporal was obliged to serve on the *Atrevida* as a private until his conduct demonstrated that his record could be cleared of so disgraceful a transgression.

In our system of organising the crews, under which we could not for any reason accept men whose safekeeping and conduct would be the constant occupation of many others, these signs of leniency were necessary rather than simply expedient and they had a considerable effect in promoting the seamen's loyalty to the service. For, apart from the fact that the recaptured seamen were good men and generally liked by their companions, there was no doubt in their minds that compassion was an inseparable part of the authority entrusted to us. Thus, while we sought them tenaciously wherever they went, this was not done with any idea of being hard to them, as we were almost obliged to be by our mission, or of subjecting them to harsh military discipline. At the same time a court martial was convened on board the *Atrevida* to examine the crime committed in Valparaíso by a marine from the garrison there who had badly wounded a seaman who later died on board, more because he had concealed his wound for a long time than because of its original severity.

Don Secundino Salamanca, in a well-argued defence, reminded the judges of the circumstances of the crime, such as the revulsion at enduring insults to his rank and service on the part of a marine who had voluntarily enlisted for the corvettes, when he might have calmly enjoyed a lengthy period of peace and lastly he spoke of the gallantry with which this man, alone and with inferior weapons, had defended himself against two seamen. Considering these circumstances the court sentenced him to six months in prison, counting what he had already served and an added sentence of five years of military service. In this respect, however, I asked His Excellency the Viceroy to allow this to be served in the frigate *Liebre*, so that the crews should not have among them, almost unpunished, a man who had killed another and so that he should not remain among the marines and seamen as a source of rancour which might in due time produce fatal results.

Apart from these examples as a warning to all, thanks to the daily work, the incentive of daily wages and normal military discipline, with frequent gun drills for the seamen and musket drills for the marines, as well as regular inspections of arms and clothing, guard duty, bugle calls for retreat and reveille, etc., we were very pleased to find that good order, mutual respect existed between the different ranks and among the individual members of them. Even their general health was becoming stronger in the very place where we had most feared its complete breakdown.

Month of July
On the 5th of July the frigate *Liebre*, under the command of Capitán de Navío Tomás

Geraldino,[1] anchored in Callao after a successful and well-conducted passage of four months and a few days from Cádiz. The administration of the port was immediately placed in the charge of this officer and the commanding officer's pendant was hoisted on board the *Descubierta* according to regulations. We had to work ceaselessly to put everything in order, and by means of constant attention and diligence, considerable progress was made in the hydrographic work, particularly as Alférez de Fragata and Senior Piloto José de Moraleda had arrived from Chiloé and had undertaken to make a fair copy of various port plans. When he had presented his chart and plans of the island of Chiloé to His Excellency the Viceroy, we were able to revise our own charts, tying in the outer parts and the vicinity of San Carlos and Chacao[2] to our astronomical and hydrographic work. To the practical knowledge of the whole coast of this skilled *piloto*, whom we consulted constantly, we added a detailed study of the journals of Piloto Machado, conferring with him frequently about his passages along the Istmo de Ofqui[3] as far as Cabo Corso.[4] Finally, a considerable contribution to our accuracy in coastal detail came from the various pilots with local knowledge of the South Sea, whom we cited several times, with sailing directions that we either managed to find or were given to us by them and some extremely useful information which Don Esteban Ventura Mestre, master of the merchant vessel *Galga* from the Cádiz trade, had collected in his long and informative time in these waters and which he presented to us with the greatest zeal and devotion to the service.

By the end of this month much of our refitting for the coming season's work was also completed. The corvettes were fully rigged, the water casks and sails very carefully overhauled, the equipment replaced or repaired, water and provisions replenished, and the necessary number of marines taken on, even after excluding one or two of the veterans, whether ill or fit. The right season for continuing our work was now very close and everything indicated that it was time to take up once more the duties that had to keep us for so long away from our beloved homeland. Neé and Haenke had rejoined us after excursions that had been useful as well as arduous. As Don Antonio Pineda, who was attempting a new survey of the *cordillera* in this area, was due back in a few days, we had to set a limit to our determination that there should be no failure in accuracy, quantity, order and neatness in every part of the current consignment[5] and turn our thoughts towards carrying out any essential duties that still required our presence in Callao.

---

[1] Tomás Geraldino was born in Jeréz de la Frontera in 1754 and served as a *guardiamarina* in Cádiz from 1770 and was promoted *capitán de navío* in 1780. He was thus considerably senior to Malaspina. In 1790 he was sent to Chile in command of the *Liebre* with a cargo of mercury, remaiming on the Pacific coast until 1794, on missions which took him as far north as Panamá. On return to Spain he was appointed in command of the *San Nicolás* but was killed on board her during the battle of Cape St Vincent on 14 February, 1797, when boarded by the *Captain*, Commodore Horatio Nelson: F. de Paula Pavia, *Galería biográfica de los generales de marina, jefes y personajes notables que figuraron en la misma corporación desde 1700 a 1868*, Madrid, 1873-4, ii, 44-6 and Christopher Lloyd, *St Vincent and Camperdown*, London, 1963, pp. 78-9.

[2] A town on the north coast of Isla Chiloé about 13 miles ENE of Ancud (Puerto San Carlos).

[3] A low-lying isthmus (46°45′S, 74°10′W) separating Golfo San Esteban from Laguna San Rafael.

[4] For Cabo Corso see p. 123, n. 4.

[5] I.e. the consignment being prepared for forwarding to Spain, details of which are given below (pp. 229-31).

20 August

The 20th of August, therefore, was the date fixed for beginning our hydrographic tasks and the return on board of most of the officers. Don José Bustamante was to take charge of our quarters at La Magdalena, since he was greatly in need of a rest and proper convalescence after two months of almost continual fevers in this climate, which was obviously disastrous for him. I was to establish myself on board to hasten the final tasks and prepare for departure, making sure that each person on board the corvettes would have, quite apart from the necessary rest, some comfort and seclusion wherever his principal work should require him to be, whether on board or at La Magdalena. All those in charge of the natural history studies, in particular, were to remain there, as were Don Felipe Bauzá, Piloto Maqueda and the two *pilotines* to continue working on the charts and Tenientes de Navío Galiano and Concha, who were attempting to put in order, before our departure, the catalogue of stars observed in Valparaíso. Don Juan Vernacci was to be responsible for obtaining fresh rates for the chronometers and setting up the observatory, very conveniently, in one of the towers of Ciudadela del Callao,[1] where we would join him as we completed a series of triangles, extended from a base towards Islas de Pachacamac.[2]

The constant mildness of the weather, which is the principal virtue of this pleasant climate, with the air, at the end of winter, now becoming gradually less hazy, allowed these measures to be carried out almost with the timing that had been planned. By the beginning of September, the hydrographic operations were finished, the port had been sounded in detail, fixing the positions of any shoals by theodolite, the rating of the chronometers was well in hand and Piloto Maqueda had made an excursion to Ancón[3] and Los Pescadores[4] to tie in those parts which we had been unable to do with our bearings, to obtain soundings there and take a few observations for latitude. Only Don Manuel Novales, who had taken the revenue felucca[5] with Guardiamarina Ali Ponzoni to survey Las Hormigas[6] and fix their position with good observations of latitude and longitude, had the misfortune of being unable to complete his task, as he was beset by an extremely strong southerly wind which raised a very heavy sea, which prevented him from anchoring among those rocky islets and exposed him to considerable risk and discomfort in regaining the port. At the same time we made up our numbers with seamen, either from the frigate *Liebre* or from other ships, and accordingly, by 15 September, the date we had set for completing the rating of the chronometers, we could consider all was ready for our departure.

Two or three days earlier we had taken our leave of His Excellency the Viceroy and of the principal residents of Lima, who had frequently given proof of their ardent desire to be of service to the Crown. Don José Bustamante and I, having completed the current consignment, had accompanied it, with an officer, to the Viceroy, hoping

---

[1] Fortaleza del Real Felipe, built between 1747 and 1773: see J. M. Zapatero, *El Real Felipe de Callao: primer castillo del mar de sur,* Madrid, 1983.

[2] A group of islets and rocks 21 miles SE of Callao.

[3] Possibly Punta Ancón, situated about 16 miles north of Callao.

[4] A group of small islands 18 miles NNW of Callao.

[5] A small double-ended vessel carrying lanteen sails, typical of the Mediterranean.

[6] A group of small islands 34 miles WNW of Callao.

Plate 27. Llama. Anon. Museo Naval, Madrid

that he would see fit to send it, through the Ministro de Marina, to the feet of the throne, by whatever means he thought most convenient or safe. The hydrographic part of it consisted of six Mercator charts, the first two covering the east coast of Patagonia from Río de la Plata to Estrecho de Magallanes, the third, all of Tierra del Fuego and the adjacent coasts to the east and west up to 50° of latitude, and the other three covering the Pacific coast from that latitude as far as Lima, including Isla Juan Fernández and Isla San Ambrosio. The plans of the ports of Deseado, Egmont, San Carlos de Chiloé, Talcahuano, Valparaíso, Coquimbo and Arica, drawn up by us, that of Valdivia by Piloto Moraleda and those of Copiapó by Ingeniero Pedro Rico,[1] formed an extremely important collection for navigation, but for lack of time it could not include the many views of the coasts we had visited which had been so skillully taken by Don Felipe Bauzá. The collection of botanical or other natural history illustrations and some scenes, the work of the draughtsman Guio and others by Don José del Pozo, formed a body of work both useful and pleasing to the eye and finally, all the manuscript part was bound in five small volumes. The first contained the narrative of the voyage, the sailing directions and a detailed explanation of the information used in the construction of the charts. The second contained a physical and political description of the coasts surveyed. The third contained the astronomical journal, the variation of the compass, the tides, the rating and irregularities of the chronometers and a catalogue of the observations taken at sea to fix the position of the coasts. The fourth and fifth contained only the descriptions by Pineda and Neé, as it had not been possible to include those of Don Tadeo Haenke because of lack of time to copy them. His Excellency was kind enough to forward all of this by the first

[1] See p.182, n. 2 above.

Plate 28. Swordfish, by José Cardero. Museo Naval, Madrid

mailboat, leaving nineteen boxes for ships sailing directly for Europe next January. These contained the plant collections, dried and stuffed animal specimens, minerals, paintings on canvas by Don José del Pozo and our draft manuscripts, which we had prepared and were sending with all possible clarity, order and safety.

Two things had, however, conspired to cloud considerably the natural pleasure arising from the contemplation of our first year's work and from His Majesty's approval of our work at Río de la Plata, of which His Excellency Bailío Don Antonio Valdés, Ministro de Marina, had recently informed us in a letter written in March. The first of these was the resignation of the artist, Don José del Pozo, from the post in which he had agreed to serve,[1] as he was unable to endure either the discipline which alone is the principle and foundation of good order, or the tenacity and perseverance in his work which were required of him as an example to others and for the harmonious achievement of our ends. The second was the unexpected failure of chronometer number 13, even though it had not been very accurate. We found that Don Pedro Pimentel, a clockmaker of Lima,[2] was able to repair it, although at a considerable sacrifice to our budget. We no longer had any confidence in it, even as a third comparison chronometer, despite which we used it as such in the subsequent passage to Guayaquil.

During the last days of our stay in Callao there were among both crews a few instances of disorderly conduct , seemingly inevitable among seamen when, on leaving a country that they see as full of delights and attractions to face the hardships of the sea, they try to bury in fleeting pleasures the thought of the perils that await them. Kindness was most often used to restrain them and rarely severity. The officers, now almost all back on board, took turns in this worthwhile endeavour and finally we succeeded in gathering on board almost the full ship's company of this corvette by

[1] Pozo remained in Lima and opened a painting school for ladies and gentlemen in 1792. Works by him adorn various churches and religious houses in the city. J. Bernales Ballesteros, 'La pintura en Lima durante el virreinato', *Pintura en el virreinato del Perú*, Lima, 1989, pp. 66–70.

[2] Tomás Ruiz Pimentel, watchmaker to the viceregal court in Lima. *Mercurio peruano,* 23rd January, 1791, pp. 55–6.

the day before our departure, missing only one marine and another who had joined from the local permanent regiment. The *Atrevida* had no more absentees than we had, as I was informed by her commanding officer through Don Francisco Javier Viana and thus, with nothing to prevent our departure by the afternoon of the 19th, we waited eagerly for dawn the next day, having hoisted some of the boats and having cut off, as far as possible, all communication with the shore.

BOOK FIVE

FROM CALLAO TO ACAPULCO

CHAPTER 1

From Callao to Guayaquil

20 September

By nightfall it appeared that the disorderly conduct of the seamen and other lower deck men on both corvettes had completely subsided. At the dawn muster only two seamen of our corvette's complement were missing. The *Atrevida,* as I was advised by her commanding officer through an officer, was still short of six seamen and their default appeared all the worse since they were the very ones who had recently been transferred from the frigate *Liebre* after having willingly consented to the transfer.

At dawn therefore the *Atrevida* was ordered to weigh her shore anchor. We did so at the same time and remained up and down on the kedge anchor ready to take advantage of the first breath of favourable wind. However, since by nine o'clock the first light airs from south had still not come up, a number of private citizens of Lima who had established strong ties of friendship either with us or with the ship's officers were kind enough to give us the opportunity to give them one last embrace. Worthy of special mention in our account are Alférez de Fragata Moraleda and Don Estevan Mestre, captain of the *Galga* who had gone to great pains to compile and pass on to me all the most useful information for the next stretch of coast northward as far as Acapulco. The time we had for our tasks was very restricted and since, despite these restrictions, we were anxious not to omit from our surveys anything which might be of interest to navigation yet further afield, we were bound to appreciate any information that would contribute so directly to that purpose.

Of all the inconveniences certain to try the most calm and even-tempered of souls, the most difficult to tolerate on this occasion was the appearance on board of some six *pulperos*[1] from Callao. These were creditors of both crews and even of some petty officers for considerable sums of money which had been spent in addition to the six reales a day which they had been given as earned in respect either of pay or of rations. I had fixed that late hour for their visit so that they could claim their debts face to face at a time when there could be no fear of further desertions and it would not be easy for the sailors to profit from their almost infinite capacity for trickery. The sight of six burly men who were almost certainly deserters from the King's ships and whose sole purpose that day was to incite the ships' companies to renewed disorders or even desertions, prompted me to provide a lesson which was now necessary if in future the ships of the royal navy were to preserve in these waters the strict discipline which

---

[1] A South American word for storekeepers or tavern keepers.

the circumstances demanded. I therefore ordered them to be put in the stocks. It was then arranged for lots to be drawn for two men to be pressed into service in our corvette and one in the *Atrevida* as far as Guayaquil. Eventually, after being severely admonished they were put ashore with the exception of one who remained on board the *Atrevida* which had also picked up some of her own seamen who were on the quay when she was setting sail.

It was ten o'clock in the morning before we were able to set sail with a steady wind. We immediately set a course, under full sail, to pass close to the NW extremity of Isla San Lorenzo which at midday was almost half a mile distant to the north.

We had completed the comparison of the chronometers against the astronomical clock on the 16th after their rates had been established and they were checked on the 19th using absolute altitudes of the Sun measured by sextant. This showed to our great surprise that their rates over the last few days had varied considerably.[1] We judged it best to adjust our calculations to the new reading closest to the old ones, knowing, however, that comparison with the *Atrevida*'s chronometers would show us whether or not this decision had been over hasty. We therefore agreed that number 72 would be compared to chronometer 61 in the other corvette. The inconvenience of not being able to communicate easily the temperature equations for number 10 by signal had led us to prefer the results from chronometer [61] even though daily comparisons in Callao had always demonstrated the marked superiority of M. Berthoud's chronometer over the others. With regard to number 13, the latest comparisons confirmed our first impression that it had benefited greatly at the hands of the skilful artificer Pimentel whom we mentioned earlier.

Before setting sail I had suggested to the commanding officer of the *Atrevida* that on the next relatively easy passage the marines and seamen could be divided into three watches. He did so, but on board the *Descubierta* it became obvious that the seamen lacked the agility for the quick handling of the rigging and in the end I thought it preferable that they should continue in two watches as far as Guayaquil.

The main reason for our anxiety to take advantage of the first light airs was that we wanted to sight Las Hormigas that same afternoon and to observe longitudes while we were within sight of them, since the storm of 23 August had prevented Don Manuel Novales from taking the astronomical observations which he had planned. So after taking a number of bearings to determine our position at noon on the new plan, we shortened sail and steered WbyN with a strong breeze from SSE on the same course that Don Manuel Novales had followed [in the revenue felucca]. We set about establishing very carefully our dead reckoning which would then always serve as a comparison with the outward and return dead reckonings plotted by the felucca if we should find ourselves short of other observations, which seemed likely in that weather. Both of these dead reckonings (when her departure and arrival points were plotted on our plan) agreed in fixing a difference of 33′32″ between the meridians of La Hormiga Grande and Torre del Callao.[2] There were considerable discrepancies in latitude but this was because the compass had not been lined up accurately with the

---

[1] Malaspina appears to have relied unduly on absolute altitudes for obtaining fresh
[2] Another reference to Fortaleza del Real Felipe, see p. 229, n. 1 above.

felucca's keel. However, since exactly reciprocal courses had been steered, there was good reason to believe that the mean of the two latitudes would be very close to the truth. This mean was 11°55′44″.

At half past three in the afternoon we managed to sight Las Hormigas from the masthead, bearing WSW, so that we luffed up immediately towards them. We were then fortunate enough to take a set of hour angles and also to get a bearing from the mizzen top to the middle part of Isla de San Lorenzo despite very hazy conditions. Finally, towards half past four, a new set of hour angles was taken and a small base measured to calculate the distance, which was no more than two miles to the main islet. At four in the afternoon we had signalled to the *Atrevida* to take soundings, but she was unable to obtain bottom; but at five o'clock we sounded ninety-four fathoms, sand and shell.

These two tiny islets, already well surveyed and sounded by Don Manuel Novales who had sailed round them and even anchored between them, could certainly present a considerable hazard to navigation because of their low elevation and the reefs that extend from them in various places. Fortunately, however, less than a mile away there is a good depth of water; moreover it is easy to avoid them by sailing west when leaving Isla de San Lorenzo. Approaching Lima, a course should be followed close to the mainland, since there is no reason at all to sheer away from the coast. Our observations and dead reckoning placed the main islet in latitude 11°54′40″ and longitude 34′00″ west of Callao. At half past five in the afternoon, having completed our soundings, we bore away, considering that we had completed this important part of our work.

As for the following morning's tasks it seemed unwise to begin them inshore of Los Farallones de Huaura.[1] The coast had already been plotted from Punta Ancón[2] as far as Chancai.[3] Further north, at Punta Salinas,[4] the coast is fronted by these *farallones*. There would not be the least risk of error in joining [Punta Salinas] to Punta de Chancai[5] and coasting to the east along all the islets from El Marquesí or Pelado[6] would give us their situation in a reciprocal direction with much more speed and accuracy. We therefore steered some nine leagues WNW by compass and lay-to for the remaining hours of the night head to shore. There was a very strong wind from SE, the sky and horizon were thick with scud and haze and the seas were high. Dawn broke with less wind and the skies fairly cloudy. We immediately sailed close hauled to the east and shortly afterwards sighted, although at a great distance, the outermost *farallón* which is called El Marquesí or Pelado.

21 September

The wind then began to strengthen somewhat so that by noon, the weather having

[1] Islas de Huauru, a chain of five needle-shaped rocks which stretch fourteen miles westward from the coast, about 55 miles NW of Callao. The two outermost mark the approach to Callao for navigators coming from the north.

[2] Situated about 16 miles north of Callao

[3] Chancay, on the shores of Bahía Chancay, about 30 miles north of Callao.

[4] Promontorio Salinas, not to be confused with Punta Salinas, the SW extremity of Isla Puná (p. 246, n. 2 below).

[5] Punta Chancay, the southern entrance point of Bahía Chancay.

[6] Westernmost islet of Islas de Huaura. Since the middle of the last century only the name Pelado has applied: A. García y García, *Derrotero de la costa del Perú*, Lima, 1870, p. 14.

cleared considerably, we were able not only to obtain our latitude from the meridian altitude of the Sun but also to close the mainland having made a careful examination of these *farallones*, the number and extent of which seem to be less than appears on the charts of these coasts. Our latitude was 11°23'30" and longitude 37' west of Callao.

As we passed along the length of the *farallones* we began to take soundings. At half past one Marzoque[1] lay to the south of us, at the same time the *Atrevida* signalled a sounding of forty-three fathoms. We ourselves then obtained a depth of thirty-six fathoms, ooze, and by four o'clock we were in twenty fathoms. By now we had been sailing for some time at a distance of one league from the coast and as we were subsequently to do until nightfall, we had taken advantage of the very clear and favourable weather to fix in detail all the nearby coves and creeks of Huaura,[2] Guacho[3] and Zupe[4] and also Isla de Don Martin.[5] We had also sighted two villages lying in beautifully wooded valleys and in another direction the celebrated ruins of Pativilca.[6] Two sets of hours angles taken at three and five o'clock and at the same time dead reckoning calculated with great care gave promise that geodesy and astronomy would contribute gratifyingly to the accuracy of our work.

At the end of the afternoon we were close to Zupe in a depth of thirty-seven fathoms, greenish ooze. We then continued for a while under light canvas to give ourselves sea room. Finally at ten o'clock we lay head to sea under topsails, swinging head to shore at one o'clock and at two o'clock sounding, eighty-five fathoms.

Even if the warnings of the Lima pilots had not led us to suspect a south-flowing current of some strength, the set which we had experienced during the night. The dead reckoning plotted then and the previous day would have left us in no doubt.[7] But navigating with great caution during the night, by dawn we were only two or three leagues from the points we had sighted and taken bearings of the previous afternoon.

22 September

There had been a fresh sea breeze throughout the night but at daybreak it died down considerably and the weather was very hazy, as on the previous morning. On the coast we could see Granadel[8] and Xagueí[9] and further to the north Guarney, Culebra and Margón,[10] all unprotected anchorages and for the most part lacking any sort of

[1] Isla Mazorca, the highest of Islas de Huaura.

[2] Not identified, but possibly Bahía Salinas, 5 miles NNE of Promontorio Salinas.

[3] Bahía Huacho, site of present day Puerto Huacho, 10 miles NNE of Promontorio Salinas.

[4] Bahía Supe, site of present day Puerto Supe, 30 miles north of Promontorio Salinas.

[5] Islote Don Martín, 16 miles north of Promontorio Salinas.

[6] Three fortifications properly called Paramonga, built by the ancient Moche people to defend the southernmost zone of their settlement along the Peruvian coast. Some reconstruction took place after the arrival of Inca conquerors in the region in the late fifteenth century.

[7] Possibly Malaspina was under the influence of a counter current close inshore of the north-flowing Peru Current.

[8] Bahía Gramadal, 57 miles NNW of Promontorio Salinas, so called by seventeenth century sailors because of the abundant couch-grass: D. Howse and N. J. Thrower, eds, *A Buccaneer's Atlas: Basil Ringrose's South Sea Waggoner*, Berkeley, 1992, p. 185.

[9] Punta Jagüey, the southern entrance point of Bahía Gramadal.

[10] Bahía Huarmey, 70 miles NNW, Caleta Culebras, 80 miles NNW and possibly Bahía Casma, 111 miles NNW, respectively, of Promontorio Salinas. The latter bay is close to Cerro Mongón, the highest point on the Peruvian coast (1144m high, 3½ miles inland).

amenity. Bases were measured from sunrise and, as the wind later freshened, we took advantage of this under full sail. By noon we had made substantial progress and found ourselves in latitude 10°10', longitude 1°6'50" west of Callao. The error in longitude was only 10', but the estimated latitude differed by 19' from that given by the meridian altitude of the Sun.[1]

The early morning soundings showed between sixty and fifty fathoms, ooze. They were then taken over by the *Atrevida* which had agreed to repeat them every hour, with a warning signal only if the depth fell below thirty fathoms.

The afternoon was clear and the wind favourable and strong, as usual. Accordingly we were able to continue measuring our bases under a press of sail, which by six in the evening brought us within sight of Guambacho,[2] a place occasionally visited by a ship looking for salt to take to Trujillo.[3] Various sets of hour angles and some azimuths either gave us longitude by chronometers or variation, the latter remaining about 10° NW and the former tending slightly west of the differences worked out from dead reckoning positions.

The entrances to [Bahía] Ferrol[4] formed the extremity of the coast which we could see clearly to the north at nightfall when we had to bring our work to a close. Immediately afterwards we shortened sail and at ten o'clock that night, having sailed a further four leagues, we lay-to for the remaining hours heading first seaward and then to shore, the depth increasing from forty-five to eighty-two fathoms, fine brown sand and then decreasing to fifty-seven fathoms; at which depth we found ourselves at dawn four leagues from the coast.

23 September

The first impression given by the entrances to [Bahía] Ferrol is that of a crowded group of hummocks resembling islands which are later seen to be for the most part linked to one another by quicksands or spits. Beyond these hummocks a large bight [Bahía Ferrol], bordered by sandy shores, provides very good shelter from the invariably lively seas from south. The island[5] and entrance to Río de Santa lie to the north, no more than three or four leagues distant. The whole of this coast, like that between there and [Islas de] Huaura, consists of hummocks interspersed with sandy stretches. The few islets or prominent points are generally covered with guano. Not far away and rising to a great height can be seen the *cordillera*, usually topped by clouds, snow-clad on some of the highest peaks and seeming to be covered with sand.

The immediate area of [Isla de] Santa was surveyed to our complete satisfaction and, favoured by the great clarity of the horizon and by favourable winds as on the previous days, we continued with our bases, being obliged to alter course to the west only by a shoal of twelve fathoms, sand, which we discovered at ten o'clock in the

[1] Possibly Malaspina was still under the influence of a south-flowing counter current.
[2] An extensive bay, 5 miles SE of Bahía Ferrol, known since the middle of the nineteenth century as Bahía de Samanco.
[3] A city founded by Pizarro in 1535 and situated about 4 miles inland.
[4] Paso del Medio, Paso del Norte and Paso del Ferrol are the channels of approach to the broad Bahía Ferrol, on the north side of which the important fishing port of Chimbote stands today. In the late eighteenth century it was sparsely populated by Indians from Huanchaco who lived by fishing.
[5] Islas Santa, two rocky islands separated by a very narrow fissure.

morning two miles north of [Isla de] Santa. Afterwards Morro de Cao[1] and Los Farallones de Corcovado[2] and La Viuda[3] were sighted. The depths gradually increased to
eighteen fathoms, fine sand. At noon we were at latitude 8°52' and longitude 1°40',
our dead reckoning warning of a considerable current towards the NW. Since the
short distance at which we were sailing off the coast made it difficult in the morning
and even at noon to get a horizon clear enough for taking sights, the method of
back-observations began to be preferable both for observing hour angles and meridian altitudes. Sextants, and even more a Wright's quintant[4] belonging to Don Juan
Vernacci were on this occasion most useful since the Sun, being close to the zenith,
very soon reached the angle of elevation subtended by the arc of the quintant. Soon
after midday the sight of a snow-capped peak not far from the sea[5] made Don Dionisio Galiano want to measure it accurately. By following M. Borda's[6] method of tying
the base and the angles by compass to the angles measured with the sextant between
the Sun and the object, he deduced simultaneously the variation of 8°50' and the
elevation of the peak as 2273½ *toesas* above sea level.

The afternoon was spent examining the coast between Guñape and Morro de
Carretas.[7] We observed hour angles and azimuths and clearly sighted the open countryside of Trujillo, which we planned to be close to at dawn the following day so as to
make a fairly careful survey of Puerto de Guanchaco,[8] which is quite busy because of
its proximity to the town. On these coasts they give the name of a port to any spot
that is sheltered to some degree from the heavy southern swell and where, near some
river or settlement, the basic necessities for an easy subsistence can be found or some
kind of diversion, however frivolous, is to be had. The traffic from Paita[9] to Lima,
previously by the coast and later overland, has led to the establishment of a number of
small settlements along these coasts in places where nature seemed to withhold even
the meanest of her gifts. However, Trujillo and Lambeyeque[10] are in a different class

[1] Punta Chao, 15 miles NNW of Punta Santa.
[2] Islote Corcovado, rugged with prominent cliffs on its western side, 5 miles NW of Punta Santa.
[3] Islote Viuda, 7 miles NNW of Punta Santa
[4] A reflecting instrument similar to a quadrant or sextant, with an arc subtending a fifth of a circle or
72°, hence its name. Because of its reflecting property it could measure angles up to 144° and was thus
ideal for taking lunar distances. This particular instrument was possibly the modified version of the
Hadley quadrant designed by George Wright in about 1780, in which the index mirror could be slewed
through 45° for making a back observation. For a full description of this instrument see Charles H.
Cotter, *A History of the Navigator's Sextant*, Glasgow, 1983, pp. 136-7.
[5] Probably Cerro Campana, see p. 241, n. 1 below.
[6] Jean-Charles Borda, 1733-99, inventor of a method named after him for calculating the effects of
temperature variation on a pendulum. Author of *Description et usage du cercle à réflexion* (Paris, 1778).
[7] Morro Guañape and Morro Carretas, headlands between Punta Santa and Caleta Huanchaco (see
below).
[8] Caleta Huanchaco, formerly the port for Trujillo, now a holiday resort.
[9] A pre-hispanic settlement of ancient origin, which became important in colonial times as an indispensable staging-post for ships from Panamá bound for Callao. Anson attacked and destroyed it in 1741.
After the Treaty of San Lorenzo (1790), which ended the Nootka Sound incident, it became a port of
call for the growing whale fishery. For the port see J. Schlupmann, 'Commerce et navigation dans
l'Amérique espagnole coloniale: le port de Paita et le Pacifique au XVIIIe siècle', *Bulletin de l'Institut
Français d'Études Andines*, xxii, pt 2 (1993), pp. 521-49. On whaling see W. L. Lofstrom, *Paita: Outpost of
Empire*, Mystic, 1996.
[10] A town situated a few miles inland, about 100 miles NW of Trujillo.

since their location combines abundantly fertile valleys with a pleasant climate, so that they can be regarded as even more favoured than Lima by the generosity of nature.

By nightfall we were no more than four leagues from La Campana[1] and consequently feared that, were we to lie hove-to throughout the twelve hours of night the current might set us to leeward of Guanchaco. With this in mind we bore away SSW close-hauled under topsails with a moderate breeze from SE and then from one o'clock until five lay hove-to, head to shore, in a depth of 40 to fifty fathoms, hard mud.

24 September

In the morning the weather was extremely hazy so that even at three leagues it was not easy to make out clearly the land we had seen the previous afternoon. The wind was no more than a breeze so that until eight in the morning we were unable to make out the church of Guanchaco and the nearby anchorage from which we must have been only some four miles distant. Later we bore up under full sail and began measuring bases, obtaining soundings of twenty-three to nineteen, sixteen and twenty-five fathoms, ooze.

Their Excellencies Juan and Ulloa had stopped in Trujillo during their journey from Tumbes[2] to Lima and observed its latitude with an astronomical quadrant, as they did in many other places in which they had stopped, however briefly. We regarded these observations as most useful for the greater accuracy of the work we were doing and so sought to compare them with our own. This time it was difficult for us to reconcile our own conclusions with the position of Trujillo relative to Guanchaco determined during that journey, since to the south of that anchorage we could see no trace of any large town, while in the other direction we could see close by a cluster of houses scattered amongst groves of trees.

Before long we saw another phenomenon as strange as it was entertaining. An infinite number of grebes[3] were feeding on the water, spreading SW out to sea for more than two leagues.[4] Consequently we had to cut through these files or lines of birds with the corvettes more than once. But, contrary to our expectations, this strange apparition of two such large objects did not frighten them. They waited, perfectly calm, until the last moment and then dived beneath the waves or flew close by with strange cries and no sooner had the stern of the corvette passed by than they resumed

[1] Cerro Campana, shaped like a giant bell, 6 miles NNE of Caleta Huanchaco and a prominent landmark used by Malaspina to identify the cove.

[2] A town on the SE side of Golfo de Guayaquil.

[3] *Zaramagullones* in the original MS is the Spanish for grebes, which is a bird of inland waters. It is also a name given in South America to the Olivaceous Cormorant (*Phalacrocorax olivaceus*) (Murphy, *Oceanic Birds*, p. 909) but in this instance Malaspina is probably referring to the Guanay Cormorant (*P. bougainvillii*) which ranges farther out to sea than the Olivaceous Cormorant.

[4] These and other species of marine birds helped to form the enormous guano deposits, which generated huge trade and enormous fortunes in the nineteenth century. See S. Hunt, 'Growth and Guano in Nineteenth-century Peru', in R. Corés Conde and S. Hunt, eds, *The Latin American Economies: Growth and the Export Sector, 1880–1930*, New York, 1985.

their former positions quite unconcerned. As to their number, which we could without hesitation call infinite, our astonishment was still greater when we approached Islote de Malabrigo[1] at noon and found it to be entirely covered with birds of the same species, so that although its whole surface was white with guano they made it look completely black.

Our latitude at noon was 7°54' and our longitude 2°17', depth twenty-five fathoms. During the afternoon this decreased to twenty although we were no more than two miles from the coast. We finished the afternoon's work off Pacasmayo[2] and after lying-to on alternate tacks during the night, by dawn [25th] we were off Chérrepe,[3] whose church we could see clearly at seven o'clock in the morning. From here the coast makes a distinct turn towards the west and is so low and shelving that today's work was for the most part difficult and even rather dangerous. We were more than two leagues from the coast in depths of thirteen and fifteen fathoms, sand, where we could barely make out the occasional dune and later could not even manage that because of the extremely featureless character of the coast which is liable to flooding and was all covered by a thickish haze. All sailing directions place Isla Baja de Lobos[4] almost on the same parallel as Chérrepe and some ten leagues offshore, but their description of this stretch of coast was so vague that we could only interpret from it the need to be wary of the land. So we paid no attention to the strong suspicions aroused in both corvettes at seeing it a long way off in the early afternoon.[5]

Since the weather warned us of an impending calm and the haze did not allow us to see any object at a distance and with the evening approaching, we decided to bear away to the west so that during the hours of darkness neither the calm nor the moderate swell could set us too far inshore. But we could never have imagined that the coast would curve round so sharply that Isla de Lobos[6] would already bear WSW5°W from our present position. Fortunately the horizon cleared at sunset and we could see it on that bearing some four leagues distant. The lofty hill which we could see on the mainland facing the island must have been Morro de Eten.[7]

We immediately luffed up under all sail on a course SW5°W and by about eight o'clock that night had managed to pass it at a distance of one league with the depth increasing from twenty to twenty-five fathoms and from there we sailed on briskly to the SW. At ten o'clock we hove-to head to shore, and having passed the whole night without obtaining bottom we got under way before dawn so as once again to close Isla de Lobos which was already in sight. At sunrise we were indeed only two miles from its shore in a depths from twenty to twenty-five fathoms. We were thus able to survey the reefs which surround it to our complete satisfaction.

[1] Islas Macabí, 7 miles SSW of Punta Malabrigo.
[2] Punta Pacasmayo, 19 miles NNW of Punta Malabrigo.
[3] Punta Chérrepe, 35 miles NNW of Punta Malabrigo.
[4] Islas Lobos de Afuera, a small group of barren offshore islands.
[5] Malaspina was right to be wary as Islas Lobos de Afuera are situated 64 miles WNW of Caleta Chérrepe and 33 miles offshore.
[6] Isla Lobos de Tierra, 30 miles NNW of Islas Lobos de Afuera and 6 miles offshore. An important landmark for coastal navigation.
[7] Malapina was mistaken. Punta Eten, an unmistakable isolated headland is situated east of Islas Lobos de Afuera and not opposite Isla Lobos de Tierra.

26 September

The wind was still fresh and weather pleasant. We approached the mainland and then coasted along it, steadily approaching Puntas Falsa de la Aguja and Verdadera de la Aguja[1] and later Punta Nonura,[2] sounding twenty-three, sixteen and twenty fathoms, ooze. At noon our latitude by observation was 4°52′ and our longitude by chronometers 4°5′19″ west of Callao.

In accordance with instructions given beforehand, the *Atrevida* continued to take hourly soundings, subsequently putting on more sail to come up with us, while the *Descubierta,* keeping the same sails set and steering a steady course, proceeded to measure bases with great accuracy, taking soundings only when shallow water called for the greatest care with navigation but did not require us to heave-to. Beyond Punta Nonura, which is easy to identify by the rock which they call De Bernal,[3] the coast begins to trend markedly to the east. It continues somewhat hummocky as far as [Punta] Piura[4] but then in Ensenada de Sechura[5] recedes so far and is fringed with such sandy flats that it is difficult to establish the coastline, particularly as sandbanks extend from it for more than one league.

It was four o'clock in the afternoon, off Punta de Piura, when we were able to alter course ENE to survey the head of the bay. Silla de Paita[6] and the nearby upland could be seen ten leagues off to the NNW and at one-and-a-half leagues from the coast we were still in a depth of thirty-eight and forty fathoms, fine sand. The wind, still strong from SSE, seemed to want to help us on our way but very soon, under the lee of the land and with the approach of evening, it died away to erratic light airs.

At sunset we could see from the mastheads the whole inner shore of the bay and at the same time the outermost headland in the vicinity of Paita bore NW. It therefore seemed pointless to make any attempt to go further into the bay, which the evening calm and the following offshore wind would in any case have prevented. We therefore took advantage of the occasional light airs to get clear of the bay and with the last of the strong trade wind sailed on some way so as to be close to Paita at sunrise. Throughout the night the wind stayed moderate between SSE and SE and we lay hove-to under topsails on both tacks. We obtained no bottom with seventy fathoms of line.

27 September

At dawn the weather was very hazy with fairly strong gusts of wind. We at once steered close-hauled to the NE and before long closed the coast so that at seven

---

[1] Punta Aguja, a long and level point terminating in a steep bluff, 45 m in height with Punta Falsa, a salient point with a seaward facing bluff 50 m in height, 7 miles SSW.

[2] Punta Nonura, 5 miles SW of Punta Aguja.

[3] Not identified, but perhaps named after Cristóbal Bernal, who served with Francisco Pizarro and settled in the nearby town of San Miguel de Piura. His services were acknowledged by grant of a coat of arms which includes a small ship on waves of azure and argent: J. A. del Busto Duthurburu, *Diccionario histórico-biográfico de los conquistadores del Perú,* Lima, 1973-, I, p. 244.

[4] Unidentified, but a short distance east of Punta Aguja.

[5] A wide bay between Punta Aguja and Silla de Paita about 40 miles north.

[6] Cerros Silla de Paita (the Saddle of Paita) - a group of three isloted hills resembling a saddle from north and forming an excellent landmark, situated 6½ miles SSW of Paita.

o'clock we were no more than a league from Islita de Lobos.[1] To anchor at Paita when we had the chance to survey the whole roadstead under sail would have been quite contrary to the time constraints we had imposed on ourselves. On the other hand it was opportune to make a careful survey of the area both because it was an important depot for substantial trade, particularly in foodstuffs, and because, since the southerly winds fall away at this point and the alternating land and sea breezes greatly facilitate navigation, this has become a port of call and landing place for those travelling to Lima from the coasts of Mexico or Tierra Firme.[2] The Viceroys of Peru generally took advantage of this when they were coming directly from the ports of Europe and did not regard the passage of Cabo de Hornos as sufficiently safe and smooth. Nor on the other hand could they entirely put out of their minds the early hostilities carried out by Admiral Anson in a country which they had come to believe was safe from invasion, given its poverty and its distance from well frequented seas, having apparently forgotten the exploits of the buccaneers.

Two leagues from Paita stands a perforated rock[3] which is about two leagues further on from Islita de Lobos. It is necessary to pass very close to this rock and then to keep at a fair distance from the mainland coast which flanks the bay. The rock is steep-to and entirely free of danger apart from the point near the anchorage from which a spit juts out to the north. This is always above water and extends for no more than two cables. When sailing close to this coast it is prudent to rely on the smaller sails as the sea breeze blows strongly. We took just these precautions. We also unfurled the national ensigns and got up a long range of cable, signalling to the *Atrevida* to do the same. At a very early hour we had sighted two large *balsas*[4] which were sailing close inshore towards Islita de Lobos in spite of the high wind which was blowing at the time.

By eight in the morning we had surveyed the whole of the outer coast and were steering towards the roadstead at no more than a mile from the coast in depths between thirty-five and twenty fathoms, ooze. Shortly afterwards we came across no fewer than four vessels at anchor and some fishing canoes and finally the town itself which was situated at the very back of the bay, confirming the accuracy of our work. At nine o'clock, with our survey complete and lying no more than a mile from the anchored vessels, which had hoisted their ensigns, we altered course to proceed along the coast towards Los Negrillos[5] and Cabo Blanco.

Very soon soundings of fourteen and sixteen fathoms warned us that we should navigate with some caution and as the offshore wind which we had encountered in the bay gave way to a fresh sea breeze, we quickly made considerable progress and sighted the shoals which extend a good distance seaward from the low-lying coast

[1] Isla Foca, an islet near the coast off Punta Foca.

[2] A broad reference to the isthmus area of Panamá south of New Spain.

[3] Unidentified.

[4] Large balsa-wood rafts using cotton sails for propulsion and a peculiar means of steering which eventually inspired the development of the adjustable keel. The rafts were used by native peoples of the region from pre-hispanic times as fishing vessels and cargo-carriers in the busy trade along what are now the coasts of Ecuador and central Peru: J. Ortiz Sotelo, 'Embarcaciones aborígenes en el área andina', *Historia y cultura*, xx, 1990, pp. 49-79. For an illustration of this craft see Eric Beerman, *Francisco Requena: la expedición de límites*, Madrid, 1996, p. 17.

[5] Not shown on contemporary Spanish charts.

between Paita and Los Negrillos. According to signals from the *Atrevida* we were in depths of sixteen to twenty-six fathoms. Our latitude at noon was 4°51′36″ and longitude 4°5′19″ west [of Callao]. By then we were not far from Los Negrillos whose rocky ledges, extending gradually to the north, we carefully surveyed as we passed them two or three miles offshore. The soundings then increased suddenly to forty fathoms.

By two in the afternoon we had raised Cabo Blanco. We steered towards it under full sail in a fresh breeze but could not round it before half past five, when we altered course NNE by compass so as to stay close to the coast which now makes a distinct turn to the east. At nightfall there was a moderate breeze and the prospect rather misty. We continued on for two or three leagues under shortened sail and then brought-to heading towards the coast so as to make good use of the first light and, should there be a strong sea breeze, reach the anchorage at Punta Arenas on Isla Puná.[1]

As we lay hove-to during the night, we were warned by the sound of a heavy sea breaking on the shore, and even by actually sighting the coast, that we were being set close to the land, but as the depth remained over eighty fathoms and the light breeze seemed reluctant to die away altogether, we continued to lie-to until four o'clock in the morning when, under topsails, we set a course diverging somewhat from the direction of the coast. At dawn the coast was barely a league away and according to our information appeared to be the stretch between Puntas Máncora and Mero.[2] Cabo Blanco and the coast that we had surveyed the previous afternoon were still in sight. The depth was sixty and sixty-five fathoms, fine sand.

28 September

We lost no time in starting to measure bases under full sail but, the wind dying away, we made very little progress until noon, especially as the need to examine the rather flat coastline did not allow us to bear away and set a course immediately for Islote el Amortajado;[3] latitude 3°40′ and longitude 3°41′25″ west of Callao by observation.

At noon on the 21st and 24th we had compared our chronometers with those on the *Atrevida* and we repeated the comparisons at noon today. The fact that our readings differed widely from those of number 10 led us to suspect that the adjustments [to the rates] of the last few days had been defective and that we should rather prefer those indicated by the earlier readings. Nevertheless we deferred the determination of their true rates until the forthcoming series of checks at Guayaquil, when we were confident of making a comparison using much wider and more dependable data.

The sea breeze was still gentle after noon and as a consequence the air remained so hazy that we were much delayed in raising El Amortajado and could only sight it at sunset from the main yardarm even though we were only about five leagues away. At the same time we could also make out Punta Malpaso,[4] at a distance of two leagues,

---

[1] A large island in the entrance to Golfo de Guayaquil, with Punta Arenas its SE extremity.
[2] Situated 14 and 32 miles NE of Cabo Blanco, respectively.
[3] Isla Santa Clara situated about 13 miles SW of Isla Puná; also known as Isla del Muerto because of its resemblance to a shrouded corpse when seen from east or west.
[4] Situated 65 miles NE of Cabo Blanco.

and the breakers close to it. We were in twenty fathoms, ooze. Although the night was very gloomy, the bearing of El Amortajado and our soundings left us with not the slightest concern about our track; indeed we now altered course to bring the islet right ahead with the object of avoiding Los Bajos de Payana which extend a good distance from Río Tumbes.[1]

Contrary to our expectations, at half past nine that night, when we believed ourselves to be close to that islet and were looking for it through the darkness, we suddenly found the depth decreasing from twenty fathoms, ooze, to fifteen fathoms, sand, and the *Atrevida*, according to what they told us the following day, obtained as little as nine fathoms, rock. We were left in no doubt that we had encountered the shoals. We therefore luffed round sharply and found ourselves once more in twenty fathoms, ooze, and soon afterwards in thirty fathoms. In this depth and now once more back on track we sighted El Amortajado to port at eleven o'clock, having supposed it to be half a league off, and now, considering ourselves to be on the right track, we set a course for La Puná.

It was generally agreed among the pilots that by steering NEbyE from El Amortajado one would soon reach Punta Arenas, the depth decreasing to ten fathoms, ooze. This was accordingly the course which we steered under full sail from eleven o'clock that night. But at two o'clock we found ourselves in only ten fathoms and uncertain which bank we had encountered. Although convinced that we were still a long way from La Puná, we let go an anchor and, after the appropriate signal, the *Atrevida* soon followed suit.

29 September

Thus by chance we found ourselves early next morning in an unfavourable situation, although in a most suitable position from which to take useful bearings to Puntas Salinas[2] and Arenas on La Puná, to El Amortajado and to various points on the coast of Tumbes which we had encountered by steering too far to the east.

Meanwhile in this position, the wind having died away, or rather the land breeze having set in, our first step was to send away the pinnace with a sergeant to the town of Puná[3] so that the Governor of Guayaquil could learn of our arrival and at the same time river pilots could come out to meet us at Punta Arenas. The Sun having by then risen a little above the horizon, hour angles were observed; and finally, not wishing to waste the time which the late arrival of the trade wind was obliging us to spend at anchor, we sent away a pinnace and a *pilotín* from each corvette to take two lines of soundings in different directions. At half past ten the sea breeze began to blow, so we recalled the pinnaces and instructed the *Atrevida* to remain at anchor until noon to take observations for latitude, while we would move to a suitable distance and bearing from which to measure a base using her masthead height, so that the principal features in sight could be fixed relative to each other and tied to an accurate astronomical position.

[1] Río Tumbes enters Golfo de Guayaquil through a delta south of Isla Puná with Los Bajos de Payana fronting the western part of the delta 17 miles south of Punta Salinas.
[2] The SW extremity of Isla Puná.
[3] A town at the NE extremity of Isla Puná.

By noon the *Descubierta* had obtained a suitable base for these operations, both corvettes agreeing on the angle measured. However, the Sun's close proximity to the zenith led to a large discrepancy between its observed meridian altitudes. There was, nevertheless, a good probability that the altitude taken in this corvette was very close to the true one, the determination of 3°9′30″ being given by our observations.

Having completed these operations we steered for Punta Arenas under full sail and at three in the afternoon anchored nearby in twelve fathoms, sand, after obtaining soundings of fifteen and eighteen fathoms with the same bottom, rather than the ooze which we had left as we approached Isla Puná. We subsequently learned that ships generally prefer the anchorage over ooze opposite this point but close to Tenguel[1] rather than the one we had taken, from which it was rather difficult to clear Banco de Mala.[2] The *Atrevida* anchored shortly afterwards and it was only when she dragged her anchor that she became somewhat separated from us. At that time the flood tide was running very strongly, at nearly three knots.

We would have liked to made a fresh longitude observation with the chronometers during the rest of the afternoon but the lack of any horizon to measure the altitude and the fact that even the quintant could not measure the complementary angle meant that this check could not be carried out.[3]

Throughout the night we remained in the same position. Before dawn the sea breeze died away as usual, followed by a weak offshore wind. [30th] With this wind and a rather misty outlook the Sun came out, revealing shortly afterwards our pinnace which was returning from Puná. Having mistakenly approached Punta Salinas instead of Punta Arenas, it had only reached the town at ten o'clock the previous night and after embarking local pilots it had then set off without delay on its return to the ship. Segundo Piloto Hurtado, who had been assigned to the soundings, had carried them out very properly. When some patches of clear sky appeared at seven o'clock in the morning, we made immediate use of them to observe lunar distances. The results gave a longitude 38′ less than the true one given by the chronometers. The distances measured in the *Atrevida* were even less than ours.

Until after two o'clock in the afternoon there was no decrease in the strength of the ebb tide, nor did the usually strong trade wind come up to allow us to make some further progress towards the anchorage under sail. But as the trade wind was now beginning to strengthen, we set sail, at first heading east so as to round Banco de Mala. Then, proceeding slowly in depths of seven or eight fathoms, mud, we continued on for the rest of the afternoon to close La Puná.

In its vicinity the wind slackened and after prayers we were obliged to make a short tack to a position very close to the ships anchored off La Puná. But the wind, strengthening again at this moment, caused the *Atrevida* to overtake us and her commanding officer called across to me that the chief river pilot had reached him that afternoon, so that if we were to enter the river[4] that night he would take care of

---

[1] An unidentified feature near Punta Arenas.
[2] A shoal area SE of Isla Puná, the principal channel being SE of the bank.
[3] Presumably by attempting to use the quintant's facility for taking back observations from the opposite horizon.
[4] Río Guayas giving access to Puerto de Guayaquil about 30 miles upstream.

setting the course. We immediately followed and in favourable conditions of wind and tide reached the mouth of the river, being guided in the narrows entirely by soundings and by the line of both banks which we could see clearly despite the rather murky night. Until eleven that night the favourable tide continued, so we made the best of it, sailing up the river almost far as the Punta de Piedras,[1] where we anchored close to that point in five fathoms, ooze, to await the onset of the next flood tide.

1 October

At daybreak we lowered the launch and, towed by it and the pinnace, we at once set sail to take advantage not only of the flood tide, which was beginning to run, but also of the light airs from the SW which, although feeble, were still persisting. The *Atrevida* picked this up rather later than we did and even with the launch hoisted we were lucky enough to come up with her at once. From then on we stayed together. We used the whole of that tide to continue upstream and did not anchor again until midnight, by now only two leagues from Guayaquil.

Before the flood tide lost its strength I sent Guardiamarina Ali Ponzoni in the pinnace to convey my compliments to the Governor of Guayaquil and to tell him what we needed most urgently: in particular a house close to the river in front of which we could anchor so as to apply ourselves more closely to our operations relating to astronomy and natural history. The Governor of Guayaquil was Capitán de Fragata Don José de Aguirre[2] who, in addition to belonging to our own service and having close ties of friendship with almost all the officers of both corvettes, was a most amiable character and utterly dedicated to the service of the Crown. I was therefore not at all surprised when he not only received the *guardiamarina* with great warmth but was also kind enough to come out to us in our pinnace to greet us when the tide had not yet allowed us to get under way. He was accompanied by Don J. Elizalde,[3] a resident and merchant of Guayaquil, who at the mere suggestion of the Governor had been kind enough to place at our disposal a large part of his house on the river bank, which could serve us both as lodgings and as a workplace for the various purposes of our mission.

As evening approached, the wind was blowing more strongly, while the ebb tide slackened, so that at half past five we were able to get under way and, guided by soundings, finally to drop anchor at seven o'clock opposite the town and very near the house where we were to be lodged. The *Atrevida* which anchored before us was only a cable from the shore while we were about a cable and a half, both ships later being moored in line with the tidal stream although, according to the custom of this country, a stopper was placed on both starboard cables so that the regular swinging to

---

[1] Punta de Piedra on the western side of the channel about midway between Puná and Guayaquil.

[2] Don José de Aguirre Irisarri had taken up his post in March, 1790. A former pupil of the Academia de Guardias Marinas de Cádiz, he had served with Malaspina during the siege of Gibraltar (1780-82) and had formed an acquaintance with Antonio and Arcadio Pineda at about the same time.

[3] Juan Bautista de Elizade y Echegaray settled in Guayaquil where he accumulated a substantial fortune in the cacao trade. He married Josefa de la Mar y Cortázar, sister of the future marshal and president of Peru, José de la Mar: R. Pérez Pimentel, *Diccionario biográfico del Ecuador*, Guayaquil, 1994, vii, p. 112 and Mendiburu, *Diccionario*, iii, p. 32.

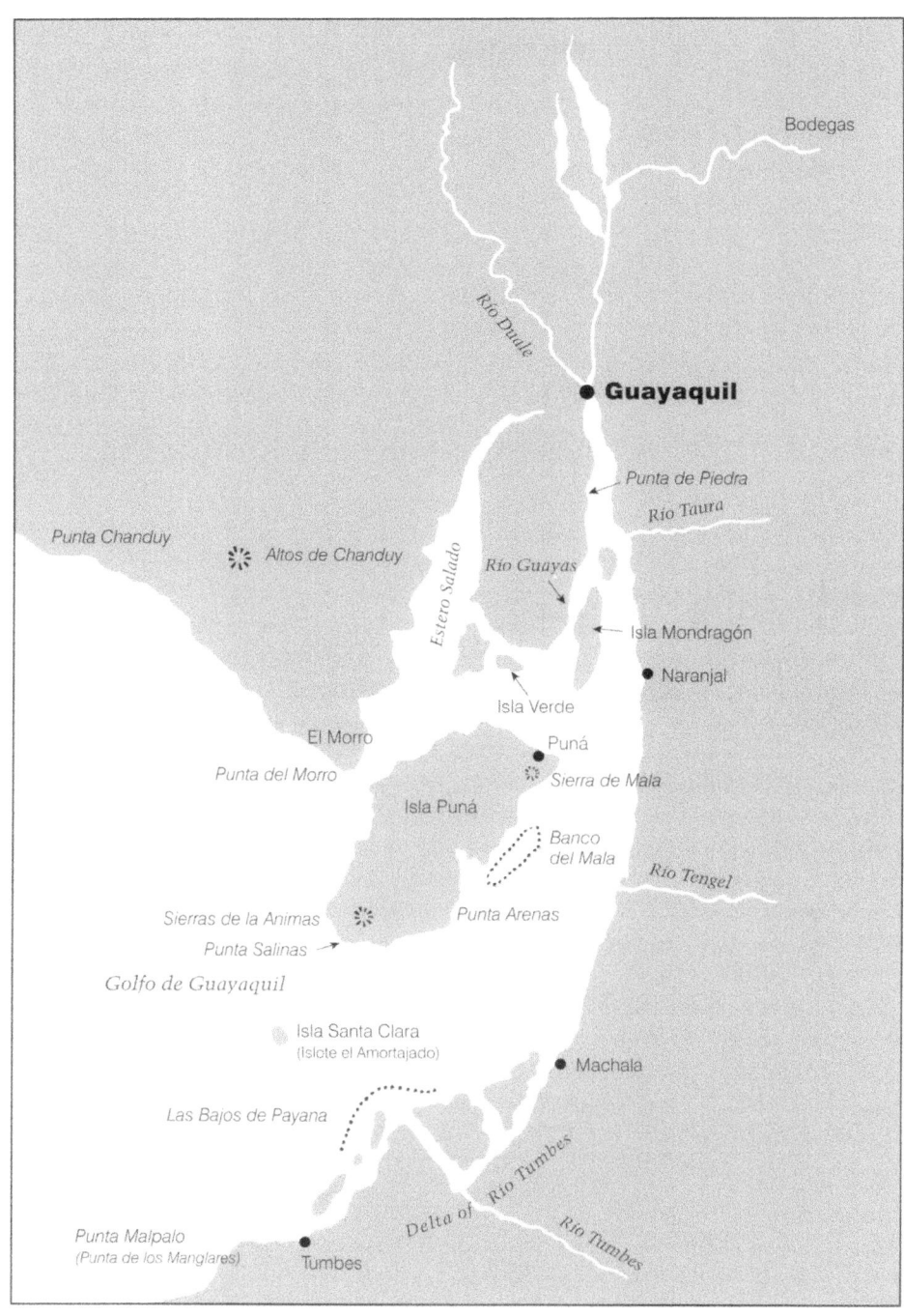

Fig. 8. Approaches to Guayaquil, September 1790

the flood and ebb tides would not cause continual twisting of the cables. The Governor later went ashore and the *guardiamarina* who accompanied him received the pleasant news that both he and his companion had been promoted to the rank of Alférez de Fragata.[1]

---

[1] Fabio Ali Ponzoni and Jacobo Murphy.

CHAPTER 2

At Guayaquil[1]

At dawn we were greeted with a spectacle as new as it was delightful, particularly to those who had not yet frequented the pleasant lands of the tropics. The shores, attractively clothed in various shades of green, the gradations of which added further contrasts to the beauty of the scene, the many birds, their songs and colouring entirely new to us, the *balsas*, the canoes, the combination of houses, trees, water and craft in almost a single grouping–all reminded the admiring spectator that the extent and variety of nature exceeds, in its marvellous beauty, what even the most vivid and rapt imagination could conceive.[2]

2 October

The astronomers began their work without a moment's delay.[3] In spite of the fact that it proved impossible to make use of the house [put at their disposal] to set up the astronomical clock and the astronomical quadrant because the upper storey floor boards caused considerable movement when people did no more than walk on them and because the proximity of the Sun to the zenith meant that the roofs blocked the view completely from an early hour. They therefore set them both up in the little square nearby so expeditiously that it was possible to calculate local apparent noon with the astronomical clock by means of equal altitudes, confirmed by comparisons with the chronometers of both corvettes.[4]

This calculation, when referred to the results from number 10 of the *Atrevida*,

---

[1] Guayaquil occupied an important place in Malaspina's vision of the sort of inexpensive empire America might become, with a chain of uniform coastal settlements in key positions. It was a relatively urbanized area of well established settlement, with a population of 36,405, including 18,154 females: 'Noticias de Guayaquil', AMN, MS 339, ff 1-24.

[2] Material considerations underpinned Malaspina's description. The forests of the region were valuable for shipbuilding and the port was well placed for commerce with the Quito region, important for textiles, foodstuffs and quinine. Guayaquil's own cacao production was increasingly important in the region's remarkable economic development.

[3] The importance of Guayaquil and its shipyard may explain the high priority given to hydrographic work and the coastal survey.

[4] 'The governor had deliberately chosen a site in advance, where we set up our oservatory and laboratory of natural history. This was one of the principal dwellings … :very close to our anchorage, but the wooden construction, universal in those parts, gave rise to various vibrations which disturbed the level of the instruments and we had to transfer them to a purpose-built building in the plaza, under the guard of a sentry.': 'Diario de Dn. Arcadio Pineda del Callao a Guayaquil, Panamá y Acapulco en 1790-1791', AMN, MS 94, f. 38v.

together with the rates of each chronometer from noon today until the same time the following day, also obtained by equal altitudes, convinced us that we had been too quick to adjust the rate according to the absolute altitudes taken on the 19th and also that the period from the 15th to the 16th in Callao had been affected by some error which our chronometers and those of the *Atrevida* indicated unanimously.

Bearing this in mind and the accuracy with which number 10, both in the daily comparisons and in the comparisons taken in Guayaquil, showed that its rate had been consistent with our expectations throughout, we decided to bring the rates of our chronometers into line with that result, by means of probable combinations. The daily equations, also combined with the comparisons to number 10 at sea, finally equated with those results and here we were pleased to find that the said comparisons indicated very precisely that the errors which at times were concealed when only three chronometers were compared could not escape notice when there were six chronometers, one of which maintained a consistent rate.

Our deductions were as follows:

	No 13	No 61	No 72
Apparent noon in Guayaquil	8^h 17^m 56^s 40	2^h 10^m 38^s 55	7^h 36^m 48^s 10
Apparent noon in Lima	8^h 7^m 15^s 25	1^h 59^m 57^s 23	7^h 26^m 8^s 14
Difference between meridians	10^m 41^s 15	10^m 41^s 32	10^m 39^s 56
Comparison equation with No 10	$- 2^s$ 52	$- 3^s$ 09	$- 1^s$ 33
Corrected difference	10^m 38^s 23	10^m 38^s 23	10^m 38^s 23
In degrees west of Callao		$2°$ $39'$ $36''$	

At the same time the *Atrevida's* chronometers gave the following results:

	No 10	No 71	No 105
Difference of meridians	$10'$ $41''$ $10'''$	11 4 40	10 50 20
Equation	$- 4$ 00	$- 27$ 40	$- 13$ 40
Corrected difference	10 37 10	10 37 00	10 36 30
In degrees west of Callao		$2°$ $39'$ $13''$	

Such consistency reassured us about the considerable difference we had with the longitude which Don Antonio Ulloa had worked out by dead reckoning from Quito, but it also allowed us to begin drawing up the chart of the coasts we had surveyed without waiting for the results of the astronomical observations of the Moon or of the satellites of Jupiter. These would vary very little or at least by an amount which would only be of significance when we were putting the finishing touches to the work.

At the same time, the astronomers took the sensible precaution of calculating the bearing and elevation at which Chimborazo[1] should be visible in these parts, according to our position and its position on Ulloa's map, if clear weather favoured us. It

---

[1] A spectacularly beautiful 6,310 metre-high dormant volcano situated 90 miles NE of Guayaquil. It was visited in 1745 by La Condamine, Juan and Ulloa, and climbed by Humboldt on 23 June 1802.

should have borne N59°4'E from us and at an apparent angle of 1°20' above the horizontal at Guayaquil, and later we compared the compass bearing where we expected to see it with the bearing where the inhabitants said it was to be seen.

The recent occurrences of unruliness and misconduct among the marines and seamen, the risk of upsetting the few remaining members of our original complement who would feel more offended for having survived more dangers and finally the unpromising news from the country where we were at the time all obliged us to devise new means of achieving our objects without violence or oppression. Accordingly, the captain of the *Atrevida* and I fully agreed upon a system for maintaining discipline and health, by keeping them constantly busy but without force; paying them a slightly larger daily allowance which would afford them some small financial relief and at the same time make them vulnerable to summary punishment, and by the daily muster; and lastly, by frequently serving wine with the rations, so that their innate desire for alcohol would be met to some degree, but they would be imperceptibly weaned from a weakness which is so detrimental to both health and discipline.

Giving them some sense of freedom, albeit remote, in their choice of food, would also help to make their rations more acceptable. Our intervention would only be necessary when required by neglect or disorder and meanwhile the serving of fresh food seasoned to their taste would save an equivalent amount of stored provisions for our coming passages. This did not, however, involve the preparation of different foods at any one meal. Following the same system as in Lima, a single pot of stew was prepared for the seamen and another for the marines, hot meals were served three times a day at regular times to allow for the proper allocation of work and finally, after the distribution to each man of the same ration as in Lima after the dawn muster, they were allowed to decide the daily amount to be spent on food, dividing the remainder equally among themselves, either daily or weekly.

For the time being the work was concentrated on cleanliness and the manning of the boats. The latter was important for our daily visits to the observatory, for our natural history studies and for travelling back and forth as well as for the many excursions required to carry out a thorough survey of the the river and indeed for the replenishment of our supplies of water which had to be brought from a considerable distance as the tides carried a long way up-river.

Such concern in the measures taken for the better regulation of the crew may seem excessive and inappropriate, if one does not take into account the seaman's natural carelessness, the Spaniard's very strong and sensitive feelings and the ravages that even minor forms of misconduct can wreak among visiting Europeans in these regions. The simpler, even unrecognizable, we could make our system for people who are by nature opposed to any monotony, so much more should be our attachment to it and indeed we were encouraged in our purpose by the pleasing sight of flourishing health and a look of natural satisfaction on the faces of both crews.

At the same spot where the first altitudes had been observed we set up the observatory tent and surrounded it with a cane fence, taking the precaution of assigning a marine by day and a soldier from the fortress by night to ensure the safety of the astronomical quadrant and the astronomical clock. Don N. Elizalde, whose house was close by, allowed us the use of suitable rooms where we could attend to the drawing, painting, and all the other concerns of natural history in great comfort, as well as a balcony from where, for purposes of comparison, we could fire pistol shots at noon to signal to both ships the time by the astronomical clock.

The brief excursions by the naturalists on this day soon gave them an idea of the magnificence of nature in these parts. Everything promised a great wealth of new acquisitions and aroused their ardent enthusiasm for the progress with their studies, particularly in botany.

The Governor, whom we all visited together this day, took great pains to contribute to our plan of operations, as may be imagined. He gave orders for every effort to be made, both in the immediate area and in the neighbouring provinces, to collect a variety of specimens for our studies in natural history. The principal residents, out of their own kindness no less than by the inspiration of this example, united in passing on to us all the information they had and anything else they knew of which might excite our curiosity.[1] Finally, a meeting every evening at the Governor's house of the most distinguished citizens of the city afforded us the relaxation and social intercourse, without which life at sea and its attendant drudgery could hardly fail to dull our faculties.

3 October

The next day was spent in arranging our principal excursions which took place on the mornings of the 4th and 5th, according to plan. Very little progress was made in the astronomical work, because the proximity of the new Moon and an expected change in the weather bringing generally cloudy skies, with some showers, prevented any kind of lunar observations. The first observations of the satellites of Jupiter would not be possible until the 17th.

4 and 5 October

According to the plan, our scientific excursions were to embrace the following aims: Tenientes de Navío Tova and Robredo from the *Atrevida* and Piloto Sánchez from the *Descubierta* were sent in a small local cutter chartered for the purpose to follow the coast from Naranjal by way of Tengael[2] and Machala[3] as far as the mouth of the Tumbes. It was to be left to those officers to decide whether to go upstream to [the town of] Tumbes itself, but it was strongly recommended that they should examine Los Bajos de Payana. Chronometer 61 and the sextants would provide the principal data during this excursion, as the tides generally made the method of measuring bases very inaccurate and difficult. The *Descubierta*'s launch, supplied with a fortnight's rations and under the orders of Don Juan Vernacci and Segundo Piloto Hurtado, was

---

[1] José de Aguirre supplied, among other documents, *Descripción de Guayaquil* by Francisco Requena (1774). The customs chief, Miguel García de Cáceres, presented statistics on ten years of exports, a memorandum on the organization of the customs office, the accounts and ladings for shipping for the year 1789, the breakdown of tribute received in 1790, a note on income and expenditure of the regional administration and a census of the entire province made in 1789, with details by race, class, sex, age-group and civil status.

[2] Río Tengel midway between Naranjal and Machala.

[3] Perhaps named after the Machala tribe of Isla Puná.

sent up-river as far as Bodegas de Babahoyo.[1] Vernacci was in charge of the Ramsden astronomical quadrant and Don José Bustamente of chronometer 105, which would not only enable them to make the necessary astronomical observations, but also to repeat the geometrical measurement of Chimborazo from accurate bases, which would allow us to make a closer approximation to the important determination of the meridian of Quito. Don Antonio Pineda and Don Luis Neé were also to set out in the launch with some good guides in an attempt to reach Chimborazo itself, the former to study the physical nature of the terrain and the latter botany and to make the best possible use of the fortnight allowed them. Don Antonio Pineda had been mainly entrusted with continuing to study lithology of the Andes and with the barometric experiments.[2]

These two expeditions left at daybreak on the 4th, with favourable tides. Early on the 5th the *Atrevida*'s launch set out in charge of Alférez de Fragata Murphy and Piloto Maqueda to survey Isla Puná between Puntas Arena and Salinas, making the appropriate observations with chronometer 71 and sextants and afterwards to determine the extent of Banco de Mala. Each of these three expeditions carried a theodolite, a compass and a lead line.

The examination of the physical conditions and botany of the surrounding area and an excursion to Montes de Taura, where the best timber was found,[3] were left to Don Tadeo Haenke. The remaining officers occupied themselves on both sides of the river with the combined tasks of hydrography, fishing and hunting, while Teniente de Fragata Francisco Javier Viana of the *Atrevida* was in charge of studying the tides.

6 October

Our only remaining concern now was the replenishment of water supplies, which was entrusted to each corvette's *bombo*;[4] these were to be sent upstream daily on the flood tide to fill up at the next low tide. This procedure, the usual one in this country, provides perfectly fresh water which keeps well on board, but it involves the inconvenience of working over three tides, which seemed to us an excessive loss of time given the uses to which the water was to be put. Accordingly it was decided that the

---

[1] Situated about 35 miles NE of Guayaquil.
[2] Possibly determining the height of Chimborazo using an aneroid barometer.
[3] Wood for the shipyard came from Montes or Sierra de Taura, 17 miles SE of Guayaquil, and Bulubulu in the judicial district of Yaguachi. Wood for export was also in demand in various Pacific ports, bringing in 25,554 pesos yearly. In 1797 Neé presented to the crown a 'Relación' of useful timbers, fibres and resins noted in the course of the expedition; in the section on Guayaquil he describes the shipyard and lists fifty-five varieties of tree useful in naval construction: 'Viajes que hizo por tierra durante la Expedición alrededor del Mundo D. Louis Neé. Botánico de ella.', AMNCN, 'Expediciones. Caja Grande N13, carpeta 17. From Don Francisco Ventura de Garaycoa, a government official in Guayaquil, Arcadio Pineda obtained a further report: 'Noticias que se desean sobre las maderas y construcción de Guayaquil', 25 October 1790, AMN, MS. 120, ff. 333-7. On the forest generally and its impact on the sensibilities of members of the expedition see J. Pimentel Igea, 'La riqueza forestal de las costas del Pacífico: noticias e informes sobre maderas de la expedición Malaspina (1789-94)', in M. Lucena Giraldo, ed., *El bosque ilustrado: estudios sobre la política forestal española en América*, Madrid, 1991, pp. 45-62.
[4] For *bombo* see p. xxv, n. 1 above.

watering boats would go upstream during the last two hours of the flood tide, would wait until two hours into the ebb to begin taking water, and would return to the corvettes in the last two hours of the same ebb tide.

We had intended to enlarge the launches and prepare them for longer expeditions and this task seemed too easy and cheap not to undertake in this country. However, I thought it better to have one done first and then the other, so that any faults that showed up either in the construction or the rigging could be more easily rectified in the second. We would also manage to save a great deal of time by having the artificers of both corvettes working on one ship; the quantity and excellence of the cedar from Realejo,[1] Amapala[2] and San Blas[3] would allow us to alter the *Atrevida*'s launch with equal facility in any of these ports where the objectives of hydrography or natural history might detain us for ten days or a fortnight. We therefore gave immediate orders for collecting the necessary timber, setting up the forge, employing sawyers and putting two seamen to work with them so that they could gradually acquire knowledge of this skill. Teniente de Navío Valdés was put in charge of all of this work and of liaison with the paymaster of the *Descubierta* so that the most productive method could be followed, according to the dockyard regulations, both in the choice of workmen and in the use and purchase of timber.

7, 8 and 9 October

The weather on the following days was generally thick, and at times showery. The tides were strong and high and the heat was considerable at times, when the *chanduy*,[4] an intermittent breeze from the WSW or SW, gave way to calms or the land breeze. At three o'clock in the afternoon of the 7th a distinct earth tremor was felt even on board, striking fear into the hearts of the few who had experience of this scourge of nature.[5] On the morning of the 8th Valdés, Haenke and Arias made an excursion with the help of the tides to Río Daule,[6] returning on board at midday, as pleased with the beauty of the place they had visited as with the acquisition of new birds and plants for their natural history studies. We also had on board a live reptile, a cayman or crocodile,[7] the description of which, together with other anatomical matters, occupied Haenke's meticulous attention.

[1] Situated 55 miles SE of Golfo de Fonseca in present day Nicaragua.

[2] Situated on an island in Golfo de Fonseca in present day Honduras.

[3] A major eighteenth century Spanish naval base 140 miles NW of Guadalajara in present day Mexico.

[4] Also the name of a town on the coast of Ecuador 46 miles NW of Punta Salinas.

[5] Pineda noted that 'During earthquakes people display no anxiety, as the houses are well built and the struts of which the walls are made are so well secured that those dwelling inside remain calm ... while ... the water boils in the river, the fish leap, the birds screech and the monkeys howl excessively': Descripción de Baba[h]oyo y Bodegas de Babaollo', AMN, MS 339, ff. 11-13.

[6] The drainage area of Río Daule covered a judicial district of some 5,000 inhabitants, settled in villages and estates, which grew tobacco, cacao, sugar cane and cotton. The woodland in the humid zone along the riverbanks was rich in timber.

[7] The *Crocodylus acutus* of the coastal region belongs to the family of Crocodylia. The cayman, which is concentrated today in the Amazon region, belongs to that of the Alligatoridae. Lázaro Spallanzani (1729-1799), who drew up the *Instructions* which Malaspina and his colleagues were to follow in their zoological researches, particularly enjoined the study of crocodiles, on which Linnaeus's information was incomplete.

A party of men from both crews was permitted to re-mast a merchant vessel which had just been careened, and since their conduct had so far been good, they were given an increasing number of opportunities for legitimate amusement, without trouble or disorderly behaviour.

10 October

On Monday afternoon we were pleased to have both the launches back, after they had been to Babahoyo and La Puná. Don Juan Vernacci had made observations for latitude and longitude at Babahoyo, acquired important information, taken soundings in various parts of the river and assisted the measuring work of Don Antonio Pineda, who had then set out for Chimborazo. He was not able, however, to get a view of that mountain because of the continuous cloud cover which we had also experienced here and so had still been unable to link our work with the determination of the meridian of Quito as we had hoped.

Murphy and Maqueda, although also impeded by the lack of clear skies, had done work of no less importance in the survey that they had been ordered to make of Banco de Mala and the coast from the Puná anchorage to Puntas Arenas and Salinas. Their results were confirmed by various observations for latitude and longitude and chance had favoured us in that their observation at Punta Arenas, when compared with the one obtained when we were anchored there, gave a basis for checking the rate of chronometer 71, which had shown an alteration in comparisons with number 10.

11 October

The astronomers were somewhat less successful this night than they had been on the previous ones. They were able to calculate the latitude of the observatory as 2°12′18″ by means of meridian altitudes of stars to the north and south, during a brief clearing of the sky and horizon. This agreed exactly with the observations of Señores Juan and Ulloa and the small difference was exactly what was to be expected from the different positions of the two observatories, one in the old town and the other in the new.[1] The rating of the chronometers proceeded, however, at rather longer intervals. Because of the clouds, the morning altitudes seldom corresponded with those of the afternoon. The method of absolute altitudes was only used when there was no alternative.

12 and 13 October

On the following days our natural history collections made good progress. The surgeon and some officers of the *Atrevida* had taken a canoe. Don Fernando Quintana and I used a boat to make different and useful excursions, while the Mayor of Bava, an area in the interior of this province, had sent a large consignment of birds and of

---

[1] The town was conventionally divided into four quarters: Ciudad Vieja, Ciudad Nueva in the centre, el Bajo or lower town and the Astillero or shipyard area.

medicinal plants.[1] The aim was to make drawings of all the lesser-known species, and José Cardero, one of the officers' servants, was most useful in this endeavour by undertaking everything that was not purely botanical;[2] he had also taken, from a suitable position, a pleasant view of the environs of Guayaquil, with the aid of the *camera obscura*.

On the night of the 12th, in preparation for calculating the longitude by the occultation by the Moon of several stars, the astronomers had obtained the latitude of the observatory with greater accuracy as 2°12'4"S by fresh meridian altitudes of stars, the positions of which were well established in [Mayer's] catalogue and as the weather had begun to look more promising, the calculations had been prepared in advance so as to miss no opportunity for getting on with hydrography or astronomy.

14 October

The weather did indeed seem more favourable on the night of the 14th. A steady calm and a particularly clear atmosphere seemed to increase the brilliance of the various heavenly bodies which appeared on the horizon. The soft, cool *chanduy* made the temperature all the more agreeable after the heat of the afternoon and the very peace and silence of the night added fresh enchantment to nature's harmonious scene.

It did not take long for Don Dionisio Galiano to make different comparisons between Mayer's 798 and the Moon with the Dollond equatorial.[3] The occultation of the same star by the dark sector of the Moon was observed at the same moment by myself, Concha and Vernacci. The latter then observed its emersion, as well as that of another smaller star, number 797 in the same catalogue, whose occultation had taken place before daylight had faded sufficiently.

These factors left us now with no worries about determining the exact longitude of Guayaquil as soon as we could be certain of the position of the star and the error in the lunar tables for the time of the observation. Having made the calculation provisionally, we found the longitude to be 81°40'45" west of Paris, while the longitude by

[1] By order of the governor, local authorities and savants put a collection of some 200 animal species at the expedition's disposal. Pineda showed great interest in testing the theories of Buffon by observing the capacities of birds. He described forty-one species in Guayaquil, but took only one to Europe, the *Crax rufus,* now called the Crested Guan *(Penelope purpurascens)* of which the flesh was prized for flavour and which, said Pineda, 'would be a useful acquisition in Europe' because it was easily domesticated and resembled the peacock: 'Descripción de Aves, Quadrúpedos y Peces del Puerto de Guayaquil', AMNCN, leg 2, carpeta 5. Años 1790-1791. All Pineda's descriptions include a classification according to the system of Linnaeus, where known, and comparative notes with species identified by Linnaeus, Georges Buffon, George Edwards (1693-1773), author of *A Natural History of Birds,* London, 1743-51, or Mathurin-Jacques Brisson (1723-1806), author of *Ornithologie,* Paris, 1760.

[2] José Cardero drew 13 specimens in Guayaquil: 12 varieties of birds and an anteater (previously thought to have been drawn in Mexico). Meanwhile, José Guío produced 24 botanical drawings. Neé reported from Guayaquil to José Celestino Mutis: 'The botanical draughtsman [Guio] ... :is good and patient. He knows the principles of botany and is well able to identify the parts of a plant including the means of propagation. The drawings I have had from him so far are uncluttered except with what is essential for systematic classification. With an accompanying methodical description, they suffice to know the plant in question thoroughly.': ARJB, Archivo Mutis, paquete 44-45, ff. 168-9.

[3] It is not clear exactly what comparisons Galiano was making, but they were in connection with determining the longitude of Guayaquil as described below.

Plate 29. One of the corvettes off Guayaquil with the volcano Chimborazo in the distance, by José Cardero. Museo Naval, Madrid.

the chronometers, which depended on the satellite [of Jupiter] observed in Lima, exceeded this by about 32′; this led us to suspect that the significant error affecting the tables would be added to by a slight error in the longitude of Lima and another one in the right ascension of the star. However, since this point, like Coquimbo, had been placed by very reliable observations, and as we would be able to adjust the partial differences of the chronometers, setting them indiscriminately either east or west by means of the daily equations, the difference, although large, gave us no cause for alarm.

17 October

Despite these considerations Don Dionisio Galiano sought further confirmation of their truth on the night of the 17th by a comparison between the Moon and φ Aquarii as they crossed the meridian; on this occasion the longitude calculated by the differences in right ascension was 81°50′54″ and served to dispel any doubts as to the first results, while at the same time various opportunities were taken to compare with the large astronomical quadrant, with which the latitude had been observed, the eccentricity of the theodolites and the Ramsden astronomical quadrant, and to observe variation by the same theodolites.

Meanwhile the other branches of the expedition had been equally successful. Don Felipe Bauzá had already compiled the chart of the recently surveyed coast between Lima and Guayaquil. Tenientes de Navío Tova and Robredo had just returned and joined their work with our survey of the afternoon of the 28th near Río Tumbes. Each day brought new riches for our natural history collections, for which purpose Don Tadeo Haenke had already spent six days on his excursion to Montes de Taura, and had sent very valuable consignments from there, and finally we knew that Pineda and Neé had reached Chimborazo and Volcán de Tunguragua,[1] despite all the obstacles in their path.[2]

The reconstruction of our launch was, at the time, the only task which did not show as much activity as the others. The shortage of seasoned timbers, the local carpenters' lack of skill, and the inflexibility of the single planks needed for the wale-timbers would have been enough in themselves to make progress very slow, even without the addition of another matter that significantly affected the situation.

[1] Situated 30 miles east of Chimborazo; *tungurahua* means inferno. The mountain appeared to the travellers who approached from the northwest as an enormous, smooth cone; columns of smoke would be visible. Pineda noted 'the most horrific aspect of wreckage, desolation and ruin.': 'Expedición al volcán de Tunguragua', AMN, MS. 120, ff. 342-56.

[2] Neé reported 'We made the strenuous journey to Chimborazo and the volcano of Tunguraguas [Tungurahua], through Bodegas, el Caracol, Pozuelos [Pozuelo], the wearisome mountain of San Antonio, Guaranda, la Mocha, Pelileo, los Baños. On the way back I climbed on foot from Chimborazo, half-way to Guaranda, and descended – again on foot – the mountain of San Antonio in order properly to collect specimens of the very rare plants which are found along those paths.': 'Relación de los viajes que Neé hizo en la Expedición' (Madrid, Mayo 17, 1796), AMN, MS 2296, ff. 269-270v. He 'found many species published in the flora of Perú, and various species unknown to its authors. The sight of such riches kindled his desire to be useful to science.': A. J. Cavanilles, 'Materiales para la historia de la botánica', *Anales de historia natural*, ii, 1800, 4, pp. 3-57, at p. 50.

Don Juan Villalengua, until now Presidente de la Audiencia de Quito, had just been appointed Regente de la Audiencia de Guatemala,[1] and therefore was obliged to move, with his family, from this posting to Sonsonate[2] or Realejo. At first I had been averse to any idea of taking him with us because of our deviations from the direct route, the need to heave-to almost every night, and the cramped quarters on board both corvettes. However, the situation of this family was such that all these inconveniences seemed trifling when compared with the fact that they would be greatly delayed in reaching their destination, would have to take their chance, probably at substantial expense, on some ship that happened to come by, with the risks that that would entail and, finally, would lack a surgeon and other attention for the chronic illness from which his wife, although very young, was suffering.

Consequently the commanding officer of the *Atrevida* and I offered him the use of the very limited space and comfort that we could provide; we would divide the baggage and family between the two ships, giving him the use of the upper part of the poop from the steering wheel aft and leaving it to him to enclose, occupy and furnish it however he wished, as long as they did not count on using any other part of the ship, however small, for the family or baggage as that would be absolutely impossible in our present circumstances. As for the eating arrangements, we were happy to be able to share with people of distinction in His Majesty's service the usual provisions with which we had been provided by the generosity of the King and the shortness of our passages between ports.

This proposal was sure to find favour, nor could we avoid making it; but in a country such as this the construction of anything intended to last would naturally encounter extended delays if we had to rely upon men as inexpert in their craft as they were unwilling to work. So it was necessary for our own carpenters to direct the work, which inevitably slowed down progress on the launch.

On the morning of the 17th I was able to set off myself on an excursion towards El Morro,[3] intending to make my way from there to the coast to examine the trend of the coast towards Chanduy. In this way we would avoid having to make a subsequent survey with the corvettes. It was also my intention to reach Punta de Santa Elena,[4] where, as well as carrying out the geodetic operations, I would take an observation for latitude to link this area with the other parts already charted from astronomical observations. I asked Pilotín Hurtado to accompany me with a theodolite, a chain and a sextant. I also had a *punque*, a small local craft,[5] dispatched [downstream] by way

---

[1] The Kingdom of Guatemala was established as an administrative unit in 1542, with its Audiencia (supreme court) having jurisdiction over a much wider territory than present-day Guatemala. The post of Regente de Audiencia was created in 1766. As the highest magistrate in the land, the Regente became the *de facto* link between the magistrates in the Audiencia and its president, the Governor General of the Kingdom.

[2] A town in present day El Salvador close to the border with Guatemala.

[3] A small hamlet, with a nearby hill, 9 miles north of Punta del Morro, a point on the mainland opposite Isla Puná.

[4] A prominent headland, whose western extremity is now called Puntilla de Santa Elena, 70 miles NW of Punta Salinas.

[5] Presumably the *punque* was the small local cutter in which Tova, Robredo and Sánchez were sent to follow the coast as far as the mouth of Río Tumbes: see p. 254.

of La Puná and Río Salado[1] for our return; and since I was to have the pleasure of Teniente N. Rocafuerte's company from Punta de Santa Elena, I took to the road with mules.

The poor endurance of these animals did not allow us to reach the hamlet of El Morro, which lies some two leagues from the coast, until three o'clock on the morning of the 18th. However, as I spent the whole time on our geodetic work I was able to complete it by the afternoon, although I had to abandon the idea of reaching Punta de Santa Elena and limited myself to taking bearings to Punta Chanduy, which is only seven leagues away. Two bases measured between El Morro and the sea linked the coast in the vicinity of Punta Lacumbe[2] with the smaller headland itself and from both positions I took bearings to the extremities of La Puná and to Punta Chanduy, thus linking these operations at the same time to those of Guayaquil, and extending them as far as necessary to the west. From El Morro we could see Amortajado and Cerros de Taura and Guayaquil, the bearings to which would serve as new points of comparison to tie all the work together.[3]

At dawn, the following day, there was nothing left to be done apart from the examination of the channel between the mainland and La Puná with the *punque*, which had arrived the previous afternoon.

So at four o'clock in the morning I set out on horseback for the landing place and at five I was able to set sail with a favourable tide. For the rest of the day I had either the wind or the tide with me and was able to make a detailed survey of the many islands and shoals which obstruct and limit the width of this channel, as well as continuing to survey, by means of transits and bearings, the stretch of the river between La Puná and Punta de Piedra along which we had sailed by night.

Having anchored for some hours at Punta de Piedra to avoid the contrary tide, I was finally able to return to the corvette at three o'clock in the morning.

20 October

The next day Don Dionisio Galiano informed me that on the afternoon of the 18th there had been a very clear view of Chimborazo, leaving them in no doubt of its height or of our longitude, as deduced from that of Quito, according to the observations both of Their Excellencies Juan and Ulloa[4] and Messieurs Bouguer and La

---

[1] Although there is a Río Salado shown on contemporary Spanish charts on the Tumbes coast opposite the southern end of Isla Puná, it is much more likely that the *punque* was to return through Estero Salado which leads to Guayaquil through an inlet west of Río Guayas.

[2] Not identified.

[3] In a region where the clergy still complained of Indian superstition and pagan survival, the expedition was able to clarify a legend recorded as early as 1550 by Cieza de León, according to which giants from the sea had been destroyed by divine fire in punishment for sodomy, leaving only their bones. Measurement and analysis proved that the bones in question were of animal origin; they were all removed to the Gabinete de Historia Natural in Madrid, to eliminate a focus of superstition: A. Pineda, 'Cuaderno de aves de Guayaquil y la de los huesos fósiles', AMNCN, Pineda, leg. 2, carpeta 4.

[4] Malaspina clearly had on board Antonio de Ulloa y de la Torre-Guiral, *Relación histórica del viaje a la América Meridional ...*, Madrid, 1746, Jorge Juan, *Observaciones astronómicas y physicas*, Madrid, 1746 and Pierre Bouguer, *La Figure de la terre, déterminée par les observations de MM. Bouguer & de la Condamine, de l'Académie royale des sciences, envoyés par ordre du Roy au Pérou pour observer aux environs de l'equateur*, Paris, 1749.

Condamine.[1] The height had been measured with both astronomical quadrants. The true position was calculated by the angle measured between the summit and an object not far off, which was referred the next morning to the Sun with a calculation of its azimuth. The results were as follows:

	Toesas	Ours	Difference
By Juan and Ulloa	3380	3161·7	− 218·3
By Bouguer and La Condamine	3217	3161·7	− 55·3[2]

The longitude of Guayaquil could be calculated by the method indicated on both the Spanish and French charts which gave the bearing and distance of Chimborazo from Quito, as follows:

	Longitude west of Paris		
According to Juan and Ulloa			
By observations of [Jupiter's] satellites in Quito	81°	59'	34"
By observations of [Jupiter's] satellites from Cayambe[3]	81	12	4
Mean of four positions determined by a lunar eclipse observed in Yaruqui[4] and in Paris by Le Monnier[5]	82	6	0
Mean of seven positions determined by the same eclipse in Paris by Grandjean de Fouchy[6]	82	4	38
According to Bouguer			
Mean of various positions determined by a lunar eclipse and some immersions and emersions of a satellite of Jupiter observed in Quito	81	38	49
According to us			
By observation of the first satellite [of Jupiter] in Lima, calculated with the chronometers	82	12	30[7]

[1] Pierre Bouguer (1698-1758), son of a professor of hydrography, won a series of prizes for precocious scientific work from the Académie Royale des Sciences before his appointment in 1735 to the expedition planned by the academy to measure the length of a degree at the equator as part of a larger project for determining the shape of the earth. Charles-Marie de La Condamine (1701-74), one of the leading members of the expedition, was a member of the academy's staff who had won acclaim for his account of a Mediterranean voyage of 1731-2. Despite dissensions and local opposition in Ecuador, the expedition's destination, the job was complete by 1742. La Condamine returned home via the length of the Amazon and published an account on his arrival in 1745. Malaspina, however, used Bouguer's account, published four years earlier, see p. 262, n. 4 above.

[2] Assuming the *toesa* = 6.3946 feet (p. 210, n. 4 above), Juan and Ulloa's height = 6588m, Bouguer and La Condamine's height = 6270m and Malaspina's height = 6173m, as opposed to the present-day accepted height of 6310m.

[3] A vantage point 2,864 m up Monte Cayambe, 40 miles NE of Quito.

[4] A town 22 miles NE of Quito.

[5] Pierre Charles Le Monnier (1715-99) accompanied Maupertuis to the Arctic on the expedition intended to complement the efforts of Bouguer and La Condamine at the equator. He was the teacher and later the enemy of Lalande.

[6] Jean-Paul Grandjean de Fouchy (1707-85), Secretary of the Académie des Sciences, 1743-76.

[7] Palau et al., *Diario de Viaje de Malaspina*, p. 172, based on AMN, MS 753, give this longitude as 82°00'45" west of Paris, having been computed from observations made in a European observatory instead of, as here, predicted values given in the almanac.

By the occultation on the 14th, the errors in the position of the star and the lunar tables not yet corrected	81	40	45
By differences in [right] ascension between the Moon and φ Aquarii observed during the meridian passage on the 16th	81	50	54
By the end of the eclipse of the Moon on the 22nd, errors not corrected	81	52	10[1]

These results were indeed all the more pleasing for the speed with which we had achieved them and with a sky not very favourable for astronomy, which had already prevented two observations of the first satellite of Jupiter visible on this meridian. We had no better luck with lunar distances, although we tried to observe them in the evening, which seemed more convenient and the sky was clearer. A thick haze which obscures the skies from midnight until the following noon often also appears at nightfall, although not as thickly, and remains all night.

It seems opportune to give here a brief account of the condition of our seamen and marines. We were very pleased to note that our example was daily leading them away from unruly behaviour and in fact we felt ourselves to be able to allay our fears in this regard when we considered that there been no desertions among those who had come on board at Lima and we had also made up our complement in this port with another four or six men, giving us a good number of extra hands. There had been no brawls, even in a country like this, almost no absences from work or musters, and an unusual degree of obedience, cleanliness and good health – achievements which we noted with all the more satisfaction when we paid attention to the dictates of solidarity or remembered the past anxieties and desertions.

To help maintain this tranquility, any absence for a single night was punished, although lightly; the coxswains of both launches were promoted to acting quartermasters, and Quartermaster Palmero, who had served as the coxswain of the *Atrevida's* launch when we surveyed the seaward coast of La Puná, was restored to his original position now that he was fully convinced of, and had repented of, his past misdeeds, most reprehensible in a warrant officer who had been especially appointed to this expedition.

21 October

By the morning of the 21st Pineda and Neé had also returned on board. They had reached the foot of Chimborazo by way of Guaranda[2] and then climbed to the summit of Tunguragua, enriching the studies of botany, lithology and physics with their detailed investigations.[3]

---

[1] These results depend on the position of the Moon depicted in the *Nautical Almanac*. Better results could have been obtained if the actual position of the Moon had been observed in a European observatory.

[2] A town, 2,694 m high, near the foot of Chimborazo.

[3] From the point of view of the contribution of the expedition to the development of eighteenth-century scientific thinking, this excursion was perhaps the most important of those made from Guayaquil. As it took place in the driest season, when there was little standing water around, Arcadio Pineda took pains to gather information about winter conditions before rejecting Buffon's theory of an environment made hostile by excessive inundation: 'Apuntes de Guayaquil 1790', AMN, MS 120, ff. 338-41. He arrived at a nicely balanced view: the rainy season was hostile indeed but 'this horrible scene of unrelieved inundations, of insects, of heat and wastes, is followed by another which is highly delightful: the waters recede, the countryside is covered with agreeable vegetation of various kinds –

The foot of Tunguragua, which abounds in nature's gifts and shelters and maintains a considerable number of Indian and mestizo families,[1] gives off thick smoke, composed largely of water vapour, from various vents or craters. Don Antonio Pineda examined the vent near the snowline, finding that the volcano produces, through fissures as wide as a palm and sometimes six or eight *varas* in length, flower-like crystalline deposits of antimony and other metallic substances which adhere to nearby rocks. In the year 1772 there was a fearful eruption and, at first appearance, the traces of a larger and older eruption were visible. There remained no doubt that Chimborazo was an extinct volcano,[2] such an abundance was there of pumice stone, lava, rock affected by heat and pozzolana,[3] although the latter is ashen in colour.[4] Both travellers praised highly the hospitality of the inhabitants of the country they had passed through, particularly that of Don Baltasar Carriedo, a resident of Quito and owner of most of the area surrounding Tunguragua.[5]

It was difficult, or I might say impossible, to continue our triangulation from Guayaquil to La Puná because of the various points and bays which form both the principal and navigable arm of the river and the many shallow inlets which, either at high water or at all states of the tide, surround parts of the islands which scatter this frequently flooded area. Fortunately, however, Altos de Puná[6] and Taura can be made out from nearly everywhere so that, having determined their positions accurately by trigonometry, it would then be easy to use bearings to fix any intermediate points and then determine the details of the coastline, linked to each other by means of transits. As for the depths and positions of the different shoals, we thought it best to depend principally on our two inward and outward passages, according to the reports of the most reputable and expert local pilots, who take soundings constantly when

mixed cover, grassland and woodland – and distinctive colours, which are immediately colonized and enlivened by innumerable handsome birds in combinations which delight the eye. The insects diminish; the livestock returns to its old haunts and pastures. Even wild animals allow man to approach and produce a wonderful illusion that you are in paradise.'

[1] The census included all mixed race individuals under the name of *libres*, who amounted to 43·3% of the population of the province.

[2] The scientific focus of this excursion was to test rival theories of the origins of stratification: was it the result of deposits made by floods or of volcanic activity? The work reinforced Pineda's ideas on the volcanic origin of stratification, without altogether displacing the flood-related theory: A. Pineda, 'Expedición al volcán de Tunguragua', AMN, MS 120, ff. 342-56.

[3] A volcanic ash named after Pozzuoli, a town near Naples.

[4] Pineda was zealous in examining what he identified as sites of special interest, describing them, collecting rock samples for detailed inspection on board and recording the indigenous names of rocks, strata and soils. He was particularly interested in the thermal springs Rancud (Badcung) y Baños, which he described and analysed. Though he repeatedly cites 1772 as the year of the most recent eruption, it was in fact on 23rd April, 1773.

[5] Pineda also made ethnographic observations which led him to a view different from Malaspina's on the vexed question of whether tropics bred lazy people. Heading upstream past Guayaquil he wrote of natives lounging in their hammocks: 'they waste their time in the indolence induced by a fierce climate, where perspiration is copious and lassitude incessant.' The worst offenders were 'generally blacks and mulattoes'. The inhabitants of the village of Pozuelo, 'because the banana needs no cultivating, live as if in paradise, never do any work and are perhaps the laziest people in America.': Eduardo Estrella, ed., *La expedición Malaspina, 1789–94*, Tomo VIII: *Trabajos zoologicos, geologicos, quimicos y fisicos en Guayaquil de Antonio Pineda Ramírez*, Madrid, 1996, p. 34.

[6] Presumably Sierra de Mala 5 Miles SW of the NE extremity of Isla Puná on Malaspina's survey.

bringing large vessels through them to careen, or when sailing after launching or careening; and lastly according to the partial examinations our pinnace could make whenever occasion demanded on the way down to La Puná.

22 and 23 October

With this in mind, Don Felipe Bauzá measured a base on one of the streets of Guayaquil to be used as a reference on parts of the coastline where an exact measurement of a base might not be possible. Bearings were taken to different points, particularly Altos de Guayaquil and the hill opposite, to which Don Secundino Salamanca had made his way beforehand with a party of axemen from both corvettes to fell trees and clear brush from the site chosen to place the theodolite to obtain an all-round view of the most important features.

[Bauzá] continued his work on the following days and, since Altos de Puná and various other heights towards Babahoyo and El Morro could be seen from this point, we were able to make a detailed study of the layout, extent and position of the major buildings of the city and could then consider this principal part of our mission to be complete.

24 and 25 October

So as to be able to make our departure on the afternoon of the 28th, as I wanted to, it was necessary to speed up the construction of the launch considerably, omitting the final work on the spars, rigging, sails and various tools until we had more time. All those occupied with natural history were required to bring their scientific research to an end. The astronomical work now had to be limited to completing rating the chronometers and in any case the generally cloudy skies would not have allowed any further progress in this field.

As for the fitting out of the upper cabins in both corvettes, which , as I have said, we had left to the Chief Magistrate of Guatemala to use as he wished, it was also necessary to leave the accommodation on the *Descubierta* unfinished, as it would have taken a long time to complete it and I could occupy it myself in these pleasant climes and vacate the cabin and part of the wardroom in its place.

A coastal packet from Panamá was anchored in the river at the time, which on a previous passage had been set off course partly by the currents but mainly by the *piloto*'s lack of skill. After making a landfall at the Gálapagos they had sailed among the islands, believing at first that they were on the coast of the mainland. For various reasons given by the revenue officers, the Governor of Panamá had no choice but to take that *piloto* and his log into custody, replacing him, for the safety of the return trip, with another native of Panamá who had a great knowledge of the Chocó coast,[1] where he had sailed since his earliest years. I considered the acquisition of this pilot a matter of no small importance for the best completion of our work so I immediately enrolled him as a member of the *Descubierta*'s crew. As for the log and the *piloto*'s reports on the passage of the *Copacabana* to the Gálapagos, I left them in the charge of

---

[1] A region on the west coast of Colombia, occupying the drainage areas of Ríos San Juan and Baudó.

Plate 30. View of the old and new town of Guayaquil, by José Cardero. Museo de América, Madrid

Governor José Aguirre to be sent to Madrid as part of the documents about our mission. According to the accounts of several of the passengers, there was a large number of islands, some of them so big that they formed a strait twenty leagues long. Most of them lacked water, as shown by their aridity, and the soil, largely composed of pumice (of which they gave samples to Don Antonio Pineda), showed that they were the remains of various volcanoes, probably destined by nature to be eternally a desert.

These accounts were unanimous in describing the constant calms and showers in the vicinity of the islands, saying that they were 160 leagues from the coast, their latitude being approximately between one degree north and one degree south of the equator. This information and a long-standing rumour among these pilots that the Gálapagos extend well to the east, obliged me to be cautious about how to make the best use of our time and not to let insignificant matters compromise the allowances of time and destinations that we had proposed for the important parts of the coast to the north.

The winds that we would encounter as far as Isla de la Plata[1] would in the end determine the most prudent course to take, and I alerted the commanding officer of the *Atrevida* to this and warned him that in embarking provisions and even in the distribution of water he should take all precautions, as if for a long voyage and deprived of all hope of assistance.

I had no intention of separating the corvettes because our tasks in Panamá (possibly to be extended as far as Portobelo)[2] would be lengthy, difficult and delicate; because it would probably be necessary to separate on the coast of Nueva España [New Spain], the products and situation of which must be assumed to be of great interest to the crown and lastly because arriving at San Blas any earlier than the expected date would be very advantageous for our coming campaign to the north.

26 and 27 October

Being so close to the city, and having received, from the moment of our arrival, every mark of favour from the Governor and, after his example, from the principal residents, both in scientific and in social matters, the commanding officer of the *Atrevida* and I could hardly fail to show our gratitude by inviting all the people whom we had visited most to dine on board with the Governor and his wife during these two days. The corvettes had been painted and cleaned beforehand. Their interior decoration and the dinner itself showed, without ostentation, the gratitude we deeply felt.

On the afternoon of the 27th, having calculated local apparent noon by equal altitudes and completed rating the chronometers, the astronomical and geodetic instruments were brought on board; the stuffed birds and animals, and Haenke's and Neé's plant collections were packed in crates and finally we put the launch, now almost completed, into the water, and hoisted it shortly after prayers. Its cost was approximately [blank] pesos. Its dimensions were [blank].

The dawn muster on the 28th, at which we began to issue sea-going rations, showed that only one marine and one seaman were missing from this corvette, both

---

[1] A small island about 15 miles off the coast of Ecuador and about 55 miles north of Puntilla del Santa Elena.

[2] An important town on the north coast of Istmo de Panamá.

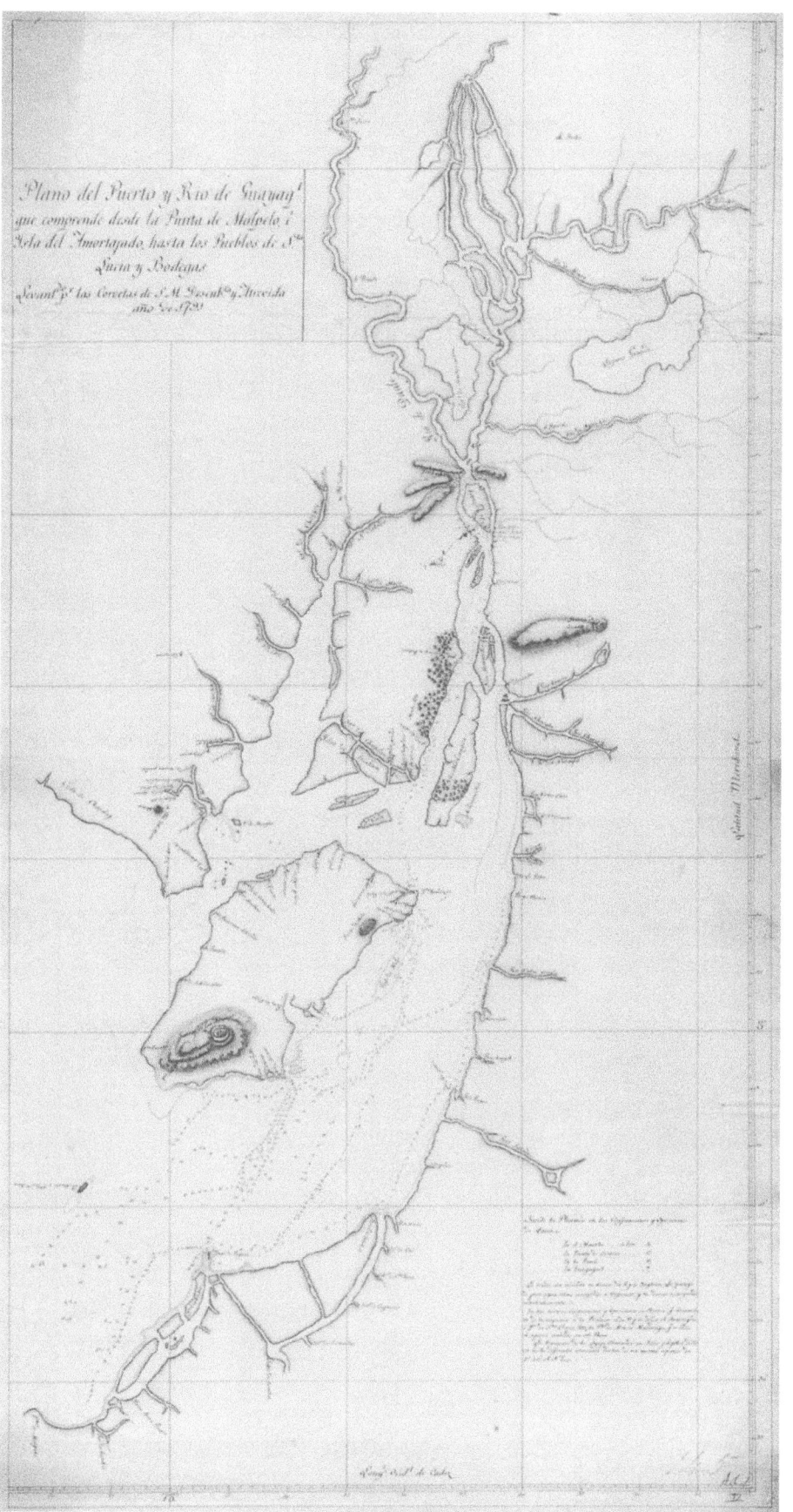

Plate 31. *Plano del Puerto y Rio de Guayaq¹... Levant⁰ pᵗ las Corvetas de S.M. Descubᵗᵃ y Atrevida año de 1791*; manuscript copy of Malaspina's survey obtained in South America by Captain Sir Thomas Staines, HMS *Briton*; UKHO, 145. Reproduced by permission of the Controller of Her Majesty's Stationery Office and the UK Hydrographic Office.

Plate 33. Young girl of Quito, by Felipe Bauzá.
Museo de América, Madrid

Plate 32. Indian woman of Quito, by Felipe Bauzá.
Museo de América, Madrid

of whom had been taken on at Lima. Six seamen were missing from the *Atrevida*, although these had already been replaced by other volunteers. On this occasion it was particularly important to be sure of the exact numbers of our crew, as we had to send to Lima for the Comisionado de Guerra y Marina, Don José de Tagle, the payment for the debts incurred by the complements of both ships in Callao.

Thus, with no further fear of losses or desertions, we were able to hand over the appropriate payment for these debts to the Governor, for forwarding to Lima. At the same time I begged him to arrange to have delivered to His Excellency the Ministro de Marina the plan of Callao and the vicinity of Lima and the chart of our last passage from Callao to Cabo Blanco, which we were sending in draft form as a precaution rather than as an official consignment as we had been unable to make a fair copy of it from lack of time. The chart was accompanied by a document drawn up by the astronomers with all the elements necessary for its accurate completion so that the errors that we knew that it contained could be corrected by the work of an observatory in Europe. This new work was enclosed with the consignment for Lima.

CHAPTER 3

From Guayaquil to Panamá

28 October

The tide today was not due to ebb until noon, so it was not until ten in the morning, when the flood had slackened somewhat, that both corvettes were able to weigh their northern anchor, leaving the other up and down, and eventually at one in the afternoon, with local river pilots on board both ships, we got under way under topsails alone.

As the breeze is usually from SW and the navigable channel is narrow and encumbered with shoals, it is usual and almost always necessary, to let oneself be carried down to La Puná by the tide, lying-to in mid-channel and using the topsails only to close one shore or the other, as required. At times it is best to go stern first, so as to have more time to seek out the channel, when the proximity of the trees, together with the stronger tidal stream, creates the amusing spectacle of the vessel proceeding stern first. The calm water, the pleasant surroundings, the breeze- against us, it is true, but slight-and the confidence of being able to avoid any danger with a kedge, make this passage to say the least enjoyable, even if it is necessarily very slow. Some parts are so shallow that they are no longer navigable at low water, in which case it is necessary to drop anchor and wait for high water at the beginning of the ebb, even if this means sacrificing part of a favourable tide. With the Moon in the last quarter, the tidal stream was at the time very weak.

For all these reasons, and principally because of the lack of depth between the southern extremity of Santay and Punta Gorda,[1] we were obliged to drop anchor off Isla Lobo[2] at half past four in the afternoon, having made only two leagues. However, the time was not entirely wasted, as Don Felipe Bauzá set off with the pilot during what remained of the afternoon to survey some distant shoals and take bearings of a point on the western shore.

When the next favourable tide began to make at half past ten that night we weighed anchor again, and, with a fair breeze, sometimes making short tacks and sometimes longer ones, we managed to anchor between Punta Miel and Punta León,[3] not far from the mouth of the Taura.

---

[1] Isla Santay a large island in mid-stream opposite Guayaquil; Punta Gorda is not shown on modern charts nor on a copy of Malaspina's survey; see Plate 31.

[2] Probably Isla Zono shown on Malaspina's survey, 7 miles south of Guayaquil and as a drying bank on modern charts.

[3] Not shown on Malaspina's survey nor on modern charts.

29 October

As the flood tide would continue for most of the morning, we made use of the time by sending the carpenters to the nearby shore to cut some mangroves. Don Felipe Bauzá took the pinnace to examine some shoals and take some more bearings, while we worked at hoisting the launch inboard,[1] since its weight and stowage with the other boats was causing us some concern.

At the time of leaving Guayaquil the commanding officer of the *Atrevida* had sent an officer to inform me that the total number of his seamen was only forty-three. For this reason we called over the pinnace from that corvette that same morning to send her two men who were superfluous to our complement, one of whom, recently recruited in Guayaquil, was very knowledgeable about the whole of the coast of Santa Fé[2] and Nueva España [New Spain].

By eleven o'clock, the tide now ebbing quite fast and the *bombo* having returned with the carpenters, we were able to weigh anchor once more and continue slowly on our way. This took us to the south of Punta de Piedras, allowing Don Felipe Bauzá to go in the pinnace to survey the isolated rock which makes this part of the channel very narrow,[3] although this is fully compensated for by the considerable depth alongside the rock and alongside the trees on the shore.

The *chanduy*, or fresh WSW breeze, did not allow us to use all the ebb tide during the afternoon, since with the weak stream it was not easy to make headway against the opposing breeze. We anchored at half past two in the afternoon, and remained there until the next tide.

This was to have carried us as far as Isla Verde but with the difficulty of keeping in the middle of the channel we were apprehensive of the shoals near Isla Mondragón,[4] [30th] and at two in the morning, after only two hours of the ebb, we anchored again between Punta de Alcatraces[5] and Isla Mondragón.

We did not waste the following morning while the tide obliged us to remain at anchor. A boat from each corvette was sent to cut firewood on the shore nearby and Don Felipe Bauzá measured a base at Punta de Alcatraces and linked the bearings taken in Guayaquil with the various important points in sight, surrounding this outer reach of the river.

At midday we were able to get under way, first seeking the centre of the channel and then keeping in it and, by half past four in the afternoon, we had almost cleared Isla Verde.[6] With the help of the sea breeze, and then under full sail, we were able to tack to starboard and then tack again at five o'clock to reach La Puná anchorage at sunset, where we dropped anchor some distance from the town, because of the increasing effect of the flood tide, which was already perceptible, so as to be in a better position to continue under way during the night.

---

[1] When the launch was hoisted before the *Descubierta* left Guayaquil it appears that it was left outboard. On approaching the open sea it would be a seamanlike precaution to bring the launch inboard when other boats could be nested inside it.

[2] The coast of central America between Panamá and Realejo.

[3] Roca San Rita, with a depth of 6 feet over it, 2½ miles south of Punta de Piedra.

[4] An island, 12 miles in length, on the eastern side of the main channel of Río Guayas.

[5] Punta Alcatráz, on the western side of Río Guayas, 10½ miles SSW of Punta de Piedra.

[6] An island on the western side of Río Guayas opposite the southen end of Isla Mondragón.

Towards midnight, at almost the same time as the tide turned in our favour, a fair breeze set in from WNW and with this, under full sail, we soon got under way in company, keeping in mid-channel between Banco de Mala and the coastal flats. We sounded five to seven fathoms, soft mud, and as we did not drop anchor until five in the morning when we were obliged to do so because of the calm and the contrary tide, by dawn we were on the parallel of Altos de las Salinas[1] on Isla Puná, with Punta Jambelí bearing S5°E by compass.

31 October

At nine o'clock next morning favourable light airs from NW began to blow and with the tide being almost slack, we set sail with the intention of being at a better heading for tacking when the sea breeze set in and so set a course towards Punta Salinas and El Amortajado. At the time we had to consider observing distances from the Sun to the Moon and obtaining a mean value for comparison with the results obtained by the chronometers and other methods we had used to obtain the longitude of Guayaquil. Thirty sets observed shortly before noon and calculated with the greatest care not only with the tables of refraction and parallax but also by Bordá's method, gave a longitude 30'26" further east than the chronometers, and therefore very close to our deductions according to the observations made in Guayaquil.

The irregularity in the rates of our chronometers, shown clearly by our daily comparisons, obliged us to compare them at noon by means of pistol shots with those of the *Atrevida,* whose officers assured us that theirs remained very exact. This precaution demonstrated that we should rely on the very accurate results of chronometer 72, the rates of numbers 61 and 13 having varied considerably. Our observed latitude was 3°2'S of the equator.

Up to noon we had made very little progress since weighing anchor and getting under way, as the various light airs from NW, north and east that we tried to use were so weak, but at one o'clock in the afternoon a fresh breeze set in from WSW and west, which we hauled to under full sail on the port tack, then tacking again to SSW and obtaining depths of seven fathoms near the flats off La Puná. Both the local pilots returned to port in their boats, which we were towing, and we hoisted the pinnace, the only boat we still had in the water.

The depth soon increased to sixteen fathoms, then gradually decreasing to ten as we approached the mainland. The ooze indicated that we were keeping in the channel and the bearings taken to the many surrounding points, among which we could already see El Amortajado, allowed us to link our present work with that of Tova, Robredo and Murphy and even the work done from the corvettes on the way here. The weather was agreeably mild and the sea breeze had only raised a slight choppy sea.

The following night was somewhat more difficult because of the need to tack continuously to keep within the channel between the coast of Tumbes and La Puná. We chose to do this rather than anchoring, as is usual on this passage, because, with no

---

[1] Sierras de la Animas near the southern end of Isla Puná on Malaspina's survey, now Cerro Zambapala, a 296-metre height.

further concern about the effects of the tide, we wanted to make as much progress as possible in the time available, although the breeze was gentle and variable. We tacked repeatedly, therefore, taking care to approach the coast of La Puná before dawn so as to be in a better position to use the NW land breeze. We were keeping to a bottom of ooze and generally obtaining depths of ten fathoms at the end of each tack, which were of no more than two leagues, often having to change tack rapidly as the depth decreased from fifteen to nine fathoms in a few minutes.

1 November

At dawn we saw with pleasure that our efforts had not been in vain. We were in depths of seventeen fathoms, ooze, between El Amortajado and Islotes de Payana,[1] which extend as far as [Río] Tumbes. [From here] the view of Alto de las Salinas provided us with a new connection by bearings with our position for Guayaquil. El Amortajado bore N70°W by compass, three or four leagues distant.

The breeze then dropped completely and we though it would be necessary to lay out a kedge, as the tide was setting us fast towards Punta de los Manglares,[2] but fortunately at eight in the morning a moderate breeze from NNW allowed us to sail, close hauled, to the west and stand off the coast of Payana. The bases we measured on this occasion dispelled, by means of the observed latitude, some doubts about the true latitude of El Amortajado, which had arisen because the Sun was too close to the zenith when we made our first observations from the corvettes. Indeed, having observed our latitude to be 3°20'30" at noon and having noted previously the times of various other observations, we ascertained that the latitude of the southern tip of El Amortajado was 3°14', when previously we had supposed it to be 3°18', and its longitude 0°32'40" west of Guayaquil.

Soon after midday we tried to use some breaks in the cloud to repeat our observations of lunar distances. These, in fact, were not completely reliable, because of the position of the sails[3] and the heavy cloud, but they agreed among themselves, and the results, very different from those of the previous day, were some 18' to 20' to the west of those of the chronometers; this revived our suspicions that the difference found in Guayaquil between one set of results and the other should not be attributed entirely to the chronometers.

The moderate sea breeze did not set in until three o'clock in the afternoon, but as El Amortajado was then bearing N20°W we were able to haul off on the port tack under full sail and finally to consider ourselves clear of the vicinity of Guayaquil and therefore of any risk of running aground. The soundings, which at the beginning of the afternoon remained at sixteen to twenty fathoms, ooze, then began to increase considerably, so that by nightfall we were in forty fathoms, sand.

After sunset the sea breeze began to abate and threatened, if we were not careful to keep a safe distance, to set us too close to the west coast of La Puná which we knew to be fringed by reefs. We quickly tacked to starboard, soon finding the advantage of this as a gentle breeze set in again from WNW and NW, giving us a safer and a better

---

[1] Small islands off the delta of Río Tumbes, protected by the dangerous Los Bajos de Payana.
[2] Punta Malpelo.
[3] Presumably Malaspina meant that the set of the sails made the observations difficult.

position from which to take advantage of the sea breeze the following day. We sounded, obtaining depths of up to thirty-two fathoms, sand.

2 November

We could not alter course during the following morning, as the wind remained light from the fourth quadrant, obliging us to remain on the starboard tack. The coast between Tumbes and Cabo Blanco could be seen clearly, but it was not easy for us to make out its position exactly, particularly as this part of the mainland had not yet been plotted on our charts, which we had been tying in every day continuously since Guayaquil. Latitude and longitude alone could not determine the position of this part of the coast. At noon the former was 3°28' and the latter, according to number 72, was 53'24 west of Guayaquil.

Soon after midday the sea breeze set in fresh from WSW and SW, and with this we got under way immediately, so as to be able to continue our hydrographic work the following morning on the coast between Punta de Santa Elena and Chanduy or Tambo, to which I had taken bearings from El Morro. The weather remained gloomy and after nightfall the wind increased considerably and a heavy swell rose, so that we had to shorten and trim sail further, being unsure of the exact position of the coast we were making for and knowing that it had many shoals and a current setting onshore.

At midnight we considered our position satisfactory and, so as to hold our station without moving too far, we kept the courses hauled amidships, and above them the topsails lowered, and lay-to on the port tack for the rest of the night. The *Atrevida*, which had fallen a little astern, was thus able to join us. We obtained no bottom with eighty and ninety fathoms of line.

3 November

The following morning the pleasant weather allowed us to steer NE under full sail towards the Chanduy coast, which indicated clearly that the unsettled conditions of the previous night had been caused by the new Moon rather than by our being in an area which belied the tranquil reputation of the Pacific Ocean. At noon, still out of sight of land despite having a reasonable horizon, we were convinced that Punta de Santa Elena did not extend as far to the west as shown on the charts, which had no doubt been influenced by a natural fear of the dangers in this area. Our latitude was 2°37' and our longitude 1°14' [west of Guayaquil].

We therefore continued steering NE under full sail, taking the precaution of posting a lookout at the masthead and sounding as we went, as the horizon to landward had become shrouded in haze and the very short day prompted us not to waste what was left of it.

By noon we had compared our chronometers with those of the *Atrevida* by means of pistol shots, confirming the uniformity of the rates of numbers 72 and 10 and prompting us to adopt new rates for numbers 13 and 61, the rates of which were very different. After synchronizing the minor alterations by means of daily equations, we managed to bring the results according to those chronometers into agreement with those of numbers 10 and 72, with the following changes:

	Rate in Guayaquil	New rate
Number 61 . . . . . . increased . . . . . . .1′ 7″ 56‴		1′ 12″ 29‴
Number 13	19″ 15‴	30″ 15‴

At two o'clock in the afternoon land was not yet in sight, and we were afraid of being set too far onshore during the night from lack of wind and by a low but rising swell and finding ourselves unable to clear Punta de Santa Elena. It seemed wiser, therefore, to luff up to NNE on a course which would not stand us into danger while still giving us the best course towards the coast, which we intended to survey. Shortly after three o'clock the haze dissipated and we saw Punta del Carnero[1] a short distance ahead with the coast stretching away on either side from Punta de Santa Elena to Chanduy and Sierras del Tambo.

The Chanduy coast and Altos de Tambo bore ENE and Punta de Santa Elena N18°W. We began to measure bases and two sets of hour angles were observed on the meridians of Punta de Santa Elena and the nearby hills; and as a rather high sea and a more moderate breeze threatened to run us foul of the many dangers with which this coast is encumbered, we hauled our wind to the NW under full sail.

Having referred the astronomical position of Sierras del Tambo obtained from our data, when combined with the operations on the following day, to that which I had determined on the recent excursion to El Morro, we saw that the latitude and longitude agreed exactly. This tied in with our work and gave us complete confidence in it, dispelling any suspicion aroused by the considerable difference found in the chart drawn up by Ingeniero Requena[2]. That gentleman does, however, deserve the praise, and I must say gratitude, that we accord to his hydrographic work. In 1770, without accurate instruments and without those aids which are so necessary for work of this kind, although generally perceived to be as costly as they are useless, and suffering the lethargy inevitable in these climates and situations, he surveyed all the surrounding areas. He made observations of latitude, took particular care over soundings and, not content to limit the mission entrusted to him in the King's name to the vicinity of Guayaquil, he extended his investigations to use all the work of Bouguer and Condamine in the vicinity of Cabo Pasado, drawing up a Mercator chart which extended as far as Río de las Esmeraldas.[3]

By sunset we had covered most of the distance to Punta de Santa Elena and were almost sure of being able to clear it on the same tack despite having a heavy sea against us, and this we did at nine o'clock. Then, with little breeze, we steered NE for a distance which would allow us to keep in sight the end of the previous afternoon's survey,[4] and at two o'clock we hove-to, waiting eagerly for daylight to begin surveying once more.

4 November

The day dawned somewhat hazy, but as we were only a short distance offshore and

[1] Situated 8 miles SE of Puntilla de Santa Elena.
[2] Francisco Requena y Herrera (1743-1824); for a biographical sketch see Beerman, *Francisco Requena*.
[3] Río Emseraldas enters the sea 27 miles ENE of Punta Galera.
[4] Malaspina clearly intended to keep within visibility range not within sight as the marks would not be visible during the night.

not far from Punta de Santa Elena itself, the summit of which bore S26°E, we were soon able to begin running bases, much assisted by the light SW breeze then blowing. Since this was our revered Sovereign's saint's day,[1] we kept the ensigns hoisted on board both corvettes and issued the crew with a larger ration of sauerkraut and wine, a new and timely cause for rejoicing.

Favoured by the breeze, although prevented by thick cloud from observing no more than two sets of hour angles in the morning and the latitude at noon, we found our work nicely linked by way of Colonche[2] and Islita Salango[3] as far as Isla de la Plata and Cabo San Lorenzo. At sunset the former bore N63°W, one-and-a-half leagues distant, and the latter N7°E, true. The soundings taken by the *Atrevida* during the day showed that we were sailing in depths of twenty to thirty fathoms and the moderately fresh breeze, blowing steadily, promised us good progress in our work. We chose to pass between Isla de la Plata and the coast, as there appeared to be no danger and in this way we would not have to go far offshore. After making good a suitable distance during the night, we lay-to on alternate tacks until dawn.

This was approximately the position from which I had intended to cross over to the Gálapagos,[4] but as the time approached more and more reasons influenced me to abandon my original idea. The wind was tending too much to the west and the local pilot warned me of fearsome calms in the Gálapagos during this month and on the coast of Gorgona[5] the following month. The time left before we were due to arrive at San Blas[6] was very short while our work in Panamá would certainly be time consuming; and finally the advantages of fixing the position of these islands would not match the disadvantages we would incur, particularly as the settled and favourable SW weather since the new Moon almost demanded that we make use of it for the much more important work to be done on the somewhat redoubtable coasts of Chocó. This decision did not mean that we gave up all hope of seeing the Gálapagos. They lay as far to the south of the coast of Nueva España as they were now to our west, so if a more suitable occasion arose, we could carry out the proposed survey at no great loss.

5 November

These considerations did not make me hesitate for an instant about the course to be followed on the morning of the 5th. Under full sail and steering NEbyE, we continued measuring bases with a starting point at Cabo San Lorenzo which linked with the finishing point of our work the previous afternoon.

We found ourselves close to the anchorage of Manta,[7] a place frequented for a small trade in provisions and manufactured goods, particularly jipijapa hats,[8] as well as

[1] Saint Charles Borromeo (1538–84), Archbishop of Milan.
[2] Situated on the north bank of Río Nuevo, 23 miles ENE of Puntilla de Santa Elena.
[3] Isla Salango, 37 miles NNE of Puntilla de Santa Elena.
[4] Archipiélago de Colón, though still widely referred to as the Galapagos Islands.
[5] The mainland coast adjacent to Isla Gorgona.
[6] The principal Spanish eighteenth-century naval base in New Spain some 400 miles NW of Acapulco; see Michael E. Thurman, *The Naval Department of San Blas*, Glendale, California, 1967.
[7] Situated 14 miles ENE of Cabo San Lorenzo.
[8] Fine woven straw hats.

by shipping from the northern coasts bound for La Puná or Guayaquil, which put in here because with the winds, tide and current against them they risk running short of water and provisions, which are more easily replaced here than at any other place on the coast.

Steering east to approach the coast which, after Manta, turns sharply to follow a [northerly] direction in the first quadrant, at ten o'clock we had reached the meridian of the town of Monte Christi[1] where we observed longitudes with all the more eagerness and satisfaction because we were able to use them to refer our present longitude to that observed by Bouguer during the lunar eclipse of the 26th of March 1736. No mistake could arise unless it had been directly set down by that astronomer in his treatise on the figure of the earth. Because of the fact that Monte Christi lay thirteen or fourteen leagues west of Portobelo, or Panamá,[2] whose longitude is assumed to be 82°10' west of Paris, we were able to infer that the longitude of Monte Christi was 82°50' and that of Cabo San Lorenzo, according to the difference of 13'15" shown by our chronometers, was 83°15'10" if that of Guayaquil was assumed to be 82°15',[3] according to our own set of longitudes observed independently of the calculations made in Guayaquil, either by lunar observations or by the different determinations made at Chimborazo by their Excellencies Juan and Ulloa. We could therefore support our previous conjectures that half the difference found in Guayaquil between the chronometers and direct observations would be the not surprising effect of the almost daily errors found in the lunar tables, adding, as suggested, some part of these to the position of the occulted star.[4]

After a few passing showers and brief calms the conditions soon began to look very promising. At noon, with fine weather and a slight to gentle breeze from SW, Sierras del Balsamo[5] and Punta Charapotó[6] were bearing approximately east at a short distance, our latitude being 46'37 south of the equator, and our longitude 46' west of Guayaquil.

We had entrusted the soundings to the *Atrevida,* as usual, so that the charts would not lack this essential information for inshore sailing and so that our bases could be measured without interruption. The soundings remained at forty to forty-five fathoms all day and, after observing another two sets of hour angles which convinced us of the existence of an easterly current, we were able to fix the end points of our work this day on the parallel of Punta Ballena[7] leaving Cape Pasado to the south, three or four leagues distant.

These coasts are generally deserted, apart from an occasional settlement near the rivers, as they are subject throughout most of the year to strong and frequent squalls, particularly from Cabo San Francisco, and to these are added the currents, which set

[1] Malaspina is probably referring to Cerro de Montecristi, 11 miles east of Cabo San Lorenzo rather than than the nearby town of the same name.

[2] Malaspina appears to imply that Portobelo and Panamá lie on the same meridian whereas Portobelo lies about 9' west of Panamá.

[3] All these longitudes are west of Paris.

[4] At Guayaquil the astronomers observed the occultation of several stars by the Moon.

[5] Not identified on modern charts.

[6] Situated 32 miles NE of Cabo San Lorenzo.

[7] Situated 15 miles NE of Cabo Pasado.

fairly hard to the north or south according to whether it is the season of the trade winds or the *vendavales*. The former is the usual time for passages from the coasts of Mexico, Guatemala and Panamá to Guayaquil and Paita and the *vendavales* are naturally preferred when sailing from Peru to the ports mentioned and those further north.

As this account includes neither the description of the coast, which is to be found in sailing directions, nor that of its political state, which can only be included in general terms referring to a single province or to an entire kingdom of the many which compose the extensive Spanish colonies, we will content ourselves here to venture the following opinion. While these coasts may have been of some importance when the conquistadors of Peru frequented them in their tedious and dangerous inshore passages, since shipping has opened these waters up and almost circumvented the obstacles of Cabo de Hornos, making the galleon fairs in Portobelo redundant,[1] one must presume that each day they will become poorer and more deserted, since the small amount of gold that the Chocó rivers yield will not be sufficient incentive for people to frequent them.

At nightfall, suspecting that the wind would drop considerably, as it had done the previous night, we thought it wiser to continue on our course for the proposed distance, so we steered NbyE at four or five knots, although, to make sounding easier, we kept all the small sails furled. On this heading, and lying-to on alternate tacks from ten o'clock, we remained at the same distance of three or four leagues from the coast and in depths of forty-five to fifty-five fathoms, sand, which later changed to ooze. It seems, as we discovered daily, that on this coast the depth changes from thirty to sixty and seventy fathoms in the space of a few miles and at times a single league.

6 November

Fortunately, apprehensive of the effects of the currents, we had been on our guard against them this night, as on previous nights, and at dawn Punta Pedernales,[2] although fairly high could hardly be seen so that when we began to measure bases at sunrise we could scarcely observe two bearings to the point chosen to link it with our work of the previous afternoon. The wind at that time was a strong SSW breeze, although much lighter than it had been during the night when it had been accompanied by some drizzle and quite a noticeable swell.

The coast now in view was the stretch which, by way of Punta Portete[3] and Ensenada de Punche,[4] leads to Cabo San Francisco which, by ten o'clock in the morning, bore N46°E by compass. This feature, which lies ln latitude [blank] and longitude [blank], deserves the closest attention by those bound for Paita from Acapulco or Panamá, since beyond it currents set strongly to the north while the land breezes lose much of their strength. Thus making a landfall or not to the south of [the cape] will

[1] These were yearly or twice-yearly fairs organized at authorized ports such as Nombre de Dios and later Portobelo in Darién where goods brought from Spain with and under the protection of the galleon fleet could be legally sold.
[2] Situated 35 miles NE of Cabo Pasado.
[3] Situated 10 miles SSE of Cabo San Francisco.
[4] About midway between Punta Portete and Cabo San Francisco.

determine the success of the voyage, saving the mariner from being driven towards [Isla] Gorgona by the currents and calm conditions. If that happens, it is best to return to Panamá and make the passage again, further offshore, such a route being expedient, and indeed necessary, during the *vendavales*.

At noon our latitude was 49'24" north of the equator and our longitude 15' west of Guayaquil, both very different from our dead reckoning position, which the last twenty-four hours placed us 20'S and 16'W of this, indicating a current setting approximately NEbyN at a rate of at least one knot. At the time Cabo San Francisco bore S25°E and Punta Galera[1] S 87° E. We were not able to repeat our observations in the afternoon, as we had hoped to do. We had only just been able to take a set of hour angles at two o'clock before the sky began to cloud over completely, accompanied by drizzle. Consequently the rest of the afternoon's work was limited to the measurement of bases until the southern entrance point of Río de las Esmeraldas bore S19°E, when we obtained no bottom with one hundred fathoms of line, after sailing in depths of ten to seventeen fathoms, sand, since two or three in the afternoon. We had previously suspected and were now quite sure that the coastal flat between Punta Galera and Gorgona extends as far as two leagues to seaward in places, making these coasts, as the sailing directions show, more than a little hazardous. For this reason it was not easy to survey the bay beween Río de las Esmeraldas and Punta Manglares,[2] an area much subject to flooding, with countless rivers, and inaccessible in parts. We felt, however, that we should at least attempt it and so we spent most of the night hove-to, but the drizzle and constant cloud cover made it very difficult to carry out what we had intended.

7 November

However, we were very pleased to see, when the Sun appeared, that the clouds and haze were clearing and, soon after it had risen, not only did the pleasant weather reveal the final positions fixed the previous afternoon, but we also had in view the entire bay as far as Punta Manglares which adjoins another low stretch of coast extending towards Isla del Gallo[3] and Punta de Salaonda.[4] Morro Tumaco,[5] standing alone, is conspicuous because of a few trees which pleasingly crown its summit and Isla del Gallo itself is only separated from the coast by a strait or narrow channel.

With a fair breeze we had reached the parallel of Isla del Gallo by noon, when Punta de Salaonda bore N68°E; our latitude being 1°57' and our longitude 54'30 east [of Guayaquil]. Accordingly we made full sail on a heading of NbyE so as to keep the coast in sight until, at four in the afternoon, we finally sighted Gorgona. Leaving the greenish-coloured water through which we had been sailing since midday behind we obtained no bottom with seventy fathoms of line and, with Gorgona bearing N46°E, we steered towards it, measuring our bases in its direction.

As the afternoon was generally clear, and the wind continued strong and

---

[1] Cabo Galera 10 miles NNE of Cabo San Francisco.
[2] Cabo Manglares (= cape of mangroves), 80 miles NE of Punta Galera.
[3] Situated 32 miles NE of Cabo Manglares.
[4] Not identified.
[5] Situated 15 miles NE of Cabo Manglares.

favourable, we made good progress in our work until five o'clock and were also able to obtain another two sets of hour angles which confirmed the existence of an easterly current.

At sunset, with the coast of Salaonda still in sight, the mid-point of Gorgona bore N46°E, eight to ten leagues distant, depth thirty-five fathoms, shell.

From Gorgona the coast recedes for a considerable distance, as far as Bahía de San Buenaventura,[1] which is here the true easternmost point of the Pacific Ocean,[2] offering rapid and probably easy communication with Ciudad de Santa Fé,[3] the capital of the viceroyalty of that name. This part of the coast, as well as being subject for most of the year to almost continual downpours and squalls, is also subject to currents which generally drive ships towards the coast and equally make it difficult to get out to sea. Our progress until now, however, had been so good on a truly rugged stretch of coast that we felt obliged to risk spending a day examining it with the detailed care that we intended and which we had successfully exercised so far. To do this, it was important that we should be fairly well inshore by dawn, while still keeping Gorgona in sight, so as not to have more easting to make than could be covered under full sail by midday. Then using the afternoon sea breeze, which the previous days' experience led us to believe would be a bit stronger and freer, to make westing again towards Cabo or Punta Chiramirá,[4] which eventually connects with Cabo Corrientes [far] to the north.[5]

We continued to make good a course during the night with the intention of passing some three leagues from the [northern] extremity of La Gorgona, taking due care to sound and obtaining no bottom with seventy fathoms of line. Once we had made a sufficient distance to be sure we had done so, the night being extremely murky and overcast with rain, we luffed up to the east and finally hove-to on the same tack at three o'clock, as it seemed prudent to wait for the dawn in this position.

8 November

Neither at dawn nor at sunrise the following day had the rain or cloud cleared. We therefore considered it impracticable to attempt to survey a coast that consisted of nothing but mangroves and was so shallow that we were fearful of approaching it, especially as the current was setting us strongly to the east.

In this situation, although we were naturally very cautious, we nevertheless made the initial assumption that the weather had spent its force in constant heavy rain overnight and would perhaps soon improve. Accordingly, while we took the necessary precautions of sounding and keeping a masthead lookout, we promptly came up into the wind under full sail on a SE heading to close the coast so that we could take bearings of it if the Sun appeared, however briefly, and finally to plan with determination our subsequent passage so as to achieve, as far as possible, our objectives under difficult circumstances. On this heading the *Atrevida* very soon hailed us

---

[1] Bahía Buenaventura.
[2] Malaspina presumably means the North Pacific since much of the coast of Chile lies further east.
[3] Santa Fé de Bogatá, now just Bogatá, or its viceroyalty.
[4] Punta Charambirá, 90 miles NE of Isla Gorgona.
[5] Situated almost 75 miles north of Punta Charambirá.

to say that they had sounded forty fathoms, ooze, and shortly afterwards our lookouts sighted, between showers, some mangroves three or four leagues to the ESE.

In this position, with the rain still falling, we thought it better to shorten sail somewhat and not to sail any closer to the coast until we had a better chance of surveying it. However, we soon changed our minds again as a little before eight o'clock, with a slightly fresher breeze from SW, the sky began to clear and soon afterwards we were able to observe hour angles and take a bearing of SW5°S to the Isla Gorgona in the distance. To the NE the trend of the coast, from ESE and SE, showed that we were not far from Bahía de San Buenaventura and with a gentle breeze from SW we could expect to reach it soon.

We altered course to NE and, sailing only three or four leagues offshore, we were soon out of sounding in depths which must have been over seventy fathoms. We observed hour angles again and at last at noon, helped more by the current than the breeze, we had a clear view of the head of the bay and the mouths of some of the rivers which discharge into it, while the extremities of the Isla Palmas[1] bore [blank]. Our latitude was 3°33', 18' further north, and our longitude 2°20', 24' further east, than by dead reckoning.

Having carried out our intentions completely with regard to the survey of the coast, we had now only to steer towards the west again so as not to run any risks from the effects of the currents during the night and to make good some distance towards our work on following day. We therefore altered course to NNW and passed Isla Palmas at a distance of two leagues in depths of forty to thirty-five fathoms, soft clay. At two o'clock we could already see the outer mangroves off Punta Chiramirá, bearing N19°W, and the point itself was visible at four o'clock, but we could not clear it on our tack in the fourth quadrant as the tide was setting us hard onto the coast. So when the wind shifted a little to the west, with a passing shower, we went about under full sail to a heading of south and SSW. By sunset the depths had increased to thirty-three fathoms and the point was bearing N12°W, thanks more to the northerly currents than to our tacking which had been to little effect because of the lack of wind.

At eight in the evening we could no longer obtain bottom and, as the breeze had swung once more to the SW with a few showers, we tacked to the NW again and set a course so as to have Punta Chiramirá bearing in the second quadrant the next morning.

As we lay-to from two in the morning and had covered very little distance, we were able to counteract the strength of the currents on this occasion. Soon after sunrise, since we had made a few miles to the east and the rain and heavy cloud that had accompanied us constantly from ten o'clock the previous night had cleared, we sighted once again the end-points of the previous day, although at a considerable distance, and were able to resume our survey.

From Punta Chiramirá to Cabo Corrientes much of the coast is of mangroves, dotted at intervals with small hills which increase in height towards Cabo Corrientes. We obtained no bottom with one hundred fathoms of line three leagues offshore.

---

[1] Situated in the NW approaches to Bahía Buenaventura.

Some of the many rivers which flood the Chocó area still flow into the sea here, although less frequently than on the coasts of Palmas[1] and Gorgona.

This morning we were quite happy with our survey. We had measured bases and observed two sets of hour angles and were in a position to link them to the astronomical observations with much satisfaction since, on these coasts, the effects of the currents on the aberrations of the bases is considerable. On the other hand, as the closeness to the equator made any errors in latitude, however considerable, unlikely to interfere with the reliability of observed times, we could also correct any errors that arose from the currents, thanks to the agreement of many bearings with the observed longitude, and so maintain the reliability of the latitude.

As the weather continued clear, we had Cabo Corrientes in sight by noon, bearing N10°W by compass, although a long way off; our latitude was 4°42' and our longitude 2°25', the former being 30' to the north and the latter only 10' to the east of our dead reckoning position. Cabo Corrientes rises to a considerable height above the sea and from a long way off looks like an island with a single peak; but close-to, some *farallones* can be seen joined to it and inland there are other high sierras, not far from it. Sailors fear it because of the currents and the violent squalls which occur there almost daily. To the south it protects a large bay[2] of medium depth of six to eight fathoms, ooze, which offers shelter to smaller vessels bound for Chocó which, having finally doubled the cape, are afraid that if the wind drops they will immediately lose that considerable advantage.

All afternoon we continued steering towards the cape, sailing some three leagues from the coast without obtaining bottom with sixty fathoms of line; we obtained further observations for longitude and because the breeze was light at nightfall we were still a considerable distance from the cape, which was now bearing N7°E by compass.

At noon, as on the 6th, we had taken care to repeat the comparison of our chronometers with those of the *Atrevida*, by means of pistol shots. A considerable variation in the temperature must have been the cause of the many daily alterations in their rates, which were not easily found on a first examination but could nonetheless give rise to quite significant errors. On the *Atrevida* they had noted considerable changes in chronometer number 71 and pocket chronometer 105, which we also found in ours; the daily comparison told us that these should be attributed to 72 rather than 61, particularly as the results according to the latter, with its new rate, agreed with those of number 10. It was becoming increasingly evident that these comparisons should be made often enough to avoid the risk of significant errors, or at least to ensure that we were not accused of failing to do everything possible to determine the truth.

If the rain had been heavy and frequent on previous nights, it was much worse tonight, particularly after ten o'clock, when we believed ourselves to be abeam of Cabo Corrientes and four leagues off, as the shadow of the land appeared to indicate. With the violence of the rain the wind, which we had used to tack to the NW under topsails and foresail, slackened and shortly before dawn light airs and further

[1] Presumably the coast adjacent to Isla Palmas.
[2] Bahía Cuevita.

downpours were followed by a squall with several bolts of lightning. Their flash, the sound of thunder and the smell of sulphur showed how close to the corvettes they had struck.

10 November

Nevertheless, the weather abated somewhat and with a slight to gentle SW breeze now prevailing, and with Cabo Corrientes bearing [blank], at eight o'clock in the morning we were able to see a great part of the coast beyond it as far as Las Anegadas.[1] We were able to take two sets of hour angles, which, when referred to local apparent noon and a very reliable meridian altitude of the Sun, dispelled any doubts about that day's work. The considerable difference of 40' to the north that we found between our observed latitude and dead reckoning could not fail to surprise us. We signalled the *Atrevida* to ask for the results of their observations and as they were the same as ours we could only wonder at the considerable effect of the currents, which, however, only altered our dead reckoning longitude by 7' to the west.

As soon as the Sun had passed the meridian, the rain began to set in again, preventing us from making a thorough survey of the coast, particularly since after Las Anegadas it is low and deeply embayed as far as Morro Quemado.[2] On the other hand, as rain is more usual than clear weather on this coast, we were determined not to wait for better weather, which probably would not materialize, as the currents would set us off course even if we attempted to stay on a particular parallel. Our plan, therefore, was limited to a gradual approach to Punta Garachiné,[3] tacking throughout the night so as to make as little leeway as possible, and sailing during the day at a distance from the coast which would not compromise our safety but would allow us to make use of any clearing in the weather to fix one or other of the points in sight.

At half past four in the afternoon we did indeed sight a short stretch of the coast beyond Las Anegadas, but before sunset it was lost from view again and at nightfall we were left in no doubt that we would be having a night of heavy and continuous rain.

In accordance with our plan and still wishing to stand somewhat further offshore in the hours of darkness, we hauled our wind to the south at midnight, tacking at four o'clock once more to NWbyW when the breeze set in again from SW. The coast in these parts runs NWbyW by compass as far as Morro Quemado and there is no bottom to found with the lead until very close inshore.

11 November

Although it had rained heavily during the night and the previous afternoon, it did not slacken the following morning. Sometimes carrying a press of canvas and at other times under easy sail, we steered sometimes NE and at other times NW, as the circumstances and weather dictated. At eight in the morning, after an hour of calm,

---

[1] Not identified

[2] Malaspina appears to have misidentified Morro Quemado, whose location on contemporary Spanish charts is confusing, but it is probably either Punta Cruces or Cabo Marzo, situated 23 miles north and 37 miles NNW, respectively, of the prominent Punta San Francisco Solano.

[3] The eastern entrance point of Bahía de Panamá and the southern entrance point of both Ensenada de Garachiné and Bahía San Miguel.

since the breeze began veering to the west and WNW, we chose a course in the third quadrant which would keep us at a reasonable distance offshore while somewhat delaying our approach to Punta Garachiné. Taking these precautions, we sighted land four or five leagues distant at half past seven, at nine o'clock and at eleven. It seemed to be the stretch of coast extending from Morro Quemado to the heights of [Punta] San Francisco Solano.[1] At noon, however, the rain was still falling and at the same time we had increased both our observed latitude and longitude.

Despite this we were more successful the following afternoon. After a calm and light airs from the west, the breeze became moderate and backed to SW, clearing the horizon and allowing a clear view of the coast from Puerto de Piñas[2] to Punta Garachiné. In fact we were unable to make any longitude observations, but since we could estimate very approximately our distance from the various points in sight and, having measured a base, could link the southernmost of these with Punta Garachiné, whose position we were to determine astronomically, we now felt with much satisfaction that the day had not been wasted. Having seen almost all the coast from Cabo Corrientes and having kept our dead reckoning for a little over twenty-four hours, there would be no significant errors. At sunset Punta Garachiné bore N5°E by compass, five or six leagues distant. We warned the *Atrevida* to prepare her anchors and to compensate for the effect of the currents we lay-to on alternate tacks from the early hours of the night, only obtaining a depth of sixty fathoms, ooze, at four in the morning.

During that night, which was very clear at times, Don Dionisio Galiano obtained from meridian altitudes of β Andromedæ and Achernar[3] our latitudes as 7°49′ and 7°46′ at 10:20 and 10:50 respectively. To some extent this confirmed our work of the previous days.

12 November

However, it was not necessary to use these results the next morning as the day had dawned fairly clear, apart from a few showers which obscured the coast at times, so we were able to take a set of hour angles and then observe our latitude as 7°54′30″, with Punta Garachiné bearing N41°E and the Isla Galera[4] N35°W. Although dawn found us only four leagues from the point, our progress was extremely slow until noon, as the breeze was very light all morning and the tide was against us. However, when the tide changed in the afternoon we made good progress, even though the breeze dropped until it was barely perceptible. At sunset Punta Garachiné bore east by compass. We could then see a considerable stretch of the coast of Darién and parts of [Archipiélago de las] Perlas, Galera and San Telmo.[5]

The entrance to Golfo de Panamá must be navigated with care because of a rocky outcrop, known as Bajo de San José,[6] which lies almost halfway between Punta

---

[1] A prominent point about 50 miles NNE of Cabo Corrientes.
[2] Bahía Piña about 36 miles SSE of Punta Garachiné.
[3] α Eridani.
[4] The SE island of Archipiélago de las Perlas.
[5] San Telmo, off the SE side of Isla del Rey.
[6] Banco San José, 14 miles west of Punta Garachiné.

Garachiné and Isla Galera, particularly as the ebb tide sets towards it quite fast and the flood tide sets towards Golfo de Darién,[1] also much encumbered with shoals. Because of these dangers it is best to anchor if the wind drops. We did this at eight o'clock in the evening, anchoring in twenty-four fathoms, ooze, as we had the tide against us and a very light breeze directly against us.

It was calm during most of the night and it was only at three o'clock when the tide began to flood that we felt a few light airs from NNE, which we did not hesitate to use, in spite of their being feeble, by setting sail immediately and ordering the *Atrevida* to do the same a little beforehand as she lay close to leeward of us.

13 November

Twice we were threatened with having to anchor, before and after a flat calm, but we maintained our course in depths of thirty fathoms, mud, and finally, as the breeze freshened a little from NE, we went about to NNW again and continued making for the coast, so that at nine o'clock Punta Garachiné bore SEby E, when we had a clear view of Isla del Rey and, on the opposite side [of the fairway], the *fallarones* of [islas] el Pelado[2] and Majagual[3] and all the northern coast of the gulf.

With the approach of the sea breeze at ten o'clock, the land breeze dropped entirely and for some time we remained at the mercy of the ebb tide, which was setting us onto the islands; but soon we felt light airs from NW and set a course, close hauled, to NE, altering course to north as the breeze freshened slightly and backed to WNW.

Our latitude by observation was 8°24′ and our longitude, according to hour angles taken in the morning, was 1°9′30″. Those observed in the afternoon on the meridian of Isla el Pelado gave it as 1°14′ east of Guayaquil. Soon after midday we began to observe lunar distances to the Sun, when the result of the mean of fifty-nine sets of observations was only 3′ east of our longitude according to the chronometers.

At three in the afternoon, when we were close to Isla el Pelado, in depths of seventeen to eighteen fathoms, mud, we tacked to the south to stay in the channel between this rock and Islas de las Perlas, as our course for Panamá would have to pass close to Isla Pacheca, the northernmost of these islands, in order to avoid a bank that ran a long way out from Punta Mangle. With the tide behind us, and on a southerly heading, we soon approached Isla Pacheca which was already bearing west two miles distant by sunset. Accordingly we went about to the fourth quadrant.

The night was exceedingly calm and pleasant. We continued on alternate tacks, remaining in depths of nineteen fathoms, ooze, and as the *Atrevida* fell well behind us, either because of the currents or because of an imperfectly understood signal, we bore away at midnight to join her and then hove-to with the foretopsail braced aback until daylight, our intention being to avoid the necessity of anchoring if becalmed at either end of the navigable channel.

14 November

At sunrise we found, as was to be expected, that we had made very little progress

---

[1] Bahía San Miguel; present day Golfo de Darién indents the north coast of present day Columbia.
[2] A small isolated island, 5 miles offshore, 36 miles NW of Punta Garachiné.
[3] Situated 2 miles offshore and 6 miles ENE of Isla el Pelado.

during the night. Farallón Pelado bore east and we could not yet consider ourselves clear of [Isla] Pacheca; the light breeze, directly against our course, did not allow for any progress on either tack. The depth remained at twenty fathoms, ooze, as on the previous night and as we began to feel puffs of the sea or westerly breeze towards noon, we hauled our wind to the north to make as much westing as possible.

In the morning we had observed longitudes with the chronometers and in the afternoon we repeated the observations of lunar distances to the Sun, the mean of which was 15' west of the chronometers. The latitude at noon was 8°36'48" by observation and the weather extremely fine and pleasant.

At four o'clock we were sailing on the port tack, as previously mentioned, making for the mainland, when the depth suddenly decreased from seventeen to ten fathoms, leading us to believe that we were very close to the bank off Punta Mangle. We went about immediately and tacked towards Isla Pacheca, finally tacking inshore again at sunset.

Anyone who knows the natural laziness of the local sailors and the type of craft they sail will not be surprised that merely the passage from Punta Garachiné to Islas de Perico takes them, with some risk, as long as ten or twelve days.

At nightfall the breeze, which had been very light almost all afternoon, dropped still further, the horizon became squally, and at eight o'clock we were struck by a squall from the NW, with much lightning, some thunder, fairly heavy rain and gusts. We lowered the topsails to half mast, and, as we were then in a depth of fifteen fathoms, we wore ship and hauled the wind on the starboard tack. Within less than an hour the sky had begun to clear when the wind veered a little to the north, allowing us to set a course to the west under full sail. At eleven o'clock Isla Pacheca was abeam to leeward, with depths of sixteen and seventeen fathoms. [15th] At dawn the following morning, in close company with the *Atrevida*, we could see Islote Chepillo bearing ENE distant one-and-a-half or two leagues. Islas Perico and Taboga, Altos de Panamá,[1] several of Islas de las Perlas, together with the whole coast beyond from Panamá to Chepillo, presented a scene which was all the more pleasant and delightful in that the variations of height and the general luxuriance of growth combined with the great clarity of the sky and horizon to impart a brilliance to the scene which Don Felipe Bauzá carefully captured in a well executed panorama.[2]

At dawn the breeze dropped considerably, but as it remained in the fourth quadrant; steering WSW, we were able to make some way towards the anchorage. The depths had gradually increased to thirty and thirty-five fathoms.

At noon our latitude was 8°40', but because of an error we were unable to calculate our longitude by chronometers. Another fifty-two sets of lunar distances, referred to the meridian calculated by the chronometers the day before, gave a result only 1' to the east of it.

The sea breeze did not set in until four in the afternoon, so until then our progress was very slow but we were then able to make for the anchorage under a press of canvas, continuing to tack as necessary, being sure of our safety both from

---

[1] Heights in the vicinity of Panamá.
[2] This drawing does not appear to have survived.

the soundings and the very clear night, until finally at two in the morning we dropped anchor in eight fathoms, ooze, two miles NE of Isla de Perico. The ledges and shallows fringing the shores of Panamá and the lack of shelter from SW winds, lead all shipping to prefer an anchorage NE of the two islets of Perico and Flamencas,[1] which cannot supply water or firewood but are well cultivated, very steep, and, besides, barely one league from Panamá.

16 November

No sooner had dawn broken than both corvettes set sail once more to approach the islets, the breeze remaining moderate from NW, and finally, at ten o'clock, we anchored in a suitable position, moored so as to lie NW/SE. From here Isla Perico bore SW three or four cables distant, and the depth was five fathoms, ooze, at low water.

---

[1] One of Islas de Perico.

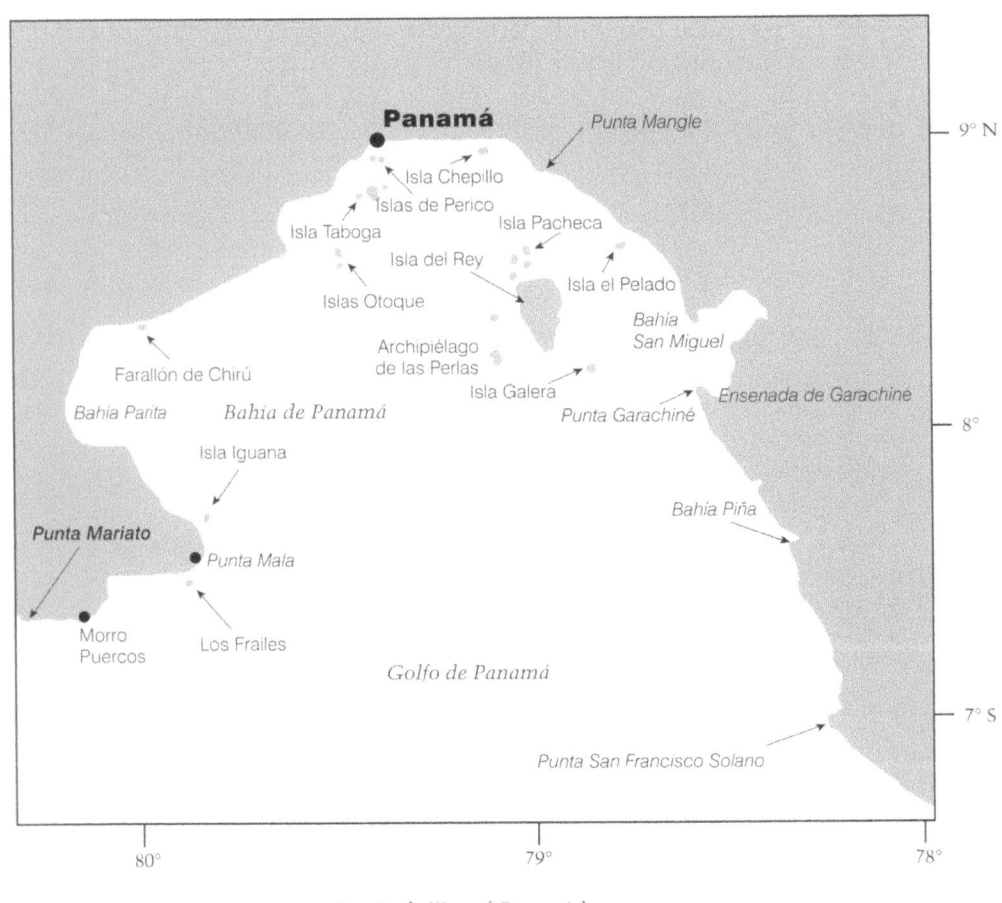

Fig. 9. Approaches to Panamá, November to December 1790

CHAPTER 4

At Panamá

[16 November]

At the time of our departure from Guayaquil all the most recent news from Europe led us to believe that the present disputes in the north and England's increasing naval armament[1] might lead our court to break off relations. So it seemed best to do nothing about our scientific operations until I had certain knowledge of the true current state of Europe, which of course would most probably reach me in this country where letters sent via Portobelo usually took no more than fifty to sixty days. With this in mind, having entrusted to Don Dionisio Galiano the responsibility of taking the corvette to the appropriate anchorage, I went in the pinnace at dawn to see the Governor and Capitán General of the province, Brigadier de la Real Armada José Domas y Valle.

This officer quickly dispelled any apprehensions about an imminent outbreak of hostilities and indeed helped us as far as he could to continue our work. I was very soon able to set up the observatory in the armoury of [Castillo de] Chiriquí,[2] whose adjoining bastions afforded a view of almost the entire horizon as well as of activity on board the corvettes so that the necessary signals could be given for an exact comparison of the chronometers with the astronomical clock. Two other houses nearby were given over to taxidermy, painting and other activities essential to our study of natural history. Finally, at midday, I was able to return on board to ensure that on that same afternoon the astronomical equipment at least was taken to the city. This apparent haste was all the more necessary because, in order to provide observations on the night of the 17th which were of the greatest importance for the deduction of longitude, the astronomical clock would have to be set up and various other preparations made well beforehand if we were to achieve the reliable results that we wanted.

Strangely enough, the longitude of Panamá was still highly uncertain despite the visits of Spanish and French astronomers who went to Quito in 1743[3] to examine the figure of the earth with the measurement of a terrestrial degree.[4] Both calculated this

---

[1] In 1790 the British fleet was mobilized following the seizure of two British trading vessels in Nootka Sound by the Spanish commandant.

[2] For a view of Panamá from Castillo de Chiriqui see Plate. 36.

[3] Jorge Juan (see p. 219, n. 2 above), Antonio de Ulloa (see p. xciv, n. 1 above), Pierre Bouguer and Charles-Marie de La Condamine (see p. 263, n. 1 above).

[4] Sometimes referred to as a degree of the meridian, but more usually described as an arc of the meridian since it is unlikely that an exact degree could be measured. For a brief description of the procedure see pp. 327-9.

longitude from that of Portobelo which had been observed at the beginning of the century by Father Feuillée, but as Messieurs Bouguer and La Condamine disagreed with their Excellencies Juan and Ulloa, they assumed that there was no difference between the meridians of Panamá and Portobelo. The latter two, however, with a detailed calculation of dead reckoning, inferred that Panamá was 31' further west than Portobelo. In addition to these doubts, we needed to fix this point with all possible certainty, so that it would serve either as a principal meridian for plotting the charts of this coast accurately when referring it by chronometers to Chagres, Portobelo or any other point on the coast opposite, or as a subject of repeated and detailed examination if the earlier observations of the officers to whom it had been entrusted gave results very different from ours.

On that same morning of the 16th (as I had instructed them) the astronomers had taken the astronomical quadrant and chronometer 61 from this corvette to the nearby beach and had attempted to obtain the longitude by chronometer of the anchorage by means of equal altitudes. However, owing either to the heavy shower which interrupted their work most inopportunely, or the same downpour which prevented complete accuracy in setting up the astronomical quadrant, we had to abandon the noon sight, as we could not make the variation in rate that it assigned to the chronometers of both corvettes tally with that of the following days. Consequently, the astronomers Galiano, Concha and Vernacci moved to Panamá that afternoon, and from that night on they concentrated on the calculation of the occultation of a star by the Moon which was to occur on the following night of the 17th.

17 November
Despite this, all their efforts were wasted, as the irregularity of the astronomical clock did not allow us to determine the time of noon by means of equal altitudes, nor, having immediately substituted it with chronometer 61, could we make the desired observation during the night because of the complete cloud cover.

Also on the same day, all the officers went together to Panamá to wait on the Governor, whose affability was no surprise, as many of us knew him from his time in command of a ship in General José Solano's squadron, or from his previous appointment as senior officer of Departamento del Ferrol.

18 November
With noon on the18th referred to our chronometers by the appropriate signals, we were now able to deduce the longitude of Panamá as follows:

	No 61 Gain	No 72 Loss	No 10 Loss
Time observed at Panamá	3h 15m 6s 11	3h 48m 28s 4	3h 49m 30s 14
Ditto at Guayaquil	3h 16m 28s 15	3h 46m 50s 55	3h 48m 9s 36
Difference between meridians	1m 22s 4	1m 37s 9	1m 20s 38
Equation for comparisons. Slow	4s 42	20s 6	4s 12
Corrected difference	1m 17s 22	1m 17s 3	1m 16s 26

The mean of which gave the longitude east of Guayaquil in degrees	0° 19' 14"
Which referred to the longitude of Guayaquil from Paris according to our series	82° 00' 30"
giving the longitude of Panamá west of Paris as	81° 41' 16"
Deduced from the observations of M. Bouguer in Manta adjusted by our chronometers	81° 42' 06"
243 sets of distance from the Moon to the Sun observed on the 13th, 14th and 15th adjusted by our chronometers	81° 57' 5"
Deduced from the observations of Father Feuillée in Portobelo { according to the authority of the French Academicians	82° 10' 0"
According to Señores Juan and Ulloa	82° 41' 0"

Consequently, since our observations at Guayaquil gave us reason to believe that the longitudes carried from Lima had somehow been affected by an error to the west, we could now, without presumption, voice our suspicion that Father Feuillée's results were as dubious in Portobelo as they had been in Valparaíso, Coquimbo and Arica.

The equations of our chronometers gave no sign that they had been subject to any undue force, since not only did their changes of rate lay in the same direction as indicated by the new comparisons but these arose from daily comparisons between the three chronometers of each corvette, which we repeated every three days between number 72 of the *Descubierta* and the excellent number 10 of the *Atrevida*.

Our comparisons between the three chronometers only extended to noon on the 14th because of the unexpected discovery on the morning of the 15th that number 13 had stopped and that no amount of shaking would start it again. We could only suppose that the spring recently replaced in Lima had broken.

The observations and signals of the 16th which allowed a thorough comparison of our chronometers with number 10 and its rate from the 18th to the 19th, adjusted by the daily equations, showed that this excellent instrument had not altered its rate in the slightest from the 14th to the 18th.

With the observatory now set up and Neé and Haenke having begun their botanical excursions at the first opportunity, Don Antonio Pineda took the opportunity to air and check again the invaluable collection of dried birds and animals from Guayaquil. Because he was afraid that they might not yet be entirely free of infected material and could have deteriorated considerably in the heat of the recent passage, he had managed to find room for the cases in a deck-house, so that they would benefit from fresh, clean air. Our surprise can be imagined when we saw, upon opening the cases, that despite all our precautions the entire collection was spoilt and useless, although it consisted of some eighty pieces, both large and small.

In accordance with our proposed system of economizing as far as possible in our use of provisions, while at the same time providing both crews with as much sustenance as would be conducive towards a greater loyalty to both the navy and this expedition, I took care that same day to have the buyers accompanied by the sergeants from both corvettes. This was so that they could examine and, when appropriate, obtain any kind of provisions, so long as that did not involve a permanent commitment or sending a boat to Panamá every day. Nor should it involve an additional administrative burden on the crown, or any greater expense than the two reales a day which until now had been allowed for rations, while preventing any semblance of liberty or room for any undue advantage to be taken that might encourage spendthrift habits in the men. This method might then make the withholding of their pay bearable; fifteen or twenty days provisions were not of course worth even slightly relaxing discipline or abandoning our scientific pursuits.

It very soon became clear, however, that this method could not be adapted to the present circumstances: fresh bread was very expensive and poor, meat was neither easy to get nor nutritious, vegetables were scarce, and in the end it was necessary to send a pinnace to Panamá every day, without the slightest profit or advantage for the crews. So that night it was resolved that we would continue with the usual rations on board and that in compensation, fish and any fruit or roots we could get would satisfy both palate and health, given that both crews, or at least the majority of them, remained in good condition.

The tertian fevers[1] which had affected carpenters, caulkers and a few sailors since Lima had now all but disappeared on board the *Descubierta*. Don Fernando Quintano seemed to have found considerable relief from his rather dangerous stomach complaint by riding, recreation and the diet that he decided to follow in Panamá; so the only sick man to occupy the surgeon's attention was a seaman from this corvette who, as we left Guayaquil, had shown symptoms of an obstruction, particularly in the liver, which had not responded to the use of emollients or aperients. Don Francisco Flores found that Dr Masdevall's antimonial compound[2] was as useful in these climes as it had been in the cold of Chile and the Patagonian coast and followed it up with Peruvian bark[3] as required. The indigestion and burning fevers that inevitably afflict the sailor in the tropics soon disappeared.

With respect to the commanding officer's particular disciplinary measures for maintaining good health, it goes without saying that on board both corvettes we did not deviate from the method used so far, dividing the crew into three watches, occasionally adding wine to excellent food, taking great care over the cleanliness of the ships and crews and, above all, moderating all these specific provisions with variety, stimulation and kindness so that a man with any feelings, who would naturally look for change, would neither find himself surrounded by bleak monotony nor likely to tend towards random disorderly conduct and destruction.

As indicated in Guayaquil, one of the principal factors in the preservation of health was that of keeping the crew busy every day. When this was implemented with moderation, tempered with half a pint of wine, and mingled with rest periods and outings

---

[1] Malaria.
[2] See page 25, n. 1 above.
[3] Quinine.

on feast days, it could also distract the men from dissolute connections ashore and maintain them in that state of beneficial perspiration, hearty appetite and need for rest, which must be considered the real basis of keeping sailors healthy in these climes.

Accordingly our tasks in this port had the following aims: first, trips with both launches for specific hydrographic purposes which, in this large expanse of gulfs and numerous islands, would inevitably be lengthy and complicated; second, the daily communication by the pinnaces with the observatory and natural history workshops in Panamá; third, using the *bombos* to collect plenty of firewood, which we would have to cut on a point of the mainland to the WNW, a good league from the corvettes and beyond the northernmost point of Isla Flamencos; fourth, a weekly attention to the corvettes' black-strakes to avoid the risk of worm, heeling the ships so as to be able to clean and expose to the sunlight all the part not protected by the copper sheathing; and finally fifth, fishing, hunting, swimming, laundering clothes etc. We managed to encourage the men to engage in these useful activities not only by cajolery and hints, but also by the excellent example of the commissioned officers.[1]

Despite the fact that the weather now seemed clear and the NW breeze settled, we suspected that the rains would continue for some time, particularly as the inhabitants unanimously assured us that this year had been extraordinarily dry. So we hastened to unbend and stow all the staysails, topgallants and mizzen topsails, these being the only sails that we could fit into our cramped and already crowded sail lockers.

23 November

From the 19th to the 23rd we arranged and carried out all the scientific excursions, the completion of which would alone determine the length of our stay in port. The *Atrevida*'s launch was sent in charge of Don Secundino Salamanca, with a *pilotín*,[2] to plot with accurate soundings the three-fathom line[3] along the coast from Panamá la Vieja[4] as far as Islas de Majagual and Pelado. In particular, he was to determine the true extension of the bank off Punta Mangle and take bearings with the theodolite of the highest points to link the details of that stretch of the coast with the work to be done here in the port and he was instructed to make several observations for latitude, linking them, if possible, with the ends of one or another of the base lines.

Don Juan Vernacci, with a *pilotín*[5] from the *Descubierta*, supplied with the Ramsden astronomical quadrant, pocket chronometer 105 by Arnold and a theodolite, had orders to make for Cruces[6] and from there by river to Chagres,[7] to measure the longitude [from Panamá] to the opposite coast, so as to link our work accurately with

---

[1] Malaspina's orders for these activities survive in documents dated 16-22 November 1790: AMN, MS 278, ff. 85-6; MS 427, ff. 69-71, 158v-161; MS 729, ff. 72v-74, MS 755, ff. 49v-50, 56v.

[2] Named as Delgado in the original but subsequently crossed out.

[3] The three-fathom line could be considered as the danger line which shipping should not cross.

[4] The ruins of the old city of Panamá, which was sacked and burnt by Henry Morgan in 1671, situated about 5 miles NE of the site of the city at the time of Malaspina's visit.

[5] Named as Sánchez in the original but subsequently crossed out.

[6] An inland town on Río Chagres, where the river changes direction.

[7] Situated at the mouth of Río Chagres on the shores of the Caribbean Sea.

that which was very soon to be undertaken at the orders of Capitanes de Navío Ugarte and Villavicencio. He was required to return as quickly as possible even if it were necessary to leave the rating of the chronometer until Cruces. He was ordered to satisfy himself about the position of Chagres, limiting himself to ascertaining the trend of the coast and finding the approximate distance to Portobelo and lastly, if unable to join us by the given date or if we could not wait for them, he was ordered to make for Cartagena and from there either to join the expedition again at San Blas when it returned from the north, or, following the orders of His Excellency the Viceroy of Santa Fé and the Commander in Chief of the navy in Cartagena, to undertake hydrographic work on the adjacent coasts.

The *Descubierta*'s launch, with the other *pilotín*,[1] a native pilot, chronometer 71, two sextants and a theodolite, was put in charge of Teniente de Navío Novales, who was to survey and carefully plot all the islands which, under the name of Las Perlas or Islas del Rey, comprise that archipelago. He was to examine Bajo de San José, lying halfway between Punta Garachiné and Isla de Santelmo and to ensure that all this work was linked with the other surveys by means of bearings from high points taken with the theodolite and compass.

It seemed to me that Ensenada de Darién[2] should be omitted, as a few years earlier, at the orders of His Excellency the Viceroy and Archbishop, the *piloto* Alférez de Navío Fernando Murillo of the royal navy had done some work in it and also because it was neither of any importance for shipping nor in any case was it significant to the masters of the barges which go there almost every month, either carrying troops or in search of timber, and they offered to give us a detailed and accurate account of it.

While we were engaged in gathering the necessary materials to carry on with our tasks, Don Felipe Bauzá and Don Juan Maqueda immediately continued the very complicated task of putting in order what had already been done. Bauzá was in charge of the Mercator chart from Guayaquil to this port. The complete survey sheet of the vicinity of Guayaquil from Cabo Blanco to Punta Santa Elena was entrusted to Maqueda who was to bring together the various results, whether mine or those of Murphy, Vernacci, Tova and Robredo or of the corvettes themselves while approaching or leaving [Guayaquil]. The artist José Cardero had already taken some views of the anchorage and the city and was busy drawing interesting specimens of natural history from life, while Haenke and Neé increased their extremely rich plant collections with important acquisitions, having been spared the fatal effects of the proximity and shade of the *manzanillo*.[3] This tree, which has already been scientifically described by botanists, and whose timber is used for various civilian purposes even though the touch of its leaves is truly deadly, naturally excited the curiosity and professional pride of our botanists and there were several of our seamen there who had rashly approached these trees when sent to cut firewood. The immediate consequences of its noxious shadow were swelling in different parts of the body, violent

---

[1] Named as Inciarte in the journal but subsequently crossed out.
[2] Bahía San Miguel, see p. 287, n. 1 above.
[3] The manchineel tree (*Hippomane mancinella*), having a poisonous and caustic milky sap and acrid fruit somewhat resembling an apple. It has been said to be so poisonous that people have died from merely sleeping beneath its shade.

Plate 34. Isla Taboga, Golfo de Panamá, by José Cardero. Museo de América, Madrid

vomiting and general and acute pain throughout the body which did not subside for many hours, after causing considerable pain.

From the 21st, with the full Moon now past, rain began to set in again and on some days this was so heavy and accompanied by wind from one direction or another, although usually calm from which ever direction, that at times all communication with the shore was cut off and the astronomers missed a number of observations of the greatest importance, which I had calculated that the passage of the Moon would have given them. However, Don Dionisio Galiano informed me, on the afternoon of the 25th, that he had been able on the preceding nights to observe occultations by the Moon of Nos 88 and 243 in Mayer's catalogue and the results had allowed him to calculate the longitude of the observatory.

By the former . . . . .81° 44′ 32″ from Paris
By the latter (not very reliable) . . . . .81° 57′ 15″

The first satellite of Jupiter, the immersion of which was observed on the night of the 25th, although the clouds obscured it for a very few minutes beforehand, gave results very similar to those shown above.

The absence of our launches, the employment of both *bombos* in collecting firewood and of the pinnaces in daily travel to the city, moved me to request one of the barges from the city's complement to be crewed by our unemployed seamen and marines, both to take them to the city on their off-duty days and to expedite the transport of firewood on working days. The Governor agreed to lend us one, as I expected, but as he was very short of artificers, equipment and funds for fitting out another which was out of commission at the time, I had to agree to let our carpenters and caulkers work on it and to donate some canvas, new rope and oakum, without which it would have been impossible to make it serviceable.

26 November

On the night of the 26th we were pleased to see the *Atrevida*'s launch return. Don Secundino Salamanca had carried out the various aspects of his mission with great accuracy although a lack of water and a contrary breeze had caused them to extend their excursion to include the closest of the Islas de las Perlas, visiting especially Islas Chapera and Pacheca, where they not only found the water and food they needed but also received the greatest hospitality from the small number of negro or mulatto settlers who live there for the pearl fishing.

30 November

Until the last day of the month nothing worthy of note occurred. On board the *Descubierta* the stock of firewood was fully replenished with the use of the barge, the off-duty marines also being assigned to this task. The *Atrevida* was not so successful. She intended to use the city's barge for the same purpose with as many men as possible on the morning of the 29th but unfortunately its mooring line parted when it was almost fully laden and it was driven onto the rocks by a strong and unexpected surge, the hull split, and it foundered. The men, who had worked zealously to save the barge, had therefore had to spend a very unpleasant night on that deserted coast with no food, shelter or dry clothing. A party of our hunters was also involved in this mishap. Under Don Pedro González, the first surgeon of the *Atrevida*, they had made an interesting collection of birds for the natural history studies and this was also lost in this incident.

Surprised when the barge failed to return the following night, Don José Bustamante dispatched the *Atrevida*'s launch at dawn the next day under the command of Teniente de Navío Tova. He took their rations with him and it may be imagined how eagerly these were received by the shipwrecked men, who were then able to work on refloating the barge and moor it, although half submerged, some distance from the rocks. They all returned on board that afternoon and at noon the next day Don José Robredo was at last able, with skilful manoeuvring, to tow the barge to Isla Naos[1] and beach it there for repairs.

2 December

As well as the observations already mentioned, Don Dionisio Galiano had taken the precaution on the night of the 27th of the previous month of measuring the distance from the Moon to Regulus and to the centre of the constellation Hydra[2] with the astronomical quadrant, thus deducing the longitude of the observatory to be 81°43'22" from Paris. Nevertheless we anxiously awaited the night of 2 December when the immersion of the first satellite of Jupiter and the occultation by the Moon of λ Virginis would confirm our conclusions on this important point. Our luck held on this occasion, and we obtained the following very reliable results:

By the first satellite (according to Don Dionisio Galiano) . .81° 53' 15"
By the immersion and emersion of λ Virginis . . . . .81° 46' 21"

Adding to these highly satisfactory figures (particularly with the removal of the error in the satellite and lunar tables),[3] our various observations made as far as Lima and then in Guayaquil, that of M. Bouguer in Manta and those made by Spanish and French academicians near Quito, we were convinced that the longitude of Panamá, and accordingly that of a point on the other coast, could be accepted without fear of any significant error.

4 December

This was the date that had been fixed for the naturalists to return from their excursions. However, they had not been able to go further than Cruces since they were busy with preparing and preserving the previous collections and with various specimens of lithology, minerals and the shells which are very abundant in these parts. Accordingly, all four of them returned as did Teniente de Fragata Vernacci, to everyone's satisfaction.

This officer had very quickly reached Puerto de Chagres and determined its latitude with the astronomical quadrant, its longitude with pocket chronometer number 105 and variation with the theodolite. He had also sailed some leagues towards Portobelo and taken bearings to the entrance to that port, tying the intermediate points with

[1] The nearest of Islas de Perico to Ciudad de Panamá.
[2] Presumably Galiano observed to Alphard, the brightest star in this constellation.
[3] Presumably as a result of actual observations made in a European observatory.

Plate 35. Isla de Naos, Golfo de Panamá (detail), by José Cardero. Museo Naval, Madrid

bases and theodolite bearings. He had checked the rate of the chronometer in Cruces and determined its latitude and longitude and, lastly, had fortunately been able to observe from Chagres the immersion of the first and second satellites [of Jupiter] on the night of the 25th and the immersion of the first satellite on the night of the 2nd.

Since we had concluded our astronomical operations, all the instruments were therefore returned on board that day, leaving the astronomical quadrant and pocket chronometer number 105 (which had arrived from Chagres) so that the rating of the chronometers could continue.

By this time Don Felipe Bauzá, having also measured a base on the nearby beaches and to the north of the city, had extended his bearings from Alto de la Vigía[1] and a rounded hill somewhat further inland, towards the most distant points to east and west. Isla de Chepillo and Punta de [blank] were the two end points that we could link with this base, but we then extended it towards Pacheca, Taboga and Otoque,[2] thus linking with the work of both launches. For greater precision, the single anchorage off Perico was tied in by means of a new base, further bearings were taken to the high point of this island, and, under various officers of the *Atrevida,* soundings were taken at every place, from the anchorage to Panamá Viejo, that might be of interest to our shipping.

It was now clear to us that the unusual delay of Don Manuel Novales must be largely due to the constancy of contrary breezes which, now settled in from north and NW, had freshened daily with the approach of the end of the lunar cycle and had cleared the weather considerably. We could, therefore, undertake the passage to Isla Taboga, where I intended to replenish our water supplies and perhaps this would also facilitate our reunion with the launch. I thought it best to wait for him, however, particularly because, as will be seen, there were many other reasons for this brief delay.

7 December

On the afternoon of the 7th we saw the launch returning, which allowed us to plan our departure for the morning of the 12th. Don Manuel Novales had, as we had suspected, been delayed since the 2nd at Pacheca, beset by contrary winds which on one occasion had split his mainmast and on another had endangered him in the anchorage itself. Finally, making for Taboga under oars and using the tides, he was able to approach these islands, where Don Cayetano Valdés immediately took him in tow with the pinnace. The fruits of this excursion were the detailed survey of all the islands and of Bajo de San José, a large number of bearings taken to the furthest points of both coasts, and a set of observations for latitude and longitude (these with chronometer 71) which left us in no doubt of the true position of each point.

As already been mentioned, there were other reasons which inclined me to delay our departure a little. The Governor had asked me to attend a council of war, at which it was to be decided what course should be followed in the unfortunate circumstances currently affecting these provinces, the Viceroy of Peru having withheld

---

[1] Literally Look-out Hill, possibly either Cerro de Gavilan or Cerro de Ancón, two hills on Malaspina's survey in the vicinity of Ciudad de Panamá.

[2] A small island about 22 miles SSW of Ciudad de Panamá.

the annual grant which alone maintained them. My proposal, with everyone's agreement, was to send to Paita one of the barges with an officer of this garrison, who would submit to the Viceroy the present urgent needs of these provinces and to obtain, at least, some assurance of an early remittance so that neither public credit nor liquidity would be seriously affected. Consequently it was essential that our artificers should start to repair that barge and somehow put off the daily requirements of a vessel being refitted.

Furthermore the Governor had suggested that I would do a great service to the crown if I could convey to the ports of Nicaragua Coronel Roberto Hodgson who had recently been so harassed by the Mosquito Indians that his life was in danger. He had been forced to flee with his family to Chagres where His Majesty's orders were that he should report in person to the President of Guatemala as soon as possible. I immediately agreed to take him as far as Realejo, and as some time was needed for the *coronel* to make the journey from Chagres, it seemed that I should fix the morning of the 12th as the latest time for our stay in Perico.[1]

At the same time this allowed me to attend to the replacement of a few seamen who had deserted, to the continuing efforts to find a good local pilot for the coast from here as far as Realejo or Acapulco and finally to put in order a consignment of six boxes, five of natural history specimens and the other containing [chronometer] number 13, which was out of order, so that they could be entrusted to Jefe de Escuadra Joaquín Cañaveral, Comandante General de Cartagena de Indias, to be sent to Señor Intendente de Cádiz, whom I informed of the whereabouts of the chronometer and the boxes.

It was not possible to find any replacements for the six seamen who had deserted from the two corvettes, all of whom were among those recently recruited in Lima or Guayaquil. The only pilot who could have been useful to us as far as Realejo was put in charge of the barge going to Paita. The remittance of the six boxes to Cartagena de Indias was entrusted to Don Agustín Gana, a resident and merchant of this city, for which I advanced him the sum of eighty *pesos fuertes* on His Majesty's account so that the consignment would not involve any expense for the local Treasury accounts, none of which had any funds left.[2]

At the same time, I knew how important it was to procure a good store of tobacco for the seamen at a place where it was of good quality and which was in all probability the last place where we could find supplies of a product so indispensable to sailors. Accordingly, three *quintales* of it were distributed to each corvette and the Administración de Tabacos was promptly paid for it, with the intention once again of benefiting the local treasury rather than causing it expense.

The chief caulker, Pedro de Lamas and Seaman Gunner Francisco Ximeno, both of this corvette, had been suffering for some time from persistent illness; the former had an obstruction of the liver and the bile duct, with the early stages of a watery

---

[1] Malaspina accepted this charge in a letter of 4 December 1790: AMN, MS 427, ff. 161v-162.

[2] The Panamanian economy never fully recovered from the withdrawal of regular Spanish silver fleets and the final collapse – complete by 1740 – of the periodic fairs at Portobelo. The *Audiencia* was suppressed in 1751 and a series of fires devastated the town at intervals between 1756 and 1781. Nevertheless, such figures as are available suggest a fitful recovery from about 1770 to the late 1790s: C. Ward, *Imperial Panama: Commerce and Conflict in Isthmian America, 1550–1800*, Albuquerque, 1993, p. 157. At the time of Malaspina's visit the population was 7,831 (862 whites, 5,112 free black, 1,676 slaves and 63 Indians).

deposit in the lower abdomen, indicating an imminent dropsy, as the result of chronic intermittent fevers; the latter, with hepatitis or an inflammation of the liver causing a suppuration of the bowels, needed an operation so painful and difficult that he would need a convalescence which could hardly be provided on board. It seemed wisest to discharge both of them from the ship so that they only had their health to worry about. As well as the money granted them to cover their wages, they were given the appropriate certificates and passports so that as soon as they had recovered they could make their way to Cartagena de Indias and continue in the King's service.[1]

The chief caulker's position was made over to the caulker's mate (according to regulations), but we were unable to replace either him or the three seamen who had deserted.

11 December

When everything was ready and the Governor's orders had been received, particularly regarding the embarkation of Coronel Hodgson, who had arrived in the city on the night of the 8th, we weighed the SE anchor at dusk and lay with the NW anchor almost up and down. At the same time we landed some supplies and stores by the city barge which we had offered them because of the city's severe lack of everything they needed to maintain their barges or to test a recently invented instrument to make pearl-fishing less dangerous.

12 December

Dawn of the 12th was most pleasant, as on previous days, with a moderate NW breeze. As we intended to make for Taboga to replenish our water at the anchorage there and to add lime, which we had obtained beforehand, to prevent it from spoiling, some of the ships' water casks were sent to the anchorage in the launches under sail. Both corvettes followed under topsails and topgallants. The tide was already beginning to ebb and the breeze continued moderate and favourable, so by ten o'clock we were easily in a suitable position to drop anchor again. We had obtained soundings of ten and eleven fathoms, ooze, until a mile from the anchorage, where the depth suddenly increased to sixteen and eighteen fathoms and the spot where we let go the anchor had a depth of not less than twelve fathoms, ooze, although only two cables from the watering place.

This anchorage is one of the most attractive and inviting to a navigator, particularly during the season of the trade winds from which it is well sheltered. Six or seven fathoms are found half a cable from the beach, across which a stream discharges into the sea providing mariners, without the slightest inconvenience, crystalline water of the finest flavour and lasting quality. In a pleasant forest of useful plants such as bananas, coconuts, medlars, avocados, pineapples and tamarinds, live some hundred families of peaceful and happy people who combine the appearance of a well-ordered existence with an air of opulence quite unusual in these parts, especially in the vicinity of Panamá. The cleanness of their houses, the whiteness of their clothing,

[1] Malaspina notified the Governor of Ximeno's condition on 10 December 1790: AMN, MS 427, ff. 165–165v.

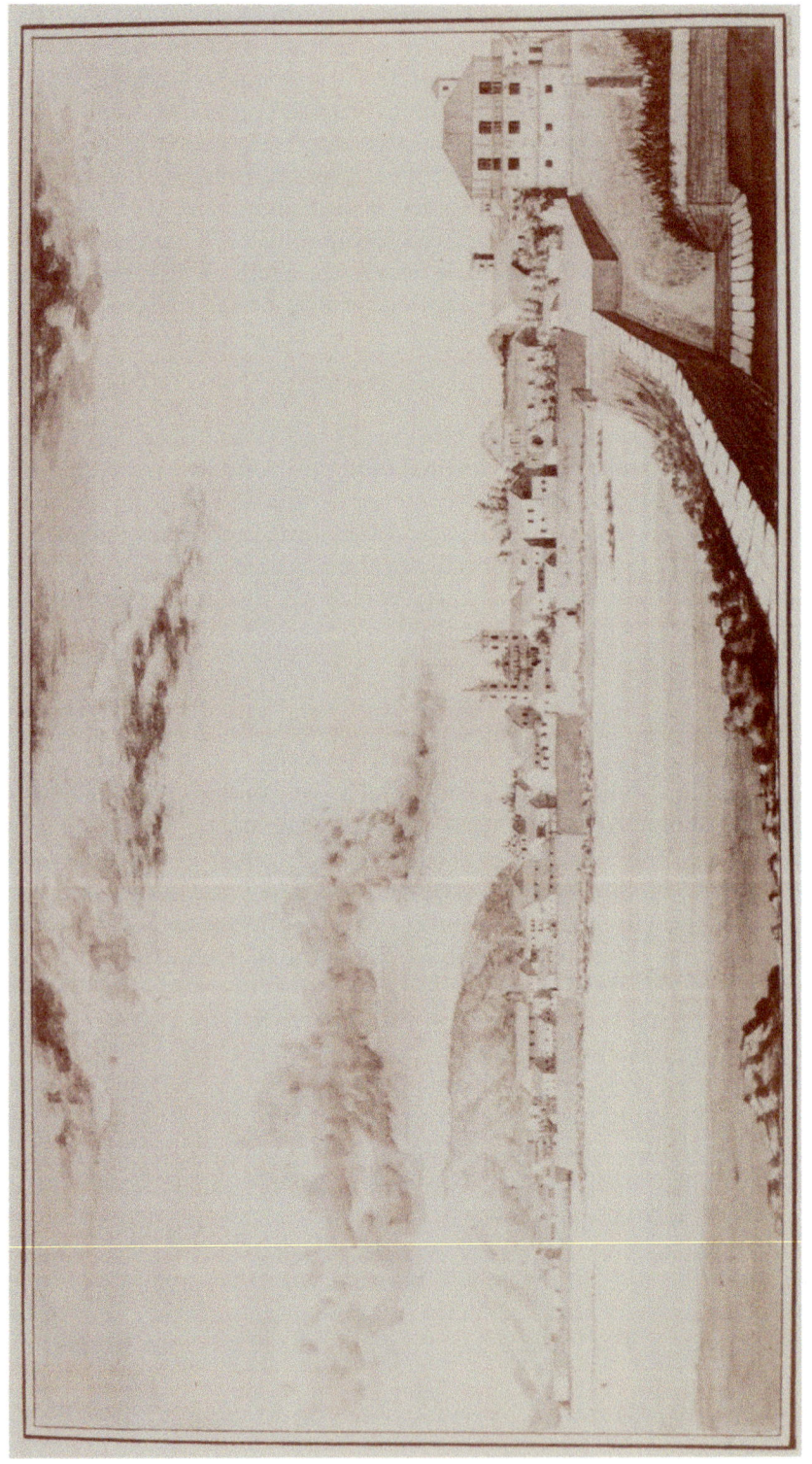

Plate 36. Panamá from the Castillo de Chiriquí, by José Cardero. Museo Naval, Madrid

particularly of the women, and their almost continual dancing add brilliance to the scene. Hardly a mile away there is abundant firewood and tasty fish abound off the beaches.

Before noon (the watering being well organized by Don Manuel Novales) first one corvette and then the other took a launch-load of water. Thanks to good work and the convenience of the place, by eight in the evening we had a considerable amount on board, although we had emptied a large number of casks into the hold so as to freshen it and clean the bilges. At the same time we replaced what we had taken on in Lima and Guayaquil with better water.

Somewhat apprehensive about our close proximity to the shore, we laid out to ESE a kedge with the help of the pinnace and later an anchor with the launch, although these were laid only half a cable apart, knowing that it would therefore not be necessary to recover a mooring laid further to seaward. The *Atrevida*, lying not so close inshore as us, was satisfied with laying out a kedge with two small hawsers to ESE.

Since we had at our disposal the whole afternoon and even part of the night, thanks to the moonlight, our naturalists took the opportunity to make botanical and lithological excursions, while Don Felipe Bauzá, having measured a short base on the beach, took advantage of the tide being out to fix a position on the sandy spit halfway between the island and the tiny islet,[1] from where bearings of great importance were taken to the heights on the mainland to the south and to the islets in our anchorage. As Haenke informed me, the lower cultivated land seemed no different from that which they had visited in Panamá, but the higher uncultivated parts, where there were other plants very different from those they had found until now, had amply rewarded them for the heat and fatigue they had suffered.

We left ashore only four seamen from each corvette for the night, to continue gradually filling the many casks we had landed. So at daybreak we were able to take on board by launch a full load of water, and during the day we were able to busy ourselves with other work especially hydrography.

13 December

For this purpose, Don Felipe Bauzá climbed at dawn to the summit of the hill[2] from where he would find it easy to continue extending his bases to distant points on the mainland to the west and to Islas del Rey. In the *Atrevida*'s pinnace Don Jacobo Murphy sounded the waters around the islet in the anchorage[3] and the northern part of the channel between Taboga and Taboguilla.[4] Piloto Maqueda was sent in the *Descubierta*'s pinnace to sound the southern part of the same channel, where, in addition to several underwater rocks between Taboguilla and Islote Sube la Vaca,[5] there was an extremely dangerous one half way between that islet and Isla Urava.[6] It showed only at low water and made the very useful passage between these islands rather hazardous.

---

[1] Islote Chama, 3½ miles SSW of Isla Taboga.
[2] Morro Taboga.
[3] Not identifiable on Malaspina's survey.
[4] Situated about 2 miles NE of Isla Taboga.
[5] Possibly one of several islets off Isla Taboguilla.
[6] Close off the SE extremity of Isla Taboga.

At three in the afternoon both pinnaces returned, having completed their missions. Don Felipe Bauzá had not been as successful because the thick haze which had shrouded the horizon since dawn had not dissipated all day and the breeze had continued very light from west and WNW so that the heat and the harmful effects of the Sun were much worse than on the previous days.

As we had received the lime before midday and had immediately put the suitable amount into the lowest tier of casks, we could now concentrate energetically on a complete renewal of the water supplies. By nightfall the task was finished on board both corvettes, the *Atrevida* having also taken on board by launch a load of firewood, which was easily cut a short distance from the anchorage.

The heat of the day and, even more, the pleasantness and convenience of the site, about half a mile from the watering place towards the higher ground, had persuaded almost all of the officers to bathe in some pools that had previously been dug for the Bishop of Panamá's recreation. The seamen and marines also bathed at the same time, either at the beach or near the watering place and as there was an extraordinary abundance of fruit on the island, particularly oranges, bananas, coconuts and limes, everybody enjoyed as much of these refreshments as they could eat, so countering the dreaded effects of the heat.

14 December

For several weighty reasons we were prevented from setting sail on the following day. Apart from the facts that Don Felipe Bauzá had to climb the hill again and that the men needed some rest since some of them were beginning to suffer from high fevers, the naturalists insisted on examining in detail the local fish which were of varied and beautiful species as yet little known in Europe.

Consequently, when the Moon had set that night and the tide was out, the two pinnaces were sent from the corvettes with, as far as possible, a reasonable hope of good conditions for the study of nature, while the rest had the opportunity to enjoy some fresh air again. Pineda and Haenke found that from the variety and novelty of their many acquisitions they had in fact much to add to the limited knowledge of natural history.[1]

At dawn this day Don Felipe Bauzá was also somewhat more successful. From the summit could be seen, and therefore fixed by means of good bearings with the theodolite, Islas Pacheca and Otoque and all the coast as far as Punta Chamé which, with the area close to Panamá, forms a fairly large bay. In this manner we avoided the possibility of the slightest error, either astronomically or geodetically, in our work on the multitude of islands which compose this truly complicated archipelago.

It was our intention to allow the men a whole day ashore to enjoy their freedom, as long as it did not give rise to disorderly conduct. At dawn, the last load of water having been taken on board by launch, the work in the hold was finished and the men were granted not only a general leave for the day but also given plenty of incentives. All the men washed their clothing and so spent half the day pleasantly in the

---

[1] Pineda's notes are collected in a series of MSS: AMN, MS 265, ff. 103-104v; MS 339, ff. 25-25v, 27-50; MS 462, ff. 115-140v; MS 2,136, ff. 1-41v, 48-116.

water, and their conduct throughout the day gave no support to the rumours among the common people of Panamá that our crews were unruly and really awful.

All the objectives which could have detained us at this watering place had now been achieved, so the rest of the afternoon was spent only in rating our chronometers to be ready for sailing early the next morning. number 71 from the *Atrevida* had been brought to this corvette when Don Manuel Novales returned so that, having taken this precaution, the rate of the three chronometers could be confirmed by daily comparisons and we could use the excellent number 10, which had not been moved, to observe longitudes aboard the *Atrevida* with the same confidence. We had also taken the precaution of adding to the observations made in Panamá for rating this chronometer; the noon observations for this day and the previous one had been taken on the beach by Don Juan Vernacci with the astronomical quadrant. As well as the fact that the observatory was almost on the meridian of this beach and that even this slight difference could be deduced with bearings, the current period, which had begun at noon on the 8th and which we extended to the 14th (discarding for the sake of greater accuracy the intermediate observations, although of absolute altitudes, on the 11th and 12th, and the equal altitudes of the 13th), included six days, during which any error would be of little significance for the mean rate. In fact, if we consulted the results of each observation in itself, we had to suspect that all the chronometers had shown considerable irregularities, as confirmed by the daily comparisons, although the short time did not allow us to refer them to number 10, which probably would have resolved all these doubts. Having paid attention to the rates of numbers 61 and 71 for this last period, as both had only been on board since the 6th and, on the other hand having adopted the mean rate for the whole period of the comparisons for number 72, we referred to the latter the comparisons of the *Atrevida's* chronometers, which, as agreed, were to be repeated every three days and referred directly to number 10.

By nightfall all the men who had been ashore were back on board, and as the excessive heat of the day had been followed by a cool, mild evening with a brilliant Moon, we chose to do the anchor work then, and so at nine o'clock, all ready to set sail, we allowed the men a full rest.

APPENDIXES

Plate 37. Antonio Valdés y Bazán. Anon. Museo Naval, Madrid

APPENDIX I

The Malaspina–Valdés Correspondence

The exchange of letters between Malaspina and Antonio Valdés y Bazán, Ministro de Marina,[1] throws much light on the objectives of the 'Scientific and Political Voyage Around the World'. Valdés was head of the Order of St John in Spain, and so took a natural interest in the career of Malaspina, a fellow member of the Order. At the time of the voyage he wrote of Malaspina that 'his knowledge, birth, nobility and elegance of person and manners, proud bearing, affability, resolute character and social gifts made him the first in our navy and a unique choice for that commission'.[2] The letters printed here demonstrate the original objectives of the voyage as set out by Malaspina and Bustamante in their 'Plan' of September 1788, and the way in which events modified those aims. The final letter in this set was written by Malaspina at Callao on 15 September 1790. Subsequent correspondence between Malaspina and Valdés will appear in later volumes of this edition.

1. Alejandro Malaspina to Antonio Valdés. Madrid, 10 September 1788.[3]
Excellency
While with these lines I am complying with your order to inform you about my plans before leaving Madrid, I am also taking this opportunity to convey my thoughts in greater detail than I was able to do when I spoke with Your Excellency. Without being thought impertinent or importunate, I hope that you may think fit to guide my future steps so that my aims, directed towards the service of His Majesty, are accomplished to the best effect.

First of all I should say to Your Excellency that the current situation in Europe leads me to think that I should return directly to the Department [of Cádiz], which I shall do at the end of this month of September if Your Excellency considers it appropriate and gives me leave to go[4]...I trust also that it will not appear presumptuous of

---

[1] Antonio Valdés y Bazán (1744-1816) joined the Real Compañía de Guardias Marinas in Cádiz in 1757. After distinguishing himself in the defence of Havana in 1762, he held numerous administrative posts in the navy. In 1781 he became director of the ordnance factory of La Cavada, and in 1783 was appointed Inspector General de la Marina. Shortly afterwards he was made Ministro de Marina, and by the time of his return to active service in 1795 he was credited with an impressive expansion of the service, and in particular with improvements in the scientific education of naval officers.

[2] See Cerezo, *Diario por Malaspina*, Pt. 1, p. 16.

[3] AMN, MS 1826, ff. 1-5v.

[4] The passage omitted here deals with the affairs of the Real Compañía de Filipinas.

me to send you the attached Plan for a Scientific and Political Voyage around the World. This Don José Bustamante y Guerra and myself undertake to carry out if we are granted a period of about eight months for the necessary assembly of resources and personnel, and for indispensable preliminary studies.

It would, however, be very distressing indeed at a time when I would rather be offering my life fighting the enemies of the Crown to be occupied with tasks other than purely military ones. In this respect I can without hesitation put myself forward either for a commission as commander of a squadron, or for the command of a single ship to operate in seas frequented by enemy forces, perhaps with the aim of observing their forces at close quarters.

The need for some respite in order to organize the results of my last voyage[1] as well as to make some progress with my professional and political studies make me think that Your Excellency will not consider unjustified a final request should I fail at this time to be assigned a commission at sea. This is to shut myself in the observatory at Cádiz for some months rather than taking up command of the Midshipmen's College (Compañías de Guardias Marinas) that at present is so ably discharged by Capitán de Fragata Don José Barrientos y Rato. It would cause me much satisfaction if Your Excellency were to accept the reports on matters of naval or national interest that I could submit as a result of such a quiet retreat.

Above all else Your Excellency can be sure that my good health and willing spirit require that I occupy myself with work, the more so because work is the only reward that I wish.

May Our Lord grant Your Excellency many years of life. Madrid 10 September 1788.

Excellency, I am &c.
Alejandro Malaspina

2. Plan for a Scientific and Political Voyage Around the World

For the past twenty years two nations, the English and the French, in noble competition, have undertaken voyages of this sort in which navigation, geography and humanity itself have made very rapid progress. The history of human society has thus been founded on much wider research; natural history has been enriched with almost endless discoveries; and possibly the most exciting victory has been the preservation of health in the course of long sea voyages through different climates while facing the most challenging labours and dangers.

The proposed voyage would aim to accomplish these objectives, and this part, which can be termed the scientific aspect, would certainly be undertaken following earnestly in the wake of Cook and La Pérouse.

But a voyage undertaken by Spanish navigators must necessarily involve two other objectives. One is the making of hydrographic charts covering the most remote regions of America and the compilation of sailing directions capable of providing safe

---

[1] To the Philippines, as captain of the *Astrea* (1786-88) in the service of the Real Compañía de Filipinas. One of the interesting features of the voyage is that Malaspina, breaking established routine, returned by way of Cape Horn.

guidance to inexperienced merchant mariners. The other is the investigation of the political status of America both in relation to Spain and to other European nations.

The commercial conditions in each province or kingdom, with individual consideration of their natural products and manufactures; their preparedness and capacity to resist an enemy invasion and conversely their ability to provide forces to attack the same enemy; the condition of the ports most suited for maintaining reciprocal trade; and finally the important activities of shipbuilding and supply of naval equipment, are aspects whose careful and secret investigation should be of some interest to the State. Our report would be adjusted to different political axioms concerning national prosperity whose acceptance or rejection must be subjected to the adjudication of reputable assessors called upon to examine them. Consequently all these tasks will have to be divided into two parts: the public one which will include not only the likely collection of curiosities for the Real Gabinete and the Jardín Botánico[1], but also the hydrographic and historical sections; the other a confidential one which will be directed to the political enquiries mentioned above. These could include, if the Government considered it appropriate, the Russian settlements in California[2] and the English ones at Botany Bay and the Liqueyos,[3] all of them places of interest whether from a commercial point of view or in the event of war.

His Majesty's Navy will be able to provide all the personnel for this commission, except for the two botanists or naturalists and the two topographical draughtsmen, for which posts it should be possible to find volunteers without undue difficulty in Madrid. As for the class of ships needed and the quality of their crews, the three principal requirements of safety, convenience and economy can easily be combined. The complement required for each of the ships can be kept down to about one hundred men. The actual detail of each vessel, such as the rigging, inboard arrangements, number of boats, their features and their equipment, and, finally, the quantity and quality of victuals, are too much to go into at present, and in any case cannot be precisely determined until such time as His Majesty shall decide on the scope of the proposed expedition.

To carry out the plan outlined below might take approximately three and a half years from 1 July 1789, the date by which the ships could depart if His Majesty were to give his approval now, either as proposed or with modifications, allowing eight months to those who will have to implement it. This is necessary for the collection of all the requisite materials as well as for the preliminary studies, the most important of which will be the acquisition of skills in practical astronomy.

The two corvettes will sail from Cádiz on the 1 July 1789 and will proceed to

[1] The Gabinete Real de Ciencias Naturales and the Real Jardín Botánico, both in Madrid.

[2] There were no Russian settlements in today's California in 1788, but for the Spaniards of Malaspina's time the name applied to a region that stretched much farther north than the present state of California. Malaspina here may have had in mind the Russian trading posts recently established in Alaska, for before sailing he had sent for all available accounts of Russian voyages to Alaska. See AMN, MS 583, f. 32v.; MS 281, f. 25.

[3] The British convict settlement at Port Jackson (near but not at Botany Bay) had been established in January 1788. There was no British settlement in the Liqueyos (Ryukyu Islands), though British merchants had for some time been interested in the possibility of establishing a base there. See Robert J. King, 'A Regular and Reciprocal System of Commerce – Botany Bay, Nootka Sound, and the Isles of Japan', *The Great Circle*, 19, No 1 (1997), pp. 1–29.

Montevideo, where a fresh rating of the chronometers, such astronomical observations as are possible, and all kinds of natural history studies, will be carried out. Various kinds of provisions will be acquired there for the maintenance of the crews, and also for experimental purposes. From there the Malvinas[1] will be surveyed and, if the Government considered it advisable, some livestock could be landed at Bahía de Buen Suceso in Estrecho de le Maire, since it seems evident that this anchorage will be the most convenient port of call for the navigation of Cabo de Hornos. From Bahía de Buen Suceso a course will be set to round Cabo de Hornos, and we should attempt to survey Cabo Victoria and some part of Archipiélago de los Chonos. Finally we will drop anchors at Chiloé, which might be towards the end of 1789. The whole of the year 1790 will be employed on the western coasts of America from Chiloé to San Blas. An effort will be made to make the navigation from Guayaquil, Acapulco &c to Lima easier than now; a search will be made for the Islas del Gallego[2], and a party will be sent from Acapulco to México City.

The first three months of 1791 will be taken up with a survey of the Islas Sandwich.[3] After following the coast of California, the voyage will continue to the north between Asia and America as far as the snows will allow. After calling at Kamchatka (if the Government thinks fit), we shall continue to Canton [Guangzhou] to sell sea-otter furs for the benefit of the crews.

Departure from this port will therefore be about October or November 1791. Advantage will be taken of this season to survey Cabos Bojeador and Engaño, and Puerto de Lampón[4] on the opposite coast of Luzón. We should then proceed to the Marianas and from there concentrate our work on charting in great detail the navigation through Estrecho de San Bernardino to Manila. From that capital we shall shape our course to survey Mindanao and after passing between Celebes and the Moluccas and north of New Holland, enter the Indian Ocean.

After coasting the whole of the western seaboard of New Holland, we shall proceed (in about March 1792) to Botany Bay. Islas de los Amigos[5] and Islas de la Sociedad[6] will then be visited, and towards October or November New Zealand from where, finally, we shall head south and then alter course to NW in order to round New Holland and head for Cabo de Buena Esperanza, aiming to return to Europe in April or May 1793.

Capitanes de Fragata Don Alejandro Malaspina and Don José Bustamante y Guerra, anxious to place all their energies in the service of the state, volunteer their services to implement this plan in the hope that they will deserve to receive for their direction and greater success, information from, and the support of, the government, as well as information from individuals in Europe and across the Americas.

---

[1] Falkland Islands.
[2] Thought to lie about 850 miles west of Archipiélago de Colón. In the event, Malaspina did not search for them.
[3] Hawaiian Islands.
[4] Cabos Bojeador and Engaño are the north-western and north-eastern extremities respectively of Luzón in the Philippines; Puerto de Lampón is now Lamon Bay on Luzón's east coast.
[5] Tonga.
[6] Îles de la Société.

As far as the officer class under our command is concerned this particular commission requires that they should all be volunteers, and that in terms of both health and capabilities they should have full confidence in each other.

Madrid 10 September 1788.

Alejandro Malaspina

3. Antonio Valdés to Alejandro Malaspina. San Lorenzo del Escorial, 14 October 1788.[1]

Your project for sailing around the world has merited the King's approval in the terms proposed by Your Honour in your letter of 10 September instant. It is His Majesty's pleasure that the project should be carried out to good effect and to this end that you should be relieved from taking up your appointment to the Midshipmen's College in the Department of Cádiz. The commandant of that body is being duly informed so that you may from this moment commit yourself at will to the scholarly pursuits and the preparations required to carry out fruitfully your commission. This I signify to you for your guidance and also, considering that the expedition must be made ready to Your Honour's entire satisfaction as regards the ships (those chosen by you), their rigging, spares and victuals, as well as the officers, *pilotos*, marines and seamen, to invite you, regarding all these aspects and any others relating to the project, to plan and propose all that you will need so that the appropriate orders can be given to ensure that all preparations are made in the manner Your Honour may consider best to achieve your design.

May God &c. San Lorenzo del Escorial, 14 October 1788.

4. Malaspina to Valdés. Cádiz, 23 December 1788.[2]

Excellency

Now that I have studied in detail the pattern of work for the first year of our commission, I shall set out for Your Excellency the extent to which I should be able to accomplish the main objectives of the expedition placed in my charge. I have assumed for this purpose a balanced pattern of adverse and favourable weather. Our work will thus be more meticulous if the latter predominates, and more assiduous and determined if the former is persistent.

After surveying the Río de la Plata, and plotted as accurately as possible its extremities and dangers, we shall sail to the Patagonian coast and there, while doing our best to keep within sight of it, shall confine ourselves to fixing certain key points astronomically; specifically, these will be Puerto Deseado, Bahía de San Julián, and Cabo Blanco. Thus disregarding other parts of the coast unless any should come into view by chance, we shall take advantage of the prevailing westerlies, even at night, and thus make use of truly precious time. This course of action will be better in that we are able to get a reliable configuration from the surveys made by the xebecs *Andaluz* and *Aventurero*, and from the charts of the surroundings of Bahía de San Julián. In latitude 48°S we shall cross over to the Malvinas, but we shall only be able to survey the

---

[1] AMN, MS 278, f. 6v.
[2] AMN, MS 2296, ff. 36-8.

western part of those islands, and we shall compare what we see with previous charts and descriptions. The English Captain McBride,[1] who sailed around these islands in a frigate, claims that the coastal charts that he drew are as accurate as those of England. Nevertheless, it would be useful to survey also the eastern extremity so as to fix its true limits with reliable longitudes; but that would involve us in the loss of a month's time, and in any case it would not be difficult to finish this work, on our return in the last year. From the Malvinas we shall return close to the coast of Patagonia. At Cabo las Vírgenes we shall compare our longitudes with those calculated by His Majesty's frigate, *Nuestra Señora de la Cabeza*,[2] and Bahía del Año Nuevo in the Isla de los Estados and that of Buen Suceso in Tierra del Fuego will be accurately surveyed.

From there it would seem most suitable to sail completely out of sight of land, more particularly because a precise survey of the southern extremity of Tierra del Fuego was perfectly carried out by Captain Cook during his second voyage.[3] It would be useful, if time is not getting short, to enter Bahía de San Francisco[4] of which Sr. Don Antonio Ulloa speaks in his letter, or Canal de Navidad,[5] surveyed by Cook himself, a short distance north-west of Cabo de Hornos. But it is not possible to combine this with the subsequent tasks that we must undertake from the moment we enter the Pacific Ocean which will of course require much time and a suitable season.

Fixing the astronomical position of Cabos Pilares and Victoria[6] will be an essential part of our tasks, to the extent that it will serve to link our work with the inland survey made by Capitán de Navío Don Antonio de Córdoba.[7] From here we shall continue to survey the coast as far as it can be done with prudence; and if circumstances and the season allow, we shall anchor in some part of the Archipiélago de Chonos to investigate if at all possible its inner configuration. From the voyage of Sarmiento[8] it is believed that from the Archipiélago de Chonos to Cabo Victoria the land must greatly resemble that of [Tierra del] Fuego, that is to say that it presents an almost infinite labyrinth of islands and rocks on which the interior fire of the volcanoes, the surrounding waters constantly agitated with the greatest violence, and the weather, a fearsome destroyer of human resources, have worked with equal intensity. As a consequence it would be rash of me to expect to be able to make a perfect survey of this area. For the general purposes of navigation it is sufficient that the west coast be well plotted, and its correct limits of latitude and longitude fixed.

[1] Captain John McBride, commander of the frigate *Jason*, who garrisoned Port Egmont in the Falklands in 1766.

[2] A reference to the surveys of Antonio de Córdoba y Lazo (see note 7 below). The name of his vessel was *Santa María de la Cabeza*.

[3] For details of this survey, carried out in December 1774, see J. C. Beaglehole, ed., *The Journals of Captain James Cook: The Voyage of the Resolution and Adventure 1772–1775*, Cambridge, 1961, pp. 585-604.

[4] 7 miles north-west of Cabo de Hornos, between the south-west coast of Isla Herschel and the south-east coast of Isla Hermite.

[5] Seno Christmas, Cook's Christmas Sound.

[6] Cabo Pilar and Cabo Victoria are the southern and northern points respectively of the Pacific entrance to Estrecho de Magallanes.

[7] Antonio de Córdoba y Lazo, who made two voyages to the Magallanes region, in 1785 and 1788, during which he improved earlier surveys of Estrecho de Magallanes.

[8] Pedro Sarmiento de Gamboa, who in 1579-80 sailed from Callao, through Estrecho de Magallanes and into the Atlantic, making careful sailing directions as he did so.

From our anchorage in Chiloé we shall explore the channel which this island forms with the mainland and, if it appears prudent to do so, we shall come out into the Mar Grande,[1] thus completing as a whole the perilous tasks of the first year and making a start with those that we shall call more pleasant, although of no less importance.

From Chiloé, or rather from Valdivia, we shall undertake the survey of the coast in a much more summary and general fashion. The two corvettes will part company and having formed the scientific parties which it proves possible to assemble, the leading corvette will land them at points on the coast which appear suitable, and will then commence its tasks from Ensenada de Arica, thus continuing the work on the coast as far as Lima. Meanwhile, separate parties will go along the coast from south to north following instructions which will combine what is useful with what is feasible. They will make all the investigations which appear most conducive to the progress of science and to the soundness of our work, as well as providing information on curiosities which Europe wants to have at this time. And finally, having arrived at the point where the leading party had begun its journey, they will wait for, or make their way to, the following corvette, which will make sure to pick them up, when work on the southernmost part of the coast has been completed.

With the help of the various settlements (on which our commission will not impose any expense, nor on the royal coffers) we shall thus achieve in two or three months an excellent survey by both sea and land of the whole coast from Chiloé to Lima. Finally, when all parties have come together in the capital, and all the plans, reports, sailing directions etc. have been coordinated, after a thorough exchange of all the useful information that we are able to draw together, the whole of the work so far done will be sent to Madrid in duplicate, so that it is not exposed to the risk of being lost on the subsequent voyages of the corvettes.

I sincerely hope that this plan for our first tasks meets with Your Excellency's approval, and, for greater clarity, I also submit in separate memoranda[2] an account of all the measures which will have to be taken in the various ports of South America which we shall visit for that purpose.

5. Malaspina to Valdés. Cádiz, 24 April 1789.[3]

Maritime information required

Excellency

Before I begin to set out the various operations of the expedition with which His Majesty has been good enough to entrust me, which will no doubt take up the whole month of June, Your Excellency may be pleased to have two preliminary studies. One will include the detailed plan of the voyage, taking into account seasonal factors and the recent clarifications regarding the principal aims of the King's service; the other will on publication serve to inform the Nation of the branches of scientific inquiry that we propose to pursue, and the means which have been and will be used to effect them. As regards the first, naturally confidential, objective it is important to decide on two points: 1st whether the exploration of the coasts of California is to be confined

---

[1] The Big Sea, that is the Pacific.
[2] Not included here.
[3] AMN, MS 583, ff. 46–46v.

solely to hydrographic charting, or should be extended to the Russian settlements with the aim at least of containing them, and finally whether it is to prepare for a further attempt on the North-East Passage according to the well-known but very imprecise manuscript of Ferrer Maldonado;[1] 2nd whether New Holland and New Zealand are to be looked at with political rather than naturalists' eyes, that is to say whether from the study of various parts of those vast regions solidly-based conclusions with relation to particular products can be developed beyond the few recently established in those areas. As regards the second objective, it would perhaps be useful to develop it in much detail, but in that case it could not escape being somewhat lengthy.

If Your Excellency were to find this work useful in whole or in part, I hope you will be kind enough to let me know so that I can undertake it in May when, being busy mainly on board with the overall preparation of the corvette, I can confirm my proposals at somewhat greater leisure.

6. Malaspina to Valdés. Carraca, 9 June 1789.[2]

Maritime information

Among the valuable documents obtained by Don José Espinosa[3] from the Archivo de Indias, worthy of first place must be the account of the voyage of Ferrer Maldonado concerning the passage or communication between the Pacific and Atlantic Oceans. This journal, summarized by the Spanish author of *Historia política de los establecimientos ultramarinos de las naciones europeas*, bears all the hallmarks of authenticity; and to the extent that I have studied the Introduction to the third voyage of Captain Cook and the voyage of La Pérouse on the coasts of California, both navigators seeking to discover this passage, it would appear that it has so far eluded all foreign searches and that there is scope for a new exploration.[4] But it would be rash of me, following behind navigators with such a reputation for daring, to vaunt myself in the eyes of Europe as undertaking this important business with greater success, so long as the public is unaware either that there is a specific order from His Majesty that we should try with equal determination to follow those who have preceded us, or that there is some certainty not taken into account so far concerning this voyage and its appearance of authenticity, which together with the explorations already carried out, leave room for a new search.

---

[1] This is the first of several references by Malaspina to the alleged voyage of Lorenzo Ferrer Maldonado through a North-West (not North-East) Passage in 1588. Maldonado's account, ignored at the time, had been rediscovered in 1781, and had recently been summarized in Volume IV of E. Malo de Luque (the pseudonym of the Duke of Almodóvar), *Historia política de los establecimientos ultramarinos de las naciones europeas*, Madrid, 1788. See Dario Manfredi, 'An Unknown Episode behind the Northwest Coast Campaign of Malaspina's Expedition', in Robin Inglis, ed., *Spain and the North Pacific Coast*, Vancouver, 1992, pp. 119-24. To describe the account as 'well-known' seems odd, and conflicts with Malaspina's remarks in his letter of 9 June to Valdés, below.

[2] AMN, MS 583, ff.47v-48.

[3] José Espinosa y Tello was a junior naval officer (*teniente de fragata*) who, among others, was commissioned to search for documents that might be of value to the Malaspina expedition. He later joined the expedition at Acapulco in March 1791.

[4] The official account of Cook's third voyage had been published as Cook and King, *A Voyage to the Pacific Ocean*, London, 1784. Malaspina had also obtained a summary of the journal of La Pérouse, covering his explorations along the northwest coast in 1786, which had reached Paris in October 1788 overland from Kamchatka. See AMN, MS 278, f.44.

Of course the journal or narrative in question require that the lower regions of America from 60° to 65° be closely explored,[1] since Ferrer himself declares that the mouth of the Strait of Anian[2] is very difficult to find; and it is also necessary that the ice proves less of an obstacle than usual, as it made a passage beyond the 71st parallel impossible in the years 1778 and '79,[3] when to follow in the wake of Ferrer Maldonado it is necessary to reach 75°. But I can confidently assert that if this passage is to be sought anywhere it must be done from the Pacific rather than from the Atlantic or Labrador region.

Without Your Excellency's express permission I have decided not to communicate a copy of this memorandum or of Ferrer Maldonado's sailing directions to the Academies of Paris or London.[4] But it would seem to me appropriate, so as not to offend the navigators who have preceded us and not to commit us to excessive effort, that at the time of undertaking this exploration (if it should please His Majesty) the foundations on which it rests should be made public. In case of overwintering in the Bahía Bucareli[5] it would be necessary to have the right equipment, and to take decked launches from San Blas.

7. Valdés to Malaspina. Madrid, 30 June 1789.[6]

The King has seen what you propose in your letter of the 9th of the month now ending regarding the statements made by Ferrer Maldonado in his voyage about the link between the Pacific Ocean and the Atlantic. His Majesty authorizes you, according to circumstances and to the information that you might acquire in the course of your expedition, to attempt to discover this passage from the Pacific. But he does not agree that you should communicate your intentions to the Academies of Paris and London, or inform them of Ferrer's route until the object has been achieved. Of which I now advise you by Royal Command for your guidance.[7]

8. Malaspina to Valdés. Callao, 15 September 1790.[8]

Excellency

Having set out in full detail to Your Excellency our progress in the first year,[9] I can now, with better knowledge and consideration, expand on what must follow very

[1] This rather curious use of 'lower' ('baja' in the original) should be seen in the context of the following reference to latitude 75°N. In Maldonado's narrative, although he sailed as far north as 75°N, the actual entrance to the passage on the northwest coast was in latitude 60°N.

[2] The Strait of Anian was the name first given in the mid-sixteenth century to the waterway thought to exist between the American and Asian continents.

[3] A reference to the unsuccessful attempts by Cook in 1778 and (after his death) by Captain Charles Clerke in 1779 to penetrate the ice-barrier beyond Bering Strait.

[4] L'Académie des Sciences of Paris, and the Royal Society of London, both with a tradition of interest in voyages of discovery and their journals.

[5] Bucareli Bay in Alaska (latitude 55°13'N), discovered by Juan Francisco de la Bodega y Quadra in the *Sonora* in 1775, and named by him Puerto y Entrada de Bucareli.

[6] AMN, MS 278, f. 53.

[7] As letters to be printed in Volume II of this edition will show, Spanish hopes that the Maldonado account would be kept secret were in vain.

[8] AMN, MS 583, ff. 76–7.

[9] In a report of the same date, not included here, in AMN, MS 1407, ff. 15–18v.

soon; this seems to me the simplest course of action, so that at a single glance His Majesty can see the tasks which we shall undertake, unless at one of the many ports we shall soon reach at predetermined intervals, an order reaches us countermanding in whole or in part the plan that I now propose.

Your Excellency will see that this does not depart much from what we had proposed at the outset. But experience, the events of the first year and constant reflection have resulted in a more detailed analysis of our future movements which, as regards the coast not yet covered by our own surveys I shall for the sake of clarity describe with reference to the chart attached to the set of plates of the third voyage of Captain Cook. Thus, when speaking of the coasts of South America I shall always refer to the charts which we now present to you.

On our departure from this port on or about the 20th instant, we shall continue to plot the coast as far as Guayaquil, where we shall remain somewhat longer than would be required solely for our hydrographic work so as to give the naturalists scope to extend their special and industrious skills in such a rich region. Then we shall cross to the Galápagos, abandoning the idea of surveying Islas de Gallego, placed more to the west than the Galápagos on Captain Cook's chart, because I have not been able to find the least trace of their existence which would justify the sacrifice of a month at a time so hemmed in by the seasons. We shall return then towards La Gorgona, Ensenada de Nicoya and Realejo; and since favourable breezes blow at this season we shall hope to catch them, and to work along the adjoining coast by way of Panamá, Amapala and Tecoantepeque as far as Acapulco.[1]

Then in the month of February 1791 we shall make all speed towards San Blas where, after strengthening the launches, and perhaps taking charge of a schooner, which may either be waiting for us there or otherwise reach us at the entrance of Prince William Sound, we shall try to undertake our surveys between the middle and end of May.

If the passage of Ferrer Maldonado exists (which on mature reflection I do not find wholly absurd) it must be precisely in Cook's River [Cook Inlet] or in Prince William Sound. Thus our searches in both places will be so thorough as to confirm or disprove the truth of that account, and it would be no surprise if we were busy with this work until the end of May (even if it turns out to be fruitless).

If this survey is frustrated we shall immediately give up the detailed examination of the coast to proceed as soon as possible through Bering Strait to reach high northern latitudes, steering always eastwards, not only because it would not be difficult if we meet with unexpected good fortune and reach at least the end of Hearne's route as described in the account of Malo de Luque,[2] but also because it seems probable from

---

[1] La Gorgona, an island off the coast of Colombia; Ensenada de Nicoya, Golfo de Nicoya on the Pacific coast of Costa Rica; Realejo, on the coast of Nicaragua; Amapala, in Golfo de Fonseca on the coast of Honduras; Tecoantepeque, an obsolete name for Tehuantepec on the Pacific coast of Mexico 300 miles ESE of Acapulco.

[2] Samuel Hearne journeyed overland from Fort Prince of Wales Fort, Churchill, to the mouth of the Coppermine River in 1771, and became the first non-Native to sight the Arctic coastline of the North American continent. He put his farthest north, within sight of the polar sea, in latitude $71°34'$N (the correct latitude is $67°48'$N). It seems odd that Malaspina does not cite here the more authoritative account of Hearne's journey that first appeared in the official account of Cook's third voyage in 1784.

Ferrer Maldonado's information that the passage is not beyond 75° latitude, between the area navigated by Captains Cook and Clerke, and Baffin Bay.

In truth, short of the strange and happy chance of the ice receding towards the Pole, I believe that the passage must either be reached in the wake of Ferrer Maldonado or it will not be found at all. If that be the case, our navigation beyond 70° would have as its only object emulation of the English, and the establishing of a new era in the study of nature as regards the advance and retreat of the ice near the Pole.

Already by the beginning of August circumstances will have decided the next most suitable step, although if this were the retracing of our course we would remain somewhat longer in those latitudes so as to surpass not only the latest date but also the length of time [previously] spent in this arduous campaign. We shall finally make land in Bahía de Avacha in Kamchatka unless our discussions with the commandant of San Blas rule out this intention. A return voyage to San Blas and Acapulco will give us scope to trace the entire coast of California with all due precision, except possibly for some part of the sea inside the peninsula.[1] The time of our arrival in either of the two ports will consequently be October or November 1791, where new orders could reach us from His Majesty so as to guide our next steps with more certainty. If it is in the mind of His Majesty that Islas de Sandwich, the Pescadores and the Carolinas should be visited, and that the Marianas should be accurately charted, we shall not reach Canton until March or April 1792, and there will scarcely be time for us to reach Manila before the stormy season sets in. But if this course appears to serve no purpose, and a stay of only fifteen days in Owihee[2] (one of Islas de Sandwich) is combined with what is necessary to fix the true astronomical position of the islands of Guajan[3] and Tinian in the Marianas, then the months from January to May 1792 could be used on the Mercator chart of the Philippines, and the following months spent in Manila on the preparations of the ships and provisions, and editing our work as we did in Lima. This second course is the one we shall follow unless orders to the contrary reach us from His Majesty.

Then in September or October of the same year we shall cross to Canton and from there to the Strait of Malacca, in fulfilment of the confidential orders which were given us in Spain. But if such a survey should appear unnecessary since there are so many of our vessels engaged in the Manila trade which pass through the Strait every year (besides innumerable foreign vessels), we would sail from Canton to survey the opposite coastline of Luzón, and then between the Moluccas until we reached New Holland, ranging until March or April 1793 along its western and southern coasts which neither Cook nor La Pérouse have surveyed.

After a short stay at Botany Bay or Port Jackson we would cross to New Zealand and then to the Friendly Islands and the Society Islands, among whose inhabitants we would spend the remaining time until in October with the favourable season we would set course direct to Chiloé. The crews having rested there briefly, and perhaps after undertaking a survey of the mainland and Archipiélago de los Chonos with the ships' launches and piraguas we shall perfect the survey of the Patagonian coast (on

---

[1] Golfo de California, east of Baja California.
[2] The island of Hawaii.
[3] Guam.

our charts) from north to south, looking for the port of Inchin,[1] the doubtful island of Santa Catalina[2] and the outer part of the Tres Morros peninsula[3], that of La Campana[4] and that part which extends from our surveys southward as far as Cabo Victoria. Then from Cabo Deseado we would coast, imitating Captain Cook, along the whole outer part of Tierra del Fuego and, after surveying Isla de los Estados through Estrecho de Maire, our corvette would first steer to disprove the alleged existence of Nueva Irlanda at 55½°, according to the information given to the Viceroy at Buenos Aires by the captain of the English brig *Hartford Packet*.[5] Then she would sail south of the Malvinas and after anchoring in the Puerto de la Soledad, would steer for Montevideo, attempting to survey the stretch between Río Negro and Cabo San Antonio. The other corvette, meeting in Puerto Deseado two launches from the Río de la Plata, under oars, decked, and with plenty of room for provisions, water and crew, would complete the exploration of Golfo de San Jorge. In March or April 1794 the two corvettes would finally join each other in Montevideo, and in the summer of the same year drop anchor in Spanish ports, the commission having been completed.

Your Excellency will realise that the survey of Golfo de San Jorge before the corvettes return to Spain is necessary. It is of utmost interest to geography, to the advancement of natural history, and even to our fisheries,[6] as well as to the security of these coasts. The fact is that the proposed plan does not involve any considerable extension of the length of time required for our voyage. But it is true that the crews and the ships themselves will be in no condition to sustain a new series of tasks, and for that reason Your Excellency will perhaps find that the Pilotos Tafor and Peña, for practical navigation, and Piloto Catalá with good instruments and a chronometer for the astronomical work, would be able to complete this survey in the launches without involving the corvettes.

In this proposed distribution of tasks it is clear that we shall have to sacrifice, for the benefit of the work being undertaken and the reputation of our nation, not only the accomplishment of completing the circumnavigation of the world, but also the survey of the Strait of Malacca, and above all the date for the completion of the voyage, which had been set for 1793. All three points require that Your Excellency sees fit to let us know if they meet with the King's approval, and whether (having advised from Manila the time of our early arrival at Puerto Deseado at the beginning of 1791) we should adhere to the same timing with the certainty of finding the launches or *sumacas* ready for the survey of Golfo de San Jorge.

Finally, if in spite of these considerations His Majesty is determined on the completion of the circumnavigation, one of the corvettes could continue its voyage by way of Cape of Good Hope, parting from the other after finishing the reconnaissance of New Zealand, and then sailing by way of Île de France [Mauritius], Table Bay, and

[1] Islas Inchin, close SW of Islas Tenquehuén in the southern part of Bahía Darwin.
[2] Probably St Katherine Island, shown on maps of the period SSW of Inchin Island.
[3] Cabo Tres Picos, the NW extremity of Isla Madre de Dios.
[4] Isla Campana.
[5] No reference to this report has been found.
[6] That is, whaling.

the islands of Annobon and Fernando de Pó,[1] while the other would follow the whole of the proposed route by way of Chiloé, the coasts of Patagonia, and Montevideo, where the corvettes would once again join company for their return to Spain.

The desire to be helpful is responsible for the excessive length and tedium of this letter. May Your Excellency see fit to excuse it etc.[2]

Aboard the corvette *Descubierta* at anchor in the port of Callao, 15 September 1790.

---

[1] The islands of Annobón off the coast of Gabon, and Bioko in the Gulf of Guinea.
[2] Valdés replied to this letter on 23 March 1791, and in general agreed with Malaspina's proposed changes to the original routes for the expedition. AMN, MS 278, ff. 109-11.

APPENDIX 2

Malaspina's Survey Methods

A major part of Malaspina's mission was to carry out detailed coastal and harbour surveys, each of which required different techniques. As Donald Cutter has pointed out in his introduction (p. xxxviii) Malaspina was an experienced hydrographic surveyor, having served under Don Vicente Tofiño during the latter's extensive survey of the coasts of Spain, as had a number of other officers from both corvettes. The survey methods Malaspina adopted were thus mainly based on those of Tofiño. For coastal surveys Malaspina adopted a method known as a running survey in which a series of 'ship stations' were established, the distance between them being computed from the time taken for the corvette to sail between the stations and the speed of the corvette as measured by log-line, a procedure that Malaspina referred to as measuring bases. The bearing between each ship station was established from the courses steered with due allowance for leeway and currents. At each ship station compass bearings were taken to prominent objects on shore and the angles between them measured by horizontal sextant angles, enabling their positions to be established relative to the *Descubierta,* a procedure that Malaspina sometimes referred to as observing triangles. To achieve the best possible accuracy in this method Malaspina placed the knots on his log-line 50⅔ English feet apart, which, in conjunction with a 30-second sand glass represents a nautical mile of 6080 feet, which is close to the present international value of 1852 metres. The use by Malaspina of English feet rather than Burgos feet, which are slightly shorter,[1] suggests that Malaspina had consulted the comments made by Phipps, who experimented with various log-lines during his voyage to Spitzbergen in 1773, a copy of whose account was held on board the *Descubierta.*[2] Bases were also sometimes measured at sea by observing the angles of elevation between the water-line and the top of the masthead of the other corvette from which the distance between the two corvettes could be calculated by simple trigonometry, a method which was also described by Phipps.[3] Horizontal sextant angles were observed simultaneously from each corvette between the main mast of the other corvette and selected objects on shore to fix them in relation to the base (see plate 38).

---

[1] 1 Burgos foot = 0·91 English feet
[2] Phipps, *Voyage towards the North Pole*, pp. 87-98.
[3] Ibid., pp. 99-107. Phipps calculated the distance between his two ships by the simple formula: distance = masthead height divided by the sine of the subtended angle, whereas Malaspina adopted a more complicated formula devised by Galiano; see p. 61, n. 3 above.

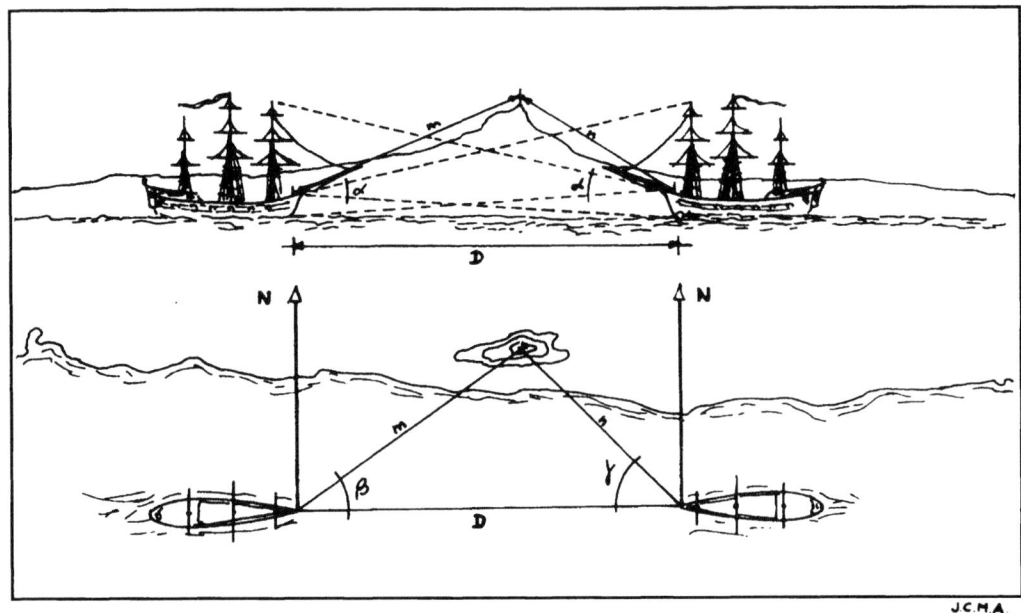

Plate 38. Measuring a base at sea by masthead heights and simultaneously fixing the position of a hill on shore by horizontal sextant angles; from Martínez-Gañavate, *Trabajos astronómicos*, p. 115.

When bases were measured by this method it was essential to exchange data as soon as possible to ensure that any points that were observed from both corvettes were correctly identified. On a number of occasions Bauzá was sent over to the *Atrevida* for this purpose.[1] However, Malaspina found this method to be unreliable when the corvettes were under way, probably because of the difficulty in acheiving simultaneous observations. To establish the geographical framework for these running surveys meridian altitudes of the Sun were observed, when possible, to obtain latitude, while longitude was obtained by observing lunar distances and by chronometer. In the latter case Malaspina had first to obtain apparent time by observing the altitude of the Sun when it was some distance from the meridian. Then knowing the corvette's latitude and the Sun's declination, it was possible to calculate the Sun's hour angle at the time of the observation. If the Sun's altitude had been observed before noon, as was usually the case, the hour angles was then subtracted from twenty four hours to obtain apparent time. This was then converted to local mean time by applying the equation of time taken out of the British *Nautical Almanac* or the French *Connaissance des temps*, both of which were held on board. The difference between local mean time and mean time at Cádiz or another meridian such as Montevideo as given by the corvette's chronometers would give her longitude west of Cádiz or Montevideo expressed as time. Malaspina referred to this operation in his journal as observing hour angles. At dusk Malaspina either lay-to or stood off and on

[1] See for example p. 80 above.

with the intention of having the final points fixed the previous evening visible at dawn to enable him to continue his survey where he had left off the night before.

In harbour bases were measured by steel chains from which a system of triangulation was then observed by theodolite. In Port Egmont, because of the terrain, Malaspina first attempted to measure a base by masthead heights, but for some unknown reason this proved unsatisfactory and Bauzá was sent ashore to measure additionally a short base by chain.[1] Soundings were usually fixed by means of compass bearings taken from the boats. Occasionally, however, two theodolites were set up on shore and angles taken to the sounding boat, the timing probably controlled by flags or pistol shots, using a method devised by Tofiño.[2]

At each place where the expedition stopped for more than a few days Malaspina landed the expedition's portable observatory, the astronomical clock, the astronomical quadrant, various telescopes and several chronometers, in order to establish the observatory's position and to rate his chronometers. The latitude of the observatory was obtained by means of meridian altitudes of the Sun and various stars, while longitude was obtained by a variety of methods, namely by lunar distances, chronometer, the occultations of the satellites of Jupiter or the occultation of stars by the Moon. At first Malaspina's longitudes were give from Cádiz, but later from the last place where he was confident that he had obtained its longitude with sufficient accuracy. The rates of his chronometers were obtained by means of equal or corresponding altitudes. Comparisons between the astronomical clock and chronometers on shore and the chronometers left on board were again carried out by means of pistol shots.

The majority of the astronomical observations taken on shore were carried out by Galiano, Vernacci and Concha, who were regarded by Malaspina as the expedition's astronomers. As well as making the necessary observations for rating the chronometers and establishing the geographical positions of the various observatories as part of the surveying programme, they also carried out observations particularly in Montevideo and Valparaíso to establish the fundamental positions of stars in the southern heaven. The results they obtained in Montevideo were forwarded by the three astronomers to the officers at the Real Observatorio de Cádiz, to Monsieur Lalande in Paris, and to Signors Oriani, Reggio and de Cesaris at the Brera Observatory in Milan.[3]

Malaspina's gravity observations

When Malaspina reached Acapulco he was joined by Teniente de Fragata José Espinosa y Tello and Teniente de Fragata Ciriaco Cevallos, who brought with them instructions from Don Antonio Valdés, dated 22 December 1790, ordering Malaspina to carry out observations for gravity with a specially designed pendulum, which the two officers brought with them. This instrument, which was probably based on one used in 1773 in Spitzbergen by Israel Lyons during Phipps's voyage,[4] was sent from

---

[1] See p. 105 above.
[2] See p. 155 above.
[3] See p. 60, n. 3 above.
[4] Phipps, *Voyage towards the North Pole*, pp. 153–79, which contains an illustration of the pendulum used by Lyons.

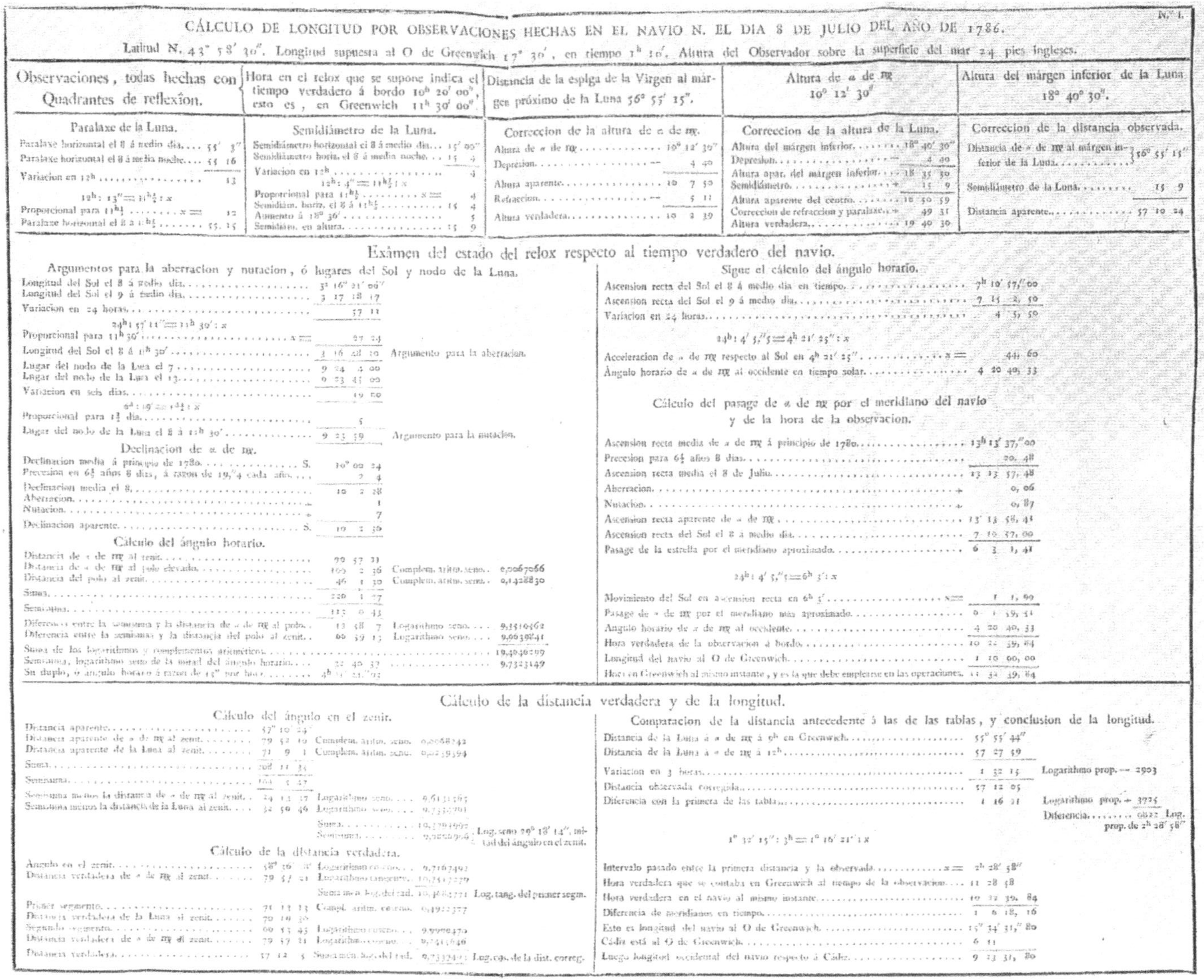

N.° I.

CÁLCULO DE LONGITUD POR OBSERVACIONES HECHAS EN EL NAVIO N. EL DIA 8 DE JULIO DEL AÑO DE 1786.

Latitud N. 43° 58′ 30″. Longitud supuesta al O de Greenwich 17° 30′, en tiempo 1ʰ 10′. Altura del Observador sobre la superficie del mar 24 pies ingleses.

Observaciones, todas hechas con Quadrantes de reflexîon.	Hora en el relox que se supone indica el tiempo verdadero á bordo 10ʰ 20′ 00″, esto es, en Greenwich 11ʰ 30′ 00″.	Distancia de la espiga de la Virgen al márgen próximo de la Luna 56° 55′ 15″.	Altura de α de ♍ 10° 12′ 30″	Altura del márgen inferior de la Luna 18° 40′ 30″.

Paralaxe de la Luna. — Semidiámetro de la Luna. — Correccion de la altura de α de ♍. — Correccion de la altura de la Luna. — Correccion de la distancia observada.

Exâmen del estado del relox respecto al tiempo verdadero del navio.

Argumentos para la aberracion y nutacion, ó lugares del Sol y nodo de la Luna.

Declinacion de α de ♍.

Cálculo del ángulo horario.

Sigue el cálculo del ángulo horario.

Cálculo del pasage de α de ♍ por el meridiano del navio y de la hora de la observacion.

Cálculo de la distancia verdadera y de la longitud.

Cálculo del ángulo en el zenit.

Cálculo de la distancia verdadera.

Comparacion de la distancia antecedente á las de las tablas, y conclusion de la longitud.

Plate 39. Example of a worked lunar distance observed on 8 July 1786 between the Moon and α Virginis (Spica); from José de Mendoza y Ríos *Tratado de navegación astronómica* Madrid 1787

London by Teniente de Navío José Mendoza y Ríos, a Spanish naval officer and a noted mathematician and astronomer, who was living there at the time.[1] As is well known, the Earth is not a perfect sphere being flattened at the poles due to the pull of gravity and the rotation of the Earth as it cooled. The purpose of these observations, as explained to Malaspina by Valdés, was to obtain information about the true figure of the Earth. In particular, observations were required in 45°S to determine whether the southern hemisphere was flatter than the northern hemisphere in the same latitude. Similar observations were being conducted in the northern hemisphere by the French. Cevallos and Espinosa had made a number of observations with the pendulum before sailing for Acapulco, probably in Cádiz. Three observers were required to carry out the observations. Two of them counted the number of oscillations in an hour, while the third recorded the time by chronometer. During the voyage observations were made at eight locations in the northern hemisphere and at seven locations in the southern hemisphere. After the voyage the results were computed in 1807 by Brigadier de la Real Armada Don Gabriel Ciscar, who obtained a value of compression of 1/321, which was published two years later by Espinosa.[2] This compares reasonably with the value of 1/298·257223563 which was adopted for the World Geodetic System (WGS) in 1984.

Malaspina was aware of a more time-consuming method of determining the figure of the Earth devised by the French astronomer Giovanni Domenico (Jean-Dominique) Cassini. In this method an accurate base was measured on land and extended by triangulation to two points on the Earth's surface in a north/south line and approximately one degree apart. From the difference in latitude between the two points obtained astronomically by meridian altitudes and the distance between them obtained through triangulation the length of a degree of the meridian at the mid latitude of the observations would be obtained. Since the length of a degree of latitude is dependent on the radius of curvature of the Earth in the meridian, two or more observations in widely different latitudes would enable the compression to be calculated. This method was referred to by the Spaniards as 'medida del grado terrestre' (measurement of the terrestrial degree), but British scientists always referred to it as the measurement of an arc of the meridian, since it was highly unlikely that the two extremities would be exactly one degree apart. In 1735 two expeditions were mounted by the French one to Lapland and the other close to the equator in the vicinity of Quito, in present day Ecuador. The later expedition, which took eight years to complete, was led by Pierre Bouguer and Charles-Marie de la Condamine, with two Spanish naval officers Antonio de Ulloa y de la Torre-Guiral and Jorge Juan ostensibly as observers, all four of whom were mentioned by Malaspina in his journal.[3]

From an earlier measurement of an arc of the meridian by Cassini near Paris the length of a sea mile (one-sixtieth of a degree of latitude) at that latitude was determined accurately. This value was used by Malaspina to determine the correct distance apart of knots on his log line.[4]

---

[1] AMN, MS 1826.
[2] José Espinosa y Tello, *Memorias sobre las observaciones astronómicas*, 2 vols, Madrid, 1809, I, pp. 190-212.
[3] Cerezo, *Diario por Malaspina*, I, p. 215. See also Taylor 'Degree of Difficulty'.
[4] Cerezo, *Diario por Malaspina*, I, p. 28.

WORKS CITED IN VOLUME I

Alcedo, A. de, *Diccionario geográfico-histórico de las Indias occidentales o América,* Madrid, 1786-9.

Alden, D., *The Making of an Enterprise: the Society of Jesus in Portugal, its Empire and Beyond, 1540–1750,* Stanford, Ca, 1996.

Almanak náutico y estado general de marina, Madrid, 1786.

Almanaque náutico y efemérides astronómicas, Madrid, 1791 to present.

An Account of Several Late Voyages and Discoveries to the North and the South, London, 1694.

Angelis, P. de, *Documentos del Río de la Plata,* Buenos Aires, n.d.

Anson, George, *A Voyage round the World,* London, 1748.

d'Après de Mannevillette, Jean-Baptiste Nicolas Denis, *Instructions sur la navigation des Indes Orientales et de la Chine, pour servir au Neptune oriental, par M. d'Aprés de Mannevillette,* Paris and Brest, 1775, pp. 2, 3.

d'Après de Mannevillette, Jean-Baptiste Nicolas Denis, *Neptune oriental,* Paris and Brest, 1745 (revised edn 1775; English version in 1782: *The East India Pilot*).

Archer, Christon I., 'Cannibalism in the Early History of the Northwest Coast: Enduring Myths and Neglected Realities', *Canadian Historical Review,* LXI, 1980, pp. 453-79.

Atienza, J. de, *Títulos nobiliarios hispanoamericanos,* Madrid, 1947.

Barber, Peter, 'Malaspina and George III Brambila and Watling: Three Rediscovered Drawings of Sydney and Parramatta', in *Malaspina '92 – Jornadas Internacionales,* Cádiz, 1994, pp. 357-69.

Beaglehole, J. C., ed., *The Journals of Captain James Cook: The Voyage of the Resolution and Adventure, 1772–1775,* Cambridge, 1961.

Beerman, Eric, *El diario del proceso y encarcelamiento de Alejandro Malaspina (1794–1803),* Madrid, 1992.

Beerman, Eric, *Francisco Requena: la expedición de límites,* Madrid, 1996.

Bellin, Jean-Nicolas, *Le Neptune français,* Paris, 1753.

[Beresford, William], *A Voyage round the World 1785–1788 by Captain George Dixon,* London, 1789.

Bernales Ballesteros, J., 'La pintura en Lima durante el virreinato', *Pintura en el virreinato del Perú,* Lima, 1989, pp. 66-70.

Borda, Jean-Charles, *Description et usage du cercle à reflexion,* Paris, 1778.

Bouguer, Pierre, *La Figure de la terre, déterminée par les observations de MM. Bouguer & de la Condamine, de l'Académie royale des sciences, envoyés par ordre du Roy au Pérou pour observer aux environs de l'equateur,* Paris, 1749.

Brisson, Mathurin-Jacques, *Ornithologie*, Paris, 1760.

Busto Duthurburu, J. A. del, *Diccionario histórico-biográfico de los conquistadores del Perú*, Lima, 1973-.

Calatayud Arinero, M. de los A., *Catálogo de las expediciones y viajes científicos españoles a América y Filipinas (siglos XVIII y XIX): Fondos del Archivo del Museo Nacional de Ciencias Naturales*, Madrid, 1984.

Campbell, L. G., Jr, 'The Military Reform in the Viceroyalty of Peru, 1762-1800', University. of Florida PhD thesis, 1970.

Carbonell, R., *Estrategias de desarrollo rural en los pueblos guaranies (1609–1767)*, Barcelona, 1992.

Carta esférica del Río de la Plata ... levantada de orden del Rey en 1789 y rectificada en 1794 por varios oficiales de su R¹ Armada, 1798.

Carta esférica de las costas de la América meridional desde el paralelo de 36°30' de latitud S hasta el Cabo de Hornos, 1798.

Carta esférica desde Punta Candor hasta Cabo de Trafalgar, 1787.

Cavanilles, A. J., *Icones et descriptiones plantarum*, 6 vols, Madrid, 1791-1804.

Cavanilles, A. J. 'Materiales para la historia de la botánica', *Anales de historia natural*, ii, 1800, 4, pp. 3-57.

Cerezo Martínez, Ricardo , ed., *La Expedición Malaspina 1789–1794*, Tomo II: *Diario general del viaje por Alejandro Malaspina*, 2 parts, Madrid, 1990.

Clavijo y Clavijo, S., *La trayectoria hospitalaria de la Armada Española*, Madrid, 1944,

Colnett, James, *A Voyage to the South Atlantic and round Cape Horn into the Pacific Ocean*, London, 1798.

Connaissance des temps, Paris, 1690 to present.

Cook, James, and King, James, *A Voyage to the Pacific Ocean for making Discoveries in the Northern Hemisphere*, 3 vols, London, 1784.

Cook, James, *A Voyage towards the South Pole and round the World*. 2 vols, London, 1777.

[Córdoba, Antonio de], *Relación del último viaje al Estrecho de Magallanes de la fragata de S.M. Santa María de la Cabeza en los años de 1785 y 1786*, Madrid, 1788.

Corés Conde, R. and Hunt, S. eds, *The Latin American Economies: Growth and Export Sector, 1880–1930*, New York, 1985.

Cotter, Charles H., *A History of the Navigator's Sextant*, Glasgow, 1983.

Croix, T. de, 'Relación que hace el excmo Señor Don Teodoro de Croix, Virrey de estos Reynos del Perú y Chile, a su sucesor el excmo Señor Fr. Don Francisco Gil de Lemos desde 4 de abril de 1784 hasta 25 de marzo de 1790', *Memorias de los virreyes que han gobernado el Perú durante el tiempo del colonizaje español*, Lima, 1859, pp. 254-9, 263-5.

Cutter, Donald C., ed., *Journal of Tomás de Suria of His Voyage with Malaspina to the Northwest Coast of America in 1791*, Fairfield, WN, 1980.

Cutter, Donald C., 'The Spanish at Hawaii: Gaytan to Marin', in *Hawaiian Journal of History*, XIV, 1980, pp.16-25.

Cutter, Donald C., *California in 1792: A Spanish Naval Visit*, Norman and London, 1990.

Cutter, Donald C., *Malaspina and Galiano: Spanish Voyages to the Northwest Coast, 1791 and 1792*, Vancouver, 1991.

David, Andrew, 'Felipe Bauzá and the British Hydrographic Office 1823-34', in *Malaspina '92*, Cádiz, 1994, pp. 235-41.

Dunmore, John, ed., *The Journal of Jean-François de Galaup de la Pérouse 1785–1788*, Hakluyt Society, 2nd ser., 179-80, London, 1994.

Edwards, George, *A Natural History of Birds*, 4 parts, London, 1743-51.

Engstrand, Iris H. W., *Spanish Scientists in the New World*, Seattle and London, 1981.

España maritima, or Spanish Coasting Pilot, London, 1812 *see* Tofiño y San Miguel, *Derrotero de las costas de España*.

Espinosa y Tello, José, *Memorias sobre las observaciones astronómicas hechas por los navegantes españoles en distintos lugares del globo*, 2 vols, Madrid, 1809

Espinosa, Juan José Martínez, *Diccionario marino español–inglés/inglés–español*, Madrid, 1849 (Editorial Naval facsimile, 1989).

Estrella, Eduardo, ed., *La expedición Malaspina, 1789–94*, Tomo VIII: *Trabajos zoológicos, geológicos, quimicos y fisicos en Guayaquil de Antonio Pineda Ramírez*, Madrid, 1996.

Falkner, Thomas, *A Description of Patagonia and the Adjoining Parts of South America*, London, 1774.

Fernández-Armesto, F., *Millennium*, New York, 1995.

Feuillée, L., *Journal des observations physiques, mathématiques et botaniques faites par ordre du roi sur les côtes orientales de l'Amérique méridionale et aux Indes occidentales*, 3 vols, Paris, 1714-25.

Findlay, Alexander G., *A Directory for the Navigation of the Pacific Ocean*, 2 vols, London, 1851.

FitzRoy, Robert, *Narrative of the Surveying Voyages of His Majesty's Ships Adventure and Beagle*, 3 vols and appendix, London, 1839.

Frézier, Amédée-François, *Relation du voyage de la Mer du Sud aux côtes du Chily et du Pérou fait pendant les anées 1712, 1713 et 1714*, Paris, 1716 (English version: *A Voyage to the South-Sea, and along the Coasts of Chili and Peru, in the Years 1712, 1713 and 1714*, London, 1717.)

Galera Gómez, A., *Alejandro Malaspina: en busca del paso del Pacífico*, Madrid, 1990.

Galera Gómez, A., ed., *La ilustración española y el conocimiento del Nuevo Mundo: las ciencias naturales en la expedición Malaspina (1789–94): la labor científica de Antonio de Pineda*, Madrid, 1988.

García y García, A., *Derrotero de la costa del Perú*, Lima, 1870.

Gibson, James R., *Otter Skins, Boston Ships, and China Goods: The Maritime Fur Trade of the Northwest Coast 1785–1841*, Seattle, 1992.

Giura Longo, R., and Rossi, P., *Con Malaspina nei mare del sud*, Bari, 1999.

González, Pedro María, *Tratado de las enfermedades de la gente del mar en que se exponen sus causas y los medios de precaverlas*, Madrid, 1805.

Grüner, V. R., *Jindy y Nyní*, Prague, 1829.

Guarda, G., *Flandes indiano: las fortificaciones del Reino de Chile 1541–1826*, Santiago, 1990.

Guerra, F., *El hospital en Hispanoamérica y Filipinas*, Madrid, 1994.

Hall-Jones, John, *Doubtfull Harbour*, Invercargill, 1984.

Hall-Jones, John, *Fiordland Explored*, Invercargill, 1990.

Hawkesworth, John, ed., *An Account of the Voyages undertaken by the Order of His Present Majesty for making Discoveries in the Southern Hemisphere*, 3 vols, London, 1773.

Hidalgo J. et al., eds, *Culturas de Chile: Prehistoria desde sus orígenes hasta los albores de la conquista*, Santiago, 1989.

Higueras Rodríguez, María Dolores, and Pimentel Igea, Juan, eds, *La expedición Malaspina, 1789–94*, Tomo V: *Antropología y noticias etnográficas*, Madrid, 1993.

Higueras Rodríguez, María Dolores, ed., *La expedición Malaspina 1789–1794*, Tomo IX: *Diario general del viaje corbeta Atrevida por José Bustamante y Guerra*, Madrid, 1999.

Higueras Rodríguez, María Dolores, 'The Malaspina Expedition (1789-1794): A Venture of the Spanish Enlightenment', in *Spanish Pacific from Magellan to Malaspina*, Madrid, 1988, pp. 147-63.

Higueras Rodríguez, María Dolores, ed., *Catálogo crítico de los documentos de la Expedición Malaspina en el Museo Naval*, 3 vols, Madrid, 1985-94.

Hipólito Unanue, J., *Guía política, eclesiástica y militar del virreinato del Perú para el año de 1793*, Lima, 1792.

Hodgen, M., *Early Anthropology in the Sixteenth and Seventeenth Centuries*, Philadelphia, 1969.

Howse, D., and Thrower, Norman J. W., eds, *A Buccaneer's Atlas: Basil Ringrose's South Sea Waggoner*, Berkeley, 1992.

Hunt, S., 'Growth and Guano in Nineteenth-century Peru', in R. Corés Conde and S. Hunt, eds, *The Latin American Economies: Growth and the Export Sector, 1880–1930*, New York, 1985.

Ibáñez, Victoria, and King, Robert, 'A Letter from Taddeus Haenke to Sir Joseph Banks', *Archives of Natural History* 23 (1996), pp. 255-60.

Ibáñez Montoya, María Victoria, ed., *La expedición Malaspina 1789–1794*, Tomo IV: *Trabajos científicos y correspondencia de Tadeo Haenke*, Madrid, 1992.

Iglesias, M. C. et al., eds, *Carlos III y la Ilustración*, 2 vols, Madrid, 1989.

Jiménez de la Espada, Marcos, 'Una causa de estado', in *Revista contemporánea*, XXXI, vol. IV, February, Madrid, 1881.

Juan, Jorge, *Observaciones astronómicas y physicas*, Madrid, 1746.

Kendrick, John, *Alejandro Malaspina – Portrait of a Visionary*, Montreal and Kingston, 1991.

King, Robert J., *The Secret History of the Convict Colony: Alexandro Malaspina's report on the British settlement of New South Wales*, Sydney, 1990.

King, Robert J., 'A Regular and Reciprocal System of Commerce – Botany Bay, Nootka Sound, and the Isles of Japan', *The Great Circle*, 19, No 1 (1997), pp. 1-29.

La Caille, Nicolas-Louis de, *Caelum australe stelliferum*, Paris, 1763.

Lamb, W. Kaye, ed., *George Vancouver, A Voyage of Discovery to the North Pacific Ocean and Round the World*, Hakluyt Society, 2nd ser., 163-6, London, 1984, I, pp. 36-7.

Lastes, J. B., *Historia de la medicina peruana*, Lima, 1951.

Lloyd, Christopher, *St Vincent and Camperdown*, London, 1963.

Lofstrom, W. L., *Paita: Outpost of Empire*, Mystic, 1996.

Lucena Giraldo, Manuel, and Pimental Igea, Juan, *Los 'Axiomas políticos sobre la América' de Alejandro Malaspina*, Madrid, 1991.

Mackay, David, *In the Wake of Cook*, London, 1985.

Manfredi, Dario, 'Alejandro Malaspina. Una biografia', in Blanca Sáiz, ed., *Alejandro Malaspina: La América imposible*, Madrid, 1994, pp. 19-133.

Manfredi, Dario, *Alessandro Malaspina e Fabio Ala Ponzone – Lettere dal Vecchio e Nuovo Mondo (1788–1803)*, Bologna, 1999.

Manfredi, Dario, 'An Unknown Episode behind the Northwest Coast Campaign of Malaspina's Expedition', in Robin Inglis, ed., *Spain and the North Pacific Coast*, Vancouver, 1992, pp. 119-24.

Markham, Sir Clements, *Early Spanish Voyages to Magellan Strait*, Hakluyt Society, 2nd ser., 28, London, 1911.

Markham, Clements R., ed., *Narratives of the Voyages of Pedro Sarmiento de Gamboa to the Straits of Magellan,* Hakluyt Society, 1st ser., 91, London, 1895.

Marshall P. J., and Williams, Glyndwr, *The Great Map of Mankind: Perceptions of New Worlds in the Age of Enlightenment*, Cambridge, Mass., 1982.

Martínez Espinosa, Juan José, *Diccionario Marino español-inglés /inglés español*, Madrid, 1849 (Editorial Naval facsimile 1989).

Martínez-Gañavate Ballesteros, Luis Rafael, ed., *La expedición Malaspina*, Tomo VI: *Trabajos astronómicos, geodésicos e hidrográficos*, Madrid, 1994.

Masdevall y Terrades, José, *Relación de las epidemias de calenturas pútridas y malignas,* 1785.

Mason, S., *Historia de las ciencias*, 5 vols, Madrid, 1986.

Mayne, R. C., *Practical Notes on Marine Surveying and Nautical Astronomy*, London, 1874.

Meares, John, *Voyages made in the Years 1788 and 1789, from China to the North West Coast of America*, London, 1790.

Mendiburu, M. de, *Diccionario histórico-biográfico del Peru*, Lima, 1878.

Mendoza y Ríos, José de, *Tratado de navagación astronómica*, Madrid, 1787.

Mercurio peruano, 3 February, 1791

Millán y Maraval, Francisco, *Descripción del Río de la Plata*, 1770.

Minguijón, Salvador, *Historia del derecho español*, 3rd edn., Madrid, 1943.

Molina Otárola, R., 'Los mecanismos del despojo del territorio Mapuche-huilliche de Osorno,' in M. Orellana Muermann and J. G. Muñoz Correa, *Comunidades indígenas en su entorno*, Santiago, 1962, pp. 23-44.

Moreyra Paz Soldán, M., 'Peralta astrónomo', *Estudios históricos,* Lima, 1995, iii, p. 536.

Morse, W. I., *Letters of Alexandro Malaspina (1790–1791).* Published as Supplement to *The Chronicle*, 240, Boston, Mass. 1944.

Muñoz Garmendia, M., ed., *La expedición Malaspina*, Tomo III: *Diario y trabajos botánicos de Luis Neé,* Madrid, 1993.

Murphy, R. C., *Oceanic Birds of South America*, 2 vols, New York, 1936.

Nautical Almanac and Astronomical Ephemeris, London, 1767 to present.

Nodal, Bartolomé and Gonzalo, Garcia de, *Relacion del viaje … los capitanes Bartholome Garcia de Nodal y Gonzalo de Nodal … al descubrimiento del estrecho nuevo de S. Vicente, que hoy es nombrado de Maire, y reconocimi° del de Magallanes,* 1821 (English translation in Markham, *Early Spanish Voyages*, pp. 169-272).

Novo y Colson, Pedro, ed., *Viaje político-científico alrededor del mundo por las corbetas Descubierta y Atrevida … desde 1789 a 1794*, Madrid, 1885.

O'Donnel, Hugo y Duque de Estrada, *El viaje a Chiloé de José de Moraleda (1787–1790)*, Madrid, 1990.

Orellana Muermann, M. and Muñoz Correa, J. G., *Comunidades indígenas en su entorno*, Santiago, 1962.

Orozco Acuaviva, A., López de Cózar, J. L. and Cabrera Afonso, J. R., 'El "Diario medico-chirúrgico" de la corbeta *Atrevida*,' *Malaspina '92*, Cádiz, 1994, pp. 115-25.

Orte Lledó, A., 'El posicionamiento astronómico de las costas de América en la expedición Malaspina', *La ciencia española en ultramar: Actas de las primeras jornadas sobre España y las expediciones científicas en América y Filipinas*, Madrid, 1991, pp. 83–96.

Ortiz Sotelo, J., 'Embarcaciones aborígenes en el área andina', *Historia y cultura*, xx, 1990, pp. 49-79.

Ortiz Sotelo, J., and Castañeda Martos, A., *Diccionario biográfico-marítimo peruano*, Lima, 1993.

O'Scanlan, Timoteo, *Diccionario Marítimo Español*, Madrid, 1831 (Museo Naval facsimile, 1974).

Pagden, A., *Lords of All the World: Ideologies of Empire in Spain, France and Britain, 1500–1800*, New Haven, 1995.

Palau M. and Sáiz, B., eds, *Moxos. Descripciones exactas e historia fiel de los indios, animales y plantas de la provincia de Moxos en el virreinato del Perú por Lázaro de Ribera, 1786–1794*, Madrid, 1989.

Palau, Mercedes, Zabala, Aránzazu, and Sáiz, Blanca, eds, *Viaje científico y político ... Diario de viaje de Alejandro Malaspina*, Madrid, 1984.

Palau Baquero M. and Orozco Acuaviva, A., eds, *Malaspina '92: Primeras jornadas internacionales*, Cadiz, 1994.

Paula Pavia, F. de, *Galería biográfica de los generales de marina, jefes y personajes notables que figuraron en la misma corporación desde 1700 a 1868*, Madrid, 1873-4.

Pérez Pimentel, R., *Diccionario biográfico del Ecuador*, Guayaquil, 1994.

Phipps, Constantine John, *A Voyage towards the North Pole undertaken by His Majesty's Command 1773*, London, 1774.

Pimentel Igea, Juan, *La física de la monarquía: ciencia y política en el pensamiento de Alejandro Malaspina (1754–1810)*, Madrid, 1998.

Pimentel Igea, Juan, 'La riqueza forestal de las costas del Pacífico: noticias e informes sobre maderas de la expedición Malaspina (1789-94)', in M. Lucena Giraldo, ed., *El bosque ilustrado: estudios sobre la política forestal española en América*, Madrid, 1991, pp. 45-62.

Pimentel Igea, Juan, ed., *La expedición Malaspina, 1789–94*, Tomo VII: *Descripciones y reflexiones políticas*, Madrid, 1995.

Pino Díaz, F. del, 'Los estudios etnográficos y etnológicos de la expedición Malaspina', *Revista de Indias*, 1982, p. 417.

Raynal, A. F., *Histoire philosophique et politique des deux Indes*, Paris, 1770.

Relación del viage hecho por las goletas Sutil y Mexicana en el año de 1792, Madrid, 1802.

Requena, Francisco, *Descripción de Guayaquil*, 1774.

Ruíz, H. and Pavón, J., *Systema vegetabilium florae peruvianae et chilensis*, 3 vols, Madrid, 1798-1802

Sáiz, Blanca, *Bibliografía sobre Alejandro Malaspina*, Madrid, 1992.

Sáiz, Blanca, *Alejandro Malaspina: la América imposibile*, Madrid, 1994.

Sanfeliú Ortiz, Lorenzo, ed., *62 meses a bordo: La expedición Malaspina según el diario de Antonio de Tova y Arredondo*, Madrid, 1943, 1988.

Schlupmann, J., 'Commerce et navigation dans l'Amérique espagnole coloniale: le port de Paita et le Pacifique au XVIIIe siècle', *Bulletin de l'Institut français d'études andines*, xxii, pt 2 (1993), pp. 521-49.

Schultes, R. Evans et al., eds, *The Journals of Hipólito Ruíz, Spanish Botanist in Chile 1777–88*, Portland, 1998.

Shadwell, Charles F. A., *Notes on the Management of Chronometers and the Measurement of Meridian Distances*, London, 1861.

Silva Vargas, F., *Tierras y pueblos de indios en el reino de Chile: esquema histórico-jurídico*, Santiago, 1962.

Sotos Serrano, Carmen, *Los pintores de la expedición de Alejandro Malaspina*, 2 vols, Madrid, 1982.

South America Pilot, Vol. 3, Taunton, 1987.

Spate, O. H. K., *Monopolists and Freebooters*, London and Canberra, 1983.

Spate, O. H. K., *The Spanish Lake*, Canberra, 1979.

Tables Requisite, 2nd edn, London, 1781.

Taylor, David, 'Degree of Difficulty: Measuring the Earth proved to be a Triumphant Fiasco', *Mercator's World*, 4, 3, May/June 1999, pp. 18-25.

Thrower, Norman J. W., ed., *The Three Voyages of Edmond Halley in the Paramore 1698–1701*, 1 vol. and portfolio, Hakluyt Society, 2nd ser., 156-7, London, 1981.

Thurman, Michael E., *The Naval Department of San Blas*, Glendale, California, 1967.

Tofiño y San Miguel, Vicente, *Derrotero de las costas de España*, 2 vols, Madrid, 1789 (republished in London in 1812 as *España Maritima, or Spanish Coasting Pilot*, with further editions in 1813 and 1814).

Tofiño y San Miguel, Vicente, *Atlas marítimo de España*, Madrid, 1789.

Torres Lanzas, P., *Catálogo de mapas y planos, virreinato del Perú (Perú y Chile)*, Madrid, 1985.

Ulloa y de la Torre-Guiral, Antonio de, *Relación histórica del viaje a la América Meridional ...*, Madrid, 1746.

Valgoma, Dalmiro de la, *Real Compañía de Guardias Marinas. Catálogo de pruebas de Caballeros Aspirantes*, Madrid, 1943.

Vargas Ugarte, R., *Historia general del Perú*, Lima, 1966.

Vargas Ugarte, R., *Historia de la Iglesia en el Perú*, Burgos, 1961.

Vattle, E., *Le Droit des gens ou principes de la loi naturelle appliqués à la conduite et aux affaires des nations et des souverains*, Paris, 1758.

Vericat, José, 'Fuentes del pensamiento socio-político de Malaspina', Palau and Orozco, *Malaspina '92*, pp. 13-18.

Viana, Francisco Javier de, *Diario del viage explorador de las corbetas Descubierta y Atrevida en los años 1789 a 1794*, Montevideo, 1958.

Ward, C., *Imperial Panama: Commerce and Conflict in Isthmian America, 1550–1800*, Albuquerque, 1993.

White, John, *Journal of a Voyage to New South Wales* [1790], Sydney, 1962.

Williams, Glyndwr, ed., *A Voyage round the World in the Years MDCCXL, I, II, II, IV by George Anson*, Oxford, 1974.

Zapatero, J. M., *El Real Felipe de Callao: primer castillo del mar de sur,* Madrid, 1983.

Zulueta, Julian de, and Higueras, Lola, 'Health and Navigation in the South Seas: the Spanish Experience', *Starving Sailors: the Influence of Nutrition upon Naval and Maritime History*, National Maritime Museum, 1981, pp. 93-6.

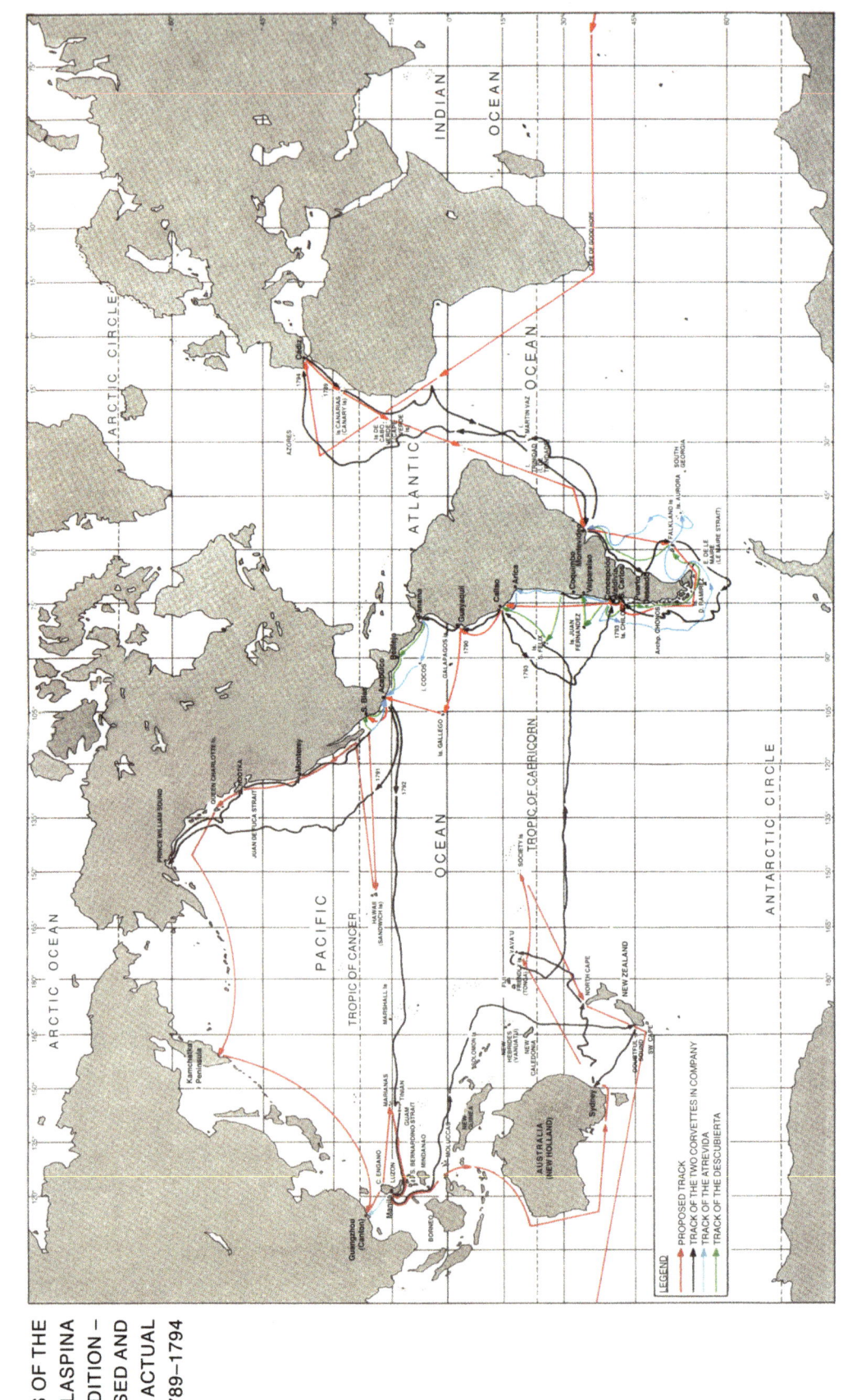

TRACKS OF THE
MALASPINA
EXPEDITION –
PROPOSED AND
ACTUAL
1789–1794